Boundary Element Methods

Proceedings of the Third International Seminar, Irvine, California, July 1981

Editor: C. A. Brebbia

Seminar sponsored by the International Society for Computational Methods in Engineering

With 232 Figures

CML PUBLICATIONS

Springer-Verlag Berlin Heidelberg GmbH 1981

Dr. CARLOS A. BREBBIA
Computational Mechanical Centre
125 High Street
Southampton
England, SO1 0AA

ISBN 978-3-662-11272-4 ISBN 978-3-662-11270-0 (eBook)
DOI 10.1007/978-3-662-11270-0

Originally published by Springer-Verlag Berlin Heidelberg New York in 1981
Softcover reprint of the hardcover 1st edition 1981

2061/3020 – 543210

CONTENTS

INTRODUCTORY REMARKS

C.A. Brebbia

1. HISTORICAL NOTE

The great interest that finite element methods have attracted in the engineering community since the beginning of 1960's has had two important consequences. 1) it stimulated an impressive amount of work in computational techniques and efficient engineering software; 2) substantial research into basic physical principles such as variational techniques, weighted residuals, etc. was originated.

The first of the above points came as a natural consequence of the emergence of new and powerful computers which were able to solve engineering problems involving large amounts of numerical storage and manipulation. The development of mathematical techniques and basic principles came instead in response to the need of extending finite element modelling, assessing convergence and accuracy and understanding the relationship of finite elements to more classical variational principles and weighted residual techniques.

These techniques can be traced to pre-computer times [1,2] and involve different ways of solving the governing equations of a problem, i.e, Galerkin, collocation, least-squares, line techniques, matrix progression or transfer, the combination of different techniques, etc. Fortunately they were not forgotten and they reappeared in the finite element literature, sometimes with different names, such as Galerkin finite elements, finite element strip-method, some time integration schemes, etc. Another important development in approximate analysis was the investigation of mixed principles and the realization that physical problems can be expressed and solved in many different ways in accordance with the part or equations of the problem that we need to approximate. These approximations are of fundamental importance for the computer implementation of the different numerical techniques. Mixed methods can be traced to Reissner [3] and more specifically for finite elements to

Pian [4]. An excellent exposition of mixed methods in structural mechanics can be found in the book by Washizu [5].

Integral equation techniques were until recently considered to be a different type of analytical method, somewhat unrelated to approximate methods. They became popular in Western Europe through the work of a series of Russian authors, such as Muskhelishvili [6], Mikhlin [7], Kupradze [8], and Smirnov [9] but were not very popular with engineers. A predecessor of some of this work was Kellogg [10] who applied integral equations for the solution of Laplace's type problems. Integral equation techniques were mainly used in fluid mechanics and general potential problems and known as the ' source' method which is an ' indirect' method of analysis; i.e. the unknowns are not the physical variables of the problem. Work on this method continued throughout the 1960s and 1970s in the pioneering work of Jaswon [11] and Symm [12], Massonet [13], Hess [14] and many others.

It is difficult to point out precisely who was the first one to propose the 'direct' method of analysis. It is found in a different form, in Kupradze's book [8]. It seems fair however from the engineering point of view to consider that the method originated in the work of Cruse and Rizzo [15] in elastostatics.

Since the early 1960s a small research group started working at Southampton University, on the applications of integral equations to solve stress analysis problems. Hadid's thesis [16] partly based on Hadjin's [17] work was published in 1964 and dealt with the use of integral equations in shell analysis. Unfortunately the presentation of the problem, the difficulty of defining the appropriate Green's functions and the parallel emergence of the finite element method all contributed to minimize the importance of this work. This work was continued through a series of theses dealing mainly with elastostatics problems. Recent developments in finite elements had started to find their way into the formulation of boundary integral equations, specially in the idea of using general curved elements. Finally the question of how to effectively relate the boundary integral equations to other approximate techniques was solved using weighted residuals [18]. The work at the Southampton University group culminated with the first book in 1978 for which the title "Boundary Elements" was used [19].

More recently this work has been expanded to encompass time dependent and non-linear problems [20]. Two important International Conferences were held at Southampton University in 1978 and 1980. The edited Proceedings of these conferences - the only ones so far on this topic - are now standard references [21,22].

One of the most important recent applications of boundary elements has been for the solution of time dependent problems. This has now opened the way for solving more complex problems, such as wave propagation, Navier-Stokes flow, elastodynamics, etc. The first boundary integral solution for the diffusion equation was due to Rizzo and Shippy [23], who used the Laplace transform to remove the time dependence. This approach was discussed by Liggett and Liu [24] where some of its disadvantages are pointed out. Chang et al. [25] in 1973 solved the problem in the time domain using space and time piecewise constant values for the variables. This approach has been extended by Wrobel and Brebbia [26] who applied it to solve complex temperature problems. Brebbia and Walker [20] proposed a step by step finite difference scheme for the time derivatives. The use of a time dependent fundamental solution implies that one does not need to perform any finite differences on the time derivative and produces an accurate and efficient solution specially when higher order time interpolation functions are employed [27], [28].

The first applications of boundary elements in elastodynamics can be traced to Rizzo and Cruse [30], [31] and the technique is now widely accepted for solving soil-structure interaction problems. Cruse's work has been of fundamental importance for the development of boundary elements as he has contributed to its application not only in elastostatics and elastodynamics but also in fracture mechanics [31] where boundary elements are much more accurate than finite elements.

The boundary element method generally offers the advantage over more classical finite elements, of working with smaller systems of equations and obtaining more accurate results. Because of these advantages the technique is of interest for solving a range of non-linear problems as well as linear ones. The first formulation of boundary elements for plasticity was presented in 1971 by Swendlow and Cruse [32] in a purely theoretical paper which laid the foundations for all subsequent work. However, they did not present the expressions for initial stresses or strains which are needed in stepwise plasticity analysis. These expressions were presented by Mendelson [33] in 1973 for two- and three-dimensional bodies although the plane strain formulation was correctly presented by Mukherjee [34] only in 1977. None of the above authors gave the proper expressions for strains and stresses at internal points. It was Bui [35] who in 1978 published the correct expressions for strains and stresses for three dimensional bodies and finally in 1979, Telles and Brebbia [36] presented the complete formulation for boundary elements applied to two- and three-dimensional problems. Applications relevant to engineering practice are presented in reference [37] and more can be found in this book.

It may be convenient in many applications to combine boundary elements with finite elements, as some subregions of the problem may be more amenable to solution using the latter. This may be achieved in either of two ways: (1) by considering the whole problem using an equivalent BEM approach or (2) converting the BEM subregions into equivalent FEM. The two approaches are described in detail in Brebbia and Georgiou [38]. As the FEM is very well established, with computer codes readily available for its implementation, the consideration of the BE subregion as an additional equivalent finite element matrix seems most attractive. The formulation of this 'equivalent' matrix used to model the BE regions presents however certain problems; at sharp geometric discontinuities there are also discontinuities of surface tractions which require special attention and the equivalent FE matrix formed is not inherently symmetric, unlike the classical FE approach. A technique which overcomes these problems and provides an acceptable FE type formulation using the BEM method has been suggested by Georgiou and Brebbia [40]. The technique is based on an idea advanced by Chaudonneret [39] in 1978.

Ways of symmetrizing the BE matrices have been suggested by Mustoe et al. [41], Margulies [42] and others and are thoroughly discussed in the present book.

2. REVIEW OF THE PAPERS PRESENTED

The 3rd International Conference on Boundary Element Methods held in California during July 1981 constituted a forum for the presentation of the most advanced research in BEM. When comparing the Proceedings - published in this book - with those of previous meetings it is evident that rapid advances have been made in important fields. Regarding applications of BEM, this book contains several important papers dealing with time dependent and vibrations problems, fluid flow and material non-linearities. Several types of non-linearity are discussed, i.e. elastoplasticity, viscoelasticity and no-tension type materials. A substantial number of papers deals with stress analysis, including treatment of body forces, higher order models, crack problems, cyclic symmetry and sliding and other types of conditions. One of the papers even extended the BEM to deal with shell analysis opening a promising new area of research. Geomechanical applications are also well represented and for these problems the advantages of using BEM which can include radiation as well as material damping are obvious.

One of the most interesting set of papers is that dealing with the systematic formulation of boundary element method and mathematical properties such as accuracy, convergence versus order and number of elements. The topic of numerical integration is discussed by several authors and some promising ideas have been put forward.

Many papers deal with the coupling of BE and FE solutions. This is a particularly important field in view of the large number of available FE codes and the convenience of combining both techniques in many practical applications.

Several of the papers presented here refer to potential and fluid flow problems. Symm [43] discusses different ways of finding the capacitance of a quadrilateral based on the direct boundary element formulation. The boundary element solution to the eddy current problem [44] produces a Helmholtz type equation which can be transformed into integral equations. The problem is assumed to be steady state and the results compared against the finite element solution. The BEM is capable of achieving high accuracy with fewer equations than the FEM.

The paper by Curran et al. [45] discusses the solution of time dependent diffusion problem using integration on time as well as in space. Although the method is only applied to one dimensional space dependent problems, the authors conclude that it can be expanded to solve two and three dimensional problems as well. Groenenboom [46] explains in detail how the boundary element in time and space can be applied to solve the wave equation. He discusses the problem of radiating boundary conditions and presents results for a three dimensional rectangular box. His work appears to be the first successful attempt to use BEM to solve the transient wave propagation problem using time as well as space dependent fundamental solutions. Williams [47] uses an integral equation technique to solve finite-amplitude waves and presents several computational techniques to determine the unknown position of the free surface.

Pignole [48] dealing with linear elastic problems, discusses how direct and indirect methods relate to each other. The basic formulation of the BEM is also discussed in the paper by Patterson and Sheikh [49], who proposes moving the singularities to points outside the domain of the problem. In this way the resulting system of equations tolerates higher order singularities in the solution. A series of numerical results demonstrates that the technique gives accurate results. The paper by Danson [50] describes in detail, and in a consistent way, how body forces can be taken to the boundary. The work is particularly important for the efficient implementation of stress boundary element programs and different types of body forces are discussed. The problem of comparison between different BEM and FEM solutions for practical problems is discussed by Seabra Pereira et al. [51]. The authors present the computer performance of different BEM models including CPU times using a general purpose BEM program which has constant, linear and parabolic 2-D and 3-D elements. The advantages of using BEM against FEM are also pointed out and the use of parabolic BEM for stress concentration problems is advocated. Patterson and Sheikh |52| propose using non-conforming BEM other than constants, to avoid the problem of defining the normal at corner points or having errors propagating into the system due to a change in the nature of the boundary conditions. They conclude that interelement continuity is not necessary with BEM and demonstrate this through several numerical examples.

Another integral formulation called by the authors the displacement discontinuity method (DDM) is presented by Dunbar and Anderson [53]. The DDM is a type of boundary integral method wherein the fundamental solution used is the static displacement discontinuity on a finite segment in an infinite or semi-infinite medium. The technique is explained in detail in the paper together with its possible linkage with finite elements. Dubois [54] presents a generalized way of obtaining second order BE models for structural analysis. The method relates displacements, rotations, stresses and couple-stresses and has the advantage of ensuring complete compatibility in complex structural analysis.

The important field of using BEM to solve crack problems is discussed in paper [55]. Several element types are discussed and the BEM is found to be an efficient and accurate tool to solve the problem. Chaudouet [56] presents a way to include cyclic symmetry and sliding between structures in BEM codes. The paper presents some interesting industrial applications.

Antes [57] extends the BEM for the solution of circular cylindrical shells. He develops a new reciprocal theorem and deduces an original boundary element formulation in what appears to be the first application of BEM to shell analysis.

BEM has been particularly successful to study contact problems for which they offer over the advantage over FEM of easily and accurately coupling normal and tangential tractions. The paper by Andersson [58] describes how BEM can be applied to two-dimensional contact problems with friction. Several BEM examples are presented and results compared against other solutions. The problem of indentation of a rubberlike material bonded to a rigid cylinder and indented by another rigid cylinder is studied by Batra [59] using BEM. Results are presented for the pressure distribution over the contact surface, the shape of the indented surface and the stress distribution at the bond surface.

Wong and Hutchinson [60] use BEM to determine the natural frequencies and mode shapes of thin elastic plates of uniform thickness with arbitrary boundary conditions and arbitrarily shaped edges. The complete theory is presented in the paper but numerical results are not given as the computer code is not yet fully operational.

Part III of this book deals with geomechanic problems. Ottenstreuer and Schmid [61] apply BEM to study soil-foundation interaction to obtain a better representation of radiation damping. The soil is assumed to be an elastic homogeneous space and the classical fundamental solution for the steady state case is used. Material damping behaviour is taken into account by introducing a complex modulus of elasticity. The paper by Wood [62] places particular emphasis upon the coupling of FEM to soil BEM to solve soil-structure interaction problems. He describes two computer codes which incorporate BEM in the areas of raft foundation and pile/wall analysis and design. Several interesting practical applications are described. Predeleanu [63] appears to be the first to apply Biot consolidation theory to BEM. The theory is extended to the general class of viscoelastic bodies defined by convolutions. The fundamental singular solution is similar to the one used in the coupled theory of thermoviscoelasticity.

BEM is increasingly becoming an accepted method of solution for non-linear material problems. The paper by Brunet [64] deals with its application to cyclic plasticity and presents a computational procedure for the solution of such classes of non-linear problems employing different types of constitutive relations. This work is closely related to the paper by Telles and Brebbia [65] which presents further developments on the application of BEM to plasticity, namely the extension of the theory to deal with half plane problems. They also show a series of examples of practical importance. The other paper in this section of the book written by Venturini and Brebbia [66] deals with the extension of BEM to no-tension materials such as those present in some applications in geomechanics. The no-tension solution is achieved using an iterative process which consists of applying at each step a series of initial stresses to compensate the tensile stresses, much in the same way as it is done with the FEM. Finally some problems are solved to illustrate the application of BEM to no-tension materials. The examples were selected to compare the BEM solution with already published finite element results and point out the advantage of using the former.

Several important papers are presented in the section dealing with numerical techniques and mathematical principles. Jaswon [67] points out how the scalar and vector field theories used in potential and elastostatics boundary elements can be unified and the analogy between the Green's formula of potential theory and Somigliana's formula of elasticity. **Wendland's** [68] paper discusses the important topic of convergence of boundary integral methods and finds an expression for the asymptotic error estimate. The work by Gourgeon and

Herrera [69] tries to establish a systematic approach to the theoretical foundation of the boundary element method with special reference to the bi-harmonic equation. The problem of convergence of the boundary element solution is again discussed in the paper by Pina, Fernandes and Brebbia [70]. A series of numerical tests are carried out using linear and parabolic elements for two different problems and the BEM results are compared against several FEM solutions. Several meshes are tried to assess the order of convergence and the effect of mesh grading on the magnitude and order of error convergence is also assessed.

The work of Futagami [71] presents the first attempt to combine the BEM with linear programming to develop a new optimization technique. The method is applied to optimal control problems in heat conduction phenomena. He recommends the extension of the technique to transient problems and developing stochastic boundary elements as well as non-linear programming. The paper by Walker [72] discusses the use of approximate fundamental solutions and alternative formulations He advocates using reference points outside the domain to eliminate having to evaluate singular integrals. Walker also discusses the effect of taking approximations to the fundamental solutions. It is interesting to notice that all his numerical work was carried out using a microprocessor. The problem of numerical evaluation of the boundary integral equations is addressed again by Vable and Sikarskie [73] who developed a special algorithm in which the unknown vector is represented as a combination of a Fourier series and piece wise linear function. The authors point out the numerical advantages of their technique which requires much less solution time and could be very efficient for certain non-linear problems. The paper by Caldwell [74] discusses the difficulties associated with the numerical solution of integral equations and their relationship with Gauss quadrature formulae. This is of course an important topic in BEM research. A new type of BE program for two dimensional elasticity is described by Bolteus and Tullberg in reference [75]. They allow for discontinuity in the tractions and propose a method for coupling BE and FEM solutions, dealing in detail with the integration needed to find the influence coefficients and producing some guidelines for 'optimal' integration.

Section VI of these proceedings deal with the topic of coupling the BEM and FEM. This is a topic of practical importance in view of the large number of existing FE codes. The paper by Felippa [76] reviews critically several techniques for coupling FE and BE solutions. He particularly assesses three different techniques relevant to his physical problem, i.e. a three-dimensional structure submerged in an acoustic field and impinged by a pressure shock wave. He stresses the usefulness of BE techniques for discretizing unbounded homogeneous domains governed by linear equations.

Hartmann [77] reviews in a very clear manner the difference between BE and FE matrices and the reasons for the lack of symmetry of the former. He then proposes simple ways of drawing stiffness matrices from BE and coupling them to FE solutions. The paper by Volait [78] describes how stiffness matrices can be obtained using BE, based on an energy minimization technique and introducing possible discontinuity of tractions simply by choosing nodes for the model tractions inside the element. The topic of numerical integration is discussed as well. Beer and Meek [79] propose a coupling of the two methods to analyse circular excavations in infinite domain with the region of plasticity confined to the FE zone. Comparisons are also carried out between coupling FE with BE and with the so called 'Infinite' Finite Elements. The paper by Dendrou and Dendrou [80] also is concerned with the applications of FE/BE techniques in geomechanics. Finite elements are used to describe the liner of an underground excavation and BE to represent the soil. They conclude that the proposed methodology offers both accuracy and cost efficiency. The paper by Katz [81] explains the reasons why it is desirable to link the FE and BE techniques, review several approaches and puts forward the idea of using a method proposed in the Second International Seminar on Boundary Elements, Southampton, 1980. The paper also discusses the numerical integration of singular equations.

3. SUGGESTIONS FOR FURTHER RESEARCH

Several important topics related to boundary solutions or more specifically the boundary element method require further investigation. Boundary element method implies an approximate technique by which the problem dimensions have been reduced by one or in the case of time dependent problems by more than one. The problem becomes a boundary problem and the surface of the domain can be divided into elements over which some approximate technique can be applied. What differentiates boundary element of classical integral equations is its emphasis on elements and compatibility as well as its reliance on general principles such as weighted residuals. This is important as some classical boundary integral solutions suffer from lack of accuracy and convergence due to poor representation of the geometrical shapes. These problems appear not to have been sufficiently investigated.

It is also evident to finite element practitioners that the degree of complexity assured for the geometry and the one for the functions representing the boundary variables are interrelated. We know from variational techniques research that trying to represent the geometry with higher order functions while assuring simple functions for the variables will introduce considerable errors in the solution. This is an important point which seems to have escaped the attention of many researchers applying integral equations.

Although current boundary element practice is heavily dependent on using fundamental solutions based on solving the governing equations under concentrated source and no boundary conditions, it is of the utmost importance to point out that other solutions can be used as well. These other solutions are illustrated by those used in such methods as hybrid finite elements and trigonometric type solutions. Analytical functions which have been obtained over a close domain with certain boundary conditions could also be used as fundamental solutions provided that our domain of interest is within the domain of the analytical solutions. This of course can be easily arranged by a change of dimensions. Once these ideas are better understood we will be able to solve many complex plate and shell and other problems using the wealth of solutions already available as approximate fundamental solutions. The general field of non-linear problem solving using BEM requires further investigation.

Time dependent problems may be one of the most fruitful research topics in BEM. The method seems well suited to solve problems in time and space. FEM solutions in time and space are on the contrary, very uneconomical and have been largely abandoned. BE is now accepted as a valid and efficient way of solving parabolic problems but little is known about its applications to hyperbolic problems such as the transient wave equation. More research should be carried out to determine its accuracy and efficiency for these problems.

Finally, BEM are also well suited for the small computers mainly microprocessors, which are nowadays being used in many engineering offices. BEM solutions tend to require less matrices and data storage than FE solutions while having a large number of repetitive operations such as numerical integration which can be economically carried out by a microcomputer. BE appears ideally suited for computer aided design activities due to the smaller amount of data generally required. This is an important advantage specially when dealing with three dimensional problems and more emphasis should be put into developing efficient CAD systems based on BE.

REFERENCES

1. Kantorovich,L.V. and Krylov, V.I., Approximate Methods of Higher Analysis, Interscience, New York (1958).

2. Courant, R. and Hilbert, D. Methods of Mathematical Physics, Interscience (1953).

3. Reissner, E. A Note on Variational Principles in Elasticity Int. J. Solids and Structures, 1, pp.93-95 and 357 (1965).

4. Pian, T.H.H. and Tong, P., Basis of Finite Element Method for Solid Continua, Int. J. Numerical Methods Engng., 1, pp.3-28 (1969).

5. Washizu, K., Variational Methods in Elasticity and Plasticity, 2nd edn, Pergamon Press, New York (1975).

6. Muskhelishvili, N.I., Some Basic Problems of the Mathematical Theory of Elasticity, P. Noordhoff, Ltd., Groningen (1953).

7. Mikhlin, S.G., Integral Equations, Pergamon Press, New York (1957).

8. Kupradze, O.D. Potential Methods in the Theory of Elasticity, Daniel Davey & Co., New York (1965).

9. Smirnov, V.J. Integral Equations and Partial Differential Equations, in A Course in Higher Mathematics, Vol. IV, Addison-Wesley (1964).

10. Kellogg, P.D., Foundations of Potential Theory, Dover, New York (1953).

11. Jaswon, M.A., Integral Equation Methods in Potential Theory, I, Proc. R. Soc., Ser. A, p.273 (1963).

12. Symm, G.T., Integral Equation Methods in Potential Theory, II, Proc. R. Soc., Ser. A, p.275 (1963).

13. Massonnet, C.E., Numerical Use of Integral Procedures in Stress Analysis, Stress Analysis, Zienkiewicz, O.C. and Holister, G.S. (eds), Wiley (1966).

14. Hess, J.L. and Smith, A.M.O., Calculation of Potential Flow about Arbitrary Bodies, Progress in Aeronautical Sciences, Vol.8, Kuchemann, D. (Ed), Pergamon Press (1967).

15. Cruse, T.A. and Rizzo, F.J., A Direct Formulation and Numerical Solution of the General Transient Elasto-Dynamic Problem, I, J. Math. Analysis Applic., 22 (1968).

16. Hadid, H.A., An Analytical and Experimental Investigation into the Bending Theory of Elastic Conoidal Shells, Ph.D. Thesis, University of Southampton (1964).

17. Hajdin, N., A Method for Numerical Solutions of Boundary Value Problems and its Application to Certain Problems in the Theory of Elasticity, Belgrade University Publication (1958).

18. Brebbia, C.A., Weighted Residual Classification of Approximate Methods, Applied Mathematical Modelling Journal, Vol.2, No.3, September (1978).

19. Brebbia, C.A., The Boundary Element Method for Engineers, Pentech Press, London, Halstead, New York, (1978).

20. Brebbia, C.A. and S. Walker, Boundary Element Techniques in Engineering, Butterworths, London,(1979).

21. Brebbia, C.A. (Ed.), Recent Advances in Boundary Element Methods, Proc. 1st Int. Conference Boundary Element Methods, Southampton University, 1978, Pentech Press, London (1978).

22. Brebbia, C.A. (Ed.), New Developments in Boundary Element Methods, Proc. 2nd Int. Conference Boundary Element Methods, Southampton University, 1980, CML Publications, Southampton (1980).

23. Rizzo, F.J. and Shippy, D.J. , A Method of Solution for Certain Problems of Transient Heat Conduction, AIAA J., 8, No. 11, pp.2004-2009 (1970).

24. Liggett, J.A. and Liu, P.L., Unsteady Flow in Confined Aquifers: A Comparison of Two Boundary Integral Methods, Water Resources Res. 15, No. 4, pp.861-866 (1979).

25. Chang, Y.P., Kang, C.S. and Chen, D.J., The Use of Fundamental Green's Function for the Solution of Problems of Heat Conduction in Anisotropic Media, Int. J. Heat Mass Transfer, 16, pp.1905-1918 (1973).

26. Wrobel, L.C. and Brebbia, C., The Boundary Element Method for Steady-State and Transient Heat Conduction, 1st Int. Conf. Numerical Methods in Thermal Problems, Swansea, Taylor, C. and Morgan, K. (Eds), Pineridge Press (1979).

27. Wrobel, L.C. and Brebbia, C., A Formulation of the Boundary Element Method for Axisymmetric Transient Heat Conduction, Int. J. Heat Mass Transfer,

28. Wrobel, L.C. and Brebbia, C., Boundary Elements in Thermal Problems, Numerical Methods in Heat Transfer, Lewis, R.W. (ed.), Wiley, Chichester, England (1981).

29. Cruse, T.A. and Rizzo, F.J., A Direct Formulation and Numerical Solution of the General Elastodynamic Problem', I, J. Math. Analysis Applic., 22 (1968).

30. Cruse, T.A. and Rizzo, F.J., A Direct Formulation and Numerical Solution of the General Transient Elastodynamic Problem, II, J. Math. Analysis Applic., 22 (1968).

31. Cruse, T.A. and Vanduren, W., Three Dimensional Elastic Stress Analysis of a Fracture Specimen with Edge Crack, Int. J. Fract. Mech., 7, pp.1-15 (1971).

32. Swedlow, J.L. and Cruse, T.A., Formulation of Boundary Integral Equations for Three-Dimensional Elasto-Plastic Flow, Int. J. Solids Struct., 7, 1673-1683, 1971.

33. Mendelson, A., Boundary Integral Methods in Elasticity and Plasticity, Report No. NASA TN D-7418 (1973).

34. Mukherjee, S., Corrected Boundary Integral Equation in Planar Thermoelasticity, Int. J. Solids Struct., 13, pp.331-335 (1977).

35. Bui, H.D., Some Remarks about the Formulation of Three-Dimensional Thermoelastoplastic Problems by Integral Equations, Int. J. Solids Struct., No.14, pp.935-939 (1978).

36. Telles, J.C.F. and Brebbia, C.A., On the Application of the Boundary Element Method to Plasticity, Appl. Math. Modelling, 3, pp.446-470 (1979).

37. Telles, J.C.F. and Brebbia, C.A., The Boundary Element Method in Plasticity, in New Developments in Boundary Element Methods, Brebbia, C.A. (Ed.), CML Publications, England (1980).

38. Brebbia, C.A. and Georgiou, P., Combination of Boundary and Finite Elements in Elastostatics, Applied Math. Modelling, 3. pp.213-220 (1978).

39. Chaudonneret, M., On the Discontinuity of the Stress Vector in the Boundary Integral Equation Method for Elastic Analysis, Recent Advances in Boundary Element Methods, Brebbia, C.A. (Ed.), Pentech Press, London (1978).

40. Georgiou, P. and Brebbia, C., On the Combination of Boundary and Finite Element Solution, Applied Math. Modelling, To be published.

41. Mustoe, G., A Symmetric Direct Boundary Integral Equation Method for Two-Dimensional Elastostatics, Paper presented at the 2nd Int. Seminar on Boundary Element Methods, Southampton University, March (1980).

42. Margulies, M., Exact Treatment of the Exterior Problem in the Combined FEM/BEM, Paper presented at the 2nd Int. Seminar on Boundary Element Methods, Southampton University, March, 1980, (Ed. C. Brebbia), CML Publications (1980).

43. Symm, G.T., Two Methods for Computing the Capacitance of a Quadrilateral, in this book.

44. Salon, S.J., J.M. Schneider and S. Uda, Boundary Element Solutions to the Eddy Current Problems, in this book.

45. Curran, D.A.S., B.A. Lewis and M. Cross, Numerical Solution of the Diffusion Equation by a Potential Method, in this book.

46. Groenenbook, P.H.L., The Application of Boundary Elements to Steady and Unsteady Potential Fluid Flow Problems in Two and Three Dimensions, in this book.

47. Williams, J.M., Computing Strategy in the Integral Equation Solution of Limiting Gravity Waves in Water, in this book.

48. Pignole, M., On the Construction of the Boundary Integral Representation and Connected Integral Equations for Homogeneous Problems of Plane Linear Elastostatics, in this book.

49. Patterson, C. and M.A. Sheikh, Regular Boundary Integral Equations for Stress Analysis, in this book.

50. Danson, D., A Boundary Element Formulation of Problems in Linear Isotropic Elasticity with Body Forces, in this book.

51. Seabra Pereira, M.F., C.A. Mota Soares and L.M. Oliveira Faria, A Comparative Study of Several Boundary Elements in Elasticity, in this book.

52. Patterson, C. and M.A. Sheikh, Non-conforming Boundary Elements for Stress Analysis, in this book.

53. Dunbar, W.S. and D.L. Anderson, The Displacement Discontinuity Method in Three Dimensions, in this book.

54. Dubois, M., A Unified Second Order Boundary Element Method for Structures Analysis, in this book.

55. Balas, J. and J. Sladek, Method of Boundary Integral Equations for Analysis of Three Dimensional Crack Problems, in this book.

56. Chaudouet, A., Cyclic Symmetry and Sliding between Structures by the Boundary Integral Equation Method, in this book.

57. Antes, H., On Boundary Integral Equations for Circular Cylindrical Shells, in this book.

58. Andersson, T., The Boundary Element Method applied to Two-Dimensional Contact Problems with Friction, in this book.

59. Batra, R.C., Quasistatic Indentation of a Rubber Covered Roll by a Rigid Roll - The Boundary Element Solution, in this book.

60. Wong, G,K.K. and J.R. Hutchinson, An Improved Boundary Element Method for Plate Vibrations, in this book.

61. Ottenstreuer, M. and G. Schmid, Boundary Elements Applied to Soil-Foundation Interaction, in this book.

62. Wood, L.A., The Implementation of Boundary Element Codes in Geotechnical Engineering, in this book.

63. Predeleanu, M., Boundary Integral Method for Porous Media, in this book.

64. Brunet, M., Numerical Analysis of Cyclic Plasticity using the Boundary Integral Equation Method, in this book.

65. Telles, J.C.F. and C.A. Brebbia, New Developments in Elastoplastic Analysis, in this book.

66. Venturini, W.S. and C.A. Brebbia, The Boundary Element Method for the Solution of No-tension Material, in this book.

67. Jaswon, M.A., Some Theoretical Aspects of Boundary Integral Equations, in this book.

68. Wendland, W.L., On the Asymptotic Convergence of Boundary Integral Methods, in this book.

69. Gourgeon, H. and I. Herrera, Boundary Methods. C-complete Systems for the Biharmonic Equations, in this book.

70. Pina, H.L.G., J.L.M. Fernandes and C.A. Brebbia, The Effect of Mesh Refinement in the Boundary Element Solution of Laplace's Equation with Singularities, in this book.

71. Futagami, T., Boundary Element and Linear Programming Method in Optimization of Partial Differential Systems, in this book.

72. Walker, S., Approximate Fundamental Solutions and Alternative Formulations, in this book.

73. Vable, M. and D.L. Sikarskie, An Efficient Algorithm for the Numerical Evaluation of Boundary Integral Equations, in this book.

74. Caldwell, J., Solution of the Dirichlet Problem using the reduction to Fredholm Integral Equations, in this book.

75. Bolteus, L. and O. Tullberg, BEMSTAT - A New Type of Boundary Element Program for Two-Dimensional Elasticity Problems, in this book.

76. Felippa, C.A., Interfacing Finite Elements and Boundary Element Discretizations, in this book.

77. Hartmann, F., The Derivation of Stiffness Matrices from Integral Equations, in this book.

78. Volait, F., Three Dimensional Super-Element by the Boundary Integral Equation Method for Elastostatics, in this book.

79. Beer, G. and J.L. Meek, The Coupling of Boundary and Finite Element Methods for Infinite Domain Problems in Elasto-Plasticity, in this book.

80. Dendrou, B.A. and S.A. Dendrou, A Finite Element-Boundary Integral Scheme to Simulate Rock Effects on the Liner of an Underground Intersection , in this book.

81. Katz, C., The Use of Green's Function in the Numerical Analysis of Potential, Elastic and Plate Bending Problems, in this book.

SECTIONS I - VI

Section I
Potential and Fluid Flow Problems

TWO METHODS FOR COMPUTING THE CAPACITANCE OF A QUADRILATERAL

George T. Symm

National Physical Laboratory, Teddington, Middlesex, U.K.

ABSTRACT

The problem of finding the capacitance of a (curvilinear) quadrilateral is formulated in two distinct ways, each involving the solution of a boundary value problem for Laplace's equation. In each case the boundary value problem is solved by the direct boundary element method based upon Green's third identity. Some numerical results are presented.

INTRODUCTION

Given four points A, B, C, D on a closed contour L bounding a simply-connected plane domain G, the capacitance K between AB and CD may be defined (Campbell, 1975, p.27) by

$$K = \frac{1}{4\pi} \int_{AB} \phi'(q)dq, \tag{1}$$

where ϕ satisfies Laplace's equation

$$\frac{\partial^2\phi}{\partial x^2} + \frac{\partial^2\phi}{\partial y^2} = 0 \tag{2}$$

in the domain G subject to the boundary conditions

$$\phi = 1 \text{ on CD}, \quad \phi = 0 \text{ on DA, AB and BC}, \tag{3}$$

q is a vector variable specifying a point on L (in this case on the segment AB), dq is the differential increment of L at q and the prime denotes differentiation along the normal to the boundary L directed <u>into</u> the domain G at the point **q**. More precisely, K represents the direct capacitance per unit length, between the faces AB and CD, of an infinite cylinder, of cross-section G (Lampard, 1957). Solution of the boundary-value problem defined by Equations (2) and (3) by the boundary element (integral equation) method based upon Green's third identity yields ϕ' on L directly, whence K may be determined from Equation (1) by integration. We shall refer to this procedure as Method 1.

Alternatively (Gaier, 1979), the capacitance between AB and CD is given by

$$K = K(M) = \frac{2}{\pi^2} \sum_{n=1,3,5...} [n \sinh(n\pi M)]^{-1}, \tag{4}$$

where M is the conformal module of the quadrilateral ABCD, defined by

$$M = b/a \tag{5}$$

when the domain G is mapped conformally onto the rectangle R with corners 0, a, a+ib, ib (a>0, b>0) in such a way that AB and CD correspond to the horizontal sides of R. In this case, defining

$$m = a/b, \tag{6}$$

so that

$$M = 1/m, \tag{7}$$

we observe, from Gaier (1972), that m minimises the Dirichlet integral

$$D_G(\phi) = \iint_G (\phi_x^2 + \phi_y^2)\, dxdy, \tag{8}$$

which is invariant under conformal transformation, over the set of functions ϕ, continuous in $\bar{G}$ and piecewise smooth in G, such that $\phi = 0$ on AB and $\phi = 1$ on CD. Indeed, by Dirichlet's Principle (Courant, 1950),

$$m = D_G(\phi), \tag{9}$$

where ϕ satisfies Laplace's equation (2) subject to the boundary conditions

$$\phi = 0 \text{ on AB}, \quad \phi = 1 \text{ on CD} \quad \text{and} \quad \phi' = 0 \text{ on BC and DA}. \tag{10}$$

By the divergence theorem, Dirichlet's integral (8) may be written, in our notation, as

$$D_G(\phi) = - \int_L \phi(q)\phi'(q)dq \tag{11}$$

whence, from Equation (9),

$$m = - \int_{CD} \phi'(q)dq, \qquad (12)$$

where ϕ satisfies the boundary-value problem defined by Equations (2) and (10). Solution of this boundary-value problem by means of Green's third identity again yields ϕ' on L directly, whence m may be determined from Equation (12) by integration; M follows immediately from Equation (7), whence the capacitance K may be determined from Equation (4). We shall refer to this procedure as Method 2.

It is clear from Equation (7) that m is the conformal module of the "conjugate" quadrilateral BCDA. It follows that if we solve Laplace's equation (2) subject to the boundary conditions

$$\phi = 0 \text{ on BC}, \quad \phi = 1 \text{ on DA} \quad \text{and} \quad \phi' = 0 \text{ on AB and CD}. \qquad (13)$$

then

$$M = - \int_{DA} \phi'(q)dq, \qquad (14)$$

whence the capacitance K may again be determined from Equation (4). This provides an alternative form of Method 2, and a further alternative to Method 1, to which we shall refer as Method 2a.

We note at this point that the capacitance between AB and CD is the same as the capacitance between CD and AB, so that each of the methods above may also be applied to the rotated quadrilateral CDAB. In general, therefore, we have a total of six possible ways of computing the capacitance K. We shall see that the optimal choice between these six formulations depends, in general, upon the geometry of the contour L and the positions of the four points A, B, C, D.

INTEGRAL EQUATION FORMULATION AND SOLUTION

Given that ϕ satisfies Laplace's equation (2) in the domain G, Green's third identity may be written at a point p on the boundary L in the form

$$\int_L \phi'(q)\log|q-p|dq - \int_L \phi(q)\log'|q-p|dq = \pi\phi(p), \qquad (15)$$

whenever the boundary L is smooth at the point p (Jaswon and Symm, 1977). Thus, if ϕ is prescribed everywhere on L, as by conditions (3) in Method 1, Equation (15) becomes an integral equation of the first kind (with a weakly singular kernel) for the boundary normal derivative ϕ'. Similarly, if ϕ is prescribed on part of the boundary and ϕ' on the remainder, as by conditions (10) in Method 2 (or conditions (13) in Method

2a), Equation (15) yields a pair of coupled integral equations for the complementary unknowns, ϕ where ϕ' is given and vice versa. In either case, solution of the integral equations yields ϕ' everywhere on L and in particular on AB in Method 1 and on CD in Method 2 (or DA in Method 2a). The required results may then be obtained by quadrature.

This formulation is implemented numerically by dividing the boundary L into N smooth intervals (elements) in each of which ϕ and ϕ' are approximated by constants. We denote these constants by

$$\phi_i \text{ and } \phi'_i, \; i = 1,2,\ldots,N, \tag{16}$$

and apply Equation (15) at one "nodal" point q_i in each interval of L to obtain

$$\sum_{j=1}^{N} \phi'_j \int^{(j)} \log|q-q_i|\,dq - \sum_{j=1}^{N} \phi_j \int^{(j)} \log'|q-q_i|\,dq - \pi\phi_i = 0, \quad i = 1,2,\ldots,N, \tag{17}$$

where $\int^{(j)}$ denotes integration over the jth interval of L. Prescribing one of the constants (16) in each interval, by applying the boundary conditions (3) or (10) (or (13)) at the nodal points, we thus obtain a system of N simultaneous linear algebraic equations, in N unknowns, whose solution includes values of ϕ' on AB and CD (or DA).

In order to evaluate the coefficients in Equations (17), we approximate (if necessary) each interval of L by the two chords which join its end points to the nodal point within it. Then all the integrations can be carried out analytically as described in detail by Symm and Pitfield (1974) and by Jaswon and Symm (1977). Indeed, the setting up and solution of Equations (17) is all carried out here by Fortran subroutines (Symm, 1980) based upon this discretisation.

Finally, if there are S intervals on AB and our interval numbering starts from A, we approximate K, as given by Equation (1), by

$$K^{(1)} = \frac{1}{4\pi} \sum_{j=1}^{S} \phi'_j h_j . \tag{18}$$

where h denotes the approximate length of an interval as given by its two chords. In Method 2, we approximate K, via Equations (4), (7) and (12), by

$$K^{(2)} = K(M^{(2)}) \text{ with } M^{(2)} = 1/m^{(2)} \text{ and } m^{(2)} = -\sum \phi'_j h_j , \tag{19}$$

where the summation extends over those intervals of L which form CD. Similarly, in Method 2a, we approximate K by

$$K^{(2a)} = K(M^{(2a)}) \text{ with } M^{(2a)} = -\sum \phi'_j h_j, \tag{20}$$

where the summation extends over those intervals of L which form DA.

EXAMPLES

The methods presented above have been applied to a number of examples as follows.

Problem 1

To compute the capacitance K between opposite sides AB and CD of a unit square (Figure 1). In this case, it is immediately evident that M = 1.0 whence it follows from Equation (4) that K = 0.01756. Indeed, it is obvious from Method 2 that K will have this value for any quadrilateral with a similar degree of symmetry, as proved by Lampard (1957), who derived the analytic formula

$$K = \log_e 2/4\pi^2. \tag{21}$$

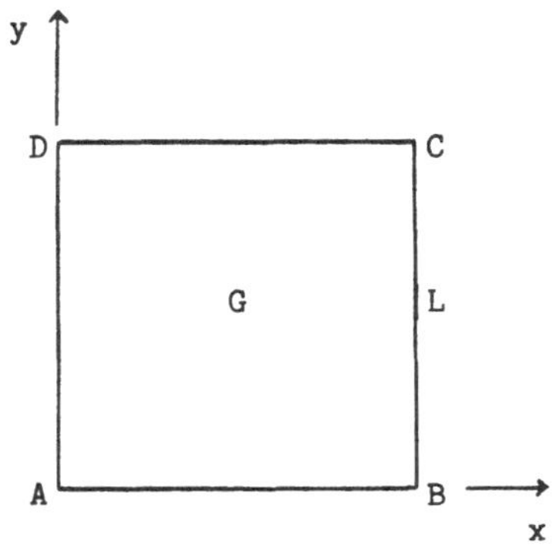

Figure 1. Domain of Problem 1

Computed values of K and M, for various subdivisions of the boundary L, are presented in Table 1.

Table 1
Capacitance K and Module M for a unit square (Figure 1)

N	$K^{(1)}$	$M^{(2)}$	$K^{(2)}$	$M^{(2a)}$	$K^{(2a)}$
16	0.01744	0.98068	0.01866	1.01970	0.01650
32	0.01755	0.99360	0.01792	1.00644	0.01720
64	0.01756	0.99793	0.01767	1.00208	0.01744
128	0.01756	0.99934	0.01759	1.00066	0.01752
64*		0.99985	0.01757	1.00015	0.01755
128*		0.99996	0.01756	1.00004	0.01756

The first four sets of results in Table 1 correspond to subdivisions of the boundary L into equal intervals and show that, for this example, Method 1 converges most rapidly. For these boundary subdivisions,

$$m^{(2)} = M^{(2a)} \tag{22}$$

by symmetry, whence, from Equation (7),

$$M^{(2)}M^{(2a)} = 1 \tag{23}$$

and $K^{(2)}$ and $K^{(2a)}$ bound K (from above and below respectively).

Improved results for Methods 2 and 2a may be obtained by noting that the corresponding boundary value problems have analytic solutions $\phi = y$ and $\phi = 1 - x$ respectively, relative to the coordinate axes shown in Figure 1. Thus both ϕ and ϕ' are constants on AB and CD in Method 2 and on BC and DA in Method 2a. The results marked * in Table 1 correspond to subdivisions which reflect this behaviour by treating these sides of the square as single intervals and dividing the other two sides into equal intervals. (Of course, in practical problems, analytic solutions of the boundary value problems for Methods 2 and 2a are not generally known. However, examination of results for successive values of N with uniform intervals will often reveal sections of the boundary L along which ϕ and ϕ' vary relatively slowly and may therefore be approximated by constants over longer intervals; cf. Problem 3 below.)

In view of the symmetry of the square, no different results are obtainable by rotating the quadrilateral in this problem.

Problem 2
To compute the capacitance K between opposite sides AB and CD of a 2 by 1 rectangle (Figure 2). In this case, it is immediately evident that M = 0.5 whence, from Equation (4), K = 0.08930.

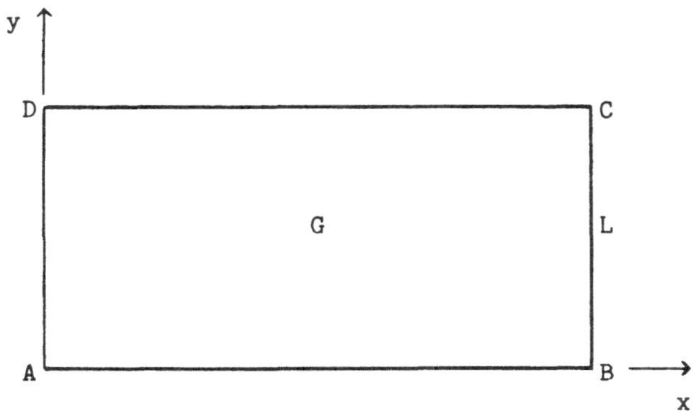

Figure 2. Domain of Problem 2

Computed values of K and M, for various subdivisions of the boundary L, are presented in Table 2.

Table 2
Capacitance K and Module M for a rectangle (Figure 2)

N	$K^{(1)}$	$M^{(2)}$	$K^{(2)}$	$M^{(2a)}$	$K^{(2a)}$
16	0.09332	0.49673	0.09034	0.52638	0.08145
32	0.09070	0.49888	0.08965	0.50852	0.08663
64	0.08983	0.49963	0.08942	0.50273	0.08845
128	0.08954	0.49988	0.08934	0.50087	0.08903
96E	<u>0.08930</u>	0.49950	0.08946	0.50070	0.08903
96*		0.49999	<u>0.08930</u>	0.50010	0.08927

In Table 2, the first four sets of results correspond to boundary subdivisions with N/4 equal intervals on each side of the rectangle. In this case, Method 2 converges most rapidly.

The results marked E in Table 2 correspond to a subdivision of the boundary L into equal intervals all round. Whilst Method 1 is much improved, the results of Method 2 are worse than before since, as in the previous example, ϕ and ϕ' are constants on CD; extra intervals here lead to oscillatory solutions for ϕ' near C and D - a typical example of "Gibbs phenomenon" (Hartree, 1952). These results may be improved, as in Problem 1, by treating AB and CD as single intervals. Such results and results of a similar treatment for Method 2a are given in the line marked * in Table 2. Here the results of Method 2 are better than those of Method 2a since the variation in ϕ, from 0 to 1, takes place over the shorter side of the rectangle in Method 2.

As in the previous example, no different results are obtainable by rotating this quadrilateral in view of its symmetry.

Problem 3

To compute the capacitance K between sides AB and CD of the L-shaped domain, formed by removing a unit square from a square of side 2, shown in Figure 3. In this case, m = 0.577350 (Gaier, 1972), whence M = 1.7321 from Equation (7) and K = 0.001756 from Equation (4).

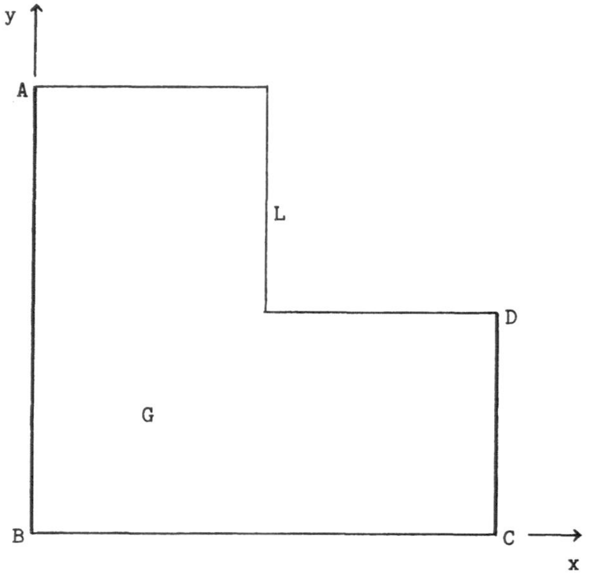

Figure 3. Domain of Problem 3

Computed values of K and M, for various subdivisions of the boundary L are presented in Table 3.

Table 3
Capacitance K and Module M for an L-shape (Figure 3)

N	$K^{(1)}$	$M^{(2)}$	$K^{(2)}$	$M^{(2a)}$	$K^{(2a)}$
48	0.001534	1.7224	0.001810	1.7348	0.001741
96	0.001621	1.7295	0.001771	1.7327	0.001753
128E	0.001764	1.7326	0.001753	1.7320	0.001756
48	0.001892	1.7278	0.001780	1.7358	0.001736
96	0.001851	1.7312	0.001761	1.7330	0.001751
128E	0.001763	1.7333	0.001750	1.7323	0.001755
128*	0.001757	1.7321	0.001756	1.7320	0.001756

In Table 3, the first two sets of results correspond to equal numbers of intervals on each side of the boundary and the third set marked E, to equal intervals all round (i.e. with extra intervals on AB and BC). These results show that

Method 2a converges most rapidly in this example.

The remaining sets of results are for the rotated quadrilateral, where the rôles of AB and CD are interchanged. These results are very similar to the previous ones for the same data. The improved results, in the line marked *, correspond to selected boundary subdivisions which take account of the behaviour of the boundary solution. In Method 1, the intervals on sides adjacent to points A and B (where ϕ is discontinuous in the rotated problem) are graded in size - reducing in length towards these points using a feature of the library software (Symm, 1980). In Method 2, sides AB and CD are each divided into 16 equal intervals and the remaining sides into 24 equal intervals each. In Method 2a, sides AB and CD are each divided into 24 equal intervals, BC into 32 equal intervals and the remaining sides into 16 equal intervals each.

<u>Problem 4</u>

To compute the capacitance K between sections AB and CD of the boundary of the L-shaped domain, formed by removing a unit square from a square of side 2, shown in Figure 4. In this case, m = 1.279262 (Gaier, 1972), whence M = 0.7817 from Equation (7) and K = 0.0351 from Equation (4).

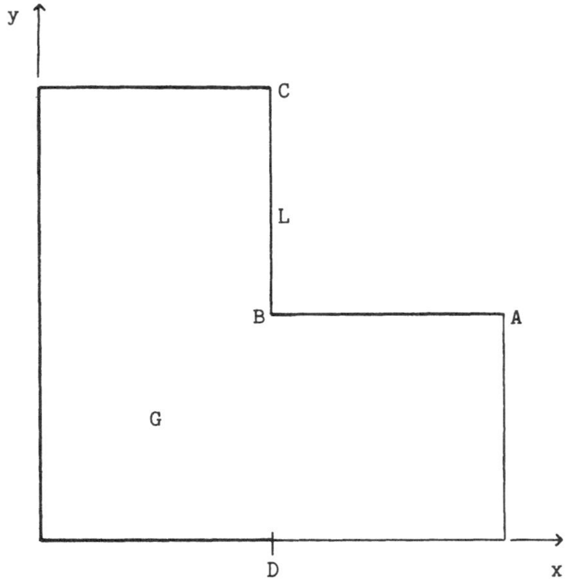

Figure 4. Domain of Problem 4

Computed values of K and M, for various subdivisions of the boundary L are presented in Table 4.

Table 4
Capacitance K and Module M for an L-shape (Figure 4)

N	$K^{(1)}$	$M^{(2)}$	$K^{(2)}$	$M^{(2a)}$	$K^{(2a)}$
56	0.0347	0.8650	0.0269	0.7053	0.0449
112	0.0349	0.8326	0.0299	0.7334	0.0410
168	0.0350	0.8199	0.0311	0.7449	0.0395
56	0.0349	0.8651	0.0269	0.7050	0.0449
112	0.0350	0.8325	0.0299	0.7333	0.0410
168	0.0351	0.8199	0.0311	0.7448	0.0395
168*		0.7856	0.0347	0.7778	0.0356

In Table 4, the first three sets of results correspond to equal numbers of uniform intervals on each of the seven boundary sections, the base of the domain G, on the x-axis, being divided into two sections at D. These results show that Method 1 converges most rapidly in this example.

The remaining sets of results are for the rotated quadrilateral, where the rôles of AB and CD are interchanged. These results are very similar to the previous ones for the same data. The results of Methods 2 and 2a are slow to converge because of boundary singularities at the points B and D. These results may be considerably improved by introducing graded intervals, as in the previous example, in the neighbourhood of these points. In particular, the results marked * in Table 4 correspond to modifications of the preceding boundary subdivision to include 12 graded intervals over a length 0.3 and 12 equal intervals over the remaining length 0.7 of each of the boundary sections BA, BC and the base section either to the left of D (Method 2) or to the right of D (Method 2a).

CONCLUSION

We have described here two methods with variations, totalling six methods in all, for computing the capacitance of a quadrilateral on a contour L. It is evident from the results obtained that which method is best will generally depend upon the geometry of L and upon the positions of the four points on it. In particular, Method 1 appears to be by far the best when, as in Problem 4, Methods 2 and 2a introduce boundary singularities at points of the quadrilateral. However, in other cases, as Problems 2 and 3 show, Method 2 or Method 2a may be the best.

Whilst the considered examples all have known analytic solutions, it is also evident that in a general problem the accuracy of the results of the various methods may be judged from the measure of agreement between them. Bearing this in mind, we observe finally that the equivalence of the various formulations may be useful in the solution of boundary value problems other than the capacitance problem, e.g. in problems of steady-state heat conduction.

ACKNOWLEDGMENT

The author wishes to thank his colleague G.F. Miller for some useful comments on the first draft of this work.

REFERENCES

Campbell, J.B. (1975) Finite difference techniques for ring capacitors. J. Eng. Maths., 9: 21-28.

Courant, R. (1950) Dirichlet's Principle, Conformal Mapping, and Minimal Surfaces. Interscience, New York.

Gaier, D. (1972) Ermittlung des konformen Moduls von Vierecken mit Differenzenmethoden. Numer. Math., 19: 179-194.

Gaier, D. (1979) Capacitance and the conformal module of quadrilaterals. J. Math. Anal. Appls., 70: 236-239.

Hartree, D.R. (1952) Numerical Analysis. Clarendon Press, Oxford.

Jaswon, M.A. and Symm, G.T. (1977) Integral Equation Methods in Potential Theory and Elastostatics. Academic Press, London.

Lampard, D.G. (1957) A new theorem in electrostatics with applications to calculable standards of capacitance. Proc. I.E.E., Monograph No. 216M, 104C: 271-280.

Symm, G.T. and Pitfield, R.A. (1974) Solution of Laplace's Equation in Two Dimensions. NPL Report NAC 44.

Symm, G.T. (1980) The Robin problem for Laplace's equation. NPL Report DNACS 32/80.

BOUNDARY ELEMENT SOLUTIONS TO THE EDDY CURRENT PROBLEM

S. J. Salon, J.M. Schneider, S. Uda
Rensselaer Polytechnic Institute

ABSTRACT

The two-dimensional eddy current problem is formulated in terms of an electric vector potential function. The resulting Helmholtz equation is transformed into a Fredholm Integral equation which involves the unknown vector potential and its normal derivative at the boundary. A set of simultaneous equations is found which approximates the integral. Results are compared to a finite element solution for a sample problem.

I. Introduction

As manufacturers of electrical equipment strive for higher efficiency, more effort has been put into obtaining efficient and accurate numerical means to predict stray losses. A large class of problems involves a slowly time-varying magnetic field in a conducting medium. In these problems, the displacement currents can be ignored when compared with the conduction current. There is extensive literature on finite difference (Roberts (1959) and King (1966)) and Finite Element (Carpenter (1975), Chari (1973), Carpenter (1977), Sato (1977), Salon (1979)) solutions to these problems. In this paper, the eddy current problem is formulated using an electric vector potential function. Only the two-dimensional case is considered, so that the eddy currents will have two components, say x and y, and be constant in the third dimension z. With the restrictions that the displacement currents be ignored and that the material is isotropic and linear, the governing differential equation is derived in section II. By means of Green's identity, this differential equation is then transformed into an integral equation. With certain restrictions placed on the forcing function term, the potential can be written entirely as a line integral on the boundary. This is shown in section III. In section IV, a set of simultaneous equations is obtained and in section V, a sample problem is worked and the results are compared to a finite element solution.

II. Electric Vector Potential

The Maxwell's Equations which apply are

$$\nabla \times H = J \tag{1}$$

and

$$\nabla \times E = \frac{\partial B}{\partial t} \tag{2}$$

along with the continuity equation

$$\nabla \cdot J = 0 \tag{3}$$

and the constitutive relationships

$$J = \sigma E \tag{4}$$

and

$$B = \mu H \tag{5}$$

We now define the electric vector potential T as

$$\nabla \times T = J \tag{6}$$

Thus T is a function similar to H and can differ from H only by the gradient of a scalar potential,

$$T = H + \nabla\phi \tag{7}$$

Substituting Equation (6) into Equation (2) and using the constitutive relationships

$$\nabla \times \nabla \times T = -\mu\sigma\frac{\partial H}{\partial t} \tag{8}$$

Assuming steady state sinusoidal conditions, Equation (8) becomes

$$\nabla \times \nabla \times T = -j\omega\mu\sigma H \tag{9}$$

We now choose $\nabla \cdot T$ to equal zero. H can be thought of as being composed of two components, an applied field, H_o, and the eddy current reaction field T. Using a vector identity, we obtain

$$\nabla^2 T - j\omega\mu\sigma T = j\omega\mu\sigma H_o \tag{10}$$

Defining

$$\alpha^2 = j\omega\mu\sigma \tag{11}$$

we obtain Helmholtz's equation,

$$\nabla^2 T - \alpha^2 T = \alpha^2 H_o \tag{12}$$

III. Boundary Element Formulation of the Electric Vector Potential

In this section, the Helmholtz equation for the two-dimensional eddy current problem will be expressed in terms of a boundary integral. We begin with Equation (12),

$$\nabla^2 T - \alpha^2 T = \alpha^2 H_o \tag{12}$$

Consider the Green's function, $G(\xi,\eta;x,y)$ which satisfies the equation

$$\nabla^2 G - \alpha^2 G = \delta(\xi - x, \eta - y) \tag{13}$$

where x, y and ξ,η are the field and source points respectively. Multiplying Equation (13) by T and Equation (12) by G, subtracting the two equations and integrating over the region R

$$\iint_R T\delta(\xi-x,\eta-y)\,d\xi d\eta = \alpha^2\iint_R H_o G\,d\xi d\eta + \iint_R (T\nabla^2 G - G\nabla^2 T)\,d\xi d\eta \qquad (14)$$

Using Green's theorem on the second integral on the right side of Equation (14) results in

$$T(x,y) = \alpha^2\iint_R H_o G\,d\xi d\eta + \int_B \left(\frac{T\partial G}{\partial n} - \frac{G\partial T}{\partial n}\right)dB \qquad (15)$$

In order to evaluate the Green's function, we return to Equation (13), which reduces to

$$\frac{\partial^2 G}{\partial r^2} + \frac{1}{r}\frac{\partial G}{\partial r} - \alpha^2 G = \delta(\xi-x,\eta-y) \qquad (16)$$

where

$$r = \sqrt{(\xi-x)^2 + (\eta-y)^2} \qquad (17)$$

The solution of Equation (16) , is

$$G(\xi,\eta;x,y) = CI_o(\alpha r) + DK_o(\alpha r) \qquad (18)$$

where I_o and K_o are modified Bessel functions of the first and second kinds order zero, respectively. From the behavior of this function at infinity we deduce that C must equal zero. The constant D is found to be $\frac{1}{2\pi}$ by integrating Equation (16) over a small disk located at r = o. The term $\frac{\partial G}{\partial n}$ in Equation (15) is evaluated as

$$\frac{\partial G}{\partial n} = \frac{\partial G}{\partial r}\frac{\partial r}{\partial n} \qquad (19)$$

where

$$\frac{\partial G}{\partial r} = -\frac{\alpha}{2\pi}K_1(\alpha r) \qquad (20)$$

and

$$\frac{\partial r}{\partial n} = -\cos\psi \qquad (21)$$

where K_1 is the modified Bessel function of the first kind order one. The terms are defined in Figure 1.

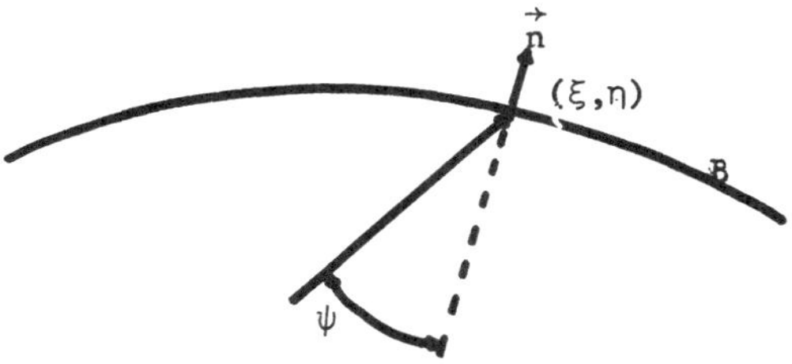

Figure 1 Section of boundary

Using Equations (15), (18), (20), and (21), we obtain

$$T(x,y) = \frac{\alpha^2}{2\pi}\iint_R H_o K_o(\alpha r)\,d\xi d\eta + \frac{\alpha}{2\pi}\int_B T K_1(\alpha r)\cos\psi\,dB - \frac{1}{2\pi}\int_B \frac{\partial T}{\partial n} K_o(\alpha r)\,dB \tag{22}$$

The surface integral involving the forcing function can be transformed into a line integral if H_o is limited to the class of functions which satisfy Laplace's Equation,

$$\nabla^2 H_o = 0 \tag{23}$$

Defining a new function g, such that

$$\nabla^2 g = K_o(\alpha r) \tag{24}$$

The first integral in Equation (22) can be written

$$\iint_R H_o K_o(\alpha r)\,dR = \iint_R \{H_o \nabla^2 g - g\nabla^2 H_o\}\,dR \tag{25}$$

Applying Green's Theorem

$$\iint_R H_o K_o(\alpha r)\,d_R = \int_B \{H_o\frac{\partial G}{\partial n} - g\frac{\partial H_o}{\partial n}\}\,dB \tag{26}$$

Solving Equation (24) for g gives

$$g = \frac{-1}{\alpha^2} K_o(\alpha r) \tag{27}$$

and

$$\frac{\partial G}{\partial n} = -\,\frac{K_1(\alpha r)\cos\psi}{\alpha} \tag{28}$$

Substituting Equation (27) and Equation (28) into Equation (26) yields

$$\iint_R H_o K_o(\alpha r)\,dR = \frac{-1}{\alpha}\int_B H_o K_1(\alpha r)\cos\,dB + \frac{1}{\alpha^2}\int_B \frac{\partial H_o}{\partial n}K_o(\alpha r)\,dB \tag{29}$$

Using this result in Equation (22) gives

$$T(x,y) = \frac{-\alpha}{2\pi}\int_B H_o K_1(\alpha r)\cos\psi dB + \frac{1}{2\pi}\int_B \frac{\partial H_o}{\partial n}K_o(\alpha r)\,dB + \frac{\alpha^2}{2\pi}\int_B TK_1(\alpha r)\cos\psi dB - \frac{1}{2\pi}\int_B \frac{\partial T}{\partial n}K_o(\alpha r)\,dB \tag{30}$$

Thus the potential at any point within the region is completely expressed in terms of an integral on the boundary. To remove the singularities which occur on the boundary, we will allow (x,y) to approach the point(x',y') on the boundary. The singularity will be integrated around on a semi-circular path, ΔB, as shown in Figure 2.

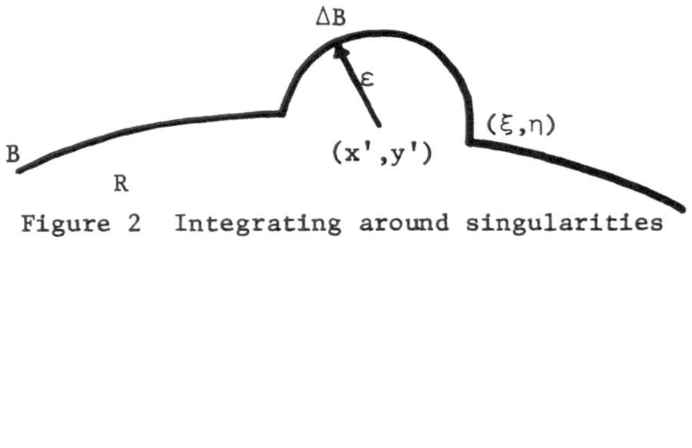

Figure 2 Integrating around singularities

Therefore T may be written as

$$T(x',y') = \frac{-\alpha}{2\pi}\int_{B-\Delta B} H_o K_1(\alpha r)\cos\psi dB \;\frac{-\alpha}{2\pi}\int_{\Delta B} H_o K_1(\alpha r)\cos\psi dB + \frac{1}{2\pi}\int_{B-\Delta B}\frac{\partial H_o}{\partial n}K_o(\alpha r)\,dB + \frac{1}{2\pi}\int_{\Delta B}\frac{dH_o}{dn}K_o(\alpha r)\,dB + \frac{\alpha}{2\pi}\int_{B-\Delta B} TK_1(\alpha r)\cos\psi dB + \frac{\alpha}{2\pi}\int_{\Delta B} TK_1(\alpha r)\cos\psi dB$$

$$+\frac{-1}{2\pi}\int_{B-\Delta B}\frac{\partial T}{\partial n}K_o(\alpha r)\,dB - \frac{1}{2\pi}\int_{\Delta B}\frac{\partial T}{\partial n}K_o(\alpha r)\,dB \qquad (31)$$

Any singularities have now been isolated into the three integrals around ΔB. Using the asymptotic property of K_o that

$$K_o(\alpha r) \to -\ell n\ r \text{ as } r \to o \qquad (32)$$

in the integrals along ΔB, the integrals are evaluated as follows:

For the first integral,

$$\lim_{\varepsilon\to o} \frac{-\alpha}{2\pi}\int_{\Delta B} H_o K_1(\alpha r)\cos\psi dB = \frac{1}{2\pi}\int_{\Delta B} H_o \frac{\partial}{\partial n}K_o(\alpha r)\,dB$$

$$= \frac{1}{2\pi}\int_{\Delta B} H_o \frac{\partial}{\partial n}(-\ell n r)\,dB$$

$$= \frac{1}{2\pi}\int_o^{\pi} H_o \frac{1}{\varepsilon}(\varepsilon d\Theta) = -\frac{H_o}{2} \qquad (33)$$

For the second integral

$$\lim_{\varepsilon\to o} 1 \int_{\Delta B}\frac{\partial H_o}{\partial n}K_o(\alpha r)\,dB = 1\int_{\Delta B} O(\varepsilon\ell n\varepsilon) = 0 \qquad (34)$$

For the third integral

$$\lim_{\varepsilon\to o} \frac{\alpha}{2\pi}\int_{\Delta B} TK_1(\alpha r')\cos\psi dB = \frac{1}{2\pi}\int_o^{\pi} T\frac{1}{\varepsilon}(\varepsilon d\Theta)$$

$$= \frac{T(x',y')}{2} \qquad (35)$$

And for the fourth integral

$$\lim_{\varepsilon\to o} -\frac{1}{2\pi}\int_{\Delta B}\frac{\partial T}{\partial n}K_o(\alpha r)\,dB = \int_{\Delta B} O(\varepsilon\ell n\varepsilon)\,dB = 0 \qquad (36)$$

An expression for T at the boundary is therefore

$$T(x',y') = -H_o - \alpha\int_B H_o K_1(\alpha r)\cos\psi dB + \frac{\alpha}{\pi}\int_B TK_1(\alpha r)\cos\psi dB$$

$$-\frac{1}{\pi}\int\frac{\partial T}{\partial n}K_o(\alpha r)\,dB \qquad (37)$$

IV. <u>Development of a Set of Simultaneous Equations</u>

In order to evaluate the integrals in Equation (37), the boundary B is approximated by N straight line segments. The values of the potential T and its normal derivative, $\frac{\partial T}{\partial n}$, are assumed to be constant on each segment and equal to its value in the center of the segment. The values of r in Equation(37) are distances between the field points and points on the boundary. Approximating the integral by a summation

$$T_i = -\pi H_o - \alpha \sum_{\substack{j=1 \\ j\neq i}}^{N} H_o K_1(\alpha r_{ij}) \cos\psi_{ij} \Delta S_j + \frac{\alpha}{\pi} \sum_{\substack{j=1 \\ j\neq i}}^{N} T_j K_1(\alpha r_{ij})$$

$$\cos\psi_{ij} \Delta S_j - \frac{1}{\pi} \sum_{\substack{j=1 \\ j\neq i}}^{N} \frac{\partial T_j}{\partial n} K_o(\alpha r_{ij}) \Delta S_j$$

$$+ \frac{\alpha T_i}{\pi} \int_i K_1(\alpha r) \cos\psi ds - \frac{1}{\pi} \frac{\partial T_i}{\partial n} \int_i K_o(\alpha r) ds. \tag{38}$$

The two integrals in Equation (38) contain the singularities. The first of these integrals is zero, since the direction cosine is zero. The second integral has been evaluated in Luke, and shown to be

$$\frac{1}{\pi} \int_i K_o(\alpha r) ds = \frac{2}{\alpha\pi} \{ \sum_{k=o}^{2} d_k \beta^{2K+1} - \ell n\beta \sum_{k=o}^{2} c_k \beta 2K+1 \} \tag{39}$$

where

$$\beta = \frac{\alpha \Delta S_i}{4} \tag{40}$$

where the d_k and c_k values are found in Table I.

Table I

K	c_k	d_k
0	2.000	0.8456
1	0.6667	0.5041
2	0.100	0.1123

Using Equation (38) at each node on the boundary gives the following matrix equation.

$$[J][T] + [K][\frac{\partial T_n}{\partial n}] = [F]$$

where

$$J_{ii} = 1 \tag{41}$$

$$J_{ij} = -\frac{\alpha}{\pi} K_1(\alpha r_{ij}) \cos\psi_{ij} \Delta S_j \tag{42}$$

$$K_{ii} = \text{(See Equation (38))}. \tag{43}$$

$$K_{ij} = \frac{1}{\pi} K_o(\alpha r_{ij}) \Delta S_j \tag{44}$$

$$F = \frac{-\alpha}{\pi} \sum_{\substack{j=1 \\ i \neq j}} H_o K_1(\alpha r_{ij}) \cos\psi_{ij} \Delta S_j + H_o \tag{45}$$

V. Sample Problem

The formulation was tested on the configuration shown in Figure 3. This is a square conductor 10 mm on each side and extending to infinity in the z direction. A uniform applied magnetic field, H_o is impinging on the conductor in the z direction. Due to symmetry, only one-fourth of the conductor must be modeled. For the boundary conditions we set T = 0 on the conductor surface and $\frac{\partial T}{\partial n} = 0$ on the two lines of symmetry. The resulting set of equations can be written as

$$[J][T] + [K][\frac{\partial T}{\partial n}] = [F] \tag{46}$$

In Equation (46), there would normally be two unknowns at each boundary element. In this example, either T or $\frac{\partial T}{\partial n}$ is specified at every point and the system of equations may be solved.

The boundary of the quarter conductor was divided into 72 equal boundary elements. Equation (46) was solved using a Gaussian elimination technique. The results were compared to a finite element solution of the same problem and the results are shown in Figure 4 for both the real and imaginary parts of the potential. The results are very close, with a small error occurring at the corner.

VI. Conclusions

A boundary element method has been presented for the slowly varying magnetic field. The results have been validated by comparison with a finite element solution of a sample problem. Very little attempt was made to optimize the number of nodes in either the finite element or the boundary element problem, but it seems clear that the boundary element method is capable of achieving high accuracy with fewer equations than the differential methods.

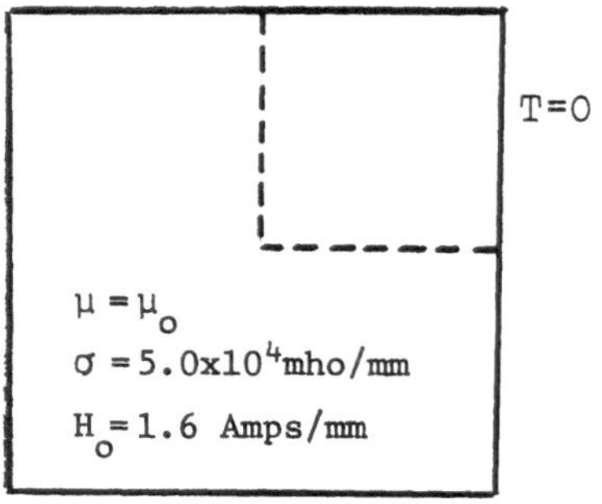

Figure 3

Sample Problem, 10 mm x 10 mm

$\frac{\partial T}{\partial n}$ = 0 on dotted lines

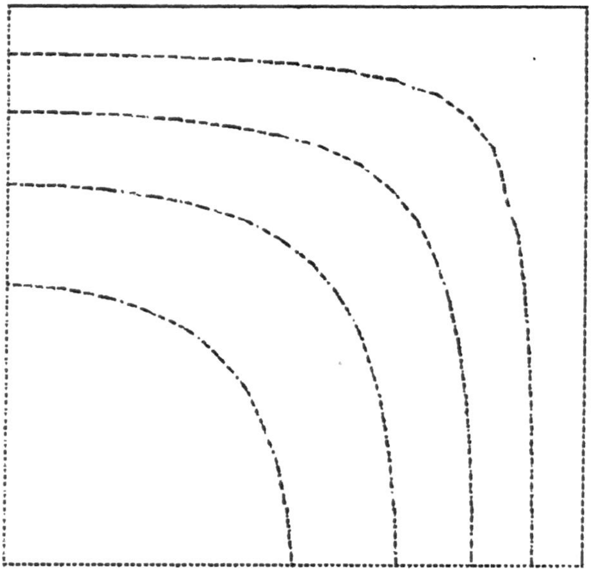

Figure 4-a

Real Part of Electric Vector Potential

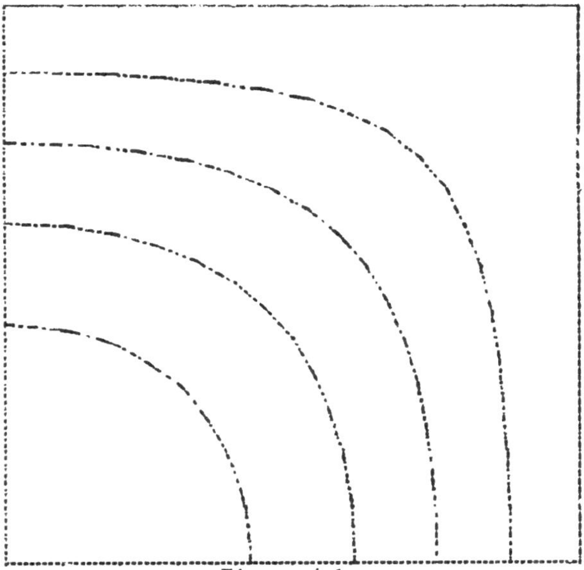

Figure 4-b
Imaginary Part of Electric Vector Potential

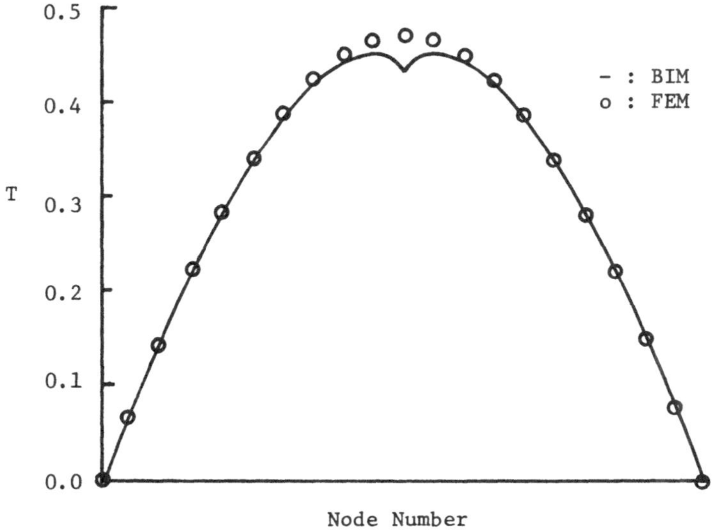

Figure 4-c
Real Part of Potential T on the boundary

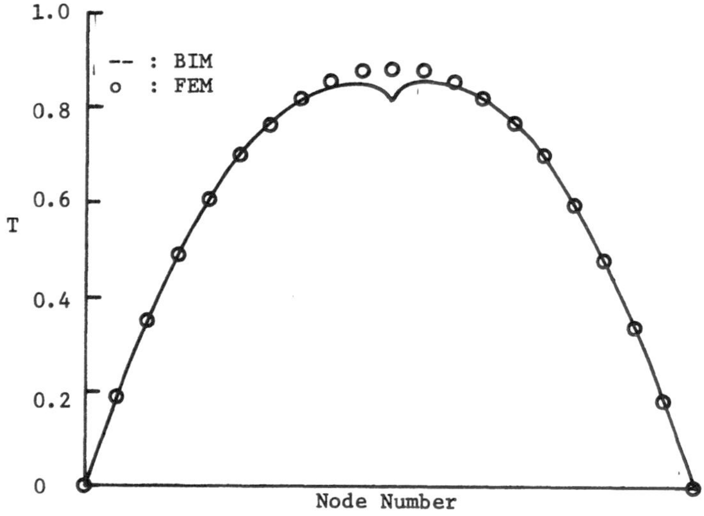

Figure 4-d
Imaginary Part of Potential T on the Boundary

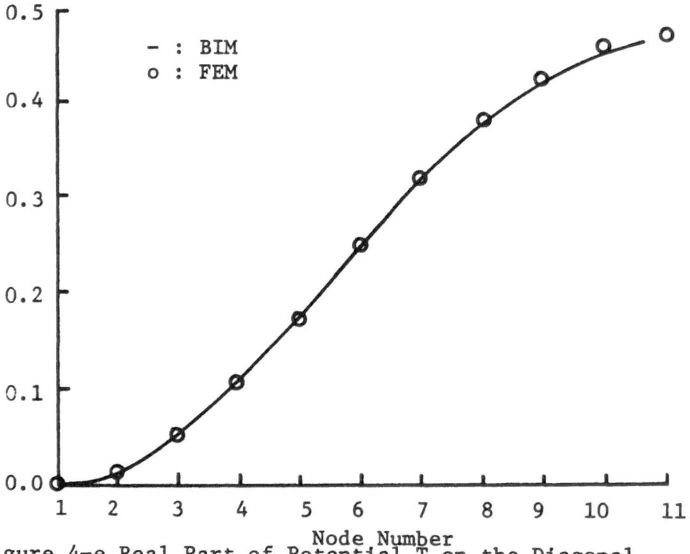

Figure 4-e Real Part of Potential T on the Diagonal

References

1. Carpenter, C.J. (1977) Comparison of Alternative Formulations of Three Dimensional Magnetic Field and Eddy Current Problems at Power Frequencies. Proceedings IEE, Vol. 124, No. 11.
2. Carpenter, C.J. (1975) Finite Element Models and Their Application to Eddy Current Problems. Proceedings IEE, Vol. 122, No. 4.
3. Chari, M.V.K. (1973) Finite Element Solution of the Eddy Current Problem in Magnetic Structures, IEEE Paper, T-73-320-9.
4. King, E.I. (1966) Equivalent Circuits for the Magnetic Field; 11-The Sinusoidally Time Varying Field. IEEE Transactions PAS-85 No. 9 pp 936-945.
5. Roberts, J. (1959) Analog Treatment of Eddy Current Problems Involving Two Dimensional Fields, IEEE, Lond. Monograph 341 M.
6. Salon, S.J. and Schneider, J.M. (1979) The Application of a Finite Element Formulation of the Electric Vector Potential to Eddy Current Losses, IEEE Paper A79-545-5.
7. Sato, T. and Sarto, S. (1977) Solution of Magnetic Field, Eddy Current and Circulating Current Problems, Taking Magnetic Saturation and Effect of Eddy Current and Circulating Current Paths into Account, IEEE Paper A-77-168-8.

NUMERICAL SOLUTION OF THE DIFFUSION EQUATION BY A POTENTIAL METHOD

D.A.S. Curran*, B.A. Lewis* and M. Cross+

* Dept of Mathematics & Computer Studies, Sunderland Polytechnic

+ Mineral Resources Research Center University of Minnesota, U.S.A.

ABSTRACT

The main advantage of using potential methods for generating numerical solutions to elliptic partial differential equations is that the dimension of the problem is reduced by one. This feature reduces, not only the number of algebraic equations to be solved, but also alleviates the problems of mesh generation. Both of these factors combine to reduce considerably the cost of computing acceptable numerical solutions. In this paper a potential method is developed which essentially extends to the diffusion equation most of the advantages currently afforded to elliptic equations.

INTRODUCTION

The numerical solution of elliptic partial differential equations by potential methods has received much attention in recent years[1-4] as they have some distinct advantages over traditional finite difference of element methods for certain classes of problems. Essentially, the potential method transforms the solution of the partial differential equation to that of a Fredholm integral equation over the boundary of the solution region. This transformation effectively reduces the dimension of the problem by one. As such, a smaller number of algebraic equations results and the costs of mesh generation are reduced since the solution is obtained entirely in terms of the boundary values. Useful accounts of these methods may be found in Jaswon and Symm[1] and Brebbia[4].

During the last ten years various extensions to the potential method have been proposed so that solutions to linear parabolic differential equations may also be solved.[5-8] However, in each case the 'parabolic' method loses the advantage of a reduction in dimension. Furthermore, they have generated their own problems with regard to numerical stability and accuracy[9]. In the present paper, a method is developed for the numerical solution of the diffusion equation in one space dimension, which retains most of the advantages of the potential methods available for elliptic partial differential equations. The solution to the diffusion equation is generated in two parts. The first is an integral over the whole solution region which involves the initial conditions of the problem only, whilst the second part is a solution of a system of Volterra integral equations involving the time coordinate only around the boundary. The solution in the region's interior is then generated from the boundary values. The method is applied to several problems and the results favourably compared with those obtained by finite difference techniques.

DEVELOPMENT OF THE METHOD

Essentially, the method involves a transformation of the diffusion equation to a standard form followed by a series of procedures to evaluate its solution both on the boundary and within the integration region.

Transformation to standard form

Consider the diffusion equation in non-dimensional form,

$$\frac{\partial u}{\partial t} = \frac{\partial^2 u}{\partial x^2} \qquad 0 < x < \ell \tag{1}$$

subject to the boundary conditions

$$u(0,t) = W_1(t), \; u(\ell,t) = W_2(t) \tag{2}$$

and the initial condition,

$$u(x,0) = f(x) \qquad 0 < x < \ell \tag{3}$$

A solution of equation (1) which also satisfies the initial condition defined by equation (3) is given by

$$u_0(x,t) = \frac{1}{2\sqrt{\pi t}} \int_0^\ell f(y)\exp\left(\frac{-(x-y)}{4t}\right)^2 dy \tag{4}$$

If a new function is defined so that

$$u_1(x,t) = u(x,t) - u_0(x,t) \tag{5}$$

then it satisfies equation (1) with initial condition,

$$u_1(x,0) = 0 \qquad 0 < x < \ell \tag{6}$$

and the following boundary conditions,

$$u_1(0,t) = W_1(t) \tag{7}$$

$$u_1(\ell,t) = W_2(t) \tag{8}$$

where

$$W_1(t) = W_1(t) - \frac{1}{2\sqrt{\pi t}} \int_0^{\ell} f(y) \exp\left(-\frac{y^2}{4t}\right) dy \tag{9}$$

$$W_2(t) = W_2(t) - \frac{1}{2\sqrt{\pi t}} \int_0^{\ell} f(y) \exp\left(-\frac{(\ell-y)^2}{4t}\right) dy \tag{10}$$

Thus, the solution to equations (1) - (3) is given by $u(x,t) = u_0(x,t) + u_1(x,t)$ where $u_0(x,t)$ is given by equation (4) and $u_1(x,0)$ satisfies the homogenous condition defined by equation (6) together with boundary equations (7) and (8).

The solution to equation (4) is easily obtained using a suitable product integration formula, i.e.,

$$u_0(x,t) = \sum_{i=0}^{m} B_i(x,t)\, f_i \tag{11}$$

where $\ell = mn$ and $f_i \equiv f(ih)$. This leads to expressions for the boundary values $W_1(t)$ and $W_2(t)$ also involving product integration formulae, i.e.,

$$W_1(t) = W_1(t) - \sum_{i=0}^{m} B_i^{(1)}(t)\, f_i \tag{12}$$

$$W_2(t) = W_2(t) - \sum_{i=0}^{m} B_i^{(2)}(t)\, f_i \tag{13}$$

Determination of $u_1(x,t)$

The solution of $u(x,t)$ leads to the need for an evluation of $u_1(x,t)$. For one dimensional problems Smirnov[10] has shown that

$$u_1(x,t) = \frac{1}{2\sqrt{\pi t}} \int_0^{t} \frac{\phi(\tau)}{(t-\tau)^{3/2}}\, x \exp\left(-\frac{x^2}{4(t-\ )}\right) d\tau$$

$$+ \frac{1}{2\sqrt{\pi t}} \int_0^{t} \frac{x(\tau)}{(t-\tau)^{3/2}}\, (\ell-x) \exp\left(-\frac{(\ell-x)^2}{4(t-\tau)}\right) d\tau \tag{14}$$

where $\phi(t)$ and $x(t)$ are the source intensities of two dipoles situated at $x=0$ and $x=\ell$, respectively. These functions may be determined by the following weakly singular Volterra integral equations of the second kind,

$$\phi(t) - \ell \int_0^t x(\tau)\, K(t,\tau)\,d\tau = W_1(t) \tag{15}$$

$$-x(t) + \ell \int_0^t \phi(\tau)\, K(t,\tau)\,d\tau = W_2(t) \tag{16}$$

where the kernel, $K(t,\tau) = \dfrac{\exp(-\ \ell^2/4(t-\tau))}{2\sqrt{\pi}\ (t-\tau)^{3/2}}$ (17)

In order to solve equations (15) and (16) numerically the integrals on the left hand side of each equation must be approximated by product integration formulae. As such,

$$\begin{aligned} \int_0^t K(t,\tau)\ \phi(\tau)d\tau &= \sum_{i=0}^{n} A_i^{(n)}\ \phi_i \\ \int_0^t K(t,\tau)\chi(\tau)d\tau &= \sum_{i=0}^{n} A_i^{(n)}\ \chi_i \end{aligned} \tag{18}$$

where the interval (o,t) has been divided into n subintervals (ik,(i+1)k) (i=0,.....,n-1,k=constant), $A_i^{(n)}$ are the series of integrals involving the kernel $K(t,\tau)$ and $\phi_i = \phi(ik)$, $\chi_i = \chi(ik)$. Using the above integration formulae, equations (15) and (16) lead to solutions at time nk for each of the point intensities as,

$$\frac{\phi_n = W_1(nk) + \ell \sum_{i=0}^{n-1} A_0^{(n)}\ \chi_i}{1 - A_n^{(n)}}$$

$$\frac{\chi_n = \ell \sum_{i=0}^{n-1} A_0^{(n)}\ {}_i - W_2(nk)}{1 - \ell A_n^{(n)}}$$

In fact, for small enough values of k, the integrals $A_n^{(n)} \simeq 0$, so that ϕ_n and χ_n may be obtained as recurrence relations, ie.

$$\phi_n = W_1(nk) + \ell \sum_{i=0}^{n-1} A_0^{(n)}\ \chi_i \tag{19}$$

$$\chi_n = W_2(nk) + \ell \sum_{i=0}^{n-1} A_0^{(n)}\ \phi_i \tag{20}$$

The evaluation of $u_1(x,t)$ is then given by:

$$u_1(x,t) = x \sum_{i=0}^{n} A_i^{(n)}(x)\, \phi_i + (\ell-x) \sum_{i=0}^{n} A_i^{(n)}(1-x)\, \chi_i \qquad (21)$$

where $A_i(x)$ involves an integral of the kernel.

PROBLEMS AND SOLUTION PROCEDURES

In order to provide a realistic test of the basic method several problems were posed and solved. For each problem the solution for u (x,t) was obtained either by analytic integration or by using a piecewise linear approximation to f(x) on (0,1). If the latter procedure was used then the interval was divided into m subintervals of constant width h so that $x_i = ih$. Then

$$u_0(x,t) = \sum_{i=0}^{m-1} \frac{1}{2\sqrt{\pi t}} \int_{x_i}^{x_i+1} f(y) \exp\left(-\frac{(x-y)^2}{4t}\right) dy$$

$$= \frac{1}{2}\ \text{erf}\left(\frac{x-x_i}{2\sqrt{t}}\right) + \text{erf}\left(\frac{x_{i+1}-x}{2\sqrt{t}}\right) P_i(\chi) \qquad (22)$$

$$+\frac{\sqrt{t}}{\pi} \exp\left(-\frac{(x-x_i)^2}{4}\right) - \exp\left(\frac{-(x_{i+1}-x)^2}{4t}\right) P_i'(\chi)$$

where

$$P_i(x) = \frac{1}{h}(x-x_i)\, f_{i+1} + \frac{1}{h}(x_{i+1}-x) f_i$$

$$P_i'(x) = \frac{1}{h}(f_{i+1} - f_i) \qquad (23)$$

with regard to the evaluation of u (x,t), the source intensities were approximated by piecewise constants so that,

$$\phi_n = W\,(nk) + \sum_{j=0}^{n-2} A_{n-j}\, \chi_{j+1} \qquad (24)$$

$$\chi_n = -W_2(nk) + \sum_{j=0}^{n-2} A_{n-j}\, \phi_{j+1} \qquad (25)$$

where

$$\phi_1 = W_1(k), \quad \chi_1 = W_2(k) \qquad (26)$$

and

$$A_{n-j} = \text{erf}\left(\frac{\ell}{2\sqrt{(n-j-1)k}}\right) - \text{erf}\left(\frac{\ell}{2\sqrt{(n-j)k}}\right) \qquad (27)$$

The solution in the interior then follows as

$$u_1(x,t) = \sum_{j=0}^{n-1} A_{n-j}(x)\ \phi_{j+1} - \sum_{j=0}^{n-1} A_{n-j}(\ell-x)\ \chi_{j+1} \tag{28}$$

where,

$$A_{n-j}(x) = \operatorname{erf}\left(\frac{x}{2\sqrt{(n-j-1)k}}\right) - \operatorname{erf}\left(\frac{x}{2\sqrt{(n-j)k}}\right) \tag{29}$$

Evaluation of equations (12), (13), (22), to (29) then defines a numerical solution to the posed problem as summarized by equations (1) to (3). Note that the solution is not required within the whole region before progressing to the next time step. The information is all carried forward in the intensity functions at the boundaries.

Problem 1 - discontinuity at the boundary

The problem here is defined by equation (1) subject to the boundary conditions,

$$u(0,t) = u(1,t) = 0$$

and the initial condition,

$$u(x,0) = 1 \ (0 \leqslant x \leqslant 1)$$

The discontinuity between the initial and boundary conditions makes it difficult for most methods to obtain accurate solutions close to the boundary. This problem is most easily solved by using the substitution $v(x,t) = 1-u(x,t)$ which leaves the form of the differential equation the same, but changes the boundary conditions to

$$v(0,t) = v(1,t) = 1$$

whilst the initial condition becomes

$$v(x,0) = 0$$

A series of results for this problem are set out in Table 1 where comparisons are shown with the Crank-Nicholson finite difference method using comparable time steps and a variety of distance steps.

Problem 2 - initial condition piecewise linear

The boundary conditions in this problem are specified as

$$u(0,t) = u(1,t) = 0$$

subject to the initial condition

$$u(x,0) = \begin{cases} 2x & 0 < x \leqslant 0.5 \\ 2(1-x) & 0.5 < x < 1 \end{cases}$$

This problem was solved using an analytic evaluation of $u_0(x,t)$ and a relevant set of results are shown in Table 2. For the Crank-Nicolson calculations a small distance step (ie. $h = 0.01$) was used to minimise space discretization errors.

Problem 3 - initial condition non-linear

The boundary condition in his problem is defined as

$$u(0,t) = u(1,t) = 0$$

subject to the initial condition

$$u(x,0) = x(1-x) \qquad 0 \leqslant x \leqslant 1$$

This problem was solved using both analytic and numerical evaluations of $u_0(x,t)$ and a relevant set of results are displayed in Table 3.

Problem 4 - sinusoidal initial condition

The boundary conditions are as specified above but the initial condition becomes

$$u(x,0) = \sin\pi x \quad 0 \leqslant x \leqslant 1$$

This problem was solved using numerical evaluations only for $u_0(x,t)$; a set of relevant results are shown in Table 4.

CONCLUSIONS

The present method compares favourably with the C.N. method for Problems 1 and 2. In Problem 1 where considerable difficulty is experienced in obtaining accurate numerical solutions because of the discontinuity between initial and boundary conditions the method produces significantly more accurate results than the C.N. method. Results for Problem 3 indicate that the use of numerical integration in space does not significantly affect the accuracy of the results, satisfactory results being obtained for relatively large integration steps. Only the results for Problem 4 are disappointing. In this problem using small space and time steps does not significantly improve the results.

Although the method discussed in the paper has only been applied to problems in one space dimension it can be easily extended to higher dimensional problems. The results from the one-dimensional problems suggest that the present method is capable of producing results at least comparable with finite difference method while retaining most of the advantages of the B.E.M.

REFERENCES

Brebbia, C.A. (1978) Bounday element method for engineers Pentech Press London.

Brebbia, C.A. (Ed) (1980) New developments in boundary element methods, CML Publications, Southampton.

Chang, Y.P. et al (1973) Int. J. Heat Mass Transfer 16, 1905

Curran, D.A.S. et al, (1980) Applied Math Modelling. 4, 398

Jaswon, M.A. & Symm, G.T. (1977) Integral equation methods in potential theory and elastostatics, Academic Press London.

Brebbia, C.A. (Ed) (1979) Recent advances in boundary element methods, Pentech Press London.

Rizzo, F.J. & Shippy, D.J. (1970) AAIAJ. 8, No.11.

Smirnov, (1964) A course of higher mathematics Vol. IV, Pergamon Press, Oxford.

Wrobel, L.C. & Brebbia, C.A. (1979) In numerical methods in thermal problems (Eds R.W. Lewis & K. Morgan) Pineridge Press Swansea, 58.

PROBLEM ONE

t = 0.005 Δt = 0.001

x	0.1	0.2	0.3	0.4	0.5
Exact	0.68269	0.95450	0.99730	0.99994	1.00000
I.E.M.	0.68269	0.95450	0.99730	0.99994	1.00000
C.N. h=0.1	0.67311	0.93179	0.98966	0.99874	0.99975
C.N. h=0.025	0.68169	0.95363	0.99668	0.99985	0.99999

t = 0.01 Δt = 0.001

x	0.1	0.2	0.3	0.4	0.5
Exact	0.52050	0.84270	0.96610	0.99530	0.99919
I.E.M.	0.52050	0.84270	0.96611	0.99530	0.99919
C.N. h=0.1	0.52343	0.83233	0.95437	0.98970	0.99632
C.N. h=0.025	0.52030	0.84220	0.96547	0.99491	0.99897

t = 0.1 I.E.M. Solution for decreasing Δt

x	0.1	0.2	0.3	0.4	0.5
Exact	0.1467	0.2790	0.3839	0.4513	0.4745
0.01	0.1488	0.2806	0.3849	0.4519	0.4750
0.005	0.1479	0.2798	0.3844	0.4516	0.4747
0.0025	0.1473	0.2794	0.3842	0.4514	0.4746
0.001	0.1469	0.2791	0.3840	0.4513	0.4745

PROBLEM TWO

t = 0.1 Δt = 0.01

x	0.1	0.2	0.3	0.4	0.5	
Exact	0.0934	0.1776	0.2444	0.2873	0.3021	
C.N.	0.0948	0.1803	0.2482	0.2918	0.3069	0.01
I.E.M.	0.917	0.1743	0.2340	0.2822	0.2968	

t = 0.1 I.E.M. Solution for decreasing Δt

x	0.1	0.2	0.3	0.4	0.5
Exact	0.0934	0.1776	0.2444	0.2873	0.3021
0.005	0.0925	0.1758	0.2420	0.2846	0.2993
0.0025	0.0929	0.1766	0.2431	0.2859	0.3006
0.001	0.0931	0.1772	0.2439	0.2867	0.3015

Solution at x = 0.5 Δt = 0.01

	Exact	C.N.	I.E.M.
t=0.01	0.7743	0.7691	0.7743
t=0.02	0.6809	0.6921	0.6796
t=0.10	0.3021	0.3069	0.2968

PROBLEM THREE

Solution for decreasing t using exact integration

t = 0.05

x	0.1	0.2	0.3	0.4	0.5
0.01	0.0470	0.0906	0.1255	0.1479	0.1557
0.005	0.0478	0.0916	0.1264	0.1487	0.1564
0.0025	0.0482	0.0921	0.1269	0.1492	0.1569
0.001	0.0485	0.0924	0.1272	0.1495	0.1572

t = 0.1

x	0.1	0.2	0.3	0.4	0.5
0.01	0.0293	0.0556	0.0765	0.0900	0.0946
0.005	0.0295	0.0560	0.0771	0.0906	0.0953
0.0025	0.0296	0.0562	0.0774	0.0910	0.0957
0.001	0.0297	0.0564	0.0776	0.0913	0.0960

Solution for decreasing t using numerical integration

t = 0.05 Δt = 0.005

x	0.1	0.2	0.3	0.4	0.5
h=0.1	0.0474	0.0908	0.1253	0.1475	9,1551
h=0.05	0.0477	0.0914	0.1261	-.1484	0.1561

t = 0.05 Δt = 0.0025

x	0.1	0.2	0.3	0.4	0.5
h=0.1	0.0478	0.0913	0.1258	0.1480	0.1556
h=0.05	0.0481	0.0919	0.1266	0.1489	0.1566

t = 0.05 Δt = 0.001

x	0.1	0.2	0.3	0.4	0.5
h=0.1	0.0481	0.0917	0.1262	0.1483	0.1559
h=0.05	0.0484	0.0922	0.1270	0.1492	0.1569

PROBLEM FOUR

Solution for decreasing Δt and varying integration steps h

t = 0.1 Δt = 0.01

x	0.1	0.2	0.3	0.4	0.5
Exact	0.11517	0.21907	0.30153	0.35447	0.37271
h=0.1	0.11250	0.21357	0.29389	0.34556	0.36339
h=0.05	0.11319	0.21489	0.29571	0.34770	0.36564
h=0.025	0.11337	0.21522	0.29617	0.34824	0.36620

t = 0.1 Δt = 0.005

x	0.1	0.2	0.3	0.4	0.5
h=0.1	0.1133	0.21521	0.29622	0.34828	0.36623
h=0.05	0.11398	0.21655	0.29805	0.35044	0.36396
h=0.025	0.11443	0.21752	0.29940	0.35199	0.37012

t = 0.1 Δt = 0.001

x	0.1	0.2	0.3	0.4	0.5
h=0.1	0.11401	0.21781	0.29842	0.35083	0.36889
h=0.05	0.11472	0.21816	0.30027	0.35301	0.36661

THE APPLICATION OF BOUNDARY ELEMENTS TO STEADY AND UNSTEADY POTENTIAL FLUID FLOW PROBLEMS IN TWO AND THREE DIMENSIONS

Paul H.L. Groenenboom

B.V.Neratoom, The Hague, Netherlands

ABSTRACT

The Boundary Element Method is applied to both steady and unsteady potential flow of a compressible liquid inside a fixed volume. For steady flow the velocity potential satisfies the Laplace or Poisson equation. This potential can be found by solving the linear equations resulting from a discretisation of the boundary integral equation.
The velocity potential for unsteady fluid flow can be found by solving the three-dimensional wave equation. The potential at a certain point and moment can be expressed in an integral over the boundary of the values of the potential at the retarded time plus a (inhomogeneous) source term.
The computer code BEREPOT (Boundary Element Retarded Potential Technique) has been developed to calculate the potential successively. This code can handle boundaries consisting of a rigid structure and non-reflecting openings representing infinitely long pipes.

This method has been applied to the flow of liquid sodium in cooling components of Liquid-Metal Fast Breeder Reactors. Special attention will be paid to the pressure wave propagation resulting from a sodium-water reaction following a tube rupture in a LMFBR steam generator. The boundary is given by (parts of) the s.g. walls and the connections with the piping system.
The validity of the potential flow model and of the solution method, the boundary conditions and other applications will be discussed. Some results will be compared with the results from analytical solutions and from other numerical methods.

INTRODUCTION

One of the main problems in the delivery of the heat transfer components to the Liquid Metal Fast Breeder Reactor (LMFBR) SNR-300 at Kalkar (F.R.G.) is the proven integrity of the

components to withstand the pressure loads that arise from sodium-water reactions in the steam generator (s.g.). Usually these loads are determined by transient pressure gradients calculated by simple one-dimensional methods as the characteristic method for water hammer phenomena. If one takes a look at a straight tube steam generator (fig.1), one may guess that such a calculation might be not too bad. However, for a helical coil tube steam generator, as in fig.2, the assumption of one-dimensional pressure wave propagation becomes highly questionable. Since these conservative calculations, in the worst case, give rise to problems in the proof of the structural integrity of the s.g., we have to look for more realistic numerical methods.
Two effects, not included in the simple methods, are expected to give a significant reduction of the structural loads:

- the r^{-1}-dependence of the pressure for spherical waves as compared to the constant pressure wave propagation in one dimension,
- the dispersion and absorbtion of pressure waves due to the elastic-plastic behaviour of the internal structures and the walls.

The boundary element method (BEM) applied to potential fluid flow may present a feasible method to perform the 3D-calculations and, in a later stage, also fluid-structure interaction may be included.

MATHEMATICAL FORMULATION

For the calculation of pressure waves and other fast transients one can derive the wave equation for the velocity potential. (See for instance:Chorin and Marsden, 1979)
The conservation equations are:

$$\frac{\partial \rho}{\partial t} + \operatorname{div}(\rho \underline{v}) = 0 \qquad (1)$$

$$\frac{\partial \underline{v}}{\partial t} + (\underline{v}.\operatorname{grad})\underline{v} = - \frac{\operatorname{grad} p}{\rho} \quad \text{(Euler)} \qquad (2)$$

The effects of viscosity have been neglected since at the time scale of the $Na\text{-}H_2O$ reaction, viscous boundary layers remain very thin and the bulk viscosity is also negligible.
The Euler equations may be rewritten as

$$\frac{\partial \underline{v}}{\partial t} + \tfrac{1}{2}\operatorname{grad} |\underline{v}|^2 + \underline{\omega} \wedge \underline{v} = - \frac{\operatorname{grad} p}{\rho} \qquad (3)$$

By taking the curl of this equation, it is found that the vorticity, $\underline{\omega}$=curl $\underline{v}$, is governed by an evolution equation

$$\frac{\partial \underline{\omega}}{\partial t} + \operatorname{curl}(\underline{\omega} \wedge \underline{v}) = (\operatorname{grad} \rho \wedge \operatorname{grad} p)/\rho^2 \qquad (4)$$

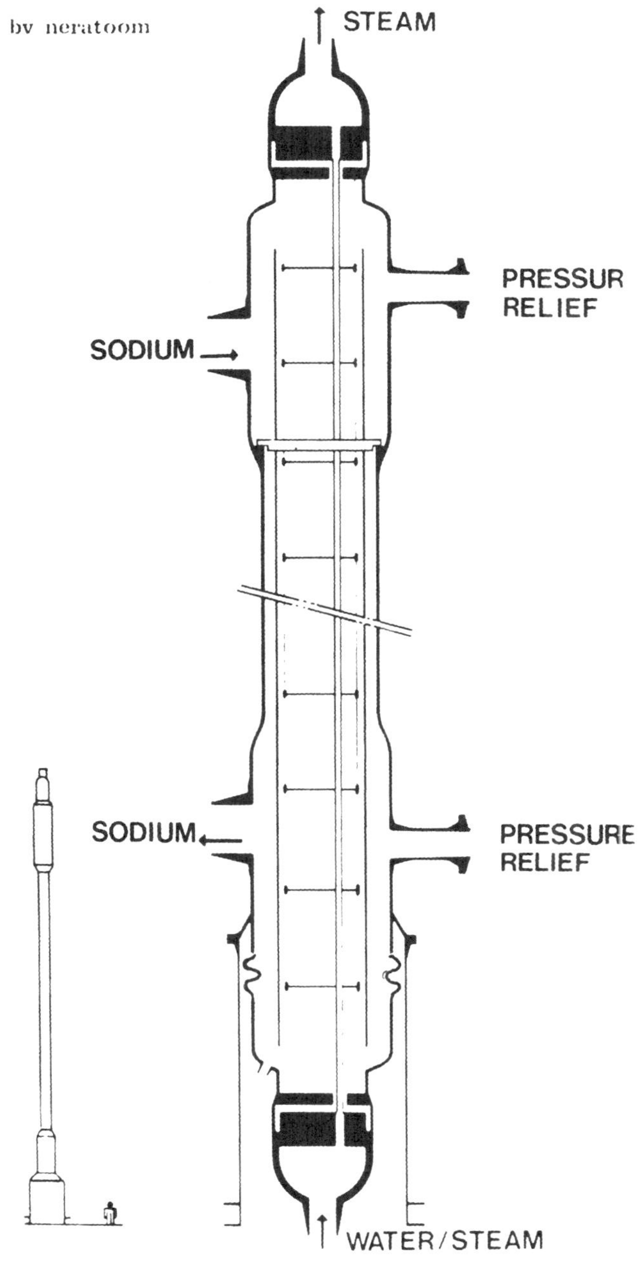

Fig. 1. Model of a straight tube LMFBR steam generator

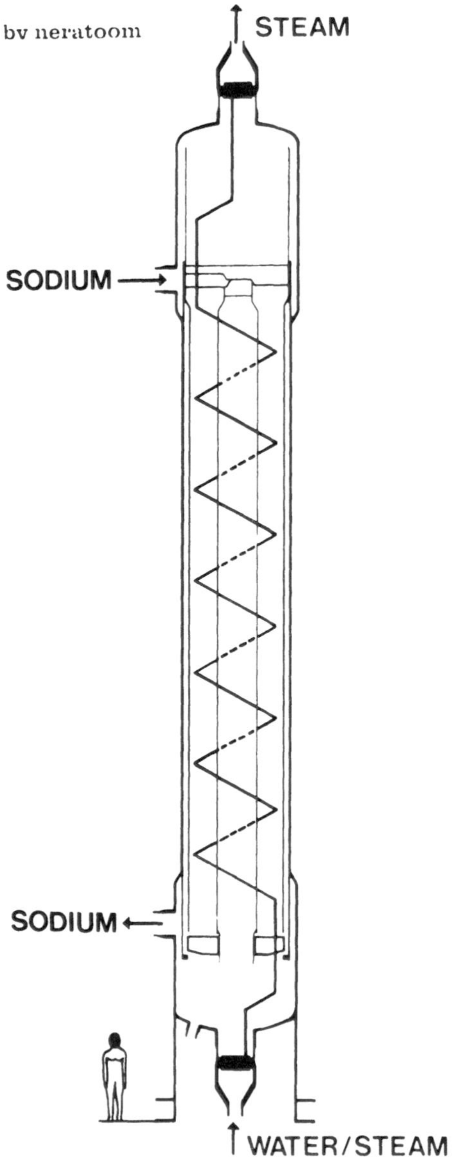

Fig. 2. Model of a helical coil LMFBR steam generator

The density, ρ, is a function of the pressure, p, and of the entropy, S. If we assume that the variation in the entropy is so small that the explicit dependence of ρ on S may be neglected, the gradients of p and ρ will have the same direction and the r.h.s. of eq.4 will vanish.
So, if there is no initial vorticity, no vorticity will be present. Consequently, the flow is irrotational and can thus be described by a velocity potential $\varphi(\underline{r},t)$ with

$$\underline{v} = \operatorname{grad} \varphi \tag{5}$$

By combining the conservation equations, it is straightforward to show that for velocities small compared to the speed of sound, which is very high in liquid sodium, the velocity potential satisfies the three-dimensional wave equation

$$\nabla^2 \varphi(\underline{r},t) - \frac{1}{c^2}\frac{\partial^2\varphi}{\partial t^2} = 0 \tag{6}$$

The pressure is related to the potential by the Bernouilli equation

$$\rho\,\frac{\partial\varphi}{\partial t} + \tfrac{1}{2}\rho\underline{v}^2 + p = 0 \tag{7}$$

For the sodium-water reaction a source term representing the expanding reaction bubble has to be included in the wave equation

$$\nabla^2\varphi(\underline{r},t) - \frac{1}{c^2}\frac{\partial^2\varphi}{\partial t^2} = \gamma(\underline{r},t) \tag{3}$$

The solutions represent the travelling waves in the sodium. Using the Kirchhoff-method (Stratton, 1941) for the scalar electrodynamical potential, this can be transformed into an explicit boundary integral equation. Since that relation is the basis of this method, a short sketch of the elegant derivation will be given.
The fundamental solution or Green's function of the wave equation is defined by

$$\nabla^2 G(\underline{r},t\,|\,\underline{r_o},t_o) - \frac{1}{c^2}\frac{\partial^2 G}{\partial t^2} = \delta(\underline{r}-\underline{r_o})\;\delta(t-t_o) \tag{9}$$

Multiplication of eq.8 by G and of eq.9 by φ and integration over the volume of interest and the timevariable yields

$$\begin{aligned}&\int_{-\infty}^{+\infty}dt \int_V d\underline{r}\{G\nabla^2\varphi - G.\frac{1}{c^2}\frac{\partial^2\varphi}{\partial t^2} - \varphi\nabla^2 G + \varphi\frac{1}{c^2}\frac{\partial^2 G}{\partial t^2}\} = \\ &= \int_{-\infty}^{+\infty}dt \int_V d\underline{r}\{\gamma(\underline{r},t)G - \varphi(\underline{r},t)\,\delta(\underline{r}-\underline{r_o})\,\delta(t-t_o)\}\end{aligned} \tag{10}$$

The Green's theorem for volume-integration reads:

$$\int_V d\underline{r}\{G\nabla^2\varphi-\varphi\nabla^2 G\}= \int_V d\underline{r}\nabla\cdot\{G\nabla\varphi-\varphi\nabla G\}=\int_\Omega dA\{G\frac{\partial\varphi}{\partial n}-\varphi\frac{\partial G}{\partial n}\}, \quad (11)$$

where $\underline{n}$ is defined as the outward normal at the surface Ω. By a similar derivation one finds that the time integration over $G\frac{\partial^2\varphi}{\partial t^2}-\varphi\frac{\partial^2 G}{\partial t^2}$ vanishes.
Application of this and of identity 11 to the l.h.s. of eq.10 yields:

$$\int_{-\infty}^{+\infty} dt \int_\Omega dA(G\frac{\partial\varphi}{\partial n}-\varphi\frac{\partial G}{\partial n}) \quad (12)$$

whereas the r.h.s. of eq.10 becomes

$$\int_V d\underline{r} \int_{-\infty}^{+\infty}\gamma(\underline{r},t)G(\underline{r},t|\underline{r_o},t_o)-c_i\ \varphi(\underline{r_o},t_o) \quad (13)$$

with c_i = +1 for $\underline{r_o}$ inside V
= $+\frac{1}{2}$ for $\underline{r_o}$ on the boundary Ω
= 0 for $\underline{r_o}$ outside V

Since, for potential problems, the field outside a spherically symmetric source is the same as that of a central point source (at $\underline{rs}$), $\gamma(\underline{r},t)$, may be replaced by $\hat{\gamma}(t)\ \delta(\underline{r}-\underline{rs})$.

The fundamental solution of the 3D-wave equation is

$$G(\underline{r},t|\underline{r_o},t_o) = \frac{\delta(t-t_{ret})}{4\pi R} \quad (14)$$

with $R=|\underline{r}-\underline{r_o}|$
and $t_{ret} = t_o - R/c$ (retardation)
Expression 13 then becomes

$$\frac{1}{4\pi}\frac{\hat{\gamma}(t_{ret})}{|\underline{r_s}-\underline{r_o}|} - c_i\varphi(\underline{r_o},t_o) \quad (15)$$

and from expression 12 one obtains:

$$\int_{-\infty}^{+\infty} dt \int_\Omega dA\{\frac{\partial\varphi}{\partial n}\frac{\delta(t-t_{ret})}{4\pi R}-\varphi\frac{\partial}{\partial n}(\frac{\delta(t-t_o+R/c)}{4\pi R})\}$$

$$= \int_\Omega dA\ \frac{1}{4\pi R}[\frac{\partial\varphi}{\partial n}]_{t_{ret}} - \frac{1}{4\pi}\int_{-\infty}^{+\infty} dt \int_\Omega dA\ \varphi\frac{\partial R}{\partial n}\{-\frac{\delta(t-t_{ret})}{R^2}$$

$$+ \frac{1}{Rc}\frac{\partial}{\partial t}(\delta(t-t_{ret}))\}$$

$$=\frac{1}{4\pi}\int_\Omega dA\{\frac{1}{R}[\frac{\partial\varphi}{\partial n}]_{t=t_{ret}}+\frac{1}{R^2}\frac{\partial R}{\partial n}(\varphi(\underline{r},t_{ret})+\frac{R}{c}\dot{\varphi}(\underline{r},t_{ret}))\} \quad (16)$$

The final expression for the potential at place $\underline{r_o}$ and time t_o is subsequently found

$$4\pi c_i \varphi(\underline{r_o}, t_o) = \frac{\hat{\gamma}(t_{ret})}{|\underline{r_s} - \underline{r_o}|} +$$

$$+ \int_{\Omega} dA \{ \frac{1}{R} [\frac{\partial \varphi}{\partial n}]_{t_{ret}} + \frac{1}{R^2} \frac{\partial R}{\partial n} (\varphi(r, t_{ret}) + \frac{R}{c} [\frac{\partial \varphi}{\partial t}]_{t_{ret}}) \} \qquad (17)$$

This is the basic equation for BEREPOT. A two-dimensional representation of the geometry is sketched in fig.3.

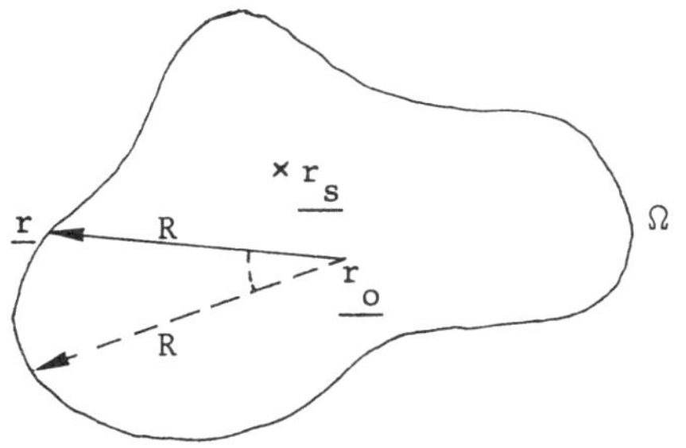

Fig. 3. Sketch of the geometry used by BEREPOT

Since the potential $\varphi(\underline{r_o}, t_o)$ is expressed solely in quantities at previous time steps, this method is explicit and requires no matrix solution. Note furthermore that neither time nor volume-integration has to be performed. The only integration involves the boundary of V.

APPLICATION TO THE NA-H_2O REACTION

According to the German licensing authorities, the present definition of the design base accident for a Na-H_2O reaction in a steam generator is an instantaneous, complete tube rupture. Also the chemical reaction,

$$Na + H_2O \rightarrow NaOH + \tfrac{1}{2} H_2 + \text{energy},$$

is assumed to take place immediately. Under these assumptions narrow pressure peaks will be formed. After that, a quasi-static pressure build-up will occur in the reaction zone and in the component. The pressure waves will propagate through the steam generator and into the connecting secondary sodium loop.

Further assumptions about the reactions are:

- isothermal reaction with $T \simeq 1150$ K
- the volume of the reaction zone is determined by the formed hydrogen (ideal gas)

Since the velocities of the sodium are small compared to the soundspeed, the pressure wave propagation and the subsequent flow may be described by a velocity potential which obeys the 3D wave equation. Therefore the theory of the previous chapter may be applied for which we will discuss a number of aspects.

Boundary conditions

The first consideration concerning the boundary is the choice of the geometry. A reasonable choice for the boundary is (a part of) the inner flow shroud enclosing the rod bundle or the structural limits of the header, depending on the assumed reaction site. For the moment these structures are assumed to be rigid, so the corresponding boundary condition is that the velocity normal to the surface vanishes;

$$\frac{\partial\varphi}{\partial n} = 0 \qquad (18)$$

In a later stage we hope to replace this condition by a more realistic condition derived from the elastic-plastic response of the structure.

The enclosure of the reaction volume is not complete; openings have to be considered. These openings give entrance to other parts of the steam generator and the connecting piping system. The corresponding boundary condition is that of an infinitely long pipe represented by the radiation condition

$$\frac{\partial\varphi}{\partial n} + \frac{1}{c}\frac{\partial\varphi}{\partial t} = 0 \qquad (19)$$

Coupling to one-dimensional codes is a future option.

Reaction zone

The source of the pressure waves and of the flow of liquid sodium is the expanding reaction bubble which is assumed to remain in spherical shape. In that case, it is easy to prove that the condition, that the velocity of the sodium at the Na-H_2 interface is equal to the growth rate of the bubble radius, may be replaced by a sourceterm, γ, as introduced in eq.8. The sourceterm is equal to the opposite of the volume-growth of the bubble.

$$\begin{aligned} \gamma(\underline{r},t) &= -4\pi R^2(t)\,\dot{R}(t)\;\delta(\underline{r}-\underline{rs}) \\ &= -\dot{V}(t)\;\delta(\underline{r}-\underline{rs}) \end{aligned} \qquad (20)$$

Treatment of pressure discontinuities

The very high pressure peaks immediately after the tube rupture have a duration in the order of 10 μs only. The corresponding size of the boundary elements is so small (a few centimeters) that a complete partitioning of the s.g. would require more data than any existing computer could store. Therefore, it is preferable to treat this as a discontinuous wave front which reflects at the rigid boundaries [Neilson et al, 1978].

Numerical aspects

Next to the choice of the source function and of the geometry, the discretisation of the boundary will primarily determine the success of the BEREPOT calculations. The time-step of the problem will have to be chosen in accordance with the time-development of the source function. Subsequently, the element size has to be equal to or greater than the distance traversed by a wave in a single time-step.
Numerical problems will arise from the fact that the retardation within a single element is not a constant and that this retardation will not be equal to an integer number of time-steps.
The second problem is solved by linear interpolation between the two nearest-by values. For the solution of the basic equation (17) it is necessary to evaluate the time derivative of the potential at the boundary. This evaluation can be done by a three-point backward difference formula (Mitzner, 1967)

$$\frac{\partial\varphi}{\partial t} = \frac{1}{\tau}\{\frac{3}{2}\varphi(m\tau)-2\varphi((m-1)\tau)+\tfrac{1}{2}\varphi((m-2)\tau) + O(\tau^3)\} \qquad (21)$$

At sharp edges or corners of the boundary the normal is not defined uniquely, so if integration points are situated at these edges, the factor c_i of eq.13 remains undetermined.
A solution of this problem has been suggested by Wrobel and Brebbia (1979).
Finally we have to check the numerical stability of the solution by applying small changes in boundary conditions, initial conditions, discretisation scheme and time-step.

TEST CASES

To demonstrate the feasibility of the boundary element method to potential flow problems, we have examined two examples.

Two-dimensional, stationary potential flow

Without sources, the velocity potential satisfies

$$\nabla^2\varphi = (\frac{\partial^2}{\partial x^2} + \frac{\partial^2}{\partial y^2})\varphi(x,y) = 0 \qquad (22)$$

The corresponding fundamental solution is

$$G(\underline{r},\underline{r}_o) = -\frac{1}{2\pi}\ln\left(\frac{1}{|\underline{r}-\underline{r}_o|}\right) \qquad (23)$$

Proceeding in the same way as for the three-dimensional, time dependent case, it is straightforward to derive:

$$2\pi c_i\varphi(\underline{r}_o)-\int_\Omega d\underline{r}\,\frac{(\underline{r}-\underline{r}_o)\cdot\underline{n}}{|\underline{r}-\underline{r}_o|^2}\varphi(\underline{r})=\int_\Omega d\underline{r}\,\frac{\partial\varphi}{\partial n}\Big|_{\underline{r}}\ln\left(\frac{1}{\underline{r}-\underline{r}_o}\right) \qquad (24)$$

with $c_i=+1,0,\frac{1}{2}$ for $\underline{r_o}$ resp. inside V, outside V or on the boundary Ω.
By division of the boundary in constant elements, this integral equation becomes a matrix equation which can be solved by standard subroutines. The boundary conditions are determined by the specific case (fig.4): homogeneous in- and outflow through openings at the top or the bottom of one wall of a rectangular (2D) box. We assume that the normal derivative of the potential at the boundary is zero everywhere except at the openings where it is assumed to be a finite constant.
The boundary has been discretised into two meshes
1) 4 elements in the x-direction, 24 for y
2) 8 elements in the x-direction, 48 for y.

The results of the simple BE-code for this case are presented in fig.5. There the values of the potential along the boundary (four walls) have been plotted. For comparison, results of a least-squares finite element method, developed in our group (De Bruijn,1980), for this case, are also given in this figure. Since the potential is defined apart from an arbitrary constant, these constants have been chosen in such a way that the curves are close to each other but do not overlap. Note the nice agreement between the three results.

Wave propagation for a point source in a 3D rectangular box
The governing equations is the one treated in the second chapter (eq.17).
The specific example of a point source in a box has been chosen because the potential can be solved analytically by the method of mirror sources. The condition of zero normal velocity at a wall near a source is satisfied by placing a source of the same (time-dependent) strength on the rear side. For two parallel walls with a source in between, one obtains a whole series of mirror sources and for a rectangular box one finds series of sources in all three directions.
Due to the finite velocity of the pressure waves, the number of sources that influence the box at a certain moment remains finite and can be found by a computer code, named SOMMIS.

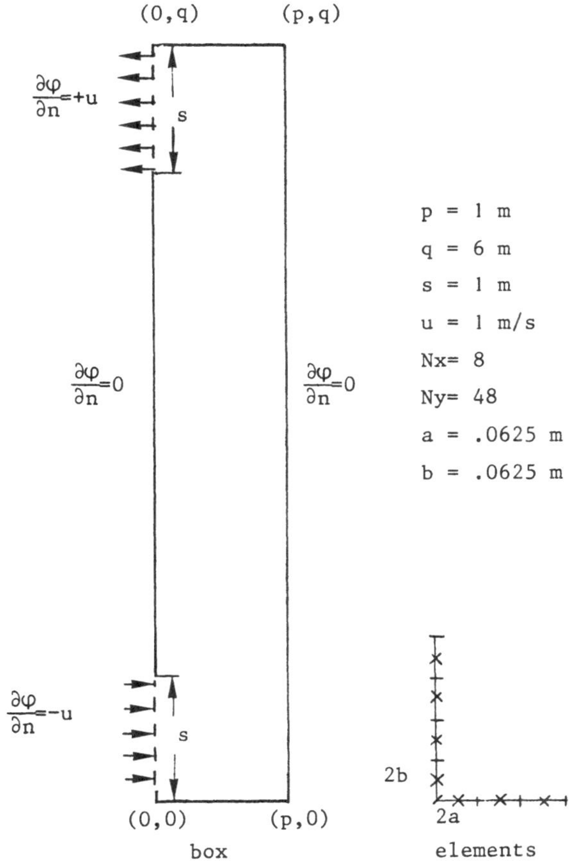

Fig. 4. Geometry for the testcase of potential flow in a 2D box

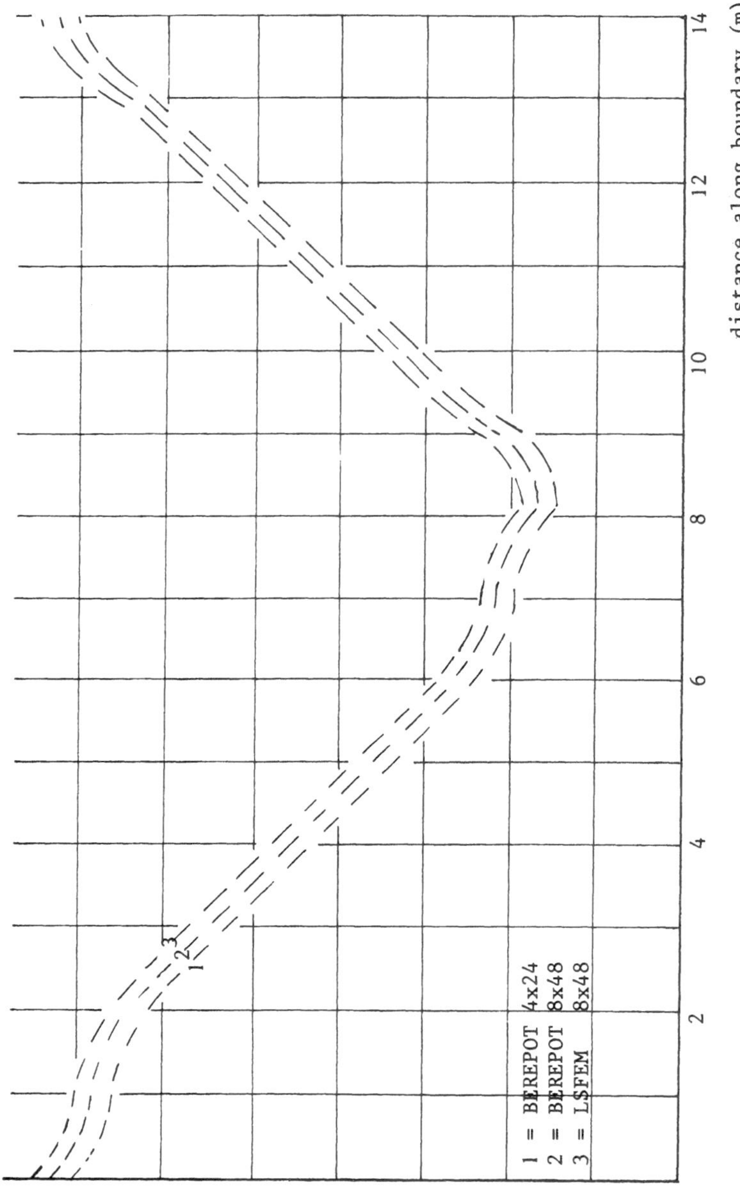

Fig. 5. Velocity potential along the boundary of the 2D box

A test version of BEREPOT has been made for this problem. The boundaries of the box, of dimensions 3x4x5, have been divided into 376 elements of dimensions 0,5 x 0,5. The potential is assumed to be constant in a single element. For the source function a short pulsed Gaussian has been taken and at two boundary points the response of the potential has been evaluated, both by the exact method of mirror sources (SOMMIS), and by the boundary element method.
In fig.6 the original pulse at the source is given and the direct influence which is calculated exactly by SOMMIS.
The calculation by the BEREPOT-code has been performed with timesteps of 2 (units of time) and with 0,5. For the largest time-step the reproduction of the exact potential is, of course, not very accurate for such a fast transient, and a possibly small delay in the response may be observed. For the shorter time-step the reproduction of the secondary and the third reflection is quite good, but at later times the solution seems to become unstable.
Still, for such a crude mesh, the reproduction of this strong transient is encouraging.
In fig.7 the results of the potential history at a boundary point closer to the source is given and a similar behaviour as in fig.6 may be observed.

CONCLUSIONS AND PROSPECTS

Although effects of viscosity and convection may cause strongly non-linear behaviour in steady fluid flow, in many cases, the potential flow model may present some indicational results (De Bruijn and Zijl,1981).
For pressure wave propagation the potential flow model is justifiable. Therefore, the application of boundary elements to the solution of the wave equation for the velocity potential, is primarily expected to be of interest for the calculation of the consequences of explosions and violent reactions like the Na-H_2O reaction.
Specific advantages of the use of the Boundary Element Method for flow problems are the explicitly time-dependent formulation and the reduction of the volume integration to the boundary. A disadvantage of the method is that at every time-step the mutual influences between all elements have to be evaluated. So, if one would like to know the pressure field inside the volume, a Finite Element Method would probably be preferable.
In the future we intend to use this method for more realistic cases such as the growth of the reaction bubble inside realistically modelled parts of the steam generators. To this end an order has been given to the Computational Mechanics Centre in Southampton to develop a BEM-computer-code for the 3-D wave propagation with more possibilities than in the presented test-cases. That code will be able to handle arbitrary geometry with both rigid walls and radiation boundary conditions and, possibly, Fluid Structure Interaction (FSI).

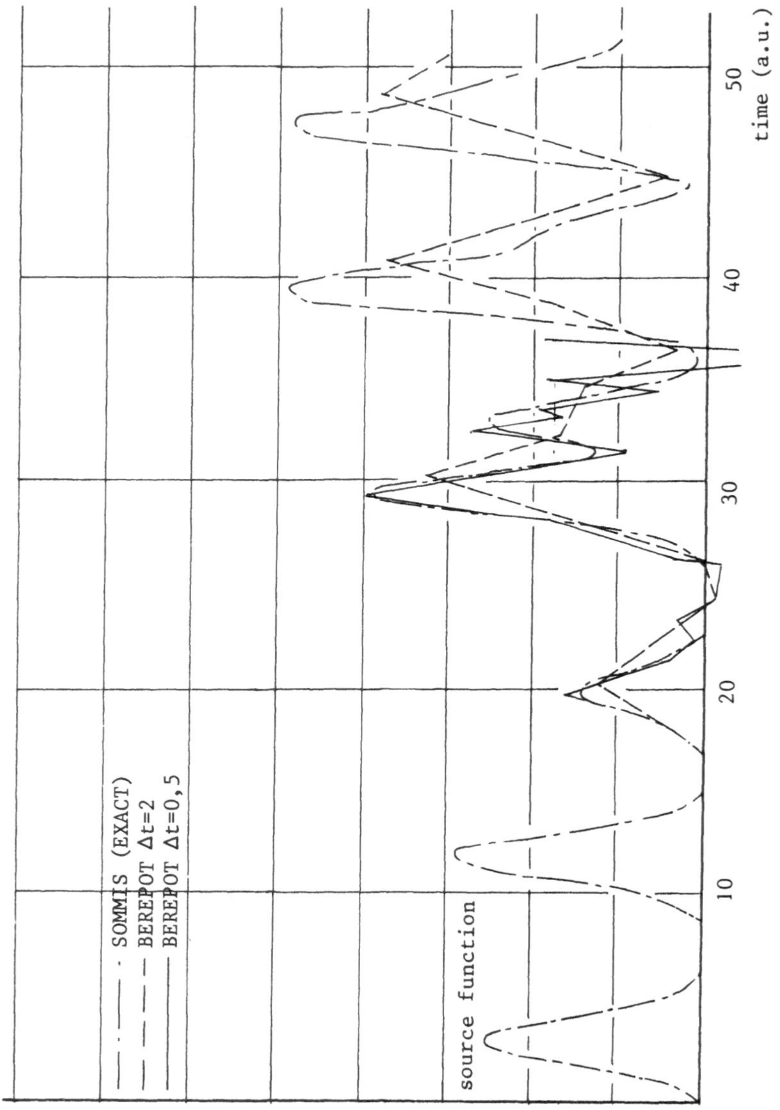

Fig. 6. Potential of a point source in a three-dimensional box

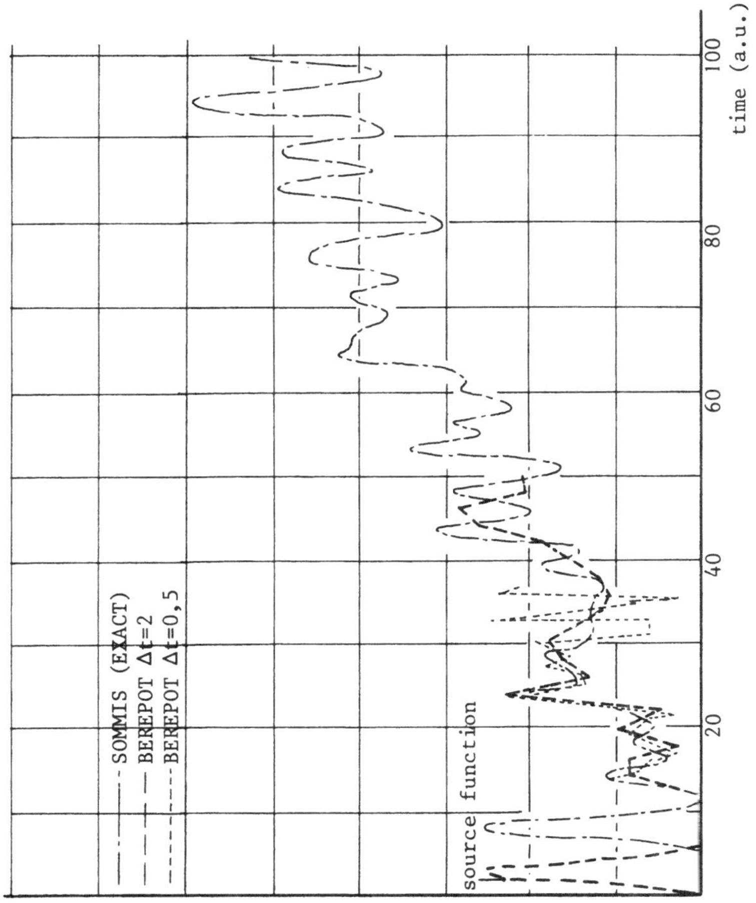

Fig.7 Potential of a point source in a three-dimensional box

Both the 3D-effects and the FSI might give a significant reduction in pressure loads on steam generator structures due to hypothetical instant tube rupture. If this method proves successful for this difficult problem, other numerical problems which can be described by the Poisson or the wave equation for a vast range of applications also come in hand.

REFERENCES

Bruijn, J.G.M. de (1980), private communication.

Bruijn, J.G.M. de and Zijl, W. (1981), Least Squares Finite Element Solution of several Thermal-Hydraulic Problems in a Heat Exchanger, Proc. of the 2nd Int.Conf. on numerical methods in Thermal Problems, Venice.

Chorin, A.J. and Marsden, J.E., (1979). A mathematical Introduction to Fluid Mechanics, Springer Verlag, New York.

Mitzner, K.M. (1967), Numerical Solution for transient scattering from a hard surface of arbitrary shape - retarded potential technique, J.Acoust. Soc.Am.42: 391.

Neilson, H.C.; Lu, Y.P. and Wang Y.F. (1978) Transient scattering by arbitrary axisymmetric surfaces, J.Acoust.Soc. Am. 63:1719.

Stratton, J.A.: (1941). Electromagnetic Theory, McGraw-Hill Co.

Wrobel, L.C. and Brebbia, C.A. (1979) in: Numerical Methods in Thermal Problems ed's: Lewis, R.W. and Morgan, K.; Pineridge Press, Swansea, p58.

COMPUTING STRATEGY IN THE INTEGRAL EQUATION SOLUTION OF LIMITING GRAVITY WAVES IN WATER

J.M. Williams

Hydraulics Research Station
Wallingford, Oxfordshire
U.K.

ABSTRACT

Progressive irrotational gravity waves in water have a limiting form in which, in their steady flow representation, the crest reaches the total energy line. The crest is then angled rather than rounded, with an included angle of 120^0, implying a singularity of order 2/3 in the complex potential plane.

An integral equation technique, employing two leading terms to represent the crest, has been successfully used by the author to generate definitive solutions of limiting waves over the whole range of depth : wavelength ratios. The paper presents a specimen solution and discusses various aspects of computing strategy, including the choice of number and position of nodal points. Conditions under which convergence fails are also briefly considered.

INTRODUCTION

Inviscid irrotational water waves of infinitesimal amplitude form part of classical hydrodynamics (Lamb, 1932). The solution depends on the depth : wavelength ratio which can vary from zero (the solitary wave) to infinity (the deep-water wave). The extension to finite wave amplitudes, when the non-linear surface condition arising from Bernoulli's equation can no longer be linearised, was begun by Stokes (1847). Stokes developed a series expansion, with successive terms giving successive Fourier components of the wave form. Third order solutions by Stokes's method have been tabulated, for example by Skjelbreia (1959), and used extensively in engineering design. More recently, and with the aid of more advanced computers, Dean (1965) has developed his so-called stream function wave theory which has also been published in tabular form (Dean, 1974).

Whereas in Stokes's method the coefficients of the first, second and third order solutions are found in sequence, each from its predecessors, Dean finds the coefficients of his component terms simultaneously by minimising the root mean square error in the free-surface condition; this is, in effect, a type of integral equation technique. Dean's tabulated solutions range from 3rd to 19th order; the higher orders extend the range of useful accuracy to larger wave amplitudes compared with the Stokes 3rd order tabulations.

A uniform wave motion may be brought to a steady state by superimposing a velocity opposed to the wave celerity. This steady motion has a total energy line above the surface to which the Bernoulli condition at the surface is referred. The highest possible wave is that whose crest just reaches the total energy line to give a stagnation point there. Stokes (1880) showed that the crest of this limiting wave would be no longer rounded but angled, with an included angle of 120^0. It follows that waves nearing the limiting form will need a large number of Fourier components to describe the profile accurately, and consequently a high-order solution whether by Stokes's, Dean's or other comparable method.

The problem has received much attention in recent years, prompted on one hand by practical design needs in offshore engineering and on the other by the availability of more powerful computers. We may mention in particular the work of Schwartz (1974) and Cokelet (1977) each of whom has developed a series expansion technique extending for some solutions to orders beyond 100. Although their results offer an important advance in accuracy over previously tabulated solutions there remains room for improvement for waves nearing the limiting height. In common with Stokes's original method, the accurary also falls away towards the shallow-water end of the range (depth : wavelength $\rightarrow$ 0).

The present author developed an integral equation solution for finite-amplitude waves in 1970 and has recently extended it to the limiting case. The method used and the results obtained for limiting waves have been set out in detail in a recent paper (Williams, 1981). The accuracy is now well within almost all practical requirements, with many field and integral properties being defined to within 1 or 2 parts in 10^6. Furthermore, this accuracy has been maintained throughout the depth : wavelength range, so that the results include a definitive description of the maximum solitary wave.

In this paper the method of solution will be described in outline only, accompanied by some specimen results for non-limiting and limiting waves. The main part of the discussion will describe several alternative computing methods for limiting waves which were considered before the final scheme was established, as well as some important features of the final scheme itself. It is hoped that this exposition will encourage others to consider similar techniques for related problems involving singularities and also to advance the associated formal numerical analysis.

INTEGRAL EQUATION METHOD FOR NON-LIMITING WAVES

The method is best introduced briefly by reference to a solution. Figure 2 presents a solution to 21st order computed by the author's method of 1970. The figure shows, in steady-flow form, a half-wavelength of the irrotational motion with symmetry about both the crest and the trough and a rigid boundary at the bed. At the free surface the flow satisfies, within the working limits of accuracy which are to be discussed, Bernoulli's equation, referred to the total energy line. The flow takes place (for a full wavelength) within the domain of the complex potential $\chi = \phi + i\psi$ given by $0 \geqslant \phi \geqslant -\lambda$, $0 \geqslant \psi \geqslant -2$; the ratio $4\pi/\lambda$ is defined as d. The space origin and parameter F^2 are chosen, as explained in Williams (1981), by preliminary consideration of the corresponding infinitesimal wave. It is sufficient to note here that

$$F^2 = (1/d) \tanh d \qquad (1)$$

and that $1/2F^2$ is in the present notation a normalised form of the acceleration due to gravity. The remaining physical dimensions defining the motion are wavelength L, mean depth h and wave height H.

To remove the difficulty of the unknown position of the free surface the complex potential χ is taken as the independent variable and is transformed in turn to the τ-plane given by

$$\chi = i(2/d)\ln \tau, \quad \tau = \rho\exp(i\theta). \qquad (2)$$

With reference to Figure 1, the flow field is bounded in the τ-plane by the concentric circles $\rho = \exp(-d) = R$ (bed) and $\rho = 1$ (surface), while the wave crest occurs at $\tau = 1$ and the trough at $\tau = -1$. Limiting cases are given by $R = 0$ (deep-water wave) and $R = 1$ (solitary wave).

It is shown in Williams (1981) that the solution of the problem may be written as

$$z = x + iy = -\chi + iF^2 + \zeta,$$

leaving $\zeta = \xi + i\eta$ as a holomorphic function, with appropriate symmetry, to be found. The boundary conditions for the imaginary part η are the bed condition

$$\eta = 0, \quad \rho = R \qquad (3)$$

and the surface condition

$$\frac{\eta}{F^2} + \alpha - \frac{1}{\left(1 - \tfrac{1}{2}d\,\frac{\partial\eta}{\partial\rho}\right)^2 + \left(\tfrac{1}{2}d\,\frac{\partial\eta}{\partial\theta}\right)^2} = 2p_s(\theta) = \frac{H\varepsilon(\theta)}{F^2} \equiv 0, \qquad (4)$$

where α is an initially unknown parameter.

p_s and ε are alternative forms for expressing the non-zero surface error which will inevitably occur in the solutions. p_s denotes a ratio of pressure to density so that $p_s(\theta)$ indicates the distribution of surface pressure which would make the solution exact. We shall also consider in the discussion the maximum modulus of p_s, over a wavelength, denoted by $\hat{p}_s$. $H\varepsilon$ expresses the error as a form of displacement, ε being the ratio of displacement error to wave height. It is also of interest to evaluate ε^*, the root mean square of $\varepsilon(\theta)$ over a wavelength, to compare our solutions with earlier results.

The solving technique consists of constructing ζ from a set of component solutions, each of which has the necessary symmetry and satisfies the bed condition, Equation 3. Thus

$$\zeta = a_0\zeta_0 + a_1\zeta_1 + \;\ldots\; + a_N\zeta_N, \qquad (5)$$

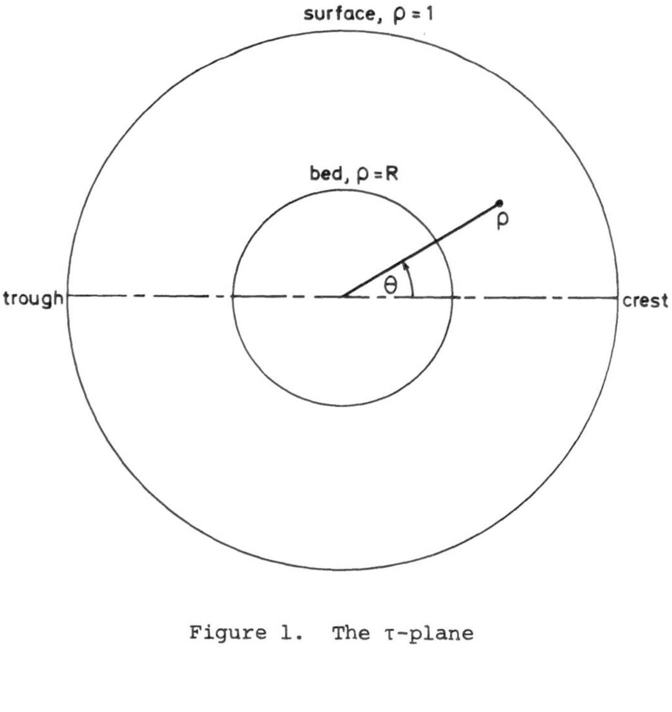

Figure 1. The τ-plane

where

$$\zeta_o = -i(1/d)\ln(\tau/R), \qquad (6)$$

$$\zeta_m = -\frac{i}{1-R^{2m}}\left[\tau^m - \left(\frac{R^2}{\tau}\right)^m\right], \quad m = 1,2,\ldots,N. \qquad (7)$$

The coefficients are computed iteratively, to satisfy Equation 4 at N + 1 nodal points, $\rho = 1$, $\theta = k\pi/N$, $k = 0,1,2,\ldots,N$. (The alternative set of nodes $\theta = (2k-1)\pi/2N$, $k = 1,2,\ldots,N$ was also used for some earlier runs.) To determine the nominal amplitude of the wave to be computed a_1 is chosen and the last unknown is then α, which governs the level of the total energy line (given by $\eta = -\alpha F^2$).

The iteration proceeds, from a trial set of coefficients, by evaluating ε at each nodal point, and hence an error vector $\boldsymbol{\varepsilon}$. Each coefficient in turn is then perturbed by a small quantity and the errors are recalculated and used to construct a rate-of-charge matrix Δ, with elements

$$\frac{\partial\varepsilon_k}{\partial\alpha}, \frac{\partial\varepsilon_k}{\partial a_o}, \frac{\partial\varepsilon_k}{\partial a_2}, \ldots., \frac{\partial\varepsilon_k}{\partial a_N}, \quad k = 1,2,\ldots., N+1. \qquad (8)$$

Corrections to the coefficients are estimated from $-\Delta^{-1}\boldsymbol{\varepsilon}$ and the cycle is repeated until convergence has been achieved to an acceptable tolerance.

This algorithm converged well and yielded useful solutions for waves up to about half of maximum amplitude over a wide range of R, with N taking values up to 21, the maximum feasible on the small computer (ICL 1901A) available at the time.

The solution shown in Figure 2, for which $\hat{p}_s$ and ε^* are each 2.5×10^{-5}, has been chosen to correspond with a published 5th order solution by Dean for which $\hat{p}_s$ is 2.2×10^{-5} and ε^* is 1.7×10^{-5}. The two solutions are thus of similar accuracy and the wave profiles are graphically indistinguishable. Dean's solution (developed in the physical rather than complex potential plane) minimises the root mean square error sampled over a large number of points, probably more than the 22 nodal points used in the author's solution. The computing effort needed in the two cases may thus be comparable but Dean's strategy has achieved a solution which is very compactly defined, with only 5 terms, after what is in effect an economization process. This topic will not be pursued in the present paper which presents compact solutions for limiting waves obtained by a different technique.

When in an integral equation solution the coefficients of the component terms have been found any other property of the flow may readily be calculated. For illustration Figure 2 includes several streamlines of the steady flow and also a corresponding set of particle paths which arise when a wave celerity is superimposed. The celerity is arbitrary unless other factors enter the problem to define it; we have used in Figure 2 the important celerity $c = \lambda/L$ which makes the space-mean bed velocity over a wavelength equal to zero. The particle paths show clearly the forward motion experienced by each particle during a half wave period. The remaining half of each particle path is a mirror image of that shown, doubling the net forward displacement in a full period.

EXTENSION OF THE METHOD FOR LIMITING WAVES

The 120^0 angle of the crest of a limiting wave transforms in the τ-plane to a singularity of order 2/3. A first step in the computation was therefore to include a leading term of this form. This has been done before by Yamada (1957 a and b), for both the deep-water and solitary waves, and Lenau (1966) for the solitary wave.

It was soon found, however, that this leading term, although valuable, was not sufficient to ensure a really high accuracy without an unreasonable overall number of terms. The next development was suggested by the work of Grant (1973) who investigated an expansion for the flow near the crest and found that the index of the second term was governed by the roots of the equation

$$\tan \tfrac{1}{2}\pi\mu = -(4+3\mu)/3\sqrt{3}\mu. \tag{9}$$

For present purposes the index of the second term in the τ-plane is the first root of Equation 9 > 2/3, namely $\mu = 1.469345741$. The series is developed further by Norman (1974) and Williams (1981).

Component ζ functions for use in the computations were therefore defined by the general form

$$\zeta_{m,A,\nu} = \frac{i\left\{\left[\frac{1}{A} - \tau^m\right]^\nu - \left[\frac{1}{A} - \left(\frac{R^2}{\tau}\right)^m\right]^\nu\right\}}{\left[\frac{1}{A} - R^{2m}\right]^\nu - \left[\frac{1}{A} - 1\right]^\nu} \tag{10}$$

For the present work with a singularity at the crest, $\tau = 1$, we take $m = A = 1$ and $\nu = 2/3$ or μ. As the singularity moves away from the crest, for near-limiting waves, A decreases from 1 and ultimately to zero, when $\zeta_{m,O,\nu}$ degenerates to ζ_m, defined in Equation 7. The remaining features of Equation 10 ensure that $\zeta_{m,A,\nu}$ satisfies the bed boundary condition, Equation 3, and is normalised in the same way as ζ_m, with $\eta = -1$ at $\tau = 1$.

Table 1. Tests with leading terms and second crest node; d=2.5

Run	N	s	q	No. of subsequent coefficients, $a_0, a_1, \ldots$	θ_c	$10^6\, \hat{p}_s$
1	15	s_0	-	15	-	5344
2	15	floating	-	14	-	5534
3	15	s_0	yes	14	-	84
4	15	floating	yes	13	-	400
5	15	s_0	yes	15	$\pi/420$	300
6	15	floating	yes	14	$\pi/420$	55
7	15	floating	yes	14	$\pi/140$	55
8	21	floating	yes	20	$\pi/420$	18
9	21	floating	yes	20	$\pi/140$	>14
10	19	floating	yes	18	$\pi/280$	12

The trial solutions for iteration thus now contained one or both of the new terms

$$s\,\zeta_{1,1,2/3} \text{ and } q\,\zeta_{1,1,\mu}. \tag{11}$$

It is shown in Williams (1981) that s has a theoretical value of

$$s_0 = \tfrac{1}{2}\left[(12F/d)(1-R^2)\right]^{2/3} \tag{12}$$

which follows from Stokes's (1880) analysis of the flow at the crest.

A series of tests was carried out in which these new terms were tried in computations of the limiting case for d=2.5, approaching the deep-water wave. The basic set of nodal points were taken as before at $\theta = k\pi/N$, $k = 0,1,2,\ldots,N$. In contrast with the non-limiting case the control of amplitude is now automatic so that no coefficient needs to be fixed at the outset. The option does exist, however, to allow the leading coefficient either to "float" or to be fixed at its theoretical value s_0. α remains as an unknown to be found.

An important innovation which developed during these tests was to introduce a further nodal point at a small distance θ_c from the crest. In this way the two leading coefficients intended to define the crest accurately were provided with two associated nodal points.

The tests are summarised in Table 1, whose final column shows the resulting value of $\hat{p}_s$, as an indication of overall accuracy. The last entry, Run 10, is the solution finally accepted.

Table 1 shows the marked improvement of accuracy from Run 3 onwards, after the introduction of the second leading coefficient q. Runs 3 to 6 then show that introducing the second crest node has a comparable restraining effect on the solution to the fixing of s to s_0; attempts to do both, however, nullify the benefit (Run 5). The marginally better accuracy of Run 6, with floating s, was preferred to the theoretically superior formulation of Run 3. It is shown in Williams (1981) that the resulting solution has a locally incorrect crest acceleration but that this has no significant consequences. Run 7 then shows that the result is comparatively insensitive to the value of θ_c while Runs 8 and 9 show the effect of increasing N. The significance of the final choice of N = 19 (Run 10) will be discussed in the next section.

The computing algorithm remained essentially as before with the program simply being enhanced to accommodate the more general form of ζ (Equation 10) and the more flexible treatment of fixed and floating coefficients and nodal point distribution. Convergence to a solution remained generally good, although with some important exceptions, to be discussed later.

The computations for this phase of the work were done on the CDC 7600 at the University of London Computer Centre. It was possible, within a reasonable run, for N to be increased up to 79. Most runs needed from 5 to 30 iterations depending on the starting values available and the rate of convergence. The time per iteration varied approximately as $N^{2.84}$ and was about 5 seconds for N = 35.

THE CHOICE OF N AND θ_c

The algorithm generated accurate solutions for the whole range of d, which are set out in detail in Williams (1981). There were, however, important features which dictated somewhat differing treatments for small, moderate and large values of d.

$0.2 \leqslant d \leqslant 0.5$

This constitutes the shallow-water end of the range, which causes greatest difficulty in most methods and where, for a given accuracy, N must be largest. Fortunately, however, the algorithm was well behaved except for the important proviso that even values of N exceeding about 40 failed to converge. N was therefore restricted to odd values which gave progressively greater accuracy as they increased. Table 2 illustrates the typical behaviour of a sequence of solutions at d = 0.2, the limit of the range computed. It is shown in Williams (1981) that this wave is, to present computing accuracy,

Table 2. Behaviour of solutions for $d = 0.2$, $\theta_c = \pi/420$

N	App. It'ns	$10^6 p_s$ for $\theta = k\pi/200$ k=0	1	2	3	4	5	6	7	8	9	10	11	12	13	14	$10^6 \hat{p}_s$
49	12	0	22	31	18	1	-7	-7	-4	0	2	2	1	0	0	-1	32
53	14	0	16	21	9	-1	-5	-4	-1	0	1	1	0	0	0	0	23
57	16	0	13	15	4	-2	-4	-2	0	1	1	0	0	0	0	0	18
63	20	0	8	9	1	-2	-2	0	0	0	0	0	0	0	0	0	11
79	30	0	3	1	-1	-1	0	0	0	0	0	0	0	0	0	0	4

Table 3. Behaviour of solutions for $d = 0.7$, $\theta_c = \pi/280$

N	It'ns	$10^6 p_s$ for $\theta = k\pi/200$ k=0	1	2	3	4	5	6	7	8	9	10	11	12	13	14	15	16	17	$10^6 \hat{p}_s$
29	8	0	0	3	7	7	5	2	0	-2	-3	-3	-2	-1	0	0	0	1	1	8
30	15	0	-1	1	5	5	4	1	0	-2	-2	-2	-2	-1	0	0	0	0	0	8
31	10	0	-1	0	3	4	3	0	0	-2	-2	-2	-1	0	0	0	0	0	0	8
32	21	0	-1	0	2	3	2	0	-1	-1	-2	-1	-1	0	0	0	0	0	0	8
33	11	0	-2	-1	1	2	1	0	-1	-1	-1	-1	0	0	0	0	0	0	0	9

identical with the maximum solitary wave. The residual surface pressure p_s has been tabulated at intervals of $\pi/200$ from the crest; beyond the range of θ shown p_s is in all cases zero to the accuracy indicated. The actual maximum, $\hat{p}_s$, generally occurs between tabulated points and is shown on the right of the table.

It was found in this range of d that the behaviour was not sensitive to the choice of θ_c.

$0.6 \leq d \leq 2.0$

This range needed moderate values of N. It was found that N could now be even without convergence being destroyed although more iterations were always needed than for odd N. A new feature was that accuracy no longer improved indefinitely with increasing N but reached a peak at a particular optimum value. Table 3 shows the behaviour of a sequence of solutions for $d = 0.7$. The position of $\hat{p}_s$ changes in this sequence from $3<k<4$ for $N = 29$ to a point between the two crest nodes in the remaining solutions.

Table 4. Coefficients of the limiting wave computed for $d = 1.3$, $N = 22$, $\theta_c = \pi/280$

s	1.82517163	q	0.52996266
a_0	-0.12288679	α	1.06815531
a_1	-1.66040564	a_{11}	0.00000833
a_2	0.12164711	a_{12}	0.00000441
a_3	0.01081024	a_{13}	0.00000232
a_4	0.00241700	a_{14}	0.00000120
a_5	0.00077108	a_{15}	0.00000060
a_6	0.00029856	a_{16}	0.00000028
a_7	0.00013036	a_{17}	0.00000012
a_8	0.00006161	a_{18}	0.00000004
a_9	0.00003070	a_{19}	0.00000001
a_{10}	0.00001583	a_{20}	0
h	1.94852	$1/2F^2$	0.754303

Apart from the value of $\hat{p}_s$, certain integrals over the wavelength which involve p_s are also important in maintaining overall accuracy, as explained in Williams (1981). The chosen solution, N = 32, was preferred on the basis of a near-minimun $\hat{p}_s$ coupled with minimum values of these integrals.

The solutions were "fine-tuned" in this way throughout the intermediate range of d. θ_c again seemed to have no critical effects.

<u>$2.5 \leq d \leq 10.0$</u>

Within this deep-water end of the range the general characteristics of the middle range remained. However, for even N convergence became very slow or sometimes failed altogether. Furthermore, with N now small the process of finding an optimum N, which had worked successfully in the middle range, no longer provided a sufficiently fine control. It was then found that θ_c could be used instead to achieve the best accuracy and in this way $\theta_c = \pi/280$ was determined to be most suitable. This value was subsequently adopted throughout the range of d since it had been found elsewhere not to be critical. The solution chosen for d = 10.0 used N = 19, with $\hat{p}_s = 12 \times 10^{-6}$. As shown in Williams (1981), this wave is identical, to working accuracy, with the theoretical extreme of the maximum infinite-depth wave.

The critical N giving best accuracy which has been found in the second and third ranges of d probably exists also in the first range. It is, however, beyond the limit of a reasonable computer run (N = 79) and would also correspond to a higher accuracy than has been generally achieved.

The algorithm has thus yielded solutions which are definitive for all practical purposes but does not in its present form seem able to give arbitrarily high accuracy. To investigate whether and how this could be achieved presents the challenging problem of reconciling the properties of the crest singularity with the needs of the numerical procedure. Some brief attempts were made to include further terms in Grant's series but it was found either that the improvement was negligible or that the iteration failed to converge.

STUDIES OF CONVERGENCE

The failure of convergence at large, and occasionally small, even values of N was discovered at a late stage of the work. The development of the algorithm had been pursued over the middle range of d, or in the deep-water range with a fortuitously odd value of N, as in Table 1. The failure does not result from a near-singular iteration matrix Δ. The algorithm had always used a perturbation of 10^{-4} for constructing the rate-of-charge matrix Δ. Some recent tests have shown that convergence can be affected by the perturbation used. The most dramatic example is for d = 10.0, N = 18, $\theta_c = \pi/420$, which failed to converge at all with a perturbation of 10^{-4} but converged strongly when it was reduced to 10^{-9}. A similar test was tried for d = 0.2, N = 54, $\theta_c = \pi/420$. The reduced perturbation produced a hesitant, intermittent convergence but further reduction to 10^{-11} gave no improvement.

These results suggest that the discretised problem which the algorithm is set to solve does not have a unique solution. It seems that, for certain even values of N, two solutions are sufficiently close to each other for the perturbation scheme to jump between them and cause an oscillating non-converging behaviour. In practical terms the strategy adopted seems completely to avoid the difficulty over the whole range of d, such that a solution is always found within the close computing tolerance which has been used. Provided the solution of the wave problem itself is unique, as is generally accepted, it should not matter which of alternative discrete close approximations is arrived at. Nevertheless this aspect of the numerical analysis would benefit from a more thorough study.

EXAMPLE OF A COMPUTED LIMITING WAVE

Figure 3 and Table 4 show the limiting wave computed for d = 1.3 with N = 22, $\theta_c = \pi/280$. The depth : wavelength ratio is

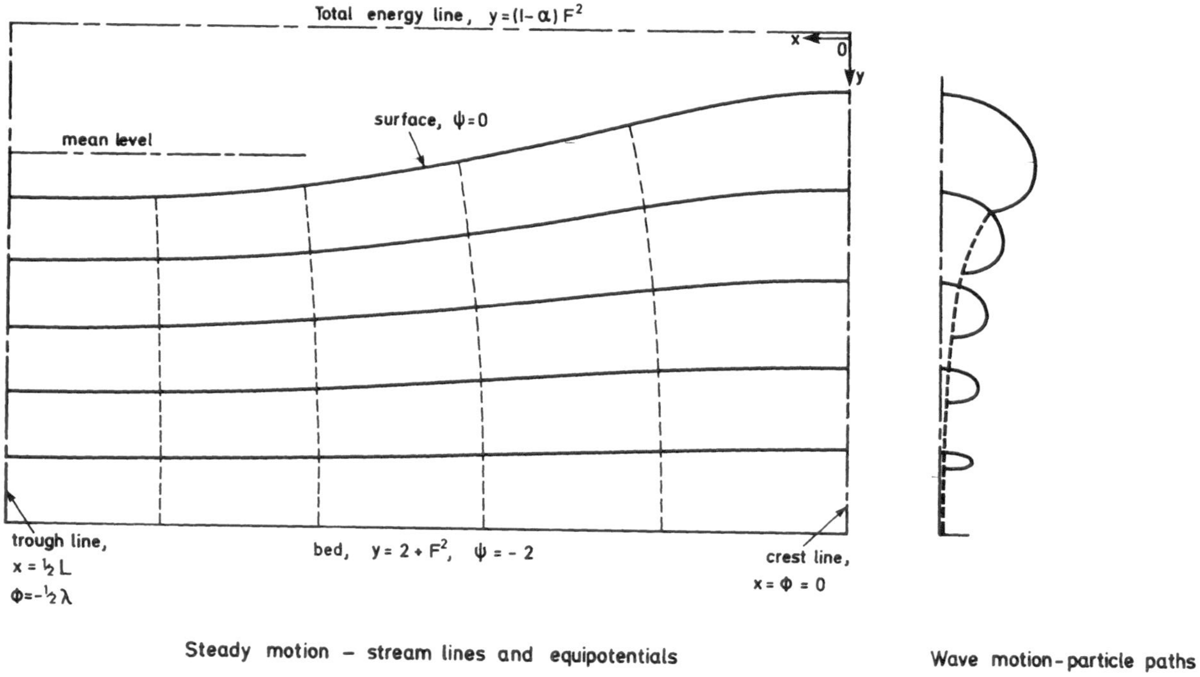

Figure 2. Computed non-limiting wave; $4\pi/\lambda = d = 1.3263$; $N = 21$; $h/L = 0.2148$, $H/h = 0.3125$, $c\sqrt{(2F^2/h)} = 0.8306$

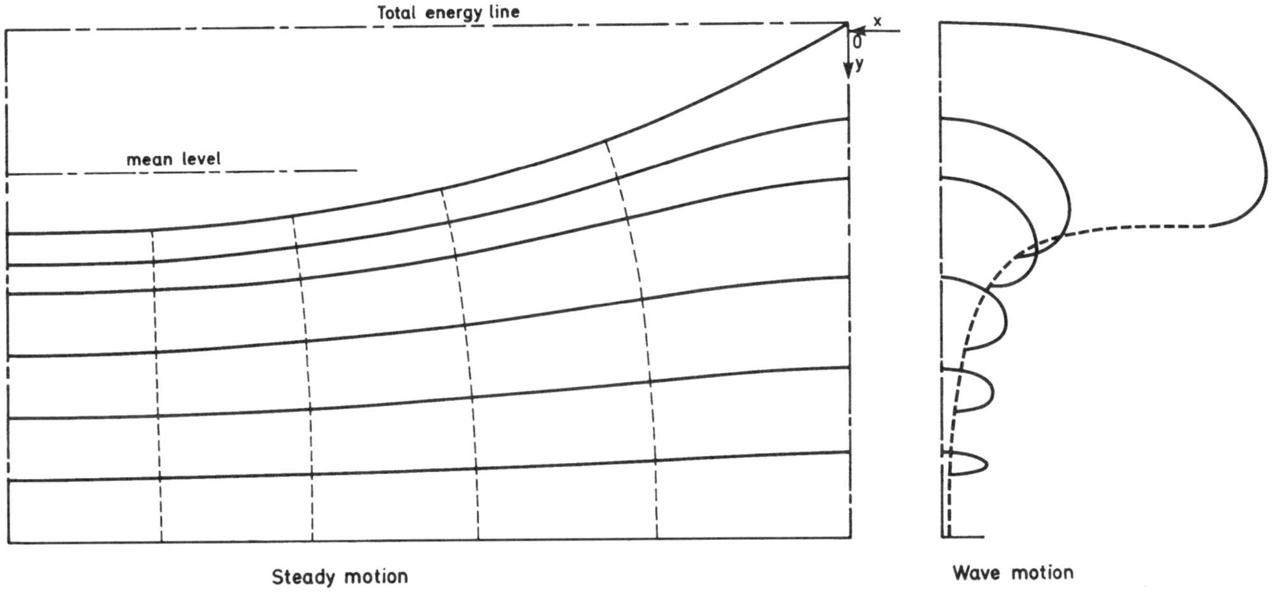

Figure 3. Computed limiting wave; $4\pi/\lambda = d = 1.3000$, $N = 22$; $h/L = 0.214772$, $H/h = 0.55257$, $c\sqrt{(2F^2/h)} = 0.87885$

0.2148, as for the non-limiting wave of Figure 2. $\hat{p}_s$ is 1.0×10^{-5} and ε^* is 1.3×10^{-6}. Despite the even N this run converged strongly in only 12 iterations.

Figure 3 shows again streamlines of the steady motion and particle paths in the unsteady motion. Note that the forward drift is now very strong at the surface.

CONCLUSIONS

Definitive solutions for limiting gravity waves have been found from an integral equation approach in a relatively compact form which permits all required properties of the flow to be calculated. The algorithm is generally well-behaved although it is not apparently able to give arbitrarily fine accuracy in its present form. The full implications of the failure of convergence under certain conditions with even N remain to be investigated. With these provisos the method constitutes a possible approach for other problems involving singularities and perhaps non-linear boundary conditions.

ACKNOWLEDGEMENTS

The work described in this paper was carried out as part of the research programme of the Hydraulics Research Station and is published with the permission of the Director of Hydraulics Research. Most of the computations were done at the University of London Computer Centre through the Institute of Computational Mathematics, Brunel University, where the author is registered for a Ph.D. degree.

REFERENCES

Cokelet, E.D. (1977) Steep Gravity Waves in Water of Arbitrary Uniform Depth. Phil. Trans. R. Soc. Lond. A, 286, 183-230.

Dean, R.G. (1965) Stream Function Representation of Nonlinear Ocean Waves. J. Geophys. Res., 70, 4561-4572.

Dean, R.G. (1974) Evaluation and Development of Water Wave Theories for Engineering Application. (2 Vol.). U.S. Army Coastal Engineering Research Center, Special Report No. 1.

Grant, M.A. (1973) The Singularity at the Crest of a Finite Amplitude Progressive Stokes Wave. J. Fluid Mech., 59, 257-262.

Lamb, H. (1932) Hydrodynamics, 6th ed., Cambridge University Press.

Lenau, C.W. (1966) The Solitary Wave of Maximum Amplitude. J. Fluid Mech., 26, 309-320.

Schwartz, L.W. (1974) Computer Extension and Analytic Continuation of Stokes' Expansion for Gravity Waves. J. Fluid Mech., 62, 553-578.

Skjelbreia, L. (1959) Gravity Waves; Stokes Third Order Approximation; Tables of Functions. Council on Wave Research - The Engineering Foundation.

Stokes, G.G. (1847) On the Theory of Oscillatory Waves. Trans. Camb. Phil, Soc., 8, 441-455.

Stokes, G.G. (1880) Considerations Relative to the Greatest Height of Oscillatory Waves which can be Propagated without Change of Form. Mathematical and Physical Papers, Vol. 1, 225-228, Cambridge University Press.

Williams, J.M. (1981) Limiting Gravity Waves in Water of Finite Depth. Phil. Trans. R. Soc. Lond. A, in Press.

Yamada, H. (1957a) Highest Waves of Permanent Type on the Surface of Deep Water. Rep. Res. Inst. Appl. Mech. Kyushu Univ., 5, 37-52.

Yamada, H. (1957b) On the Highest Solitary Wave. Rep. Res. Inst. Appl. Mech. Kyushu Univ., 5, 53-67.

Section II
Elasticity Problems

ON THE CONSTRUCTION OF THE BOUNDARY INTEGRAL REPRESENTATION AND CONNECTED INTEGRAL EQUATIONS FOR HOMOGENEOUS PROBLEMS OF PLANE LINEAR ELASTOSTATICS

M. Pignolé

Laboratoire de Mécanique Physique
Université de Bordeaux I, France

ABSTRACT

Four fundamental identities are established. They rest on addition to the interior problem (resp. exterior) of an exterior problem (resp. interior) and consideration of displacement and tension jumps across the boundary, connected with these two problems. Two of these identities express, according to these jumps, displacements and stresses everywhere in both the interior and exterior domains. The two others respectively connect the previous jumps to a boundary displacement or tension linear combination.

We obtain the true boundary integral equations by introducing boundary conditions into these identities.

Two ways of discretisation are shown.

INTRODUCTION

Boundary element methods are usually classified into direct and indirect. In the direct method, the boundary node unknowns are directly obtained by solving a boundary integral equation. Then, domain unknowns can be computed everywhere with the aid of boundary node values. In indirect methods, boundary node parameters are used. A boundary integral system has to be solved first to compute them, allowing then the calculation of boundary and domain unknowns.

The direct method [Lachat, J. C. (1975)] and indirect methods [Banerjee, P. K. (1976)] have been simultaneously developed for elastostatics. They have allowed an effective treatment of practical size engineering problems in the last few years. The equivalence between each category has been established [Brebbia, C.A. and Walker, S. (1978)]. The parameters used in indirect methods have been often called fictitious density distributions but they can be given a physical meaning [Hartmann, F. (1980)].

The main aim of this paper is to show how direct and indirect formulations ensue from four fundamental identities and how the indirect method parameters have a mathematical meaning. They are displacement and tension jumps when a complementary exterior boundary value problem (resp. interior) is associated with the initial interior one (resp. exterior). We limit our study to practical two dimensional problems. The displacements are continuous and the tensions almost everywhere continuous on the boundary. Study is analogous for three dimensional case. It may be extended to other linear operators.

A BASIC IDENTITY FOR REGULAR FUNCTIONS

Basic equations

The displacement vector u, the strain tensor ε and the stress tensor σ are functions of the point. Later on, we shall use the well known relations.

$$\varepsilon = \tilde{\nabla}\, u \tag{1}$$

$$\sigma = \lambda^{*}\,(\mathrm{tr}\,\varepsilon)\, I + 2\,\mu\,\varepsilon \qquad \text{(Hooke)} \tag{2}$$

$$\sigma = \lambda^{*}\,(\mathrm{div}\, u)\, I + 2\,\mu\,\tilde{\nabla}\, u \qquad \text{(Lamé)} \tag{3}$$

$$-\mathrm{div}\,\sigma = f \qquad \text{(equilibrium)} \tag{4}$$

$$-\left[\mu\,\Delta\, u + (\lambda^{*} + \mu)\,\nabla\,\mathrm{div}\, u\right] = f \qquad \text{(Navier)} \tag{5}$$

where the operators and the coefficients mean as follows

- ∇ gradient
- $\tilde{\nabla}$ symmetric gradient
- div divergence
- tr trace (with the meaning $\mathrm{tr}\,\varepsilon = \varepsilon_{mm}$)
- I identity matrix
- f volume forces assumed to be zero in homogeneous problems
- λ^{*}, μ constant coefficients

$$\lambda^* = \begin{cases} \lambda & \text{for plane strain} \\ \dfrac{2\,\mu\,\lambda}{\lambda + 2\,\mu} & \text{for plane stress} \end{cases} \qquad (6)$$

λ and μ are the Lamé coefficients.

<u>Connected system</u>

In spite of the common use of Equation 5, we prefer constituting a system with Equation 3 and Equation 4. For a homogeneous case, we write

$$\begin{cases} \lambda^*(\operatorname{div} u)\, I + 2\,\mu\, \tilde{\nabla}\, u - \sigma = 0 \\ \operatorname{div} \sigma = 0 \end{cases} \qquad (7)$$

or

$$\mathrm{Sys}\,(u, \sigma) = 0 \qquad (8)$$

<u>Variational identity</u>

We call ω a bounded open set
γ its smooth boundary
(ψ, Φ) some couple of a regular vector ψ and a regular symmetric second order tensor Φ

(u, σ) is a solution of Equation 8. Then

$$\int_\omega \left[\lambda^* (\operatorname{div} u)\, I + 2\,\mu\, \tilde{\nabla}\, u - \sigma\right] \cdot \Phi \, d\omega + \int_\omega \operatorname{div} \sigma \cdot \psi \, d\omega = 0 \qquad (9)$$

- $\cdot$ denotes the scalar product of two vectors or the inner product of two tensors.

Integrating by parts, we find

$$-\int_\omega \left[\lambda^* \nabla\, (\operatorname{tr} \Phi)\, I + 2\,\mu\, \operatorname{div} \Phi\right] \cdot u \, d\omega - \int_\omega (\Phi + \tilde{\nabla}\, \psi) \cdot \sigma \, d\omega$$
$$+\int_\gamma \{\left[\lambda^* (\operatorname{tr} \Phi)\, I + 2\,\mu\, \Phi\right] n\} \cdot u \, d\gamma + \int_\gamma \psi \cdot (\sigma\, n)\, d\gamma = 0 \qquad (10)$$

Where n is the outward normal

Now, let us consider σ as a function

$$u \overset{\sigma}{\to} \sigma\,(u)$$

Equation (10) may be rewritten in a shortened form

$$-\int_\omega A\,(\Phi) \cdot u \, d\omega - \int_\omega B\,(\psi, \Phi) \cdot \sigma\,(u)\, d\omega$$
$$+\int_\gamma \sigma^*(\Phi)\, n \cdot u \, d\gamma + \int_\gamma \psi \cdot t\,(u)\, d\omega = 0 \qquad (11)$$

with

$$A\ (\Phi) = \lambda^{*} \nabla\ (\mathrm{tr}\ \Phi)\ I + 2\ \mu\ \mathrm{div}\ \Phi \tag{12}$$

$$B\ (\psi,\ \Phi) = \Phi + \tilde{\nabla}\ \psi \tag{13}$$

$$t = \sigma\ n \tag{14}$$

t is the tension vector on γ.

A PARTICULAR 2-UPLE OF COUPLES $(\psi,\ \Phi)$

We seek a 2-uple of $(\psi^{(j)},\ \Phi^{(j)})$ couples, solution, in the distribution meaning, of

$$\begin{cases} A\ (\Phi^{(j)}) = \delta_y\ I^{(j)} & (15) \\ B\ (\psi^{(j)},\ \Phi^{(j)}) = \mathbf{0} & (16) \end{cases} \qquad j = 1,2$$

Where δ_y is the Dirac distribution in y, and

$$\mathbf{I}^{(1)} = \begin{bmatrix} 1 \\ 0 \end{bmatrix} \qquad I^{(2)} = \begin{bmatrix} 0 \\ 1 \end{bmatrix}$$

$$\mathbf{0} = \begin{bmatrix} 0 & 0 \\ 0 & 0 \end{bmatrix}$$

Calculation of $\psi^{(j)}$

Substituting for $\Phi^{(j)}$ from Equation 16 into Equation 15, we obtain

$$N\ \psi^{(j)} = \delta_y\ I^{(j)} \tag{17}$$

where N is the Navier operator

$$N = -\ [\mu\ \Delta + (\lambda^{*} + \mu)\ \nabla\ \mathrm{div}] \tag{18}$$

$\psi^{(j)}$ is the jth column-vector of a matrix U, an elementary solution of the N operator. U is a matrix of distributions that may be represented by functions everywhere except in x = y and aresummable on a disk around y. Theoretical justification may be found in [Schwartz, L. (1973)], details of calculation in [Pignolé, M. (1980)].

$$\psi^{(j)}_{(y)i} = U_{(y)ij} = U_1\ \delta_{ij}\ \ell n\ \frac{1}{r} + U_2\ r,_i\ r,_j \tag{19}$$

with

$$r = |x - y| \tag{20}$$

$$r,_i = \frac{\partial r}{\partial x_i} = -\ \frac{\partial r}{\partial y_i} \tag{21}$$

$$U_1 = \frac{\lambda^* + 3\ \mu}{4\ \Pi\ \mu\ (\lambda^* + 2\ \mu)} \tag{22}$$

$$U_2 = \frac{\lambda^* + u}{4\ \Pi\ \mu\ (\lambda^* + 2\ \mu)} \tag{23}$$

U has two properties of symmetry

$$U_{(y)ji}\ (x) = U_{(y)ij}\ (x) \tag{24}$$

$$U_{(x)ij}\ (y) = U_{(y)ij}\ (x) \tag{25}$$

Calculation of $\Phi^{(j)}$

Using Equations 16 and 19, we obtain

$$\Phi^{(j)}_{(y)ik} = \frac{1}{2r}\left[(U_1 - U_2)\ (\delta_{jk}\ r,_i + \delta_{ij}\ r,_k) - 2\ U_2\ \delta_{ik}\ r,_j + 4\ U_2\ r,_i\ r,_j\ r,_k\right] \tag{26}$$

Where $\delta_{\ell m}$ is Kronecker 's symbol

The 2-uple of $\Phi^{(j)}$ constitutes a third order tensor.

A PARTICULAR 2x2-UPLE OF COUPLES (ψ, Φ)

Now, we seek a 2x2-uple of $(\psi^{(jk)}, \Phi^{(jk)})$ couples, solution, in the distribution meaning, of

$$\begin{cases} A\ (\Phi^{(jk)}) = 0 & (27) \\ B\ (\psi^{(jk)}, \Phi^{(jk)}) = \delta_y\ \mathbf{I}^{(jk)} & (28) \end{cases} \qquad j,\ k = 1,2$$

with

$$\mathbf{I}^{(11)} = \begin{bmatrix} 1 & 0 \\ 0 & 0 \end{bmatrix} \qquad \mathbf{I}^{(12)} = \begin{bmatrix} 0 & 1 \\ 0 & 0 \end{bmatrix}$$

$$\mathbf{I}^{(21)} = \begin{bmatrix} 0 & 0 \\ 1 & 0 \end{bmatrix} \qquad \mathbf{I}^{(22)} = \begin{bmatrix} 0 & 0 \\ 0 & 1 \end{bmatrix}$$

Calculation of $\psi^{(jk)}$

System 27 28 may be rewritten

$$N\ \psi^{(jk)} = V^{(jk)} \tag{29}$$

with

$$V^{(11)} = - \begin{bmatrix} (\lambda^* + 2\ \mu)\ \delta_{y},_1 \\ \lambda^*\ \delta_{y},_2 \end{bmatrix} \tag{30}$$

$$V^{(12)} = V^{(21)} = - \begin{bmatrix} \mu\ \delta_{y},_2 \\ \mu\ \delta_{y},_1 \end{bmatrix} \tag{31}$$

$$V^{(22)} = - \begin{bmatrix} \lambda^* \delta_{y',1} \\ (\lambda^* + 2\mu)\ \delta_{y',2} \end{bmatrix} \tag{32}$$

Now, we know an elementary solution U of N. Therefore

$$\psi^{(jk)}_{(y)} = U_{(y)} * V^{(jk)} \tag{33}$$

$$\psi^{(jk)}_{(y)i} = \frac{1}{r} \Big[D_1 \ (\delta_{ik}\ r,_j + \delta_{ij}\ r,_k - \delta_{jk}\ r,_i) + D_2\ r,_i\ r,_j\ r,_k \Big] \tag{34}$$

with

$$D_1 = \mu\ (U_1 - U_2) = \frac{\mu}{2\ \Pi\ (\lambda^* + 2\ \mu)} \tag{35}$$

$$D_2 = 4\ \mu\ U_2 = \frac{\lambda^* + \mu}{\Pi\ (\lambda^* + 2\ \mu)} \tag{36}$$

The 2 x 2-uple of $\psi^{(jk)}$ constitutes a third order tensor D

$$D_{(y)jki} = \psi^{(jk)}_{(y)i} \tag{37}$$

D has the following properties

$$D_{(y)\,kji}\ (x) = D_{(y)jki}\ (x) \tag{38}$$

$$D_{(x)\,jki}\ (y) = - D_{(y)jki}\ (x) \tag{39}$$

<u>Calculation of $\Phi^{(jk)}$</u>

Using Equations 28 and 34, we obtain everywhere except in y

$$\begin{aligned} \Phi^{(jk)}_{(y)i\ell} = \frac{1}{r^2} \Big[& - D_1\ (\delta_{ik}\ \delta_{j\ell} + \delta_{ij}\ \delta_{k\ell} - \delta_{jk}\ \delta_{i\ell}) \\ & + (D_1 - \frac{D_2}{2})\ (\delta_{ik}\ r,_j\ r,_\ell + \delta_{ij}\ r,_k\ r,_\ell \\ & \qquad + \delta_{k\ell}\ r,_i\ r,_j + \delta_{j\ell}\ r,_i\ r,_k) \\ & - 2\ D_1\ \delta_{jk}\ r,_i\ r,_\ell - D_2\ \delta_{i\ell}\ r,_j\ r,_k \\ & \qquad + 4\ D_2\ r,_i\ r,_j\ r,_k\ r,_\ell \Big] \end{aligned} \tag{40}$$

The 2 x 2-uple of $\Phi^{(jk)}$ constitutes a fourth order tensor.

FOUR FUNDAMENTAL IDENTITIES

Let us consider a bounded open set Ω with a smooth boundary Γ and the complementary Ω' of $\bar{\Omega}$. Let us suppose that u may be twice differentiated in $\Omega \cup \Omega'$ and the interior and exterior traces $u|_{I\Gamma}$, $u|_{E\Gamma}$, $t|_{I\Gamma}$, $t|_{E\Gamma}$ exist. (u, σ) is a solution of

Equation 8 and u and σ decrease as $1/r$ and $1/r^2$ at infinity.

The displacement and tension jumps are given by

$$[u] = u|_{I\Gamma} - u|_{E\Gamma} \tag{41}$$

$$[t] = t|_{I\Gamma} - t|_{E\Gamma} \tag{42}$$

n is the outward normal for Ω.

β_y is a y centred disk the radius of which is assumed to vanish and β_R a disk such as $\bar{\Omega} \subset \beta_R$, the radius of which is assumed to tend to infinity. By writing Identity 11 for (u, σ) and a particular (ψ, Φ) in $\Omega \cup \Omega' \cap \beta_R \setminus \beta_y$, we want to obtain, when ε vanishes and R tends to infinity, four fundamental identities between

u and $\mathcal{L}(u|_{I\Gamma}, u|_{E\Gamma})$ with the 2-uple of $(\psi^{(j)}, \Phi^{(j)})$

σ and $\mathcal{L}(t|_{I\Gamma}, t|_{E\Gamma})$ with the 2x2-uple of $(\psi^{(jk)}, \Phi^{(jk)})$

$\mathcal{L}$ has the meaning of linear combination.

<u>Calculation of $\sigma(\psi^{(j)})$, $\sigma(\psi^{(j)})\,n$, $\sigma(\psi^{(jk)})$, $\sigma(\psi^{(jk)})\,n$</u>

Using Equations 3 19 and 34, we find

$$\sigma_{ik}(\psi^{(j)}_{(y)}) = \frac{1}{r}\left[T_1(\delta_{jk}\, r,_i + \delta_{ij}\, r,_k - \delta_{ik}\, r,_j) + T_2\, r,_i\, r,_j\, r,_k\right] \tag{43}$$

with

$$T_1 = D_1 \tag{44}$$

$$T_2 = D_2 \tag{45}$$

D_1 and D_2 are given by Equations 35 and 36.

$$T^{(j)}_{(y)i} = \sigma_{ik}(\psi^{(j)}_{(y)})\, n_k = (T_1\, \delta_{ij} + T_2\, r,_i\, r,_j)\frac{\partial r}{\partial n} - T_1(n_i\, r,_j - n_j\, r,_i) \tag{46}$$

The $T^{(j)}$ constitute a third order tensor.

$$\sigma_{i\ell}(\psi^{(jk)}_{(y)}) = \frac{1}{r^2}\Big[S_1(\delta_{ik}\,\delta_{j\ell} + \delta_{ij}\,\delta_{k\ell}) + S_2\,\delta_{jk}\,\delta_{i\ell} + S_3(\delta_{ik}\, r,_j\, r,_\ell + \delta_{ij}\, r,_k\, r,_\ell$$

$$+ \delta_{k\ell}\, r,_i\, r,_j + \delta_{j\ell}\, r,_i\, r,_k) \qquad (47)$$

$$+ S_4\, (\delta_{i\ell}\, r,_j\, r,_k + \delta_{jk}\, r,_i\, r,_\ell)$$

$$+ S_5\, r,_i\, r,_j\, r,_k\, r,_\ell]$$

with

$$S_1 = -\,2\,\mu\, D_1 = -\,\mu^2\, S_o \qquad (48)$$

$$S_2 = -\,2\,(\lambda^* - \mu)\, D_1 = -\,(\lambda^* - \mu)\,\mu\, S_o \qquad (49)$$

$$S_3 = \mu\,(2\, D_1 - D_2) = -\,\lambda^*\,\mu\, S_o \qquad (50)$$

$$S_4 = -\,4\,\mu\, D_1 = 2\, S_1 \qquad (51)$$

$$S_5 = 8\,\mu\, D_2 = 8\,\mu\,(\lambda^* + \mu)\, S_o \qquad (52)$$

$$S_o = \frac{1}{\Pi\,(\lambda^* + 2\,\mu)} \qquad (53)$$

$$S^{(jk)}_{(y)i} = \sigma_{i\ell}\,(\psi^{(jk)}_{(y)})\, n_\ell = \frac{1}{r^2}\left\{\frac{\partial r}{\partial n}\left[S_3\,(\delta_{ik}\, r,_j + \delta_{ij}\, r,_k)\right.\right.$$

$$\left. S_4\,\delta_{jk}\, r,_i + S_5\, r,_i\, r,_j\, r,_k\right] \qquad (54)$$

$$+ S_1\,(\delta_{ik}\, n_j + \delta_{ij}\, n_k) + S_2\,\delta_{jk}\, n_i$$

$$+ S_3\,(n_j\, r,_i\, r,_k + n_k\, r,_i\, r,_j)$$

$$+ S_4\, n_i\, r,_j\, r,_k$$

The $S^{(jk)}$ constitute a third order tensor.

Identity between u, [u] and [t] for y Ω Ω'

Using the previous procedure, we find

$$\int_\Gamma T^{(j)} \cdot [u]\, d\Gamma + \int_\Gamma \psi^{(j)} \cdot [t]\, d\gamma + \int_{\Gamma_\varepsilon} T^{(j)} \cdot u\, d\Gamma$$

$$+ \int_{\Gamma_\varepsilon} \psi^{(j)} \cdot t\, d\Gamma + \int_{\Gamma_R} T^{(j)} \cdot u\, d\Gamma + \int_{\Gamma_R} \psi^{(j)} \cdot t\, d\Gamma = 0 \qquad (55)$$

It is easy to prove that the integrals on Γ_R and Γ_E vanish, except $\int_{\Gamma_\varepsilon} T^{(j)} \cdot u\, d\Gamma$, when ε vanishes and R tends to infinity.

$$u_j(y) = \int_\Gamma T^{(j)}_{(y)}(x) \cdot [u(x)]\, d\Gamma(x) + \int_\Gamma U^{(j)}_{(y)}(x) \cdot [t(x)]\, d\Gamma(x) \quad (56)$$

noting $U^{(j)}$ as $\psi^{(j)}$.

Writing Identity 11 for (u, σ) and the particular (1, 0) couple (ψ, Φ), in Ω' and $\Omega \cup \Omega'$ we find that the hypothesis of decreasing at infinity for u and σ is equivalent to

$$\int_\Gamma t|_{I\Gamma}\, d\Gamma = \int_\Gamma t|_{E\Gamma}\, d\gamma \quad (57)$$

Identity between σ, $[u]$ and $[t]$

Using the previous procedure or using Equations 3 and 56, we find

$$\sigma_{jk}(y) = \int_\Gamma S^{(jk)}_{(y)}(x) \cdot [u(x)]\, d\Gamma(x) + \int_\Gamma D^{(jk)}_{(y)}(x) \cdot [t(x)]\, d\Gamma(x) \quad (58)$$

noting $D^{(jk)}$ as $\psi^{(jk)}$.

Identity between $(u|_{I\Gamma}, u|_{E\Gamma})$, $[u]$ and $[t]$

For a smooth boundary

$$\frac{1}{2}(u_j(y)|_{I\Gamma} + u_j(y)|_{E\Gamma}) = \int_\Gamma T^{(j)}_{(y)}(x) \cdot [u(x)]\, d\Gamma(x) + \int_\Gamma U^{(j)}_{(y)}(x) \cdot [t(x)]\, d\Gamma(x) \quad (59)$$

interpreting the first integral in the meaning of its principal value.

Identity between $(t|_{I\Gamma}, t|_{E\Gamma})$, $[u]$ and $[t]$

Assuming that the tension is continuous and that the displacement satisfies a Hölder condition at y, using the method of [Cruse, T. A. (1977)], we find

$$\frac{1}{2}(t_j(y)|_{I\Gamma} + t_j(y)|_{E\Gamma}) = \int_\Gamma S^{(jk)}_{(y)}(x)\, n_k(y) \cdot [u(x)]\, d\Gamma(x) \quad (60)$$
$$\int_\Gamma D^{(jk)}_{(y)}(x)\, n_k(y) \cdot [t(x)]\, d\Gamma(x)$$

The two integrals must be interpreted in the meaning of their principal value.

OBTAINING THE INTEGRAL REPRESENTATIONS AND INTEGRAL EQUATIONS

We consider, an initial interior boundary value problem IP,

or an exterior one EP, with boundary conditions

$$\text{IP}\begin{cases} u|_{\Gamma_u} = u_{oI} \\ t|_{\Gamma_t} = t_{oI}\end{cases} \qquad \text{EP}\begin{cases} u|_{\Gamma_u} = u_{oE} \\ t|_{\Gamma_t} = t_{oE}\end{cases} \tag{61}$$

such as indicated on figure 1.

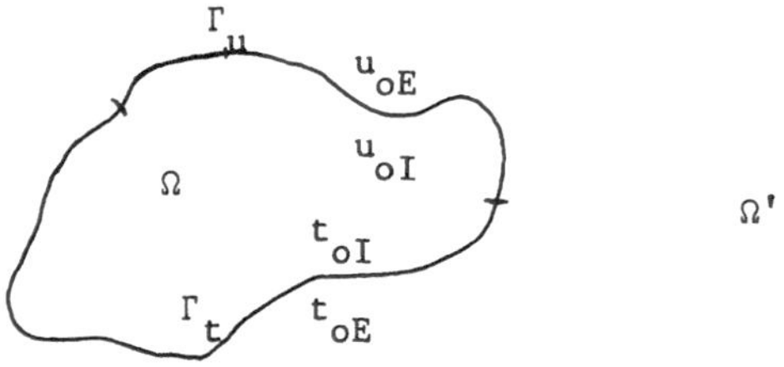

figure 1

First, we obtain particular identities from the four fundamental identities, according to the choice of an associate problem, then integral equations by introducing boundary conditions.

The direct method

We choose

$$\text{for IP} \quad u|_{E\Gamma} = t|_{E\Gamma} = 0 \tag{62}$$

$$\text{for EP} \quad u|_{I\Gamma} = t|_{I\Gamma} = 0 \tag{63}$$

From identities 56 58 and 59 and Equations 62 and 63, we obtain the direct particular identities

$$\int_\Gamma T^{(j)}_{(y)}(x)\cdot u(x)\, d\Gamma(x) + \int_\Gamma U^{(j)}_{(y)}(x)\cdot t(x)\, d\Gamma(x) =$$

$$\begin{cases} \frac{1}{2} u_j(y) & y\in\Gamma \\ u_j(y) & y\in\Omega \end{cases} \text{ for IP} \qquad \begin{matrix}(64)\\(65)\end{matrix}$$

$$\begin{cases} -\frac{1}{2} u_j(y) & y\in\Gamma \\ -u_j(y) & y\in\Omega \end{cases} \text{ for EP} \qquad \begin{matrix}(66)\\(67)\end{matrix}$$

$$\int_\Gamma S^{(jk)}_{(y)} (x) \cdot u (x)\, d\Gamma (x)$$

$$+\int_\Gamma D^{(jk)}_{(y)} (x) \cdot t (x)\, d\Gamma (x) =$$

$$\begin{cases} \sigma_{jk} (y) & y \in \Omega \quad \text{for IP} \quad (68) \\ - \sigma_{jk} (y) & y \in \Omega' \quad \text{for EP} \quad (69) \end{cases}$$

By introducing the boundary conditions into the identities for Γ, we obtain the B. I. Equation. After resolving it we know all the values of u and t on Γ and we may compute u and σ everywhere in the domain with the aid of Equations 65 and 68 for IP or 67 and 69 for EP.

Indirect methods

The particular identities contain $[u]$ and / or $[t]$. We may construct an infinite number of indirect formulations, each according to a complementary problem, using Identities 56 58 59 and 60. First, a system has to be solved on Γ, then we may compute u and / or t on Γ and u and σ in the domain. We give some examples.

Tension jump formulation For this formulation, we choose, for IP and EP

$$[u]_\Gamma = 0 \qquad (70)$$

For $y \in \Gamma$, Identities 59 and 60 become

$$u_j (y) = \int_\Gamma U^{(j)}_y (x) \cdot [t (x)]\, d\Gamma (x) \qquad (71)$$

$$t_j (y) = \int_\Gamma D^{(jk)}_{(y)} (x)\, n_k (y) \cdot [t (x)]\, d\Gamma (x) \begin{cases} + \frac{1}{2} [t_j (y)] & \text{for IP} \\ - \frac{1}{2} [t_j (y)] & \text{for EP} \end{cases} \qquad (72)$$

For y in the initial domain, we may write, from Identity 56 Identity 71 and from Identity 58 as follows

$$\sigma_{jk} (y) = \int_\Gamma D^{(jk)}_{(y)} (x) \cdot [t (x)]\, d\Gamma (x) \qquad (73)$$

The $[t]$ system to be solved is obtained by using Equation 71 on Γ_u and Equation 72 on Γ_t (for IP or EP respectively).

Next,

Identity 71 => u on Γ_t and in the domain

Identity 72 => t on Γ_u
Identity 73 => σ in the domain

<u>Displacement jump formulation</u> For this formulation, we choose, for IP and EP

$$[t]_\Gamma = 0 \tag{74}$$

In the same way, for $y \in \Gamma$

$$u_j(y) = \int_\Gamma T^{(j)}_{(y)}(x) \cdot [u(x)]\, d\Gamma \begin{array}{ll} + \frac{1}{2}[u_j(y)] & \text{for IP} \\ - \frac{1}{2}[u_j(y)] & \text{for EP} \end{array} \tag{75}$$

$$t_j(y) = \int_\Gamma S^{(jk)}_{(y)}(x)\, n_k(y) \cdot [u(x)]\, d\Gamma(x) \tag{76}$$

For y in the initial domain, we may write

$$u_j(y) = \int_\Gamma T^{(j)}_{(y)}(x) \cdot [u(x)]\, d\Gamma(x) \tag{77}$$

$$\sigma_{jk}(y) = \int S^{(jk)}_{(y)}(x) \cdot [u(x)]\, d\Gamma(x) \tag{78}$$

The $[u]$ system is obtained by using Equations 75 on Γ_u and Equation 76 on Γ_t (for IP and EP respectively).

Next,

Identity 75 => u on Γ_t
Identity 76 => t on Γ_u
Identity 77 => u in the domain
Identity 78 => σ in the domain.

<u>A mixed formulation</u> Now, we choose, for IP and EP

$$[u]_{\Gamma_u} = 0 \tag{79}$$

and

$$[t]_{\Gamma_t} = 0 \tag{80}$$

We obtain, for $y \in \Gamma$

$$\int_{\Gamma_t} T^{(j)}_{(y)}(x) \cdot [u(x)]\, d\Gamma(x) + \int_{\Gamma_u} U^{(j)}_{(y)} \cdot [t(x)]\, dx = u_j(y) \tag{81}$$

$y \in \Gamma_u$

$$\int_{\Gamma_t} S^{(jk)}_{(y)} (x)\, n_k (y) \cdot [u (x)]\, d\Gamma (x) + \int_{\Gamma_u} D^{(jk)}_{(y)} (x)\, n_k (y) \cdot [t (x)]\, d\Gamma (x) = t_j (y) \qquad y \in \Gamma_t \tag{82}$$

Equations 81 and 82 together constitute a $[t (x)]_{\Gamma_u}$ and $[u (x)]_{\Gamma_t}$ system which has to be solved at first.

Then,

$$u_j (y) = \int_{\Gamma_t} T^{(j)}_{(y)} (x) \cdot [u (x)]\, d\Gamma (x) + \int_{\Gamma_u} U^{(j)}_{(y)} [t (x)]\, d\Gamma (x) \begin{cases} + \frac{1}{2} [u_j (y)] & \text{for IP} \\ - \frac{1}{2} [u_j (y)] & \text{for EP} \end{cases} \qquad y \in \Gamma_t \tag{83}$$

$t \in \Gamma_u$ is given by an analogous expression using the kernels Sn and Dn and the jump $[t_j (y)]$.

u is given in the domain by Identity 81 and σ by

$$\sigma_{jk} (y) = \int_{\Gamma_t} S^{(jk)}_{(y)} (x) \cdot [u (x)]\, d\Gamma (x) + \int_{\Gamma_u} D^{(jk)}_{(y)} (x) \cdot [t (x)]\, d\Gamma (x) \tag{84}$$

A FEW REMARKS AS A CONCLUSION

The four fundamental identities we have written must be regarded as formal identities. Actually, u ($\in \Omega$ or Ω'), σ ($\in \Omega$ or Ω'), $u|_\Gamma$, $t|_\Gamma$ belong to suitable vectorial spaces and Γ - integrals are dualities between trace spaces. Basic elements can be found in [Nedelec, J. C. (1977)].

Identity 60, for example, must not be written in a point where t is not continuous, on the given form, but in an "integrated form", using a duality.

In this way, a variational formulation with an associated bilinear form may be written, in the place of the boundary identities between $[u]$, $[t]$ and $\mathcal{L}(u|_{I\Gamma}, u|_{E\Gamma})$ and / or between $[u]$, $[t]$ and $\mathcal{L}(t|_{I\Gamma}, t|_{E\Gamma})$.

Using the identities in the previously written form, the discretisation leads us to a boundary collocation method (BCM). Link equations must be added [Chaudonneret, M. (1978)]. Numerical examples of plane elastostatic problems may be seen in [Pignolé, M. (1980)].

Using a variational formulation, it is logical to use boundary finite elements (BFEM). For theoretical approach see [Nedelec, J. C. (1977)] and [Baldino, R. R. (1979 and 1980)].

ACKNOWLEDGEMENTS

For the helpful theoretical discussions, the author greatly thanks Professor Y. Haugazeau.

REFERENCES

Baldino, R. R. (1979 and 1980) An integral equation solution of the mixed problem for the Laplacian in R^3. Part 1 and 2. Rapports internes 48 et 54, Centre de Math. Appl., Ecole Polytechnique, Palaiseau - France.

Banerjee, P. K. (1976) Integral equation methods for analysis of piece - wise non - homogeneous three dimensional elastic solids of arbitrary shape. Int. J. Mech. Sci., 18 : 293-303.

Brebbia, C. A. and Walker, S. (1978) Introduction to boundary element methods. Recent advances in boundary element methods, Pentech Press, Plymouth and London, 293-303.

Chaudonneret, M. (1978) Calcul des concentrations de contraintes en élastoviscoplasticité. Publication O.N.E.R.A. 1978-1, Chatillon - France.

Cruse, T. A. (1977) Mathematical foundations of the boundary-integral equation method in solid mechanics. AFOSR-TR 77-1002, Pratt and Whitney Aircraft Group, United Technologies Corporation, East Hartford, Connecticut 06108.

Hartmann, F. (1980) The complementary problem of finite elastic bodies. New developments in boundary element methods, CML Publications, Southampton, 229-246.

Lachat, J. C. (1975) A further development of the boundary integral technique for elastostatics. Thesis, University of Southampton.

Nedelec, J. C. (1977) Cours de l'école d'été d'analyse numérique CEA-IRIA-EDF.

Pignolé, M. (1980) Une méthode intégrale directe pour la résolution de problèmes d'élasticité. Applications aux problèmes plans. Thèse, Université de Bordeaux I.

Schwartz, L. (1973) Théorie des distributions. Hermann, Paris.

REGULAR BOUNDARY INTEGRAL EQUATIONS FOR STRESS ANALYSIS

C. Patterson and M.A. Sheikh

Dept. of Mechanical Eng., University of Sheffield, U.K.

INTRODUCTION

THE boundary integral method is now well established as a general numerical technique available for the solution of field problems. In contrast with the finite element method freedoms need only be defined on the boundary of the domain of the problem. Once these are determined the solution within the domain is obtained using appropriate surface integrals of the boundary solution.

Central to the method is the generation of Boundary Integral Equations which properly state the problem to be solved in terms of unknown field functions on the boundary only. These equations are usually obtained using the Fundamental Solution of the given problem with the singular point located on the boundary. (The equations for the interior solution are obtained similarly, by locating the singular point within the domain of the problem). There ensues an infinite system of singular surface equations, one for each boundary point (being generated by moving the singularity around the boundary). The system is discretized by defining boundary elements, after the manner of finite elements, and the resulting finite system of singular integrals are evaluated, thereby giving a system of algebraic equations.

Two discomforting features are apparent in this, conventional, approach. Firstly, accurate evaluation of the singular integrals requires special and careful treatment in the neighbourhood of the singular point. Secondly, the class of problems for which the method is well defined may be unduly restrictive because of divergence of the equations.

In this paper it is shown that 'Regular Boundary Integral Equations can quite readily be derived which also properly state the given problem. These are obtained by the simple device of moving the singularity of the fundamental solution outside the domain of the problem. The resulting system of equations tolerates higher order singularities in the solution than previously and requires no special attention to a singular integrand.

The practicality of the method is demonstrated in two-dimensional elastostatics. A critical comparison is made of the results obtained using the new and conventional approaches for constant and linear elements.

THEORY

The governing equations of elastostatics in terms of the stress field and in the absence of body forces can be written as:

$$\sum_{j=1}^{3} \frac{\partial \sigma_{ij}}{\partial x_j} = 0 \qquad i = 1, 2, 3 \tag{1}$$

The stresses and strains are related by the constitutive relation for an isotropic body as:

$$\sigma_{ij} = \lambda \delta_{ij} \frac{\partial u_k}{\partial x_k} + 2\mu \varepsilon_{ij} \tag{2}$$

where

$$\varepsilon_{ij} = \frac{\partial u_i}{\partial x_j} + \frac{\partial u_j}{\partial x_i}$$

and δ_{ij} is the unit tensor known as kronecker delta and λ, μ are Lame's constants given as:

$$\lambda = {}^{E}/(1 + \nu)(1 - 2\nu); \quad \mu = {}^{E}/_{2}\ (1 + \nu)$$

where E is the Modulus of Elasticity and ν is the Poisson's Ratio.

The fundamental solution which satisfies the homogenous equation of elasticity theory (Equation.1) is:

$$u^*_{ij} = \frac{1}{16\pi\mu(1-\sigma)} \left[(3-4\sigma) \frac{1}{r} \delta_{ij} + \frac{(x_i - y_i)\ (x_j - y_j)}{r^2} \right] \tag{3}$$

where the so called Poisson's Coefficient $\sigma = \lambda/2(\mu + \lambda)$ and r is the distance between the points x and y i.e. the point of application of load to the point under consideration.

Let the displacement field $u_j(y)$ be twice continuously differentiable in some finite region Ω and continuously differentiable in the closed region $\bar{\Omega} = \Omega + S$. We choose a point 'x' inside Ω and remove from Ω a ball of radius ε with center at 'x'. Applying the third Betti formula [Mikhlin,1965] to u_j (y) and u^*_{ij} (y) in the region Ω with the ball removed and letting $\varepsilon \to 0$, we get

$$u_i(x) = \int_\Omega u^*_{ij} \left\{\frac{\partial \sigma_{ij}}{\partial x_j}\right\} d\Omega_y - \int_S \left\{ u^*_{ij}(x,y) t_j(y) - u_j(y) t^*_{ij}(x,y) \right\} ds_y \quad (4)$$

If equation (1) is to be satisfied the first integral on the right hand side of equation (4) vanishes and we get,

$$u_i(x) = \int_S u^*_{i,j}(x,y) t_j(y) ds_y - \int_S t^*_{ij}(x,y) u_j(y) ds_y \quad (5)$$

where

$$t^*_{ij}(x,y) = \frac{1-2\sigma}{8\pi(1-\sigma) r^2}\left[\delta_{ij} + \frac{3}{1-2\sigma}\ \frac{(x_i - y_i)(x_j - y_j)}{r^2}\right] \quad (6)$$

Equation (5) is known as Somigliana's identity which in simplified notations take the form,

$$c^i u^i = \int_S u^* t ds - \int_S t^* u ds \quad (7)$$

This equation can be used to determine the interior solution once the boundary integral equation is solved. However, if the singularity of the fundamental solution is located outside the domain $\bar{\Omega}$, r does not vanish in $\bar{\Omega}$ so that $C^i = 0$ and so,

$$\int_S u^* t ds = \int_S t^* u ds \quad (8)$$

Equation (8) forms the basis of 'Regular Boundary Integral Method' in elastostatics.

If the given boundary conditions are:

$u = \bar{u}$ on S_1

$t = \bar{t}$ on S_2

where the total boundary is $S = S_1 + S_2$, we can write an extended weighted residual equation of the form

$$\int_{S_1} \bar{u} t^* ds + \int_{S_2} u t^* ds = \int_{S_1} t u^* ds + \int_{S_2} \bar{t} u^* ds \quad (9)$$

For the two dimensional case, equation (9) gives a

system of equations, two for each singular point (one for each direction) corresponding to the boundary node under the boundary element discretization and located outside the domain of the problem. Once this system of regular integral equations is solved, interior solution can be generated routinely, using equation (7).

Constant Boundary Elements

The boundary is discretized into N elements (say) and the values of traction and displacement are assumed to be constant on each element and equal to the value at the mid-node of the element. The singular point corresponding to this mid-node 'i' is located at an arbitrary distance (taken as L/2 in our case studies, where L is the length of the element) from 'i' and along the positive normal (Figure.1).

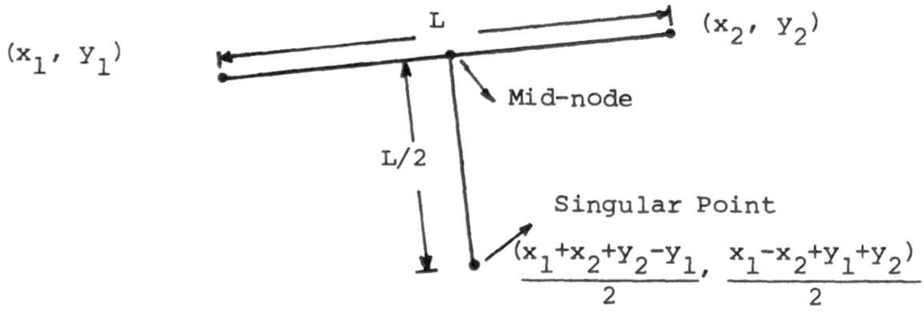

Fig.1. Location of Singular point for Constant element

Equation (8) for this singular point corresponding to the boundary node 'i' becomes in discretized form,

$$\sum_{j=1}^{N} t_j \int_{S_j} u^* ds = \sum_{j=1}^{N} u_j \int_{S_j} t^* ds \qquad (10)$$

The integrals in equation (10) can be evaluated numerically for all segments over the boundary including the one containing the node 'i'. 2N such equations (one for each direction at N points) are obtained and solved for the boundary unknowns. Once the solution over the whole boundary is obtained, interior solution can be generated using equation (7) in which case c^i equals unity.

Linear Elements

The variation of t and u is assumed to be linear within each element. For N elements equation (8) can be written as

$$\sum_{j=1}^{N} \int_{S_j} u^* \, tds = \sum_{j=1}^{N} \int_{S_j} t^* \, uds \qquad (11)$$

The values of u and t at any point of an element can be

written in terms of their nodal values and interpolation functions F_1 and F_2 as:

$$u(\xi) = F_1\, u_1 + F_2\, u_2$$
$$t(\xi) = F_1\, t_1 + F_2\, t_2 \qquad (12)$$

where ξ is the dimensionless coordinate $\xi = \frac{x}{L/2}$ (Figure.2).

$$F_1 = \frac{1}{2}(1-\xi) \; ; \quad F_2 = \frac{1}{2}(1+\xi) \qquad (13)$$

The singular point corresponding to a node 'i' is located at an arbitrary distance (taken as L/2 in the present study, where L is the length of the element) from 'i' and along the positive normal (Figure.3).

Equation (11) for any singular point corresponding to the boundary node 'i' can now be written in two parts (one for each direction) as:

$$\sum_{j=1}^{N} \int_{S_j} \begin{bmatrix} F_1 & F_2 \end{bmatrix} u^* \, ds \begin{Bmatrix} t_{1x} \\ t_{2x} \end{Bmatrix} = \sum_{j=1}^{N} \int_{S_j} \begin{bmatrix} F_1 F_2 \end{bmatrix} t^* ds \begin{Bmatrix} u_{1x} \\ u_{2x} \end{Bmatrix}$$

$$\sum_{j=1}^{N} \int_{S_j} \begin{bmatrix} F_1 & F_2 \end{bmatrix} u^* \, ds \begin{Bmatrix} t_{1y} \\ t_{2y} \end{Bmatrix} = \sum_{j=1}^{N} \int_{S_j} \begin{bmatrix} F_1 F_2 \end{bmatrix} t^* ds \begin{Bmatrix} u_{1y} \\ u_{2y} \end{Bmatrix} \qquad (14)$$

A system of 2N 'Regular Boundary Integral Equations' can now be obtained and solved as for constant elements.

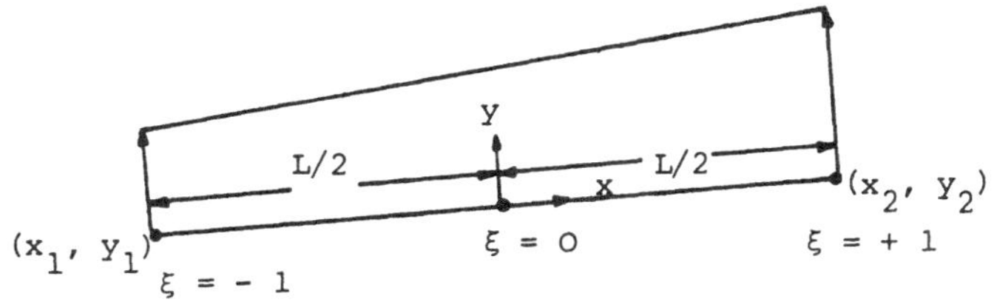

Fig. 2. Linear Element

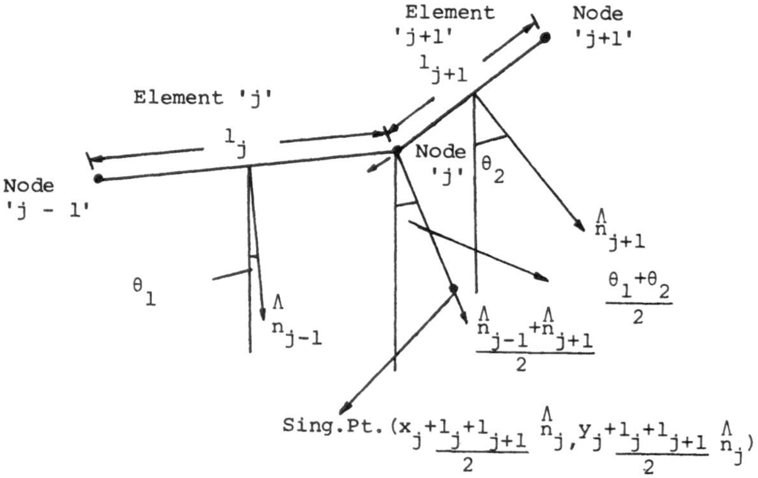

Fig.3. Location of Singular Point for Linear Element

APPLICATIONS

TWO two dimensional stress concentration problems are analysed using both the conventional and the proposed regular boundary integral methods for constant and linear boundary elements and a critical comparison of the results is made.

Notched Circular Hole in a Rectangular Sheet under Uniform Tension

The problem was first analysed without notch at the circular hole in the 150 mm X 110 mm rectangular plate as shown in Figure.4. The radius of the hole was taken as 15 mm, i.e. less than one-third of half width of the plate to produce high stress concentration at point XX. Only a quarter of the domain needs to be taken for analysis due to symmetry, (Figure.5). Boundary conditions for the problem are:

$$t_x = 0 \ , \ u_y = 0 \quad \text{along AB} \qquad (i)$$

$$t_x = 120 \text{ N}, \ ty = 0 \quad \text{along BC} \qquad (ii)$$

$$t_x = t_y = 0 \quad \text{along CD} \qquad (iii)$$

$$u_x = t_y = 0 \quad \text{along DE} \qquad (iv)$$

$$t_x = t_y = 0 \quad \text{along EA} \qquad (v)$$

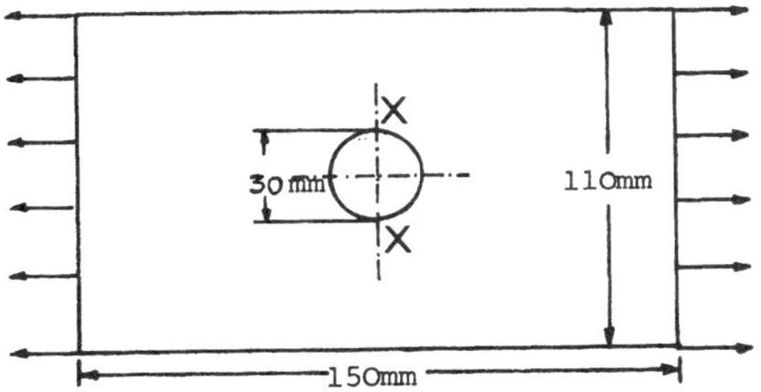

Fig.4. Rect.Plate with a circular hole under uniform tension

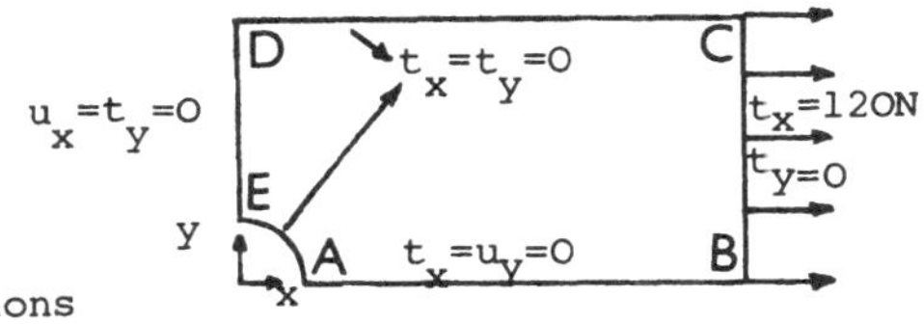

Fig.5. Boundary conditions for Quarter domain

Conditions $u_x = 0$ along DE and $u_y = 0$ along AB are specified to avoid rigid body motion. The material properties are taken as $E = 2.1 \times 10^5$ N/mm^2 and = 0.3.

The finite element and constant and linear boundary element discretizations used for the solution of the problem are shown in Figure.6.,and the stress distribution along ED thus obtained for conventional and regular integral methods is compared against finite element results [Patterson and Bagdatlioglu, 1981] in Figures.7a. and 7b., respectively for constant and linear elements. Although, as expected, the stress distribution obtained by using finite element method is better, the results for conventional and regular boundary integral methods both for constant and linear elements stay marginally close.

The problem was then made to yield even higher stress concentration by taking a semi-circular notch of radius 1.5 mm at points XX of the plate as shown in Figure.8. Again, only a quarter of the plate was considered with the boundary conditions remaining unchanged (Figure.9). The discretizations using finite element and constant and linear boundary elements are shown in Figure.10., and the stress distribution along ED obtained for constant and linear elements using conventional and regular methods is compared with finite element results [Patterson and Bagdatlioglu, 1981] in Figures.11a. and 11b., respectively. The stress levels at points XX is as much as three times than in the previous case of circular hole without notch. Again, the stress distribution given by finite elements is better but it is very similar for the conventional and regular methods in case of both constant and linear elements.

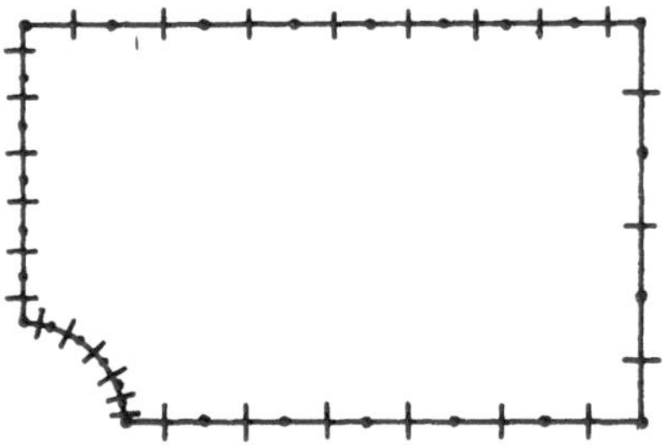

Fig.6C. Linear boundary element discretization

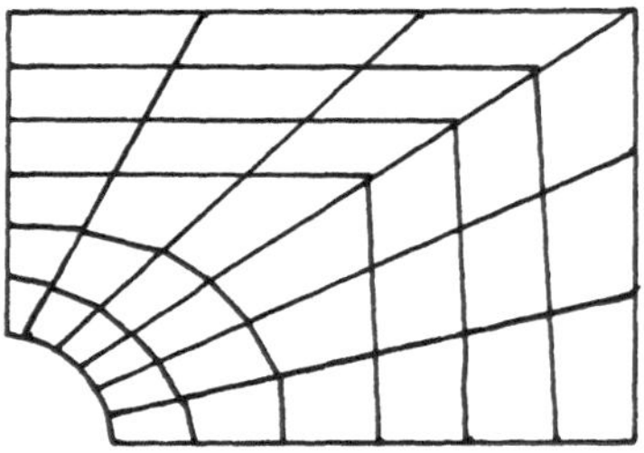

Fig.6a. Finite Element discretization

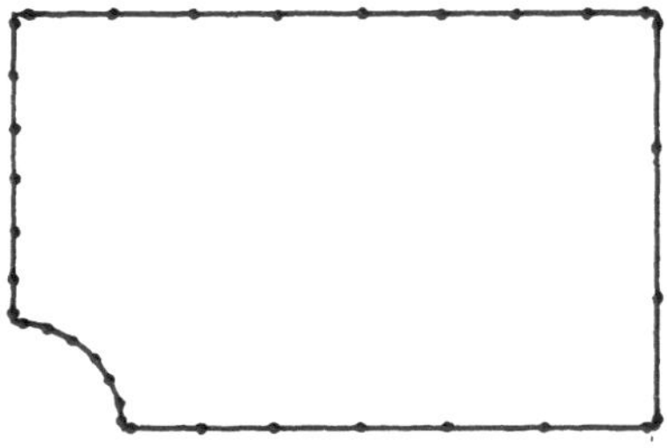

Fig.6C. Linear boundary Element discretization

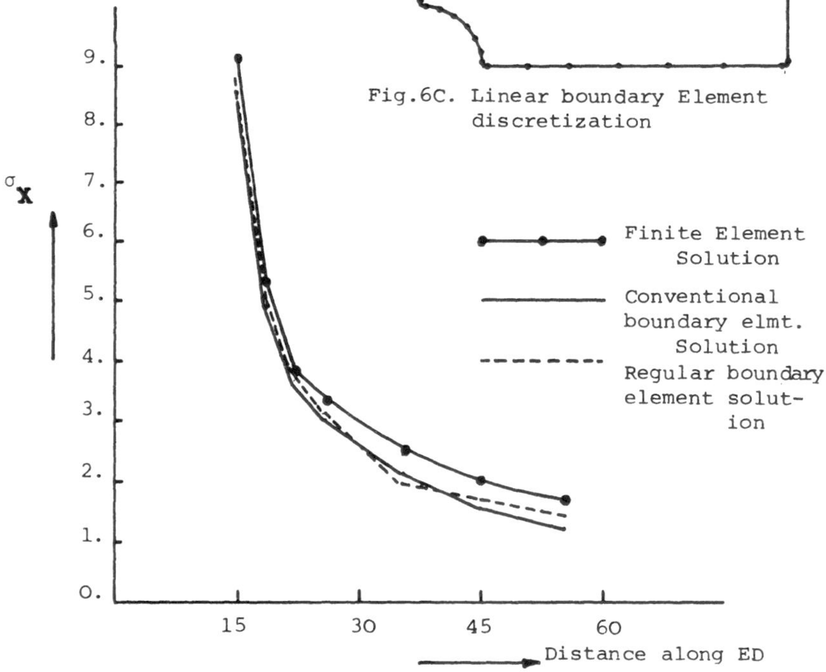

Fig.7a. Stress distribution along the edge ED of the domain using constant boundary elements

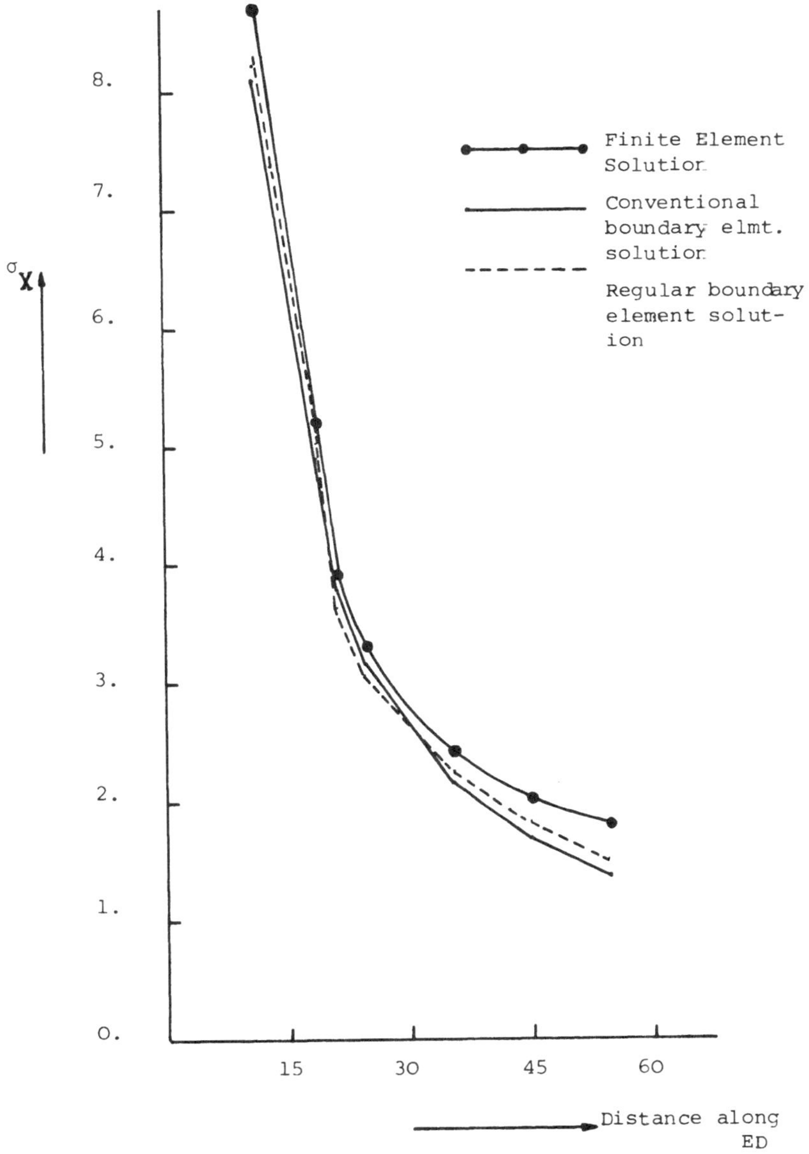

Fig. 7b. Stress distribution along the edge ED of the domain using Linear Boundary elements

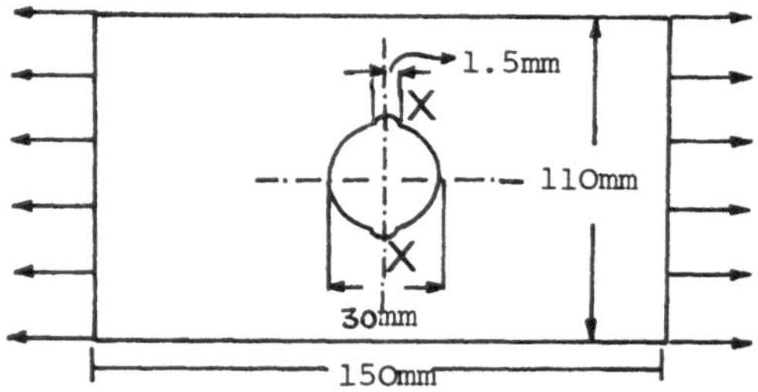

Fig.8. Rect.Plate with a notched circular hole under uniform tension

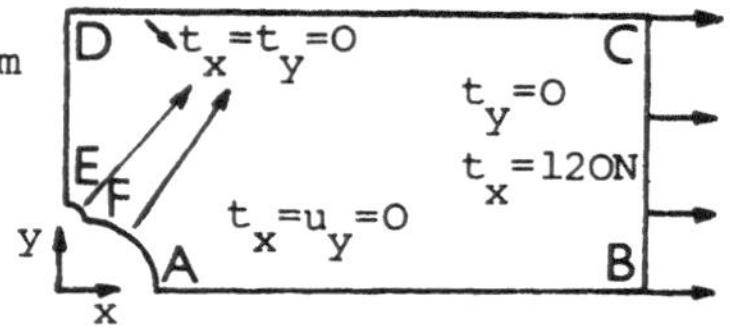

Fig.9. Boundary Conditions for Quarter domain

<u>A Lug loaded by Free Fitting Pin</u>

This is an actual design problem which has already been analysed both using photoelastic testing and as a finite element contact problem [Patterson, Wearing and Arnell, 1980]. The geometry of the Lug is described in Figure.12. Due to symmetry only half of the domain is considered and the boundary conditions are such that all the nodes on the axis of symmetry are restrained in y direction and the nodes on the edge EA are restrained in x direction as shown in Figure.13. The load distribution (Figure.14) that the pin puts on the lug was calculated by the finite element program [Patterson, Wearing and Arnell, 1980] and corresponds to 16000 lbs load on the pin of diameter 1.25 inches and diametrical clearance of 0.001 inches. The properties of the lug material are $E = 26.5 \times 10^6 \text{lbs/in}^2$ and poisson's ratio = 0.3.

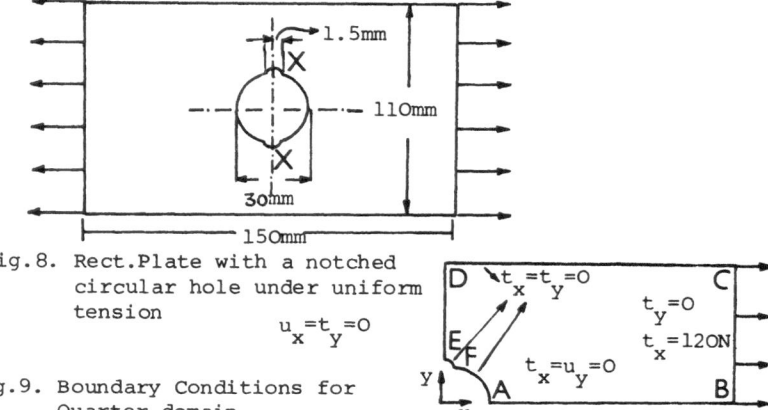

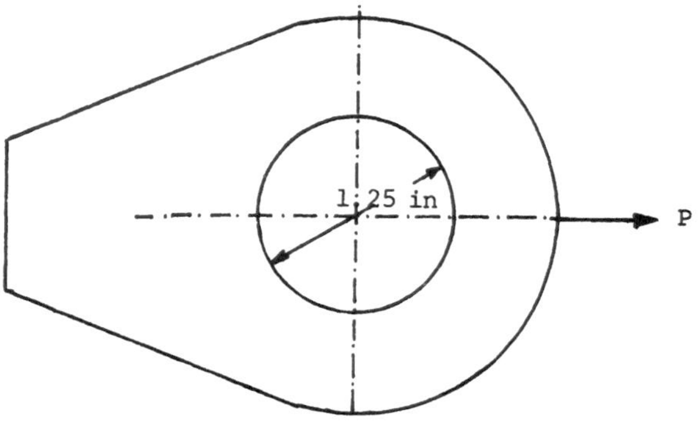

Fig.12. Geometry of the Lug

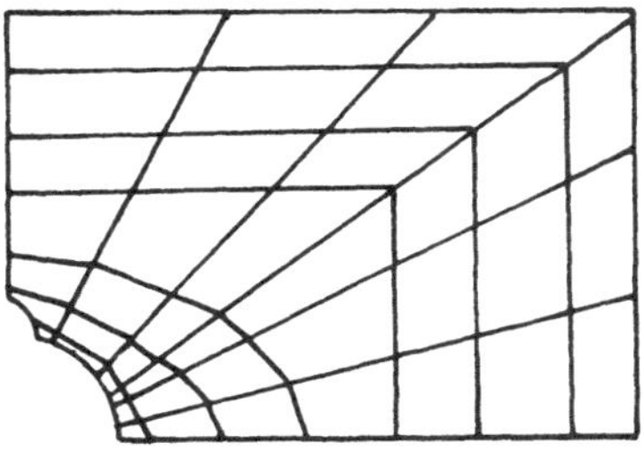

Fig.10a. Finite Element discretization

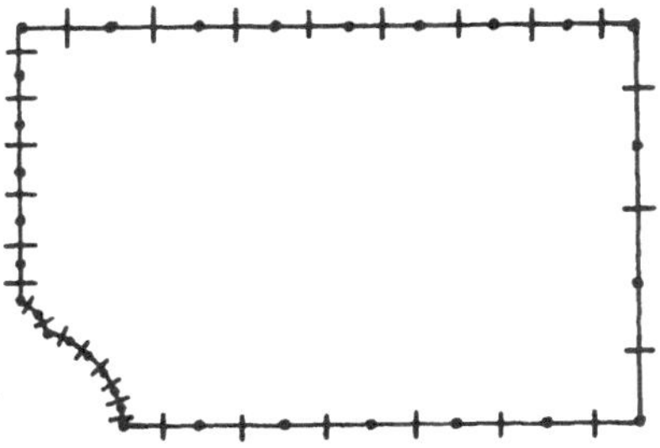

Fig.10b. Constant boundary elmt. discretization

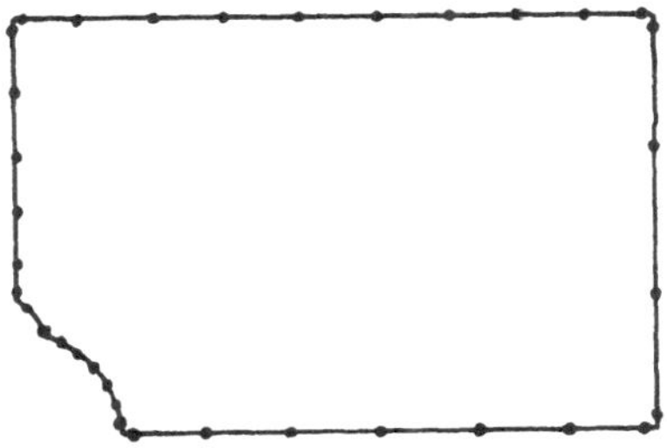

Fig.10c. Linear boundary element discretization

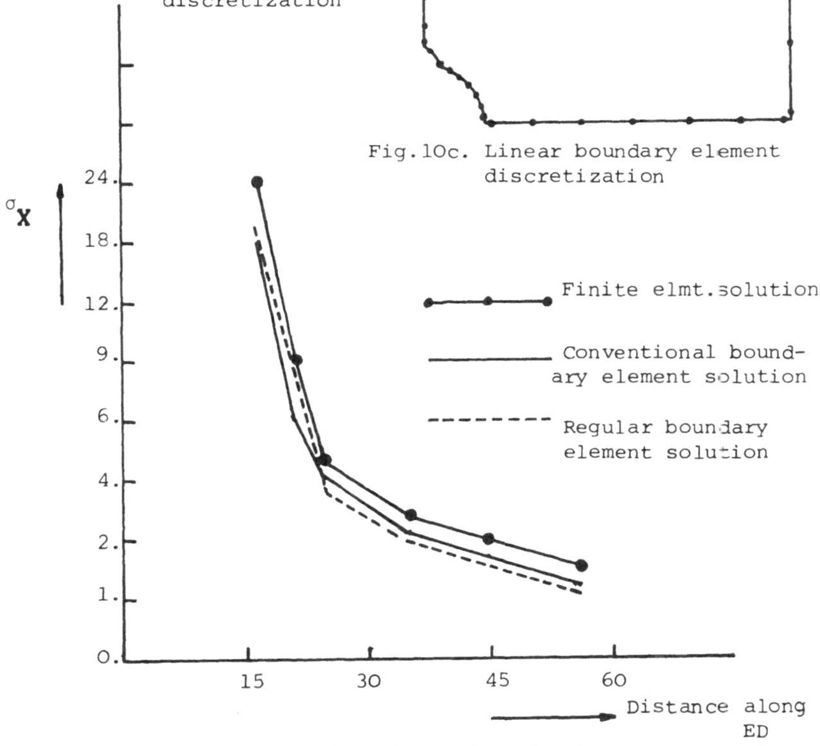

Fig.11a. Stress distribution along the edge ED of the domain

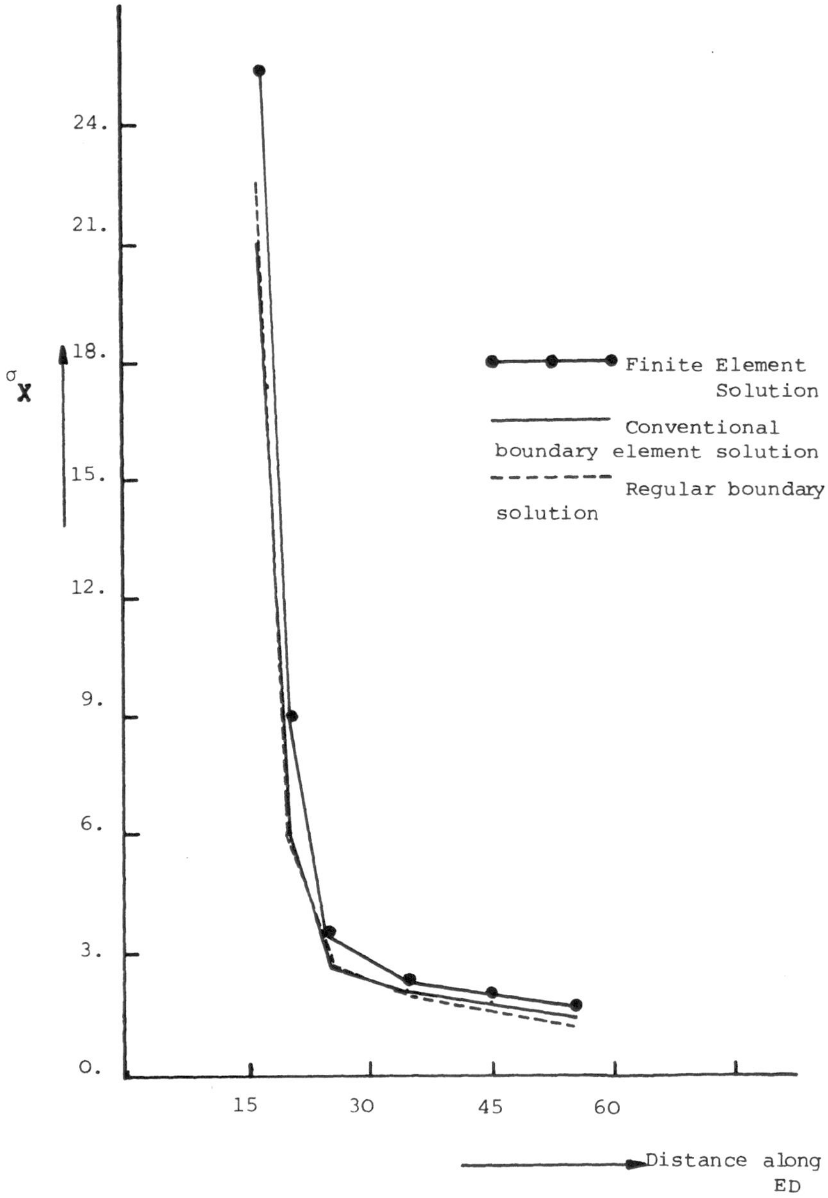

Fig.11b. Stress distribution along the edge ED of the domain using linear boundary elements

The boundary of the domain of the problem was discretized using 66 constant elements (Figure.15a) and 71 linear elements (Figure.15b) in which case two nodes (one corresponding to each side) were taken close to each other at all the five corners of the boundary to avoid any freedom constraint.

Comparison of surface stress distribution calculated by using conventional boundary integral method with that obtained from regular boundary integral method is shown in Figures.16a. and 16b., respectively for constant and linear boundary element discretizations. The stresses are calculated at various internal points along the critical section X - X' where maximum stress occurs and the stress distribution is shown in Figures. 17a. and 17b., for constant and linear elements respectively.

It is clear that although the boundary solution obtained by using conventional and regular integral methods shows a little difference, the solution in the interior is in almost complete agreement. Moreover, it should be noted that the results obtained for the constant and linear boundary element meshes used in the present study are less accurate than the ones obtained by employing photoelastic testing or a refined finite element mesh, [Patterson, Wearing and Arnell, 1980], however, they have been found to be much improved by refining these meshes. But despite of the incompleteness of the meshes used, the main purpose of comparing the conventional boundary integral method with the proposed regular boundary integral method for two different types of elements is well served.

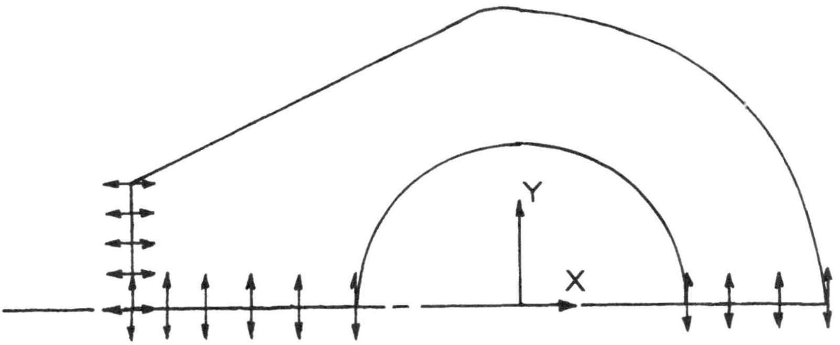

Fig.13. Boundary Conditions

Scale : 1 mm = 50 lbs-f

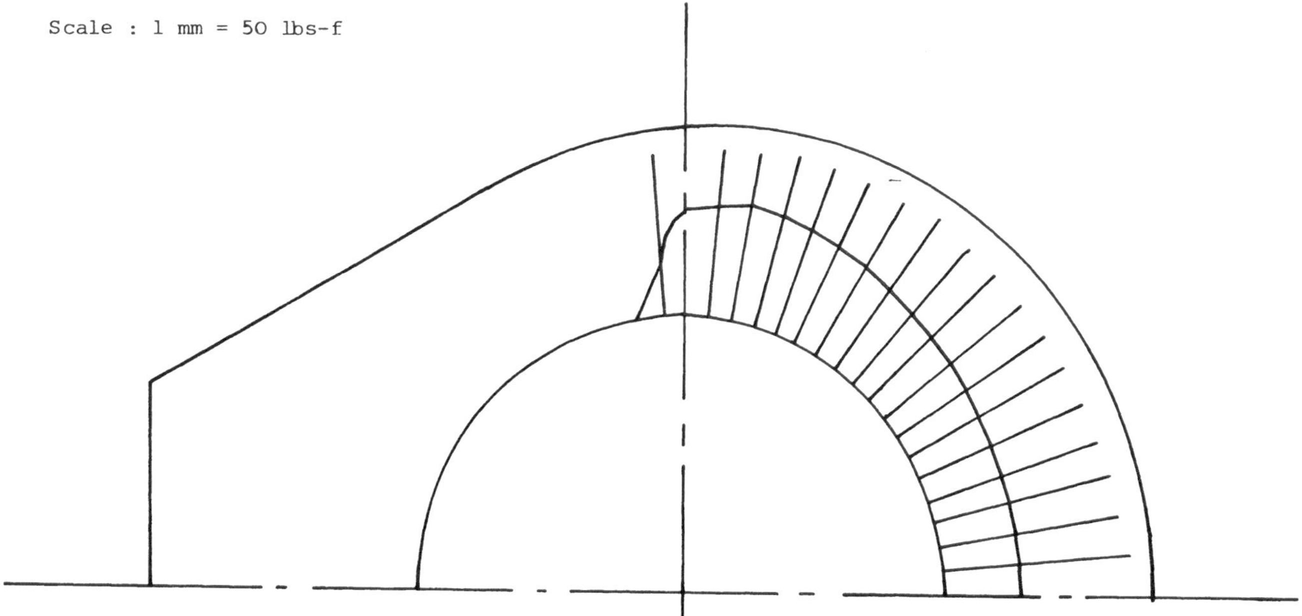

Fig. 14. Nodal Load distribution acting on bore surface at 16,000 lbs.f total pin load

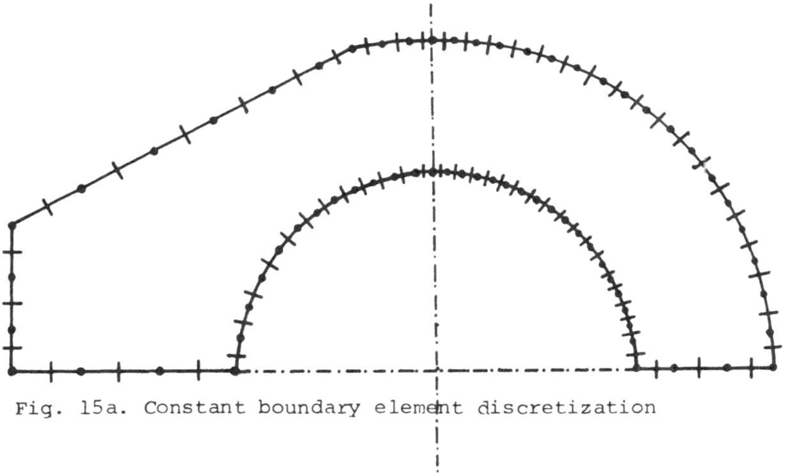

Fig. 15a. Constant boundary element discretization

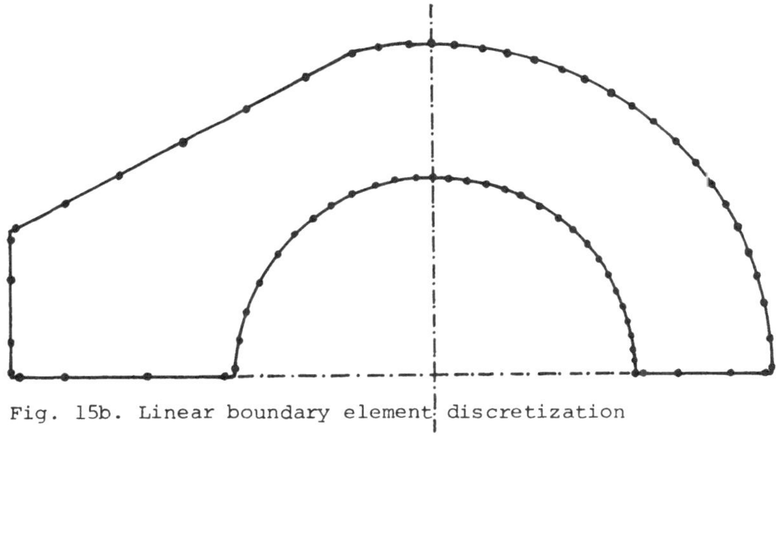

Fig. 15b. Linear boundary element discretization

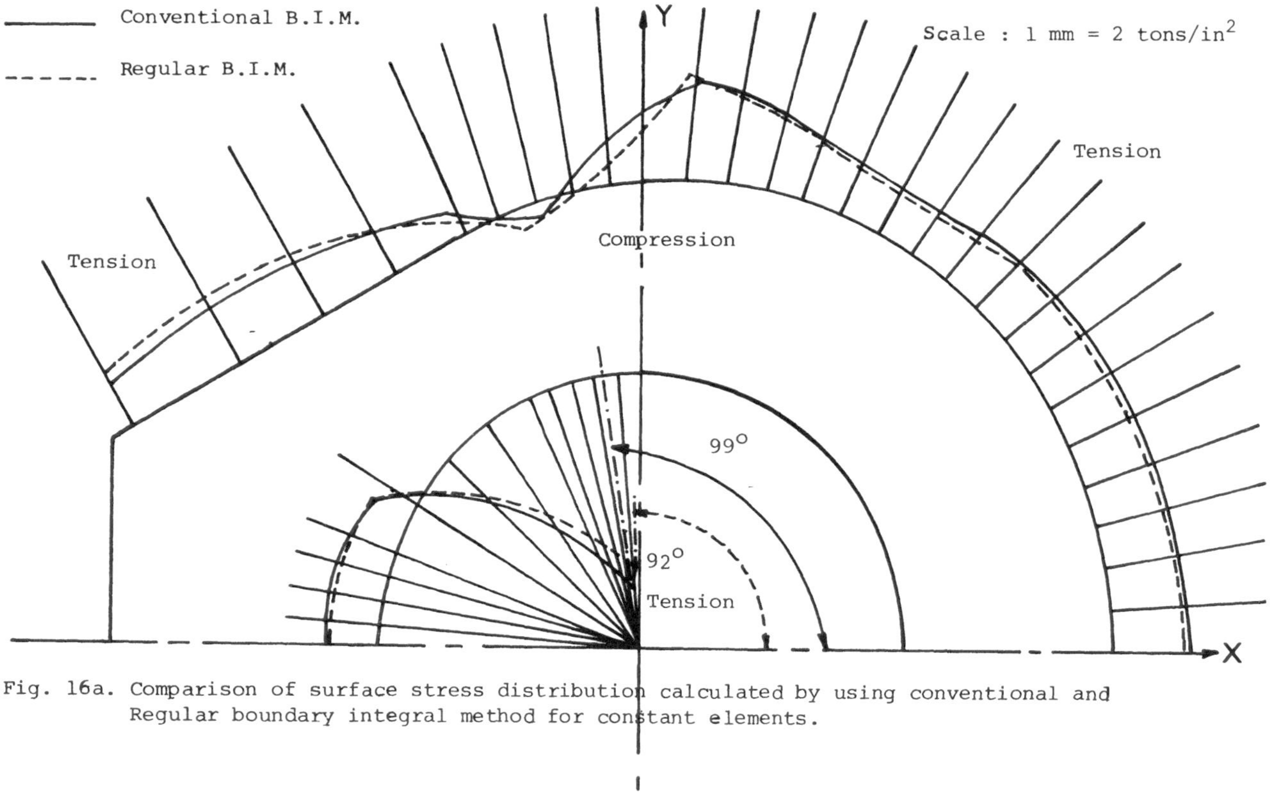

Fig. 16a. Comparison of surface stress distribution calculated by using conventional and Regular boundary integral method for constant elements.

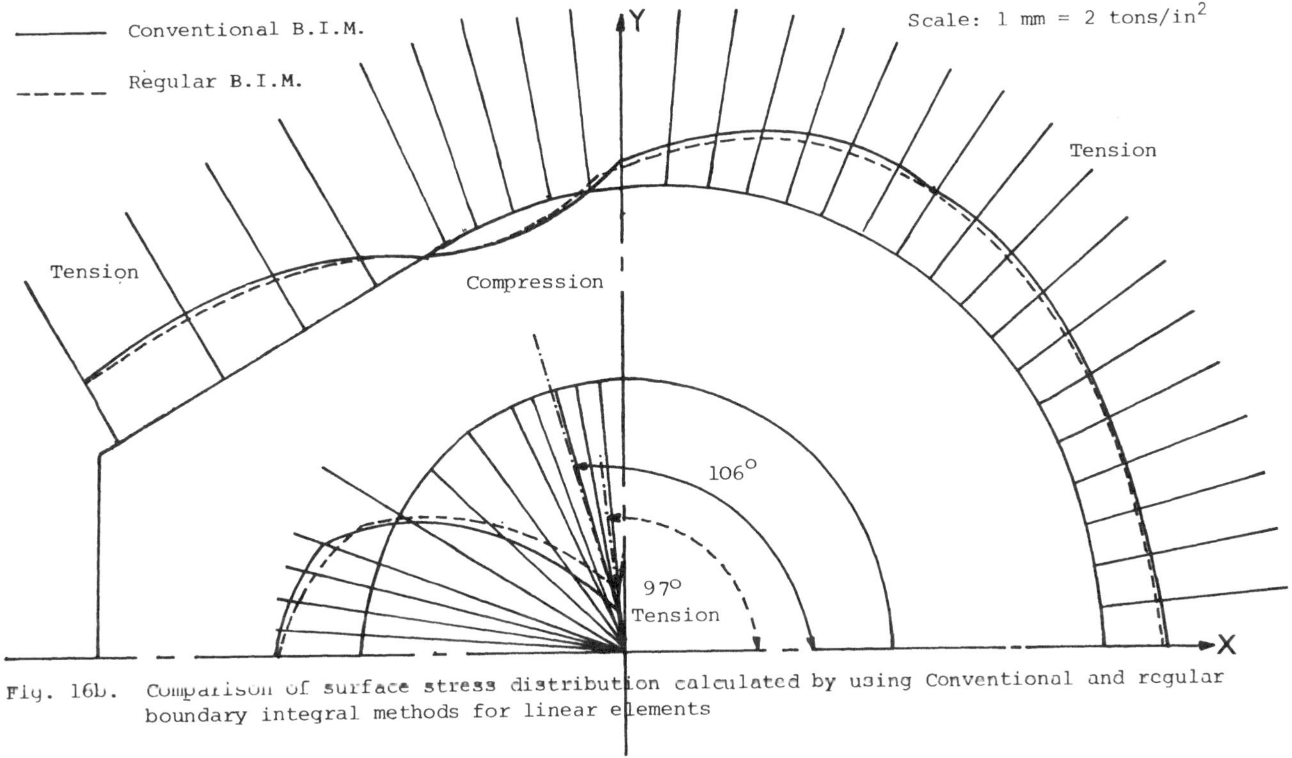

Fig. 16b. Comparison of surface stress distribution calculated by using Conventional and regular boundary integral methods for linear elements

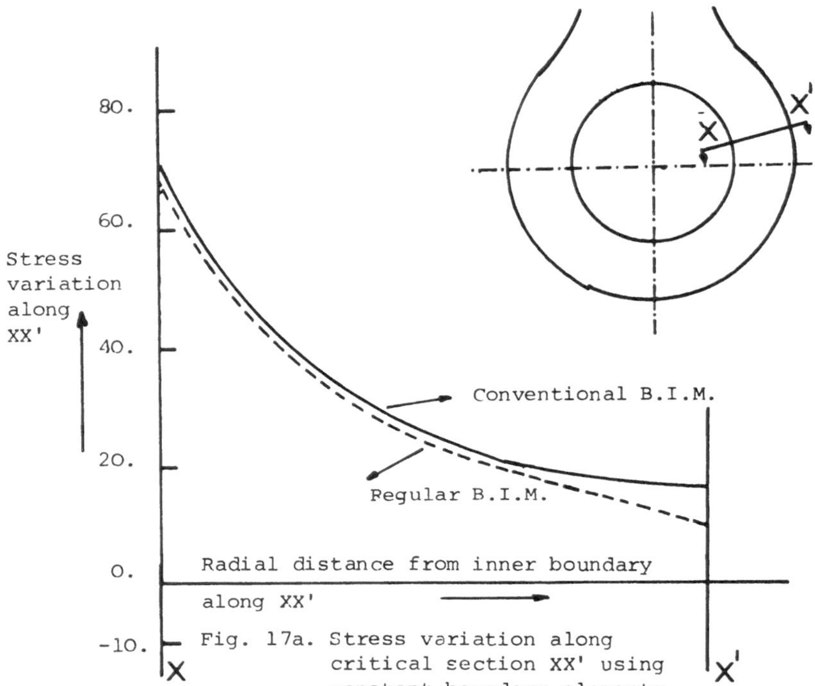

Fig. 17a. Stress variation along critical section XX' using constant boundary elements

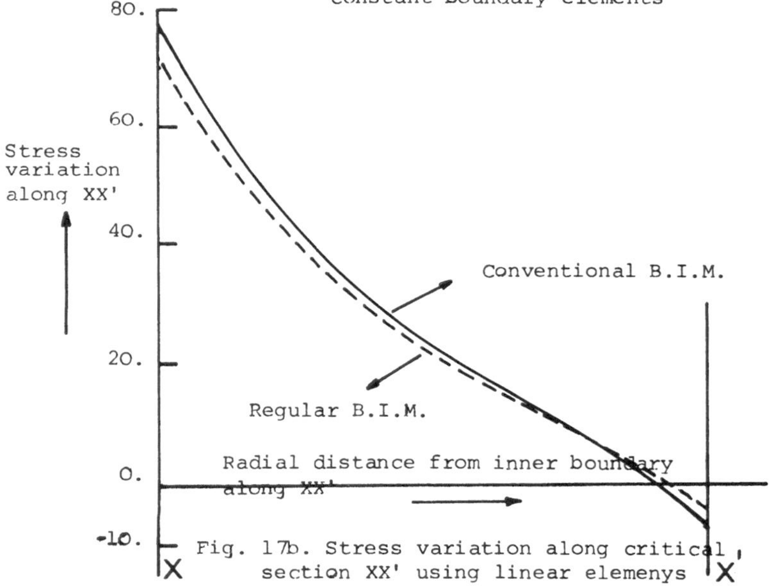

Fig. 17b. Stress variation along critical section XX' using linear elemenys

DISCUSSION AND CONCLUSIONS

THE test problems analysed were chosen to present a menu of ascending difficulty to the numerical methods leading to a typical design problem. The boundary element meshes were taken to correspond reasonably to what were known satisfactory finite element meshes but, otherwise, convergence studies were not regarded as a central issue of the study.

The simplest test was the plate perforated with a circular hole. Here the normal stresses along the load bearing ligament, Figs.7a,7b, as given by the singular and regular methods for constant and linear elements were both in good mutual agreement and agreed well with the finite element result. It is noted that the boundary element stresses are consistently below the finite element values. The effect of adding the notch to the circular hole is to enhance the stress concentration by about two and a half times. Again, the boundary element results all are in good qualitative agreement with the finite element values, Figs.11a,11b, and again they lie beneath. The free fitting pin in lug problem arose as a live design problem due to unsuspected fatigue failure of the lug. Earlier photoelastic and contact-finite element analyses had revealed a severe stress concentration on the inner face of the lug at about 106° behind the centre load point. The contact analysis revealed the surface load distribution given in Fig.14. This load distribution was used in the boundary element analyses giving the surface stress values indicated in Figs.16a,16b, for constant and linear elements respectively. For each element type the boundary method results are in good mutual quantitative agreement and show good qualitative agreement with previous values. The constant elements do not show the well developed compressive region on the outer face and the tensile zone ahead of the pin, on the outer surface, varies significantly from the linear element results. Previous values support the linear element values. On the inner surface the strong stress concentration is manifest in all cases. The linear elements again give better results than the constant variants. Using linear elements the stress concentration is better described by the conventional method being located at about 106°. Nevertheless, with the meshes used the discrepancy is in the order of a single element and both methods underestimate the finite element/photoelastic values. The interior stresses normal to the critical section XX' are given in Figs.17a,17b. This section is taken at 92° for constant elements and 97° for linear; to correspond with peak surface stress given by the regular method. The computed values are as expected, given the disrepancies in surface values already seen.

The problems analysed show that under similar conditions the regular and singular methods yield similar results. Whilst the perforated plate problem did not discriminate constant and linear elements the pin-lug problem yielded better results with linear elements. The meshes used were suggested by finite element experience and gave values which were generally marginally low in value but trends were well presented. The likely reason for this is inadequate mesh refinement which was supported by plate analyses with refined meshes.

In conclusion: An alternative to the traditional, singular boundary element method has been presented which gives regular kernel functions in the boundary element equations. In the test problems analysed results computed using the singular and regular methods were in close mutual agreement and in good general agreement with known values. The advantages of the regular method are that the familiar singular kernel of the usual method is avoided and that higher singularities can be tolerated in the solution without leading to a divergent system.

REFERENCES

Mikhlin, S.G. (1965) The Problem of the Minimum of a Quadratic Functional, Holden-Day, San Fransisco.

Patterson, C., Wearing, J.L. and Arnell, P.P. (1980). A Study of stress distribution in lugs loaded by free fitting pins. In: Stress Analysis in Fabricated Structures, Eds.C.Patterson and T.H. Richards, Applied Science Publishers, London.

Patterson, C. and Bagdatlioglu, N.I. (To be published) A Nesting Technique for Efficient Determination of Localized Stress Concentrations. In: Mathematics of Finite Elements and Applications, Ed. J.R. Whiteman, Academic Press, London.

A BOUNDARY ELEMENT FORMULATION OF PROBLEMS IN LINEAR ISOTROPIC ELASTICITY WITH BODY FORCES

D.J. Danson

Computational Mechanics Centre, Southampton, U.K.

1. INTRODUCTION

Traditional formulations of the Boundary Element Method (BEM) applied to elastostatics [1] [2] are extremely convenient when the loading on the body under analysis is limited to surface loading, since it is necessary to discretize only the boundary of the body and not the whole domain as must be done when using a technique such as the Finite Element Method (FEM). When body forces are present these have usually been handled by evaluating a domain integral [3]. Unfortunately this requires the domain of the problem to be divided into integration cells since for any practical problem the domain integral must be evaluated numerically. This greatly increases the amount of data preparation required and causes the BEM to lose much of its advantage over domain type methods. However, Cruse [4] and Cruse, Snow and Wilson [5] have shown that for certain types of commonly encountered body forces the domain integral may be transformed to a boundary integral or boundary integrals which may be evaluated at the same time as the boundary integrals involving the surface displacements and tractions. Ref. [5] is concerned exclusively with axisymmetric geometry; however, the authors' use of the Galerkin vector to achieve the required transformation from domain to boundary integrals provides the key to the present paper which relates to two and three dimensional geometry. The three dimensional case was derived in Ref. [4] without resort to the Galerkin vector. However, the Galerkin vector approach is used below both to demonstrate the power of the technique and to present the results of [4] in a slightly more general form. The two dimensional formulation is also presented.

The following commonly encountered body forces will be considered.

(i) The body force due to a body being placed in a constant gravitational field.

(ii) The body force due to constant rigid body rotation about a fixed axis.

(iii) Thermal stresses may be calculated by applying a body force proportional to the temperature gradient [7]. The domain integral resulting from this body force may be transformed to a boundary integral provided that the temperature distribution is a "steady state" one.

2. BOUNDARY INTEGRAL FORMULATION

The Somigliana identity for the displacements inside an elastic body may be derived using Maxwell-Betti's reciprocal theorem to give (Ref, [1])

$$u_k(x) = \int_S U_{ki}(x,y)t_i(y)dS_y - \int_S T_{ki}(x,y)u_i(y)dS_y$$

$$+ \int_V U_{ki}(x,y)b_i(y)dV_y \qquad (1)$$

Where
- $u_k(x)$ is the displacement at an internal point x in the k direction
- $U_{ki}(x,y)$ is the displacement in the i direction at y due to a unit point load in the k direction at x
- $T_{ki}(x,y)$ is the traction in the i direction at y due to a unit point load in the k direction at x
- $t_i(y)$ is the traction at y
- $u_i(y)$ is the displacement at y
- $b_i(y)$ is the intensity of the body force at y
- S represents the boundary of body
- V represents the domain of the body

The coordinates of the point x are x_j
The coordinates of the point y are y_j
The subscript y on dS_y, dV_y is to indicate that the coordinates y_j and not the coordinates x_j are the integration variables.

Applying Maxwell-Betti's reciprocal theorem to the thermo-elastic case gives (Ref. [4])

$$u_k(x) = \int_S U_{ki}(x,y)\ t_i(y)dS_y - \int_S T_{ki}(x,y)\ u_i(y)dS_y$$

$$+ \frac{\alpha E}{1-2\nu}\int_V U_{ki,i}(x,y)\ \theta(y)dV_y \qquad (2)$$

where α is the coefficient of linear expansion
E is Young's modulus
ν is Poisson's ratio
$\theta(y)$ is the temperature at y
$,i$ denotes differentiation with respect to y_i

Calling the domain integral in both Eqns (1) and (2) $B_k(x)$, so that for the gravitational or rotational cases

$$B_k(x) = \int_V U_{ki}(x,y)\ b_i(y)dV_y \tag{3}$$

and for the thermoelastic case

$$B_k(x) = \frac{\alpha E}{1-2\nu} \int_V U_{ki,i}(x,y)\ \theta(y)dV_y \tag{4}$$

the aim of the following is to express $B_k(x)$ as a boundary integral instead of a domain integral.

A generalized form of Gauss' theorem will be used which may be expressed as follows.

If $A_{jk\ell\ldots}$ is a general Cartesian tensor field then

$$\int_V A_{jk\ell\ldots,i}\ dV_y = \int_S A_{jk\ell\ldots}\ n_i(y)dS_y \tag{5}$$

where $n_i(y)$ is the unit outward normal at y.

3. TRANSFORMATION TO BOUNDARY INTEGRALS

We define a tensor $G_{ki}(x,y)$, called the Galerkin tensor such that the displacement kernel $U_{ki}(x,y)$ is given by

$$U_{ki}(x,y) = G_{ki,jj}(x,y) - \frac{G_{kj,ji}(x,y)}{2(1-\nu)} \tag{6}$$

It is worth noting that in the same way that $U_{ki}(x,y)$ may be regarded as three (in 3D) or two (in 2D) displacement vectors each corresponding to the direction k in which the unit load is applied, similarly $G_{ki}(x,y)$ may be regarded as three (in 3D) or two (in 2D) Galerkin vectors each corresponding to the direction k in which the unit load is applied.

Substituting Eq. (6) into Eq. (3) gives

$$B_k(x) = \int_V \left[G_{ki,jj}(x,y) - \frac{G_{kj,ji}(x,y)}{2(1-\nu)} \right] b_i(y)\, dV_y \tag{7}$$

Although Eq. (7) looks a lot more complicated than Eq. (3) it is much easier to arrange Eq. (7) in a form where it can be transformed to a boundary integral.

Gravitational loads (self weight)

A body of constant mass density in a constant gravitational field experiences a constant body force $b_i(y)$.

Thus $b_i(y)$ may be taken outside the integral in Eq. (7) and so using Eq. (5) we get

$$B_k(x) = b_i \int_S \left\{ G_{ki,j}(x,y) - \frac{G_{kj,i}(x,y)}{2(1-\nu)} \right\} n_j\, dS_y \tag{8}$$

which is the required boundary integral form.

Rotational Inertia

Let us consider a body of uniform mass density ρ rotating about some axis with angular velocity ω_i. We may without loss of generality consider the axis of rotation as passing through the origin of our coordinate system. The acceleration at a point y_i is given in vector notation as

$$\underset{\sim}{f} = \underset{\sim}{\omega} \times (\underset{\sim}{\omega} \times \underset{\sim}{y}) \tag{9}$$

By d'Alembert's principle we may replace this acceleration by a body force $\underset{\sim}{b}$ and treat the dynamic problem as a static one, hence

$$\underset{\sim}{b} = -\rho\, \underset{\sim}{\omega} \times (\underset{\sim}{\omega} \times \underset{\sim}{y}) \tag{10}$$

where ρ is the mass density of the body.

We may write Eq. (10) in tensor notation as

$$b_i(y) = -\rho\, c_{ijk}\, \omega_j\, c_{kpq}\, \omega_p\, y_q \tag{11}$$

where c_{ijk} is the permutation tensor.

Defining

$$\Omega_{iq} = -\rho\, c_{ijk}\, \omega_j\, c_{kpq}\, \omega_p \tag{12}$$

it is easily shown that

$$[\Omega_{ij}] = \rho \begin{bmatrix} \omega_2^2+\omega_3^2 & -\omega_1\omega_2 & -\omega_3\omega_1 \\ -\omega_1\omega_2 & \omega_3^2+\omega_1^2 & -\omega_2\omega_3 \\ -\omega_3\omega_1 & -\omega_2\omega_3 & \omega_1^2+\omega_2^2 \end{bmatrix} \tag{13}$$

and Eq. (11) becomes

$$b_i(y) = \Omega_{ij}\, y_j \tag{14}$$

Substitution of Eq. (14) into Eq. (7) gives

$$B_k(x) = \Omega_{ij} \int_V y_j \left\{ G_{ki,mm}(x,y) - \frac{G_{km,mi}(x,y)}{2(1-\nu)} \right\} dV_y \tag{15}$$

which after some manipulation becomes

$$B_k(x) = \Omega_{ij} \int_V \frac{\partial}{\partial y_m} \left\{ y_j\, G_{ki,m}(x,y) \right\} - G_{ki,j}(x,y) + \frac{1}{2(1-\nu)} \left[G_{kj,i}(x,y) - \frac{\partial}{\partial y_m} \left\{ y_j\, G_{km,i}(x,y) \right\} \right] dV_y \tag{16}$$

This is now in a form which can be transformed to a boundary integral using Eq. (5) and using the symmetry of Ω_{ij} to effect some slight simplification we have

$$B_k(x) = \Omega_{ij} \int_S y_j \left\{ G_{ki,m}(x,y) - \frac{G_{km,i}(x,y)}{2(1-\nu)} \right\} n_m - \frac{1-2\nu}{2(1-\nu)}\, G_{ki}(x,y)\, n_j\, dS_y \tag{17}$$

which is the required boundary integral form.

<u>Steady state thermal loading</u>
Differentiating Eq. (6) with respect to y_i and substituting in Eq. (4) gives

$$B_k(x) = \frac{\alpha E}{2(1-\nu)} \int_V \theta(y)\, G_{ki,ijj}(x,y)\, dV_y \tag{18}$$

For steady state heat conduction $\theta_{,jj}(y) = 0$ and we may write

$$B_k(x) = \frac{\alpha E}{2(1-\nu)} \int_V \theta(y) G_{ki,ijj}(x,y) - G_{ki,i}(x,y)\, \theta_{,jj}(y)\, dV_y \qquad (19)$$

Manipulating as before gives

$$B_k(x) = \frac{\alpha E}{2(1-\nu)} \int_V \frac{\partial}{\partial y_j} \left\{ \theta(y) G_{ki,ij}(x,y) \right.$$

$$\left. - \frac{\partial}{\partial y_j} \left\{ G_{ki,i}(x,y)\, \theta_{,j}(y) \right\} dV_y \right. \qquad (20)$$

which is now in a form which can be transformed to a boundary integral using Eq. (5).

Thus

$$B_k(x) = \frac{\alpha E}{2(1-\nu)} \int_S \left\{ \theta(y) G_{ki,ij}(x,y) \right.$$

$$\left. - G_{ki,i}(x,y)\, \theta_{,j}(y) \right\} n_j \, dS_y \qquad (21)$$

which is the required boundary integral form.

4. 2D BODY FORCES

The Galerkin Tensor corresponding to the 2D Fundamental Solution is

$$G_{ki}(x,y) = \frac{1+\nu}{4\pi E} \delta_{ki}\, r^2 \ln \frac{1}{r} \qquad (22)$$

where r is the distance between the force point x and the field point y
and δ_{ki} is the Kronecker delta.

Differentiating Eq. (22) twice, performing the appropriate contractions and substituting into Eq. (6) gives the Fundamental Solution

$$U_{ki}(x,y) = \frac{1+\nu}{4\pi E(1-\nu)} \left[\left\{ (3-4\nu) \ln \frac{1}{r} - \frac{7-8\nu}{2} \right\} \delta_{ki} + r_{,k} r_{,i} \right] \qquad (23)$$

where r_j is the unit vector from the force point x to the field point y^j so that

$$r_j = \frac{y_j - x_j}{r}$$

We note that Eq. (23) differs from the fundamental solution usually given by a constant. This is not of any great importance as a constant is equivalent only to rigid body motion.

Gravitational Loads (2D)

For gravitational loads we differentiate Eq. (22) and substitute into Eq. (8) to get

$$B_k(x) = \frac{1+\nu}{4\pi E} \int_S r(2 \ln \frac{1}{r} - 1) \left\{ b_k n_m r_m - \frac{n_k b_m r_m}{2(1-\nu)} \right\} dS \tag{24}$$

Differentiating Eq. (1)(with the boundary integral replacing the domain integral) and substitution in the stress displacement relations gives the stress at an internal point x. Thus

$$\sigma_{ij}(x) = \int_S D_{kij}(x,y)\ t_k(y) dS_y - \int_S S_{kij}(x,y)\ u_k(y) dS_y$$

$$+ \int_S S^*_{ij}(x,y) dS_y \tag{25}$$

where $D_{kij}(x,y)$ and $S_{kij}(x,y)$ are given in Ref. [3] and

$$S^*_{ij}(x,y) = \frac{1}{2\pi} \left[\left(1 - \ln \frac{1}{r}\right) \left\{ n_m r_m (b_i r_j + b_j r_i + \frac{\nu}{1-\nu} b_s r_s \delta_{ij}) \right.\right.$$

$$\left. - \frac{b_m r_m}{2(1-\nu)} (n_i r_j + n_j r_i) \right\} + \frac{1}{2(1-\nu)} \left\{ \ln \frac{1}{r} - \frac{1}{2} \right\}$$

$$\left. \times \left\{ \frac{1-2\nu}{2} (b_i n_j + b_j n_i) + \nu\ n_m b_m \delta_{ij} \right\} \right] \tag{26}$$

Rotational Inertia (2D)

For rotational loads we note that to keep the problem two dimensional the axis of rotation must be at right angles to the plane containing the problem. Thus if the angular speed of rotation is ω then from Eq. (13)

$$\Omega_{ij} = \rho\ \omega^2\ \delta_{ij} \tag{27}$$

Substitution of Eqns (27) and (22) into Eq. (17) gives

$$B_k(x,y) = \frac{\rho\omega^2(1+\nu)}{4\pi E}\int_S r\left\{(2\ \ln\frac{1}{r} - 1)\left(n_m r_m y_k - \frac{r_m y_m n_k}{2(1-\nu)}\right)\right.$$

$$\left. - \frac{1-2\nu}{2(1-\nu)}\ r\ \ln\frac{1}{r}\ n_k\right\} dS_y \tag{28}$$

The boundary integral equation for stresses is once again Eq.(25) with this time

$$S^*_{ij}(x,y) = \frac{\rho\omega^2}{8\pi}\left[\nu\ \delta_{ij}\ \left\{\frac{2n_m r_m y_s r_s}{1-\nu}\right.\right.$$

$$\left. + (2\ \ln\frac{1}{r} - 1)\ \left(\frac{3-2\nu}{(1-\nu)(1-2\nu)}\ n_m y_m + 2r\ n_m r_m\right)\right\}$$

$$+ 2n_m r_m(y_i r_j + r_i y_j) - \frac{y_m r_m}{1-\nu}(n_i r_j + r_i n_j)$$

$$+ \frac{1}{2(1-\nu)}\ (2\ \ln\frac{1}{r} - 1)\{(3-2\nu)(n_i y_j + n_j y_i)$$

$$\left. + (1-2\nu) r(n_i r_j + n_j r_i)\}\right] \tag{29}$$

<u>Steady state thermal loading (2D)</u>

For steady state thermal loads we write Eq. (21) as

$$B_k(x) = \int_S P_k(x,y)\ \theta(y) dS_y - \int_S Q_k(x,y)\ \theta_{,m}(y) n_m\ dS_y \tag{30}$$

where

$$P_k(x,y) = \frac{\alpha E}{2(1-\nu)}\ G_{ki,ij}(x,y) n_j$$

(31)

and

$$Q_k(x,y) = \frac{\alpha E}{2(1-\nu)}\ G_{ki,i}(x,y)$$

Differentiating Eq. (22) and substituting in Eq. (31) gives

$$P_k(x,y) = \frac{\alpha(1+\nu)}{4\pi(1-\nu)} \left\{\left(\ln\frac{1}{r} - \frac{1}{2}\right) n_k - n_m r_m r_k\right\}$$

$$Q_k(x,y) = \frac{\alpha(1+\nu)}{4\pi(1-\nu)} \; r_k \; r\left(\ln\frac{1}{r} - \frac{1}{2}\right) \tag{32}$$

Differentiating Eq.(2) (with the boundary integrals replacing the domain integral) and substition in the stress displacement relations gives

$$\sigma_{ij}(x) = \int_S D_{kij}(x,y)\; t_k(y)dS_y - \int_S S_{kij}(x,y)\; u_k(y)dS_y$$

$$+ \int_S S^*_{ij}(x,y)\; \theta(y)dS_y - \int_S V^*_{ij}(x,y)\; \theta_{,m}(y)\; n_m(y)dS_y$$

$$- \frac{\alpha E}{1-2\nu}\, \theta(x)\delta_{ij} \tag{33}$$

where $D_{kij}(x,y)$ and $S_{kij}(x,y)$ are as before and

$$S^*_{ij}(x,y) = \frac{\alpha E}{4\pi(1-\nu)r} \left\{n_m r_m \left(\frac{\delta_{ij}}{1-2\nu} - 2r_i r_j\right) + n_i r_j + n_j r_i\right\}$$

$$\text{and } V^*_{ij}(x,y) = \frac{\alpha E}{4\pi(1-\nu)} \left\{r_i r_j + \frac{\delta_{ij}}{1-2\nu}\left(\frac{1+2\nu}{2} - \ln\frac{1}{r}\right)\right\} \tag{34}$$

5. 3D BODY FORCES

The Galerkin Tensor corresponding to the 3D Fundamental Solution is

$$G_{ki}(x,y) = \frac{(1+\nu)R\;\delta_{ki}}{4\pi E} \tag{35}$$

where R is the distance between the force point x and the field point y. The symbol R is used when discussing the 3D case so that equations involving the 3D case are easily distinguished from the 2D case where the symbol r is used.

Differentiating Eq. (35) twice, performing the appropriate contractions and substituting into Eq. (6) gives the fundamental solution

$$U_{ki}(x,y) = \frac{1+\nu}{8\pi E(1-\nu)R}\left\{(3-4\nu)\delta_{ki} + r_k r_i\right\} \tag{36}$$

Gravitational loads (3D)

For gravitational loads we differentiate Eq. (35) and substitute into Eq. (8) to get

$$B_k(x,y) = \frac{1+\nu}{4\pi E}\int_S \left\{n_m r_m b_k - \frac{b_m r_m n_k}{2(1-\nu)}\right\} dS_y \tag{37}$$

The boundary integral equation for stress is Eq. (25) where for the 3D case $D_{kij}(x,y)$ and $S_{kij}(x,y)$ are given in Ref. [3] and $S^*_{ij}(x,y)$ is obtained in a similar manner as for the 2D case, whence

$$S^*_{ij}(x,y) = \frac{1}{8\pi R}\Big[n_m r_m(b_i r_j + b_j r_i) + \frac{1}{1-\nu}\Big\{\nu\delta_{ij}(n_m r_m b_s r_s - b_m n_m) - \frac{1}{2}(b_m r_m[n_i r_j + n_j r_i] + [1-2\nu][b_i n_j + b_j n_i])\Big\}\Big] \tag{38}$$

Rotational inertia (3D)

Differentiation of Eq.(35) and substitution into Eq.(17) gives

$$B_k(x,y) = \frac{1+\nu}{4\pi E}\int_S \Big[\Omega_{kj} y_j r_m n_m - \frac{1}{2(1-\nu)}\Big\{r_i \Omega_{ij} y_j n_k + (1-2\nu)R\,\Omega_{kj} y_j r_m n_m\Big\}\Big] dS_y \tag{39}$$

The boundary integral equation for stresses is once again Eq. (25) with

$$S^*_{ij}(x,y) = \frac{1}{8\pi}\Big[\Big\{n_s r_s \frac{y_m}{R} + \frac{1-2\nu}{2(1-\nu)} n_m\Big\}(\Omega_{im} r_j + \Omega_{jm} r_i) + \frac{1}{1-\nu}\Big\{\nu\delta_{ij}(r_k \Omega_{km}\frac{y_m}{R} n_s r_s - n_s \Omega_{sm}\frac{y_m}{R} + r_s \Omega_{sm} n_m) - \frac{y_m}{2R}(r_s \Omega_{sm}[n_i r_j + n_j r_i] + [1-2\nu][n_i \Omega_{jm} + n_j \Omega_{im}])\Big\}\Big] \tag{40}$$

Steady State thermal loading (3D)

For steady state thermal loading Eqns (30) and (31) apply in 3D as in 2D.

Differentiating Eq. (35) and substituting in Eqns (31) yields

$$P_k(x,y) = \frac{\alpha(1+\nu)}{8\pi(1-\nu)}\left(\frac{n_k - n_m r_m r_k}{R}\right)$$

$$\text{and} \quad Q_k(x,y) = \frac{\alpha(1+\nu)}{8\pi(1-\nu)}\, r_k \tag{41}$$

The boundary integral expression for stresses is Eq. (33) where this time

$$S^*_{ij}(x,y) = \frac{\alpha E}{8\pi(1-\nu)R^2}\left\{n_m r_m\left(\frac{\delta_{ij}}{1-2\nu} - 3r_i r_j\right) + n_i r_j + r_i n_j\right\}$$

$$V^*_{ij}(x,y) = \frac{\alpha E}{8\pi(1-\nu)R}\left(r_i r_j - \frac{\delta_{ij}}{1-2\nu}\right) \tag{42}$$

6. DIRECT COMPUTATION OF 3D BODY FORCE KERNELS

Cruse [4] has indicated how to transform the domain integral due to body forces to a boundary integral directly, without involving the Galerkin Tensor.

Gravitational loads

We have that

$$B_k(x) = \frac{(1+\nu)b_i}{8\pi E(1-\nu)}\int_V \frac{(3-4\nu)\delta_{ki}}{R} + \frac{r_k r_i}{R}\, dV_y \tag{43}$$

$$\text{Noting that} \quad \frac{\partial r_k}{\partial y_i} = \frac{\delta_{ki} - r_k r_i}{R} \tag{44}$$

this may be written

$$B_k(x) = \frac{(1+\nu)b_i}{8\pi E(1-\nu)}\int_V \frac{4(1-\nu)\delta_{ki}}{R} - \frac{\partial r_k}{\partial y_i}\, dV_y \tag{45}$$

We now make use of the following property of the Laplacian operator on the three dimensional function R(x,y)

$$R_{,mm} = \frac{2}{R} \tag{46}$$

And remembering that $R_{,m} = r_m$ we may write $R_{,mm} = r_{m,m}$ and so Eq.(45) becomes

$$B_k(x) = \frac{(1+\nu)b_i}{8\pi E} \int_V 2\delta_{ki} \frac{\partial r_m}{\partial y_m} - \frac{1}{1-\nu} \frac{\partial r_k}{\partial y_i} \, dS_y \tag{47}$$

This may now be reduced to an equivalent boundary integral by using Eq. (5)

$$B_k(x) = \frac{1+\nu}{4\pi E} \int_S b_k \, n_m r_m - \frac{r_k n_i b_i}{2(1-\nu)} \, dS_y \tag{48}$$

It should be noted that the integrand (or kernel) of Eq. (47) is not the same as that derived previously in Eq. (37). The second term is different and clearly $b_m r_m n_k$ is not in general equal to $b_m n_m r_k$.

However it is necessary only that the integral of the kernel round a closed boundary has a unique value.

Consider the integral

$$I_k = \int_S b_m r_m n_k - b_m n_m r_k \, dS_y \tag{49}$$

Transforming to a domain integral using Eq. (5) this becomes

$$I_k = b_m \int_V r_{m,k} - r_{k,m} \, dV_y = 0 \tag{50}$$

Thus we have shown that both Eq. (37) and Eq. (48) are valid integral transformations of the body force domain integral. In other words the form of the body force kernel is not unique.

We note that if we had written

$$\frac{\partial r_i}{\partial y_k} = \frac{\delta_{ki} - r_k r_i}{R} \tag{51}$$

instead of Eq. (44) and followed the reasoning through as before we would have arrived at Eq. (37).

The equivalent stress equation kernel $S^*_{ij}(x,y)$ is obtained by differentiating Eq. (48) and substituting in the stress displacement relations to get

$$S^*_{ij}(x,y) = \frac{1}{8\pi R}\left[n_m r_m\left\{\frac{2\nu b_s r_s \delta_{ij}}{1-2\nu} + b_i r_j + b_j r_i\right\}\right.$$

$$\left. + \frac{n_m b_m}{1-\nu}\left\{\left(1 + \frac{2\nu^2}{1-2\nu}\right)\delta_{ij} - r_i r_j\right\} - b_i n_j - b_j n_i\right] \qquad (52)$$

Comparison with Eq. (38) shows that the form of $S^*_{ij}(x,y)$ is not unique either.

Rotational inertia

A similar process of algebraic manipulation enables the body force kernels to be computed directly without using the Galerkin tensor. The results are, for the displacement equation

$$B_k(x,y) = \frac{1+\nu}{4\pi E}\int_S \left\{n_m r_m \Omega_{kj} y_j - R\,\Omega_{km} n_m\right.$$

$$\left. + \frac{1}{2(1-\nu)}\left(\Omega_{mm} R n_k - n_m \Omega_{ms} y_s r_k\right)\right\} dS_y \qquad (53)$$

and for the stress equation

$$S^*_{ij} = \frac{1}{4\pi}\left[\frac{\nu}{1-2\nu}\left\{n_m r_m\left(r_s \Omega_{sq}\frac{y_q}{R} - \frac{\Omega_{ss}}{2(1-\nu)}\right) + \frac{\nu}{1-\nu} n_m \Omega_{ms}\frac{y_s}{R}\right.\right.$$

$$\left. + r_s \Omega_{sm} n_m\right\}\delta_{ij} + \frac{1}{2}\left\{n_m r_m\left(r_i \Omega_{js}\frac{y_s}{R} + r_j \Omega_{is}\frac{y_s}{R}\right)\right.$$

$$- \left(n_i \Omega_{jm}\frac{y_m}{R} + n_j \Omega_{im}\frac{y_m}{R}\right) + r_i \Omega_{jm} n_m + r_j \Omega_{im} n_m$$

$$\left.\left. + \frac{1}{1-\nu}\left(n_m \Omega_{ms}\frac{y_s}{R}\left[\delta_{ij} - r_i r_j\right] - \frac{1}{2}\Omega_{mm}\left[n_i r_j + n_j r_i\right]\right)\right\}\right]$$

(54)

Once again comparison with Eqns (39) and (40) indicates that the form of the body force kernels is not unique.

Steady state thermal loads

Cruse [4] has derived in detail the 3D body force kernels for the steady state thermal case without making use of the Galerkin tensor. This time the resulting kernels are identical to those derived using the Galerkin Tensor.

7. EXAMPLES

To verify all the above theory, a fairly considerable test programme is necessary. We have three different types of body forces each in two different geometries (2D and 3D). For the 3D geometry we have two different versions of the body force kernels to test. For all these different cases both the displacement and the stress equation kernels need testing. Time has not permitted a comprehensive test of every kernel. However, the ability to handle body forces without recourse to integration over the domain is gradually being introduced into BEASY which is a general boundary element system for the solution of problems in potential theory and in linear isotropic elasticity in 2D, axisymmetric and 3D geometries. The tests on the body force kernels which have been carried out have all used constant boundary elements.

Example 1. Cylinder under thermal stresses
This example consists of a long hollow cylinder with the inner surface heated to 100°C and the outer surface held at 0°C. The details of the problem are given below

Inner radius	30 mm
Outer radius	80 mm
Coefficient of Linear Expansion	$10 \times 10^{-6}/^{\circ}C$
Young's Modulus	210 kN/mm^2
Poisson's Ratio	0.3

The infinitely long cylinder was modelled by a quarter cylinder of length 50 mm with an appropriate normal displacement boundary condition applied at either end to model the infinite length. A solution to this problem is given in Ref. [6] and Figure 1 plots values of hoop and axial stresses for the analytical solution compared with those obtained using BEASY's 3D constant element. A distinction is made between stresses obtained in the boundary solution which tests the kernels in the displacement equation, Eqns (41), and the internal point solution which tests the kernels in the stress equation, Eqns (42). The same problem was run using BEASY's axisymmetric constant element which implements the theory given by Cruse, Snow and Wilson [5]. The figures show good agreement with the analytical solution except for an internal point close to the boundary. This last error is due to inaccuracies in the integration scheme employed. It is noticeable that the axisymmetric element gives better results than the 3D element. This is no surprise as one of the integrations has been done analytically in the axisymmetric scheme.

Example 2. Rotating Disc as a 2D problem
The details of this example are:-

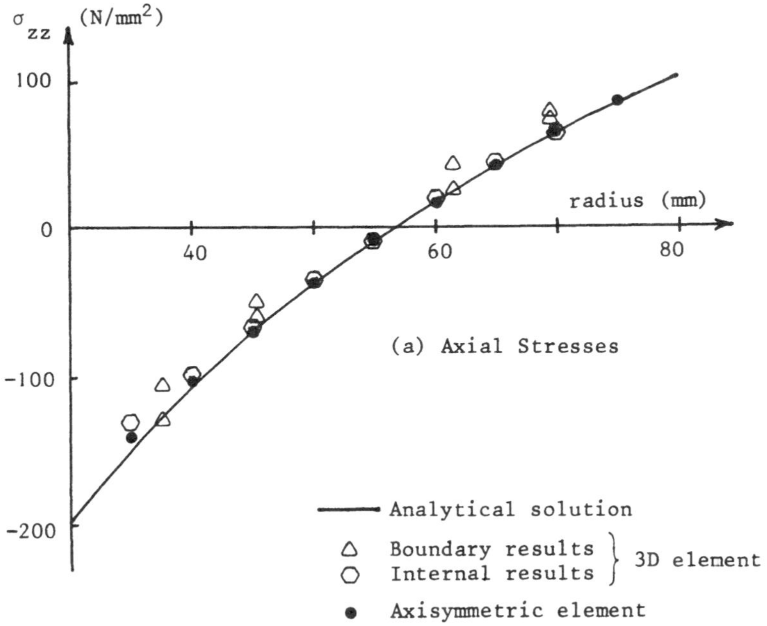

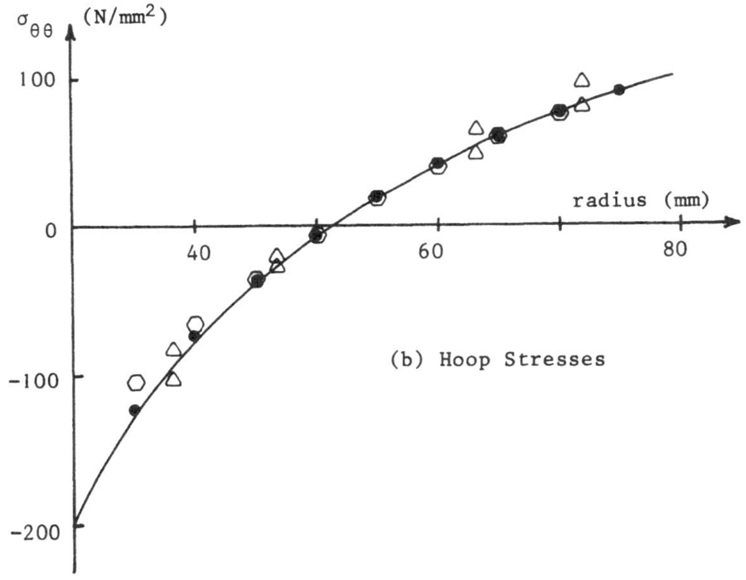

Figure 1 Cylinder under Thermal Load

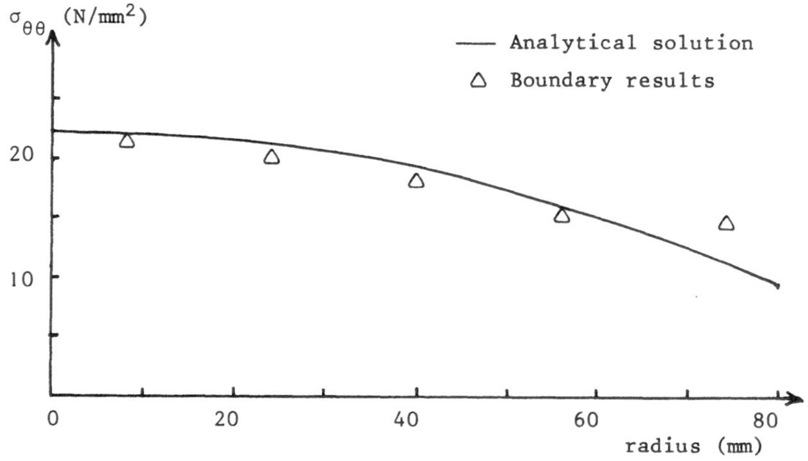

Figure 2 Hoop Stresses in Rotating Disc (2D)

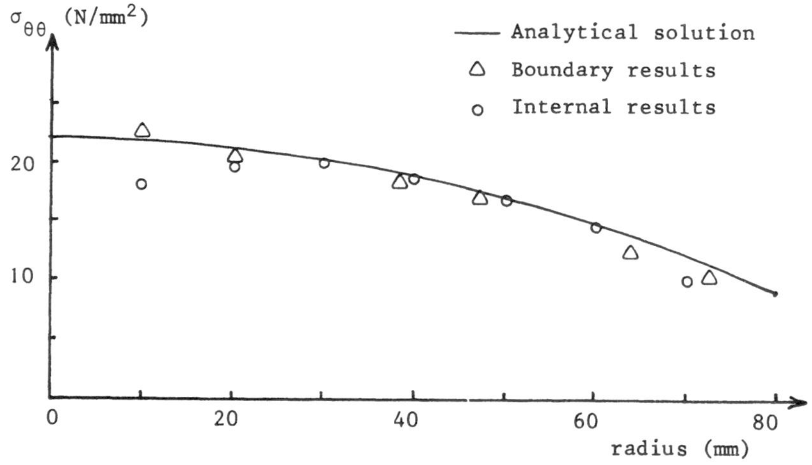

Figure 3 Hoop Stresses in Rotating Disc (3D)

Radius	80 mm
Mass Density	7.8 tonnes/m^3
Rotation Speed	10,000 r.p.m.
Young's Modulus	210 kN/mm^2
Poisson's Ratio	0.3

An analytical solution to this problem is given in Ref. [6]. Figure 2 plots the values of the hoop stresses for the analytical solution compared with those obtained using BEASY's 2D constant element. Only the boundary solution is plotted as the kernel for the stress equation has not yet been incorporated into BEASY. Thus the plotted points are a test of Eq. (28) only. Reasonable agreement has been obtained between the Boundary Element and analytical solutions.

Example 3. Rotating Disc as a 3D problem
The details of this example are:-

Radius	78.2 mm
Thickness	50 mm
Mass Density	7.8 tonnes/m^3
Rotation Speed	10,000 r.p.m.
Young's Modulus	210 kN/mm^2
Poisson's Ratio	0.3

Once again Ref. [6] supplies an analytical solution to this problem. Figure 3 plots the hoop stresses for the analytical solution along a radius which meets the axis of the cylinder at its centre ($z = 0$). The boundary element results were obtained using BEASY's 3D constant element. The boundary results plotted are actually obtained at various axial positions ($-25 < z < 25$). However, inspection of the analytical solution shows that the axial variation of the hoop stresses is very small. The internal results were obtained on the plane $z = 0$. The boundary and internal results verify Eqns. (53) and (54) respectively. As before, reasonable agreement between the analytical and boundary element solutions has been obtained.

8. CONCLUSIONS

The ability to handle three of the most commonly encountered body forces by transforming the body force domain integral to an equivalent boundary integral greatly increases the attractiveness of the Boundary Element Method for use in stress analysis. The limited tests of the method outlined above are sufficient to verify the validity of the method, although a complete test programme to test all the kernels presented here is still necessary. The disagreements obtained between analytic solutions and solutions obtained from BEASY are of similar magnitude to those obtained for problems without body forces and are due to

i) The approximations inherent in the use of constant elements.

ii) The relatively simple numerical integration schemes at present used in BEASY.

Linear and quadratic elements are in the process of being introduced into BEASY and much work has also been done to develop sophisticated integration schemes to handle the singular nature of the displacement and traction kernels. Preliminary results indicate that these developments will turn the Boundary Element Method into an extremely accurate analysis tool.

REFERENCES

1. T.A. Cruse, Numerical solutions in three-dimensional elastostatics, Int. J. Solids Struct. 5, 1259-1274 (1969).

2. T.A. Cruse, An improved boundary-integral equation method for three-dimensional elastic stress analysis. Comput. Struct. 4, 741-754 (1974).

3. C.A. Brebbia, The Boundary Element Method for Engineers. Pentech Press 1978.

4. T.A. Cruse, Boundary-integral equation method for three-dimensional elastic fracture mechanics analysis. AFOSR-TR-75-0813 May 1975.

5. T.A. Cruse, D.W. Snow and R.B. Wilson. Numerical Solutions in Axisymmetric Elasticity. Computers and Structures 1977 pp.445.

6. S.P. Timoshenko and J.N. Goodier, Theory of Elasticity Third Edition, McGraw-Hill, 1970.

7. A.E.H. Love, A Treatise on the Mathematical Theory of Elasticity. Fourth Edition, 1926.

A COMPARATIVE STUDY OF SEVERAL BOUNDARY ELEMENTS IN ELASTICITY

M.F.Seabra Pereira, C.A.Mota Soares and L.M.Oliveira Faria

CEMUL-Centro de Mecânica e Materiais da Universidade Técnica de Lisboa, Instituto Superior Técnico, Av. Rovisco Pais
1096 Lisboa Codex - PORTUGAL

INTRODUCTION

Although many elasticity problems have been solved by both analytical and numerical techniques, solutions for actual problems involving complex geometries are still relatively scarce.

For practical structural geometries, numerical methods are advantageous, but have some limitations in representing high stress gradients.

A general purpose Boundary Element program has been developed for elasticity problems, in which different elements can be easily implemented.

In this paper several boundary element solutions for elasticity problems are compared with finite element results and analytical solutions. The advantages and disadvantages of the boundary element method and of the different elements used are discussed with reference to the applications. Also, the principal features of the computer program are described.

ANALYTICAL BACKGROUND

For the sake of notation and completeness it is recalled that the boundary element method is based on the Somigliana's boundary identity, as presented by Brebbia and Walker (1980),

$$c_{ij}(P)u_i(P) + \int_s u_j(Q)T_{ij}(P,Q)ds = \int_s t_j(Q)U_{ij}(P,Q)ds + \\ + \int_V b_j(q)U_{ij}(P,Q)dV(q) \qquad (1)$$

where P,Q are points of the surface s of the domain V. U_{ij} and T_{ij} are the fundamental Kelvin solutions, c_{ij} are coefficients depending on the geometry of the boundary at the point P, and q is a point in the domain. Tensor notation is used in this expression and the indices have the range 1,2,3. When body forces are not considered equation (1) is only dependent on the surface displacements and tractions, u_i and t_i respectively.

The surface can be discretized into elements with a particular shape and a certain number of nodes, say n. Any field variable within each element is assumed to be given by

$$\theta(\xi) = M^{\alpha}(\xi)\theta^{\alpha} \qquad \alpha = 1,2,\ldots n$$

or

$$\theta(\xi,\eta) = M^{\alpha}(\xi,\eta)\theta^{\alpha} \qquad \alpha = 1,2,\ldots n \tag{2}$$

for 2D and 3D cases respectively, M^{α} are shape functions of local coordinates ξ and η, θ^{α} are the values of field variables at nodal points. If the surface displacements and tractions are introduced as field variables equation (1) becomes, for 2D problems

$$c_{ij}(P_n)u_i(P_n) + \sum_{k=1}^{n_\ell} \int_{s_k} M^{\alpha}(\xi)T_{ij}(P_n,Q(\xi))J(\xi)d\xi u_j^{\alpha} = \sum_{k=1}^{n_\ell} \int_{s_k} M^{\alpha}(\xi)U_{ij}(P_n,Q(\xi))J(\xi)d\xi t_j^{\alpha} \tag{3}$$

where n_ℓ is the number of boundary elements

u_j^{α} is the value of u_j at local node α

t_j^{α} is the value of t_j at local node α

J is the Jacobian of transformation of coordinates.

Noting that P_n refers to a particular node, then for all nodes equations (3) can be expressed in matrix form as follows:

$$[C]\{U\} + [\hat{H}]\{U\} = [G]\{T\} \tag{4}$$

or

$$[H]\{U\} = [G]\{T\} \tag{5}$$

where $\{U\}$ and $\{T\}$ are the surface nodal displacements and tractions respectively. The elements of $[\hat{H}]$ and $[G]$ can be obtained from the integrals

$$H_{ij}^{n\alpha} = \int_{sk} M^{\alpha}(\xi)T_{ij}(P_n,Q(\xi))J(\xi)d\xi \tag{6}$$

$$G_{ij}^{n\alpha} = \int_{sk} M^{\alpha}(\xi)U_{ij}(P_n,Q(\xi))J(\xi)d\xi \tag{7}$$

These integrals are normally integrated by standard Gaussian quadrature. When these integrands are singular they are calculated by rigid body considerations or by the technique described in the Appendix.

Applying the boundary conditions, equation (5) becomes

$$[A]\ \{x\} = \{f\} \tag{8}$$

where $\{x\}$ and $\{f\}$ are the vectors of unknown and known quantities respectively.

The displacements at interior points can be calculated using the following expression

$$u_i(P_n) = \sum_{k=1}^{n\ell} \int_{s_k} M^{\alpha}(\xi_i) U_{ij}(P_n, Q(\xi)) J(\xi) d\xi \; t_j^{\alpha} - \qquad (9)$$
$$- \sum_{k=1}^{n\ell} \int_{s_k} M^{\alpha}(\xi_i) T_{ij}(P_n, Q(\xi)) J(\xi) d\xi \; u_j^{\alpha}$$

and the stresses at interior points are given by

$$\sigma_{ij}(P_n) = \sum_{k=1}^{n\ell} \left(\int_{s_k} M^{\alpha} D_{\ell ij} J(\xi) d\xi t_{\ell}^{\alpha} - \int_{s_k} M^{\alpha} S_{\ell ij} J(\xi) d\xi u_{\ell}^{\alpha}\right) \qquad (10)$$

where $D_{\ell ij}$ and $S_{\ell ij}$ are given by Brebbia and Walker (1980).

BRIEF DESCRIPTION OF THE PROGRAM

Basically, the computer program consists of four main subroutines and the element subroutines. A brief flow-chart of the program is illustrated in Figure 1.

Subroutine BEM 001

This subroutine reads all the input data and sets up all the relevant array dimensions for the analysis.

Subroutine BEM 010

This subroutine performs the overall assembly of matrix [A] and vector {f}. This subroutine calls the element subroutines which calculate the integrals of expressions (6) and (7) for the boundary element with reference to a particular node. These element subroutines can be called from the element library or supplied by the user.

Subroutine BEM 030

This subroutine is an equation solver, using the Gaussian elimination technique.

Subroutine BEM 020

This subroutine calculates the displacements and stresses at internal points according to equations (9) and (10). This subroutine calls the element subroutines, which calculate the integrals of equations (9) and (10). Again, these element subroutines can be called from the element library or supplied by the user.

Element Subroutines

For each boundary element three subroutines have to be developed. The first subroutine calculates all the necessary information for each element (Jacobian, normals, etc.) at the Gaussian points. The second subroutine calculates the integrals of expression (6) and (7) for the boundary element with reference to a particular boundary node or internal point. The third subroutine calculates the integral of expressions (9) and (10) for the boundary element with reference to a particular point.

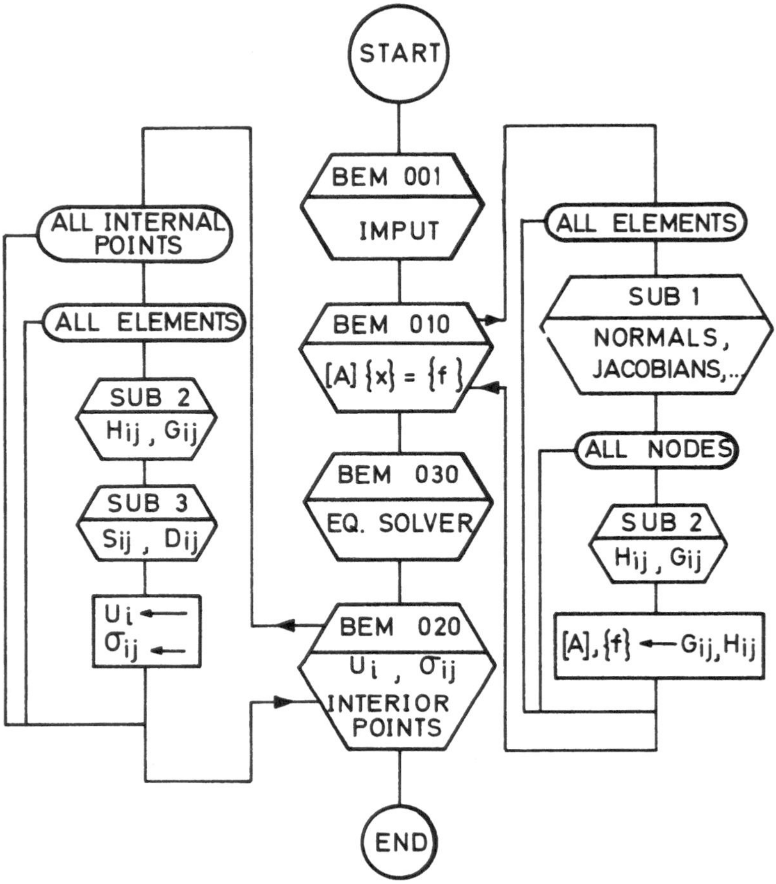

H_{ij} - coefficients of matrix $[\hat{H}]$

G_{ij} - coefficients of matrix $[G]$

S_{ij}, D_{ij} - See expression (10)

SUB1, SUB2, SUB3 - Element subroutines which are in library or supplied by the user

Figure 1. Primary flow-chart of program BEM

Element Library

The following boundary elements have been developed:

- constant, 2D element
- linear, 2D element
- parabolic, isoparametric, 2D element
- triangular, constant, 3D element
- quadrilateral, isoparametric, linear, 3D element
- quadrilateral, isoparametric, parabolic, 3D element

APPLICATIONS

The different test cases which have been chosen have, in general, severe stress gradients either because of the geometry or due to the type of loading.

Different boundary discretizations have been idealized and the following criteria were adopted: for each case a number m of nodal points is specified, yielding m/2 parabolic elements or m linear elements. Coarser meshes have been defined from the previous cases by simply supressing every other node.

The corresponding meshes for finite element analysis have been such that the outside nodes coincide with the boundary element nodes. In the finite element analysis, the isoparametric parabolic element for plain strain/stress problems has been used.

The first three case studies are 2D analyses and the last case is a 3D example. All the analyses were performed in double precision on an IBM 360/44 with a core memory of 256 kbytes.

Hole in a flat plate

The geometry of the problem is illustrated in figure 2a. In the present case the ratio a/b is 5. A constant stress σ_0 is applied to the plate at faces $x=\pm a$. Figure 2b represents the mesh used in the finite element analysis (FEM). Due to symmetry only one quarter of the structure has been modelled.

In Figure 3 the values of the σ_{xx} stresses along the y axis are compared with the analytical solution of the infinite plate case. The solution corrected for the finite width of the present proportions yields a value of 3.93 for the stress concentration factor $K_t = \sigma_{xx}/\sigma_0$.

In Table 1 a comparison of results is done jointly with an indication of the computer times involved in all analyses. A good measure of the accuracy of the various analyses is made on the basis of the σ_{xx} stresses at the point A near the hole. The discrepancy of the results is evaluated in comparison with the value $K_t = 3.93$

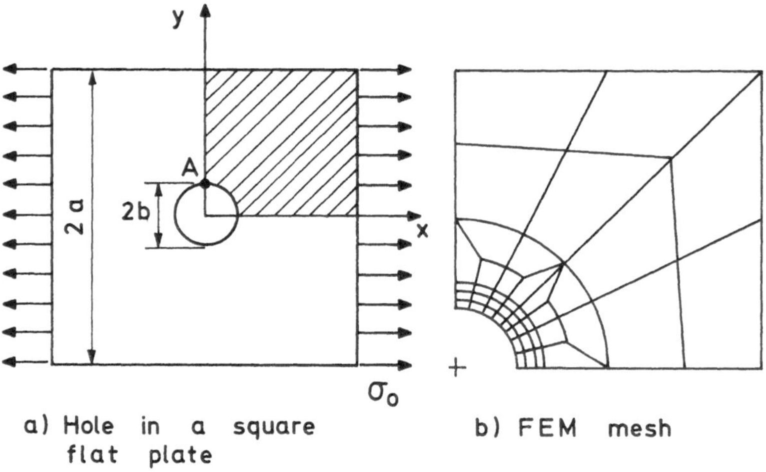

Figure 2

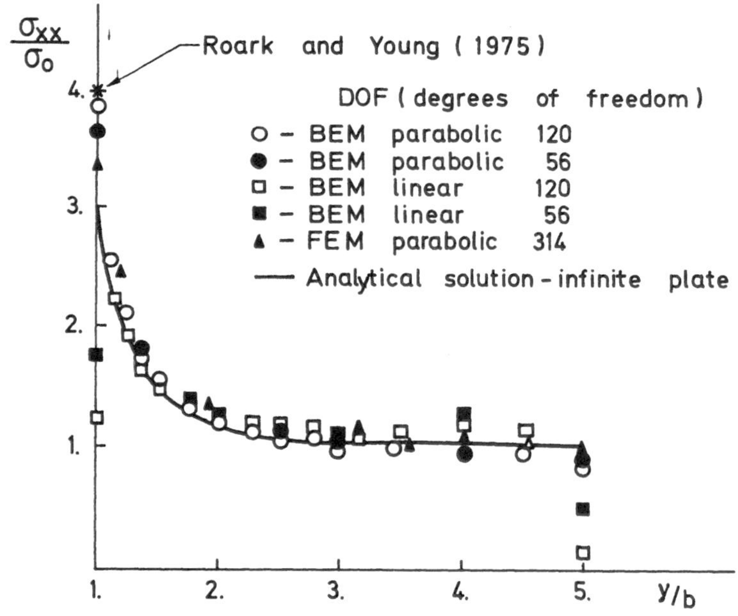

Figure 3. σ_{xx} stresses near the hole

Table 1. Comparison of results for the hole in a flat plate case

ANALYSIS	DOF	Discrepancy %	CPU (seconds)
BEM Parabolic	120	1.3	437
BEM Parabolic	56	4,7	91
BEM Linear	120	-	345
BEM Linear	56	-	68
FEM Parabolic	314	13	342

The discrepancies for the linear elements are not indicated, as the corresponding analysis yielded rather poor results. This fact is associated with the limitations of the BEM in modelling properly the stress fields at corner regions. The results show that the linear elements are more sensitive to this effect. However, in regions away from the corners all the analyses show a reasonable agreement.

In all cases a relatively small number of degrees of freedom have been used. The discrepancy of 4.7% using the parabolic element is very satisfactory regarding the rather coarse discretization of the boundary.

Circular disc compressed by two diametrically opposed forces

The geometry and loading assumed in this example is shown in Figure 4a and the FEM idealization is represented in figure 4b. Due to symmetry only one half of the structure has been considered in the BEM idealization.

Again four analyses have been made using parabolic and linear boundary elements with two different refinements of the boundary. In Figure 5 a graph of σ_{yy} stresses along the diameter perpendicular to the direction of loading is indicated. The results are compared with an analytical solution of Timoshenko and Goodier (1970).

In Table 2 a comparison of results, is presented. The computer times involved in each analysis is also indicated. A measure of the accuracy of the results is done by comparing the maximum stress σ_{yy} for x=0 for the different analyses with the analytical solution.

Stresses at interior points have been calculated in the vicinity of the loading point A. The distribution of compressive stresses in a small neighbourhood of the load point approaches that predicted by Flamant's problem which is governed by the expression $\sigma_{yy}=-2P/\pi d$ where P is the applied load and d is the distance from the load point to an interior point. In figure 6 the results obtained from the BEM and the FEM analyses are compared with the analytical solution .The typical element size near the point of application of the load is about 0.01D where D is the diameter of the disc. This element size is small enough

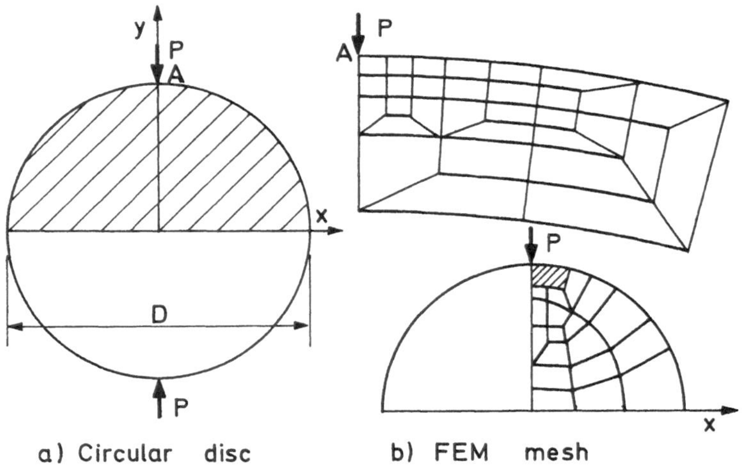

Figure 4

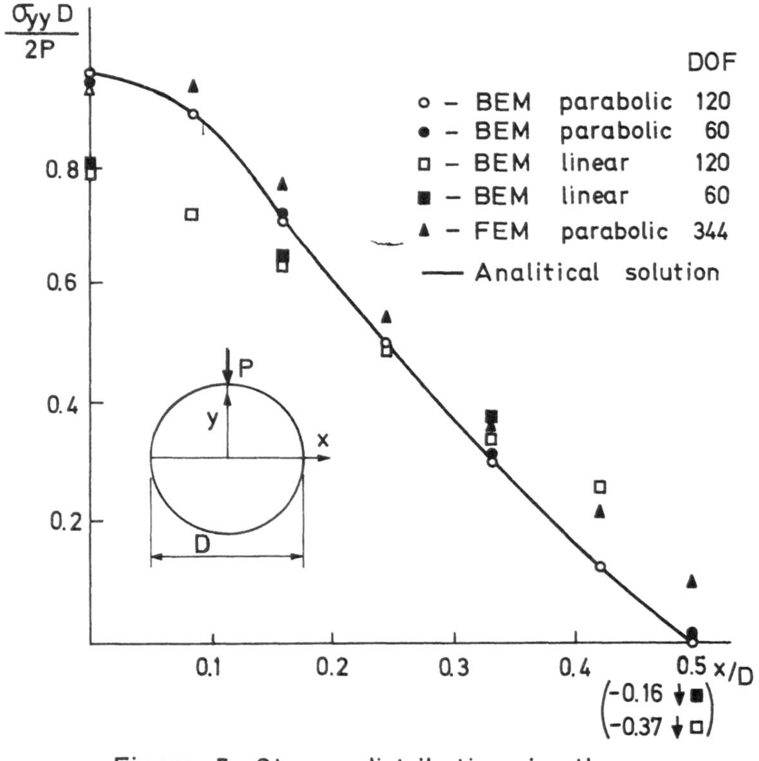

Figure 5. Stress distribution in the diameter perpendicular to loading

to allow the evaluation of stresses at interior points relatively close to the surface. The distances to which the stresses are referred are always larger than 5 times the typical element size in that region.

Table 2. Comparison of results for the disc case

ANALYSIS	DOF	Discrepancy %	CPU (seconds)
BEM Parabolic	120	0,02	442
BEM Parabolic	60	0,08	101
BEM Linear	120	19,0	350
BEM Linear	60	17	77
FEM Parabolic	344	4	100

In this figure it can be observed the inability of the FEM to model such severe stress gradients. The FEM results near the point load clearly depart from the analytical solution as well as from the BEM values. Again the parabolic boundary elements behave quite satisfactory and with an error of less than 1% for distances greater than 0.05 D from the point load.

Compact Tension Specimen (CTS)

Another example analysed is a CTS with an edge crack as illustrated in figure 7. In this case H=W and the crack length A is .25W. Due to symmetry only one half of the structure is idealised. Because of the singular stress and strain fields in the vicinity of the crack tip only parabolic boundary elements have been used.

In both, BEM and FEM discretizations elements near the crack tip have an average size of about 0.05 of the crack length. Plane strain conditions were assumed.

Two methods of deriving the values of crack tip stress intensity factors, K_I, have been used, firstly, the extrapolation of computed displacements near the crack tip, and second, an energy method based on the Fracture Mechanics plain strain relation $G_I=K_I^2(1-\nu^2)/E$, where G_I is the strain energy release rate, K_I is the mode I stress intensity factor and E and ν are the Elasticity Modulus and the Poisson's ratio respectively.

In this case the energy available for extension of the crack length dA is provided by the work done by the nodal point tractions applied in the top faces of the specimen. Variations of crack length dA of 0.05 times the initial crack length have been considered. A significant improvement on the results is obtained by displacing in the BEM and FEM elements, the mid-side node of the near tip elements to 1/4 position as described by Henshell and Shaw (1975).

The present results have been compared with a solution of $K_I/K_0=1.5$ presented by Rooke and Cartwright (1976). Table 3 shows the results and the CPU times of the various analyses

Table 3. Comparison of results for the CTS

ANALYSIS	DOF	K_1/K_0	Discrepancy %	CPU (seconds)
(1) BEM Parabolic	120	1,51	0,67	415
(1) BEM Parabolic	60	1,49	0,67	103
(2) BEM Parabolic	120	1,495	0,33	850
(1) FEM Parabolic	530	1,48	1,3	922

$K_o=\sigma\sqrt{\pi A}$; σ is the applied remote stress.

(1) Extrapolation method
(2) Energy Method

The value of the stress intensity factor obtained with the energy method involves, at the present stage, the execution of two analyses with two different crack lengths, thus increasing considerably the computing costs.

Beam subjected to an axial force

The BEM discretization of the beam is illustrated in Figure 8. Two analyses have been carried out using linear and quadratic elements. The results for the linear element have errors of 2.8% and 3.6% in the axial stresses and maximum displacements respectively. The corresponding errors for the parabolic element are 20.2% and 3.5%.

CONCLUSIONS

A general purpose Boundary Element program has been developed using several 2D and 3D boundary elements.

Results for particular elastostatics situations without body forces have been presented and the usage of different element types have been considered.

In general, the results show that the parabolic 2D element is much more efficient than the 2D linear element. The linear boundary element is particularly sensitive to corner regions where the stress results tend to degradate considerably.

The relative computing costs for the various analyses have been shown in the Tables. In situations of severe stress gradients, particularly in the vicinity of point loads the BEM method appears to offer a reasonable accuracy, of the order of 5%, for relatively modest computing facilities.

FEM analyses with similar refinement on the boundaries to the BEM discretizations involve rather larger numbers of nodes without any benefit in the accuracy of the results, yet more information is generally provided about the stress and displacement fields on the domain.

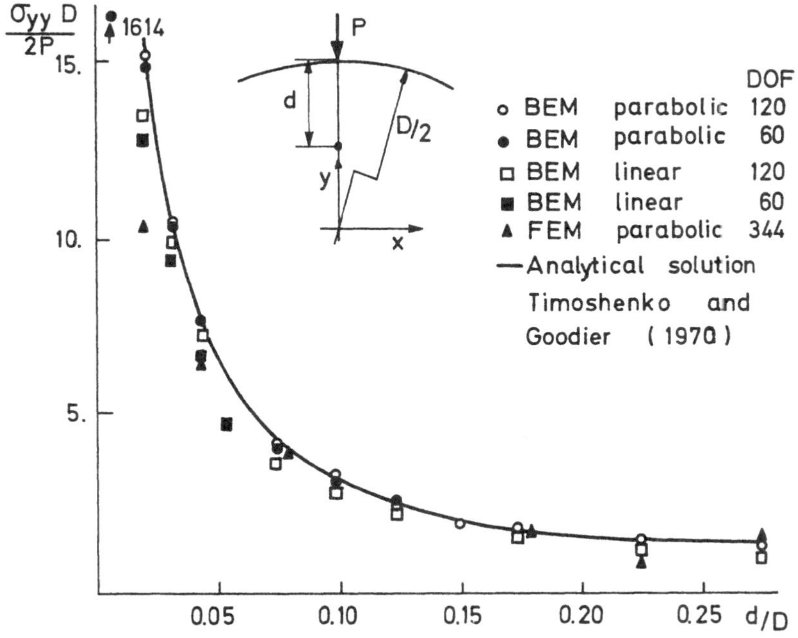

Figure 6. σ_{yy} stresses at interior points

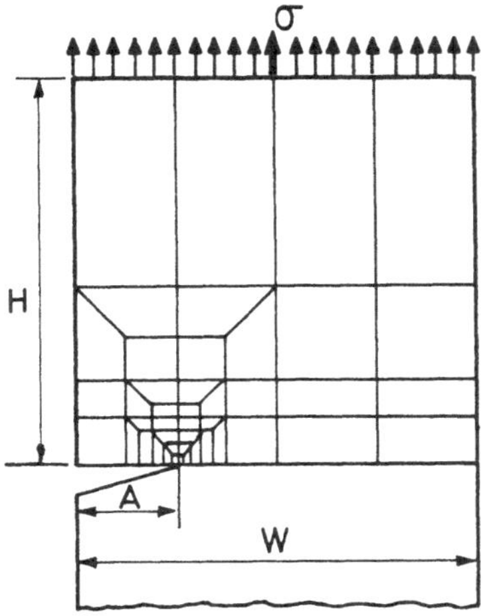

Figure 7. CT Specimen

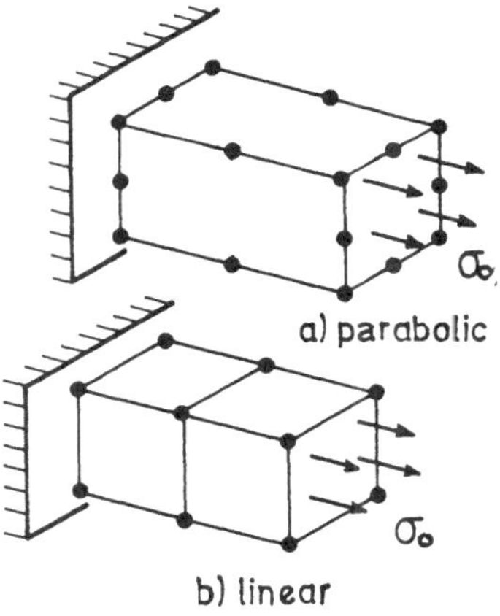

Figure 8. 3D Beam

REFERENCES

Brebbia, C.A. and Walker,S. 1980, Boundary Element Techniques in Engineering. Newnes - Butterworths, London.

Cristescu,M. and Loubignac,G. 1978, Gaussian Quadrature Formulas for Functions with Singularities in 1/R Over Triangles and Quadrangles. Recent Advances in Boundary Elements Methods, Editor Brebbia,C.A., Pentech Press, London.

Fairweather,G., Rizzo,F.J. and Shippy,D.J. 1979, On the Numerical Solution of Two - Dimensional Potential Problems by an Improved Boundary Integral Equation Method, J.Comp. Phys., Vol. 31, pg. 96-112.

Henshell,R.D. and Shaw,K.G. 1975, Crack Tip Finite Elements Are Unnecessary. Int. J. Num.Meth. in Engng. Vol. 9, p 495.

Roark, R.J. and Young,W. 1975, Formulas for Stress and Strain, 5th Ed. MacGraw-Hill, Tokyo.

Rooke,D.P. and Cartwright,D.J. 1976, Compendium of Stress Intensity Factors. Her Majesty Stationery Officer, London.

Saada,A.S. 1974, Elasticity, Theory and Applications. Pergamon Press, New York.

Stroud,A.H. and Secrest,D. 1966, Gaussian Quadrature Formulas, Prentice Hall, Int. Inc., London.

Timoshenko and Goodier 1970, Theory of Elasticity, McGraw-Hill, Tokyo.

APPENDIX: INTEGRATION OF KERNEL-SHAPE FUNCTION PRODUCTS

For 2D problems the evaluation of the singular integrals in equation (7) leads to integrals of the following type

$$\int_{-1}^{+1} \ln \frac{1}{r} f(\xi)d\xi \tag{11}$$

where r is the distance between the points P and Q and $f(\xi)$ is a product of the shape functions and the Jacobian of transformation of coordinates. This integral can be transformed as follows

$$\int_{-1}^{1} \ln \frac{1}{r} f(\xi)d\xi = \int_{-1}^{+1} \ln(\frac{1+\xi}{2r}) f(\xi)d\xi + 2\int_{0}^{1} \ln \frac{1}{\zeta} f(\zeta)d\zeta \tag{12}$$

where $\zeta = \frac{1}{2}(\xi+1)$

The first integral on the right-hand side is evaluated using standard Gaussian quadrature while the second integral is calculated numerically by formulas given by Stroud and Secrest (1965).

For 3D problems the integrals corresponding to equation (7) present a 1/r singularity and are integrated using a formula of the following type

$$\int_{-1}^{1}\int_{-1}^{1} \frac{1}{r} f(\xi,\eta)d\xi\, d\eta = \sum_{i=1}^{m} A_i\, f(\xi_i,\eta_i) \tag{13}$$

The weights A_i and the integration point coordinates ξ_i and η_i depend on the geometry of the region of integration and m is the number of integration points. These points car be calculated following the method presented by Cristescu and Loubignac (1978). For 3D linear and parabolic quadrilateral elements Gauss formulas have been developed to integrate exactly polynomials of 5th order multiplied by the weight function $1/\sqrt{\xi^2+\eta^2}$ in the regions illustrated in Figure 9.

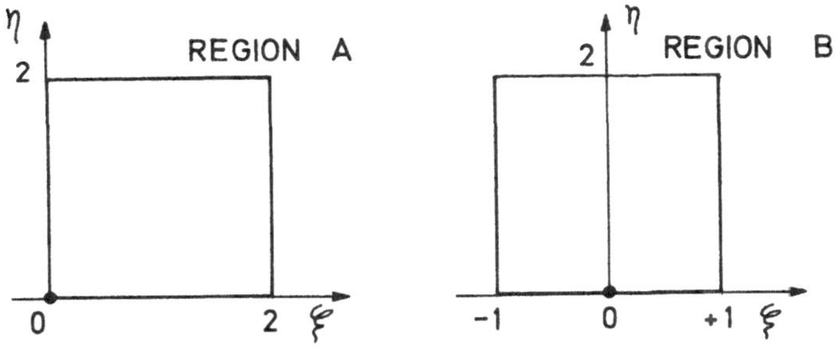

Figure 9. Regions of integration

If the singularity is located in a corner node or in a mid-side node, the integration region is transformed in the squares A or B, respectively, by applying simple linear transformation techniques to the domain of integration.

The integral (13) can be transformed as follows

$$\iint \frac{1}{\sqrt{\xi^2+\eta^2}} \frac{\sqrt{\xi^2+\eta^2}}{r} f(\xi,\eta)d\xi\, d\eta \qquad (14)$$

where the function $(\sqrt{\xi^2+\eta^2}/r)\ f(\xi,\eta)$ is suitably represented by polynomials.

The values of ξ_i and η_i have been found by calculating the common roots of orthogonal polynomials in ξ and η. Using a least square technique the corresponding weights A_i have been evaluated. The Gauss point coordinates and the corresponding weights are indicated in Table 4.

Table 4. Gauss points and corresponding weights

REGION A

	ξ_i	η_i	A_i
1	0.14110148	0.14110148	0.84568628
2	0.18488014	0.91104203	0.47062986
3	0.19800081	1.7500503	0.16899828
4	0.87593347	0.87593747	0.66675077
5	0.89746038	1.7453278	0.27317899
6	0.91104203	0.18488014	0.47062986
7	1.7118237	1.7118237	0.18744311
8	1.7453278	0.89746038	0.27317899
9	1.7500503	0.19800081	0.16899828

REGION B

	ξ_i	η_i	A_i
1	-0.74772761	0.84529169	0.49133682
2	0.0	0.87119449	0.84924859
3	0.74772761	0.84529169	0.49133682
4	-0.77589199	1.7321786	0.19429025
5	0.0	1.71964092	0.34350839
6	0.77589199	1.7321786	0.19429025
7	-0.72502542	0.15427398	0.36769444
8	0.0	0.14090886	1.5127182
9	0.72502542	0.15427398	0.36769444

NON-CONFORMING BOUNDARY ELEMENTS FOR STRESS ANALYSIS

C. Patterson and M.A. Sheikh

Dept. of Mechanical Eng., University of Sheffield, U.K.

INTRODUCTION

IN the boundary element method the familiar infinite system of integral equations is discretized by the introduction of boundary elements. As with Finite Elements one can define a heirarchy of elements here ranging from constant through linear, quadratic etc.

Hitherto, in the case of linear and higher order elements, interelement continuity of the unknown functions has been imposed in applications, undoubtedly in analogy with finite elements practice. This continuity gives rise to at least three problems [Lachat,1975]. First, at a point where surface is not smooth the normal is not defined but the freedom there (traction say) demands a valid normal. Second, at an interface where there is a change in the nature of boundary conditions (say between displacement and traction) apparently both types of freedom are constrained. Third, when the problem is partitioned into sub-regions, a convenient device on grounds of computational efficiency, there can be excessive constraint where several surfaces meet. Accommodation to the second and third problems have been devised by appropriate suppression of degrees of freedom and to the first by introduction of two freedom nodes close to the geometric singularity, on either side. The last approach is objectionable since if the nodes are not closely spaced the boundary integrals are not well discretized, while if they are the resulting algebraic equations are indefinite. This is because proximity of freedom nodes implies linear dependence in the algebraic equations. The root of the problem is imposition of interelement continuity. In the Finite Element Method adequate continuity is necessary to maintain positive definiteness of the system [Patterson,1973]. In the Boundary Element Method there is no such requirement; so that interelement continuity is simply not necessary. This point is amply borne out by the apparent success of constant elements.

In this paper a selection of two dimensional elastostatic test problems is analysed using conventional and non-conforming boundary elements and the advantages of non-conforming elements particularly at boundary discontinuities are discussed.

THEORY

THE general boundary integral equation for two dimensional elasticity observing interelement continuity and with singular point 'i' located on the boundary of the domain S, of the given problem is:

$$c^i u^i + \sum_{j=1}^{N} \left\{ \int_{S_j} t^* F^T dS \right\} u_j = \sum_{j=1}^{N} \left\{ \int_{S_j} u^* F^T dS \right\} t_j \qquad (1)$$

where c^i is the unknown coefficient at 'i'

N = Number of boundary elements in which the domain is divided,

u and t are displacement and traction respectively,

u^* is the fundamental solution for two dimensional elasticity given as,

$$u^* = \frac{1+\nu}{8\pi E(1-\nu)r} \left\{ (3-4\nu)\delta_{ij} + \frac{(x_i-y_i)(x_j-y_j)}{r^2} \right\} \qquad (2)$$

where ν = Poisson's Ratio

E = Modulus of Elasticity

δ_{ij} is the Kronecker delta

and r is the distance between points x and y, that is, the point of application of the load to the point under consideration.

Also, the Kernel t^* for plane stress and plane strain problem is:

$$t^* = \frac{1}{4\pi(1-\nu)r} \left\{ (1-2\nu) \left[\frac{n_i(x_j-y_j)}{r} - \frac{n_j(x_i-y_i)}{r} \right] \right.$$

$$\left. + \left[(1-2\nu)\delta_{ij} + 2\,\frac{(x_i-y_i)(x_j-y_j)}{r^2} \right] \frac{\partial r}{\partial n} \right\} \qquad (3)$$

where n_i, n_j are direction cosines and n is the normal to the surface of the body so that: $\partial r/\partial n = n_s(x_s - y_s)/r$

The displacements and tractions for a linear element 'j' (Figure.1.) are:

$$u = \begin{Bmatrix} u_1 \\ u_2 \end{Bmatrix} = \begin{bmatrix} F_1 & F_2 & 0 & 0 \\ 0 & 0 & F_1 & F_2 \end{bmatrix} \begin{Bmatrix} u_x^1 \\ u_x^2 \\ u_y^1 \\ u_y^2 \end{Bmatrix}$$

$$t = \begin{Bmatrix} t_1 \\ t_2 \end{Bmatrix} = \begin{bmatrix} F_1 & F_2 & 0 & 0 \\ 0 & 0 & F_1 & F_2 \end{bmatrix} \begin{Bmatrix} t_x^1 \\ t_x^2 \\ t_y^1 \\ t_y^2 \end{Bmatrix} \tag{4}$$

and the integrals in Equation (1) can be written as:

$$\int_{S_j} \begin{bmatrix} F_1 & F_2 & 0 & 0 \\ 0 & 0 & F_1 & F_2 \end{bmatrix} t^* \quad dS \quad \begin{Bmatrix} u_x^1 \\ u_x^2 \\ u_y^1 \\ u_y^2 \end{Bmatrix}$$

$$\int_{S_j} \begin{bmatrix} F_1 & F_2 & 0 & 0 \\ 0 & 0 & F_1 & F_2 \end{bmatrix} u^* \quad dS \quad \begin{Bmatrix} t_x^1 \\ t_x^2 \\ t_y^1 \\ t_y^2 \end{Bmatrix} \tag{5}$$

in which u_x^1, u_y^1 are the displacements at node 1 in x and y directions respectively,

u_x^2, u_y^2 are the displacements at node 2 in x and y directions,

t_x^1, t_y^1 are the tractions at node 1 in x and y directions,

t_x^2, t_y^2 are the tractions at node 2 in x and y directions,

F_1 and F_2 are the interpolation functions given by,

$$F_1 = (1-\xi)/2 \; ; \quad F_2 = (1+\xi)/2 \tag{6}$$

where ξ is the dimensionless coordinate $\xi = x/L/2$

The values of displacement and traction in x and y directions at any point on the element are:

$$\begin{aligned} u_x(\xi) &= F_1 u_x^1 + F_2 u_x^2 \\ u_y(\xi) &= F_1 u_y^1 + F_2 u_y^2 \\ t_x(\xi) &= F_1 t_x^1 + F_2 t_x^2 \\ t_y(\xi) &= F_1 t_y^1 + F_2 t_y^2 \end{aligned} \tag{7}$$

For a node 'i' at the intersection of elements 'j-1' and 'j', Equation (1) can be written in two parts for simplicity (for each direction) as:

$$\begin{aligned} c^i u_x^i + \sum_{j=1}^{N} [h_{x_{i,j-1}} + h_{x_{i,j}}] u_{x_i} &= \sum_{j=1}^{N} [g_{x_{i,j-1}} + g_{x_{i,j}}] t_{x_i} \\ c^i u_y^i + \sum_{j=1}^{N} [h_{y_{i,j-1}} + h_{y_{i,j}}] u_{y_i} &= \sum_{j=1}^{N} [g_{y_{i,j-1}} + g_{y_{i,j}}] t_{y_i} \end{aligned} \tag{8}$$

where

$$\begin{aligned} h_{x_{i,j-1}} &= \int_{S_{j-1}} [F_1 t^*]_x \, dS \; ; \quad h_{x_{i,j}} = \int_{S_j} [F_2 t^*]_x \, dS \\ g_{x_{i,j-1}} &= \int_{S_{j-1}} [F_1 u^*]_x \, dS \; ; \quad g_{x_{i,j}} = \int_{S_j} [F_2 u^*]_x \, dS \\ h_{y_{i,j-1}} &= \int_{S_{j-1}} [F_1 t^*]_y \, dS \; ; \quad h_{y_{i,j}} = \int_{S_j} [F_2 t^*]_y \, dS \\ g_{y_{i,j-1}} &= \int_{S_{j-1}} [F_1 u^*]_y \, dS \; ; \quad g_{y_{i,j}} = \int_{S_j} [F_2 u^*]_y \, dS \end{aligned} \tag{9}$$

That is two equations are set up for each node (one for each direction) taking into account the contributions made by the adjoining elements. This approach gives rise to problems mentioned earlier in the 'Introduction' section. Also, the determination of c^i in Equation (8) is cumbersome. To overcome these problems 'Regular Boundary Integrals for Non-Conforming Elements' are employed here. In this method, freedoms of an element are not defined at the geometric nodes i.e. $\xi = \pm 1$ but at $\xi = \pm 1/2$ (Figure. 2). Also, the singular point is relocated outside the domain and along positive normal to the boundary (taken as L/2 in the present case study where L is the length of the element) at any node 'i' [Patterson and Sheikh, 1981]. The resulting integrals are then regular over the whole domain and no special treatment is required for the evaluation of singular integrands ($c^i = 0$).

The interpolation functions take the form

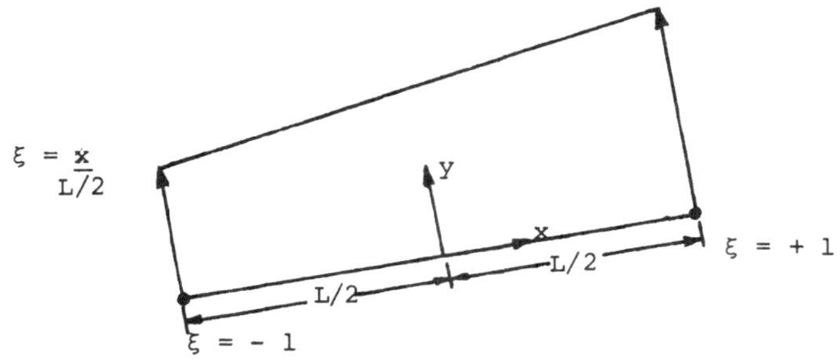

Fig. 1. Linear Element

$$F_1 = (\xi - 1/2) \; ; \quad F_2 = (\xi + 1/2) \qquad (10)$$

There are 2N freedom points defined over the whole boundary of the domain of the problem and the equations for any point 'i' similar to Equation (8) can be written as:

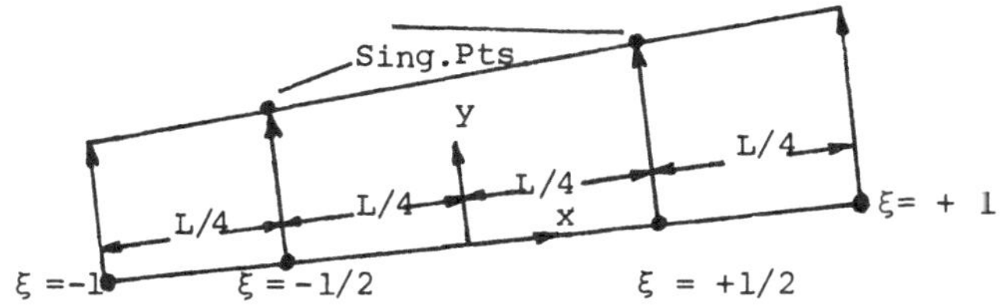

Fig. 2. Discontinuous Linear Element

$$\sum_{j=1}^{N} [h_{x_{i,2j-1}}] u_{x_i} = \sum_{j=1}^{N} [g_{x_{i,2j-1}}] t_{x_i}$$

$$\sum_{j=1}^{N} [h_{y_{i,2j-1}}] u_{y_i} = \sum_{j=1}^{N} [g_{y_{i,2j-1}}] t_{y_i}$$

$$\sum_{j=1}^{N} [h_{x_{i,2j}}] u_{x_i} = \sum_{j=1}^{N} [g_{x_{i,2j}}] t_{x_i} \qquad (11)$$

$$\sum_{j=1}^{N} [h_{y_{i,2j}}] u_{y_i} = \sum_{j=1}^{N} [g_{y_{i,2j}}] t_{y_i}$$

where

$$h_{x_{i,2j-1}} = \int [F_1 t^*]_x \, dS \; ; \quad h_{x_{i,2j}} = \int [F_2 t^*]_x \, dS$$

$$h_{y_{i,2j-1}} = \int [F_1 t^*]_y \, dS \; ; \quad h_{y_{i,2j}} = \int [F_2 t^*]_y \, dS$$

$$g_{x_{i,2j-1}} = \int [F_1 u^*]_x \, dS \; ; \quad g_{x_{i,2j}} = \int [F_2 u^*]_x \, dS \qquad (12)$$

$$g_{y_{i,2j-1}} = \int [F_1 u^*]_y \, dS \; ; \quad g_{y_{i,2j}} = \int [F_2 u^*]_y \, dS$$

Equations (12) produce a 4N x 4N system of equations which can be solved to give the solution at the boundary. Once the values of displacements and tractions in both directions are known over the whole boundary, the solution in the interior can be generated routinely.

APPLICATIONS

THREE two dimensional elastostatic problems are analysed using both the conventional and non-conforming linear elements and a critical comparison of the results is made.

Uniform Cylinder under Internal Pressure

The exact solution of the problem is well known and given by Lame's equations. The cylinder with inner radius r = a and outer radius r = b is considered to have free ends and subjected to uniform internal pressure P, with no pressure applied on the outer surface. Only a quarter of the cylinder is considered for analysis due to symmetry. The boundary conditions are specified in a way to avoid rigid body motion i.e. along AB zero displacement is prescribed in y direction and CD is fixed in x direction, as shown in Figure.3. The numerical values for the problem are assumed to be:

$$a = 1 \text{ cm} \quad ; \quad b = 2 \text{ cm} \quad ; \quad P = 1000 \text{ N/cm}^2$$

The Modulus of Elasticity is taken as 210 ,00000 N/cm^2 and Poisson's Ratio as 0.3.

The boundary of the domain was divided into 26 linear elements. No special care was taken in concentrating the integration points around the singularities to show that this is not a particular requirement when non-conforming elements are used. The problem was solved using conventional boundary integral method and regular boundary integral method for non-conforming elements. Radial displacements were calculated at five different radii across the thickness of the cylinder for four different values of θ = 0,10,30,45 degrees. As shown in Table.1., the best results are obtained at $\theta = 45^{\circ}$ in the case of conventional method whereas when regular integrals are used over non-conforming elements, not only the values are improved but there is no considerable variation in radial displacement for different values of θ. This is because the way boundary conditions are specified at corners A,B,C and D, using conventional method. We can specify either traction or displacement at A and B in x direction and at C and D in y direction, depending on the side under consideration and one generally chooses to prescribe displacement. No such problem arises due to ambiguity in the specification of boundary conditions in case of non-conforming elements being used. Results for stresses in both cases at $\theta = 45^{\circ}$ are compared in Figure.4.

Rigid Die on Elastic Half Plane

A load Q is applied through a rigid flat die as shown in Figure.5. The solution to the problem is discussed in [Saada,1974]. The pressure distribution on the die is given by,

$$p = \frac{Q}{\pi\sqrt{R^2 - x^2}} \tag{13}$$

where R = Half length of the die

x = distance from point of application of load $\underset{\sim}{Q}$ through the die, along OX direction.

Expression (13) gives $p = Q/\pi R$ as the pressure under point of application of load i.e. at x = 0 and infinite

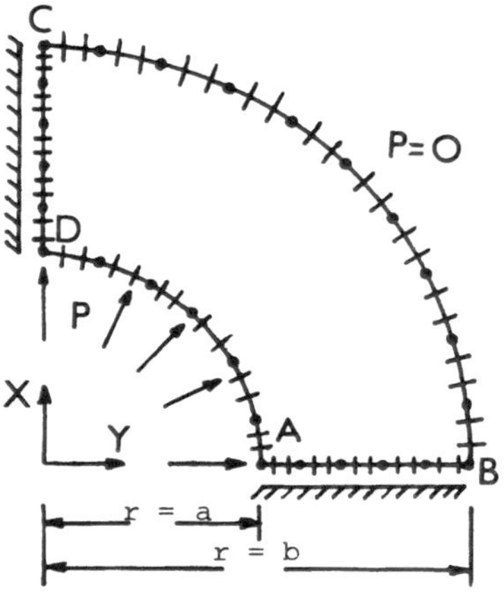

Fig. 3. Boundary Conditions and discretization of the cylinder using conventional and non-conforming linear elements

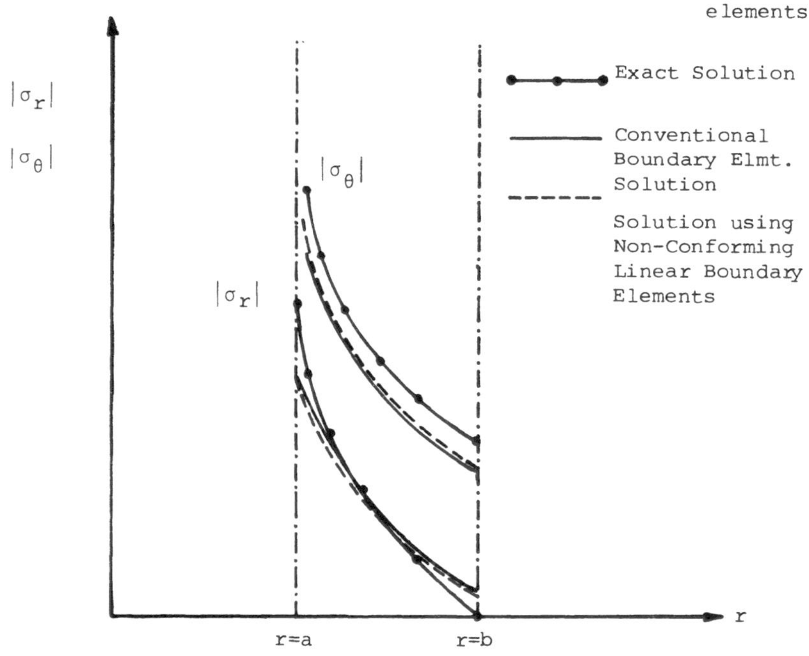

Fig. 4. Comparison of Exact Solution with Solutions obtained using Conventional and non-conforming linear elements

Table.1. Comparison of exact solution with the calculated values of radial displacement using conventional and non-conforming linear boundary elmts.

Function	Radius (cm)	Exact Sol. ($\times 10^{-4}$cm)	Theta (θ)	Conventional Lin.Elmts ($\times 10^{-4}$cm)	Non-Conf. Lin.Elmts. ($\times 10^{-4}$cm)
Radial Disp.	1.1	0.872	0°	0.801	0.825
			10°	0.802	0.825
			30°	0.810	0.828
			45°	0.828	0.830
	1.3	0.779	0°	0.712	0.745
			10°	0.716	0.745
			30°	0.724	0.746
			45°	0.740	0.750
	1.5	0.717	0°	0.657	0.689
			10°	0.660	0.690
			30°	0.667	0.695
			45°	0.681	0.696
	1.7	0.674	0°	0.620	0.651
			10°	0.621	0.651
			30°	0.627	0.652
			45°	0.640	0.653
	1.9	0.645	0°	0.595	0.618
			10°	0.598	0.618
			30°	0.601	0.622
			45°	0.612	0.625

pressure at x = R i.e. the edge of the die, which makes the solution singular at that point.

Due to symmetry only half of the die is considered and discretized as a part of a large rectangle (Figure.6.) such that the edges AB and BC are sufficiently far from the die to remain undeformed. The vertical displacement under the die is constant and taken as unity, whereas zero lateral displacement is prescribed at all points along OY. The boundary of the rectangle is divided into 40 linear elements taking a refined mesh along the die and its nearest boundaries. The problem was analysed using both the conventional and non-conforming linear elements. Figure.7., shows the pressure distribution along the die obtained for the two cases. Figure.8., displays the variation of horizontal and vertical displacements along ODC. The clarity of these results at the singular point D in case of non-conforming elements being used, is to be noted.

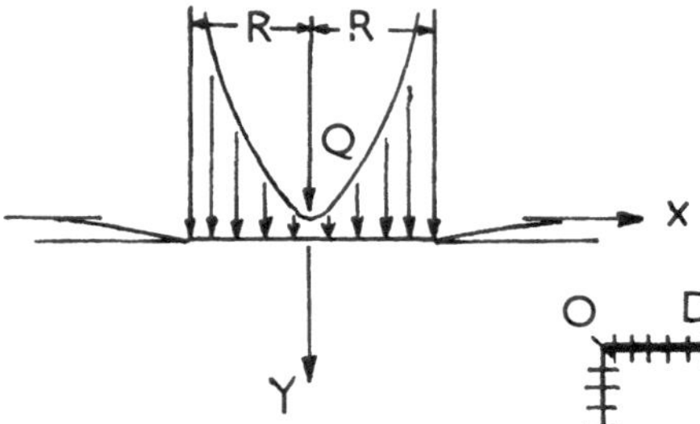

Fig.5. Rigid Die on Elastic Half Plane

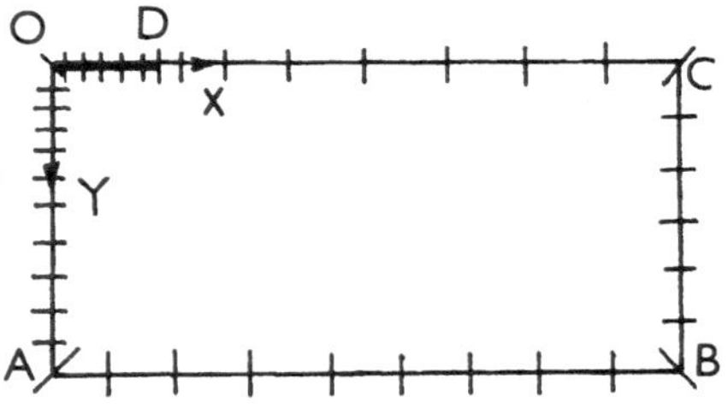

Fig.6. Boundary Element Discretization

Stress Concentration in a Plate under Uniform Tension due to a small Circular Hole

The plate with a circular hole in the center is subjected to a uniform tensile load as shown in Figure.9. The diameter of the hole is taken as less than one fifth of the width of the plate to produce high stress concentration at points XX. Only a quarter of the plate needs to be considered because of symmetry, and is subjected to boundary conditions shown in Figure.10.

The problem was solved by dividing the boundary into 36 linear elements and using conventional and regular boundary integral methods. In the latter case non-conforming elements were used with appropriate boundary conditions specified at the 'freedom points' of the element and not at the geometric

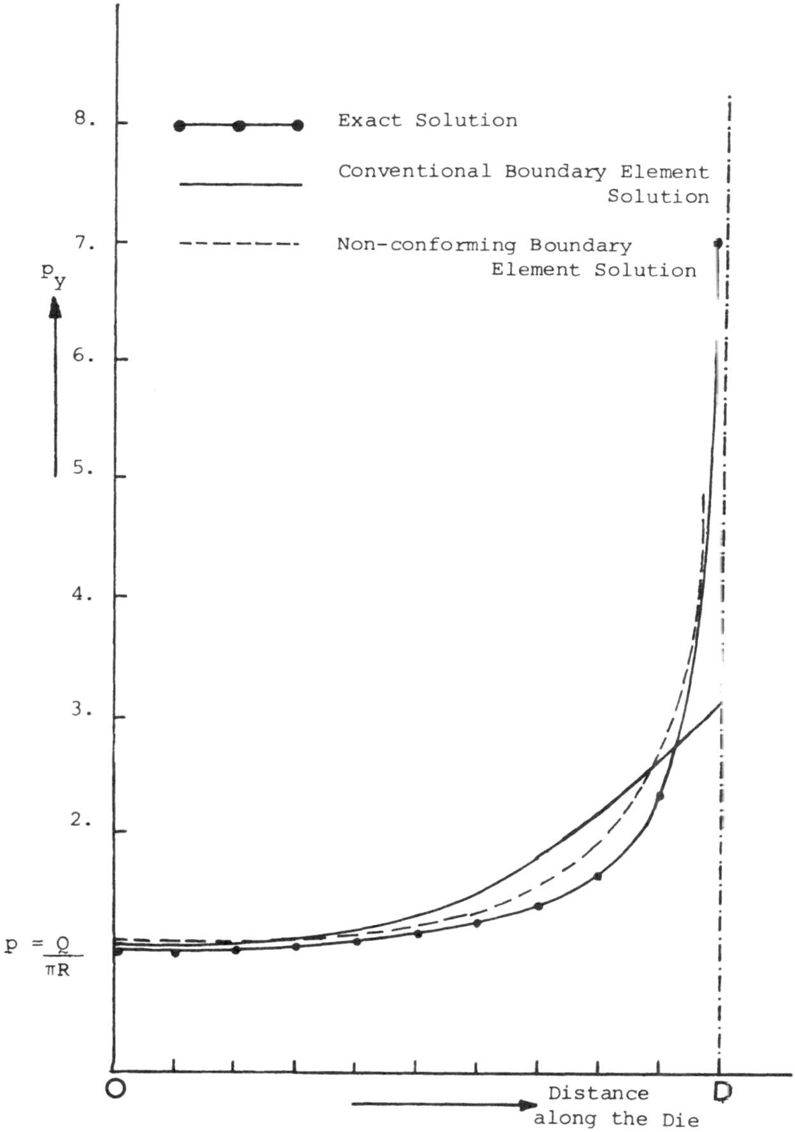

Fig.7. Pressure distribution along the die

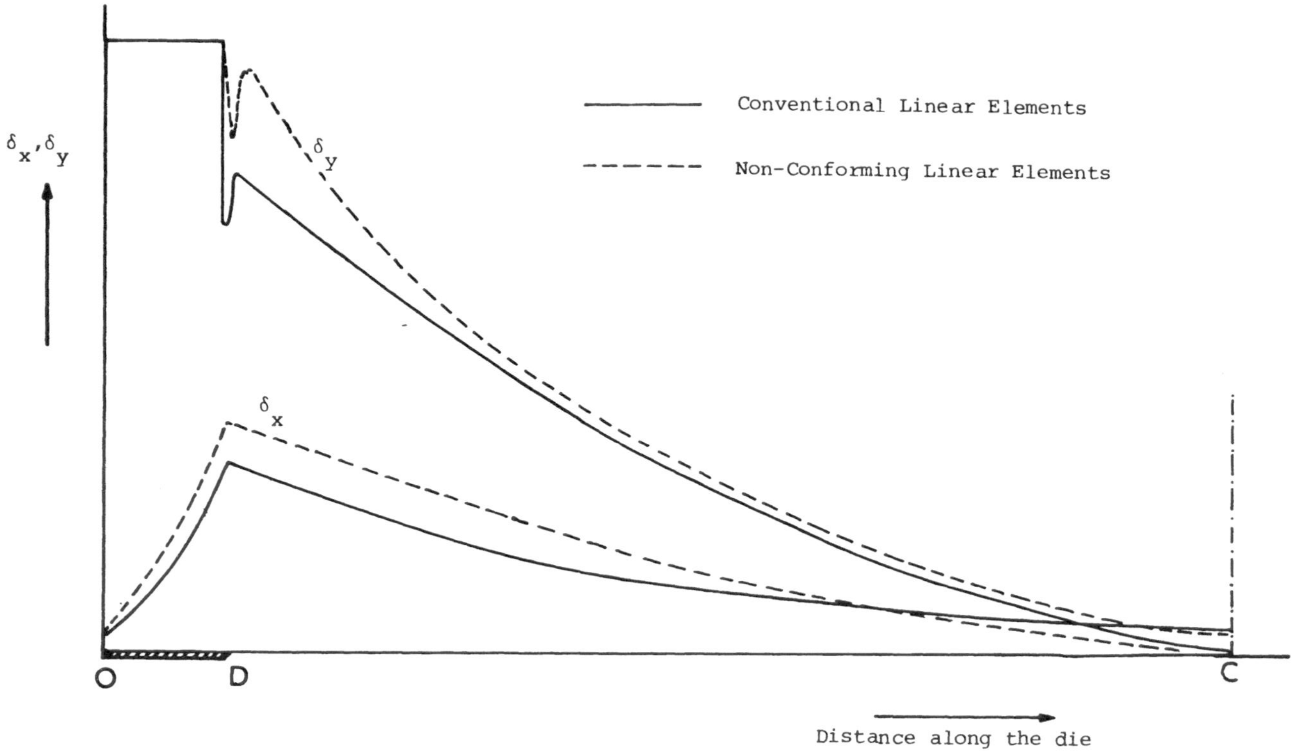

Fig. 8. Vertical and Lateral displacements along the die

nodes. The curves showing stress distribution along DE are plotted in Figure.11., for the two cases used.

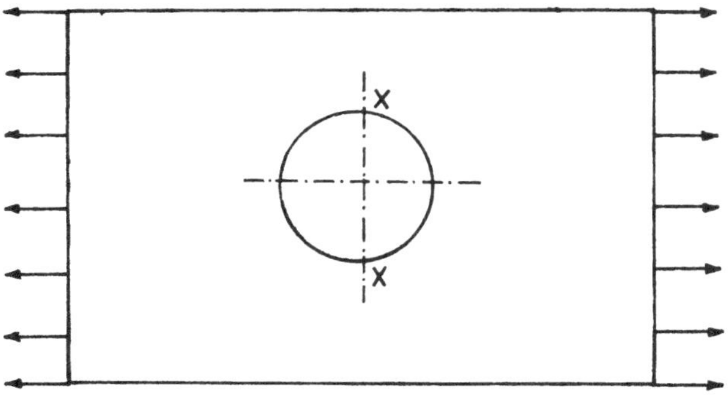

Fig. 9. Plate with a circular hole under uniform tension

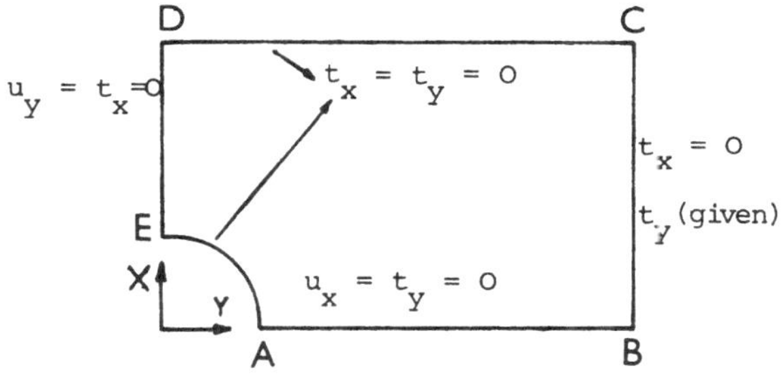

Fig.10. Boundary Conditions for quarter domain

DISCUSSION AND CONCLUSIONS

THE main objective in this study is to compare the behaviour of conforming (i.e. continuous across an element boundary) and non-conforming boundary elements. The linear element in two dimension was selected as the simplest case. Because of modelling difficulties at singular boundary points (corners) and at points where abrupt changes in boundary conditions occur it is expected that the solutions obtained, using conforming elements, will be poorer in the neighbourhood of these anomalous points than their counterparts. It is of interest to see if this degradation is local or more widespread. Although a reasonable quality of convergence was sought in each case analysed no attempt was made to assess mesh densities required for a solution of prescribed accuracy.

On examining the calculated displacements for the thick cylinder problem, Table 1, it is evident that the non-conforming elements give results which respect the axial symmetry

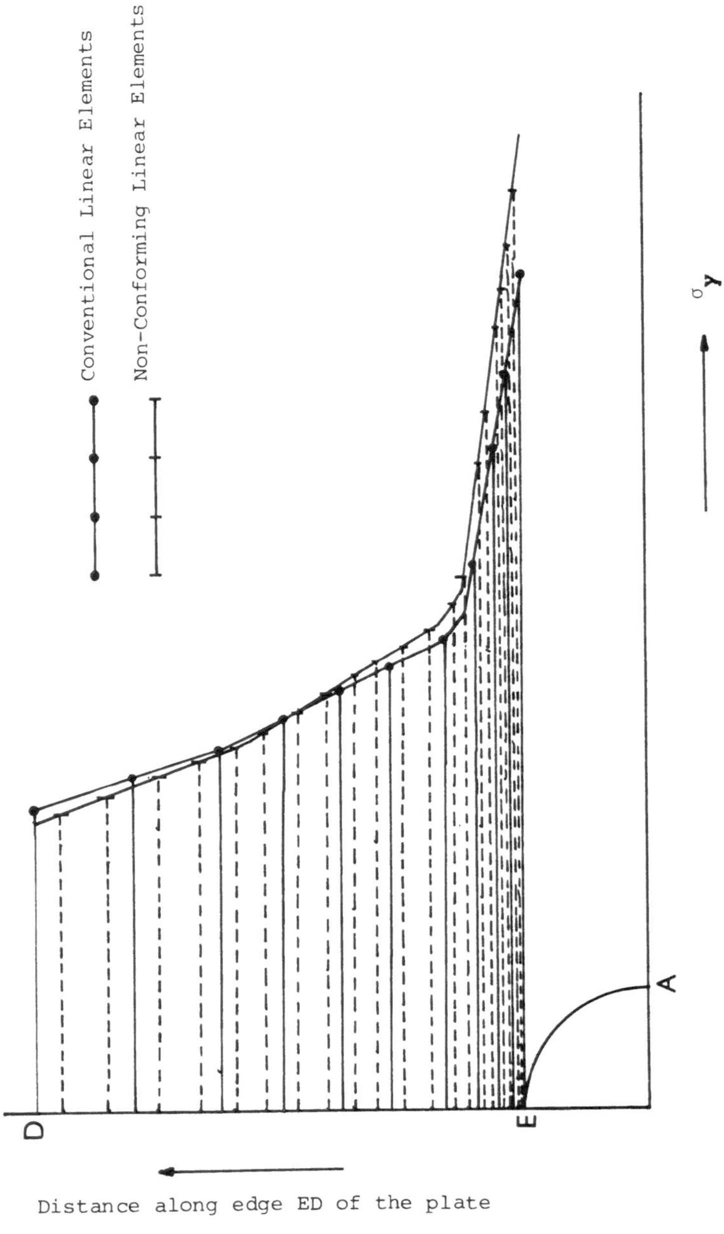

Fig. 11. Stress distribution along edge ED of the plate

to around 1% whereas the conventional elements show a pronounced degradation at the radial edge of the model which is felt even on the symmetry line (45°). The non-conforming results are about 5% low on the inner and outer edges with some improvement in the interior. The best conforming results; at 45°, are only marginally lower than their counterparts but there is a further degradation of around 5% as the radial boundary is approached. The degradation of the continous element results is due to the modelling problem at the points where the radial edges meet the inner and outer edges of the tube. On cutting the mesh at the corners and introducing two adjacent nodes, separated by 10% of element length, at each side of the corners the continuous elements gave better results, comparable with the non-conforming elements. Computed internal stresses, at 45°, are given in Fig.4, these again give similar low values with worst performance on the boundary.

The rigid die problem was selected as a severe test because of the known singular normal stress at the edge of the die with zero normal stress immediately beyond. The computed and known pressure distribution across the die are given in Fig.7. Away from the singular region both elements agree with the exact solution adequately, within 5%. At the edge of the die the continuous elements gave a normalized pressure 3.1 while the non-conforming elements gave 4.9 at the node nearest the edge. The exact value at that node is 7.1. Clearly the non-conforming results are substantially better. In Fig.8. are given the displacements at the surface under and away from the die. The horizontal motions are qualitatively satisfactory for both elements.

Neglecting the oscillation apparent at the edge of the die in both cases, the vertical motions are qualitatively satisfactory for the non-conforming elements but not for the continuous elements. A developing trough with a step at the die is not correct. At the time the analyses were performed a displacement solution was not known by the authors, however one was subsequently found [Sneddon,1951]. This shows that the oscillation at the die edge is a spurious numerical feature; this is not really surprising in view of the severe singularity. Otherwise it confirms the non-conforming results.

The perforated plate problem is a routine test. Here again, the non-conforming results are superior at a corner and give a satisfactory stress concentration factor.

In conclusion: It has already been observed that interelement continuity with boundary elements gives problems at geometric singularities, at interfaces between different load types and, in subdomain models, where several surfaces meet. In this paper it has been observed that interelement continuity is not necessary with boundary elements. In the test cases analyzed the continuous elements gave poorer results where the model was ambiguous, the degradation being felt well away from the anomaly. The non-conforming elements did not suffer this defect and gave results which were generally acceptable.

REFERENCES

Lachat, J.C. (1975) A Further Development of the Boundary Integral Technique for Elastostatics, Ph.D. Thesis, Southampton University.

Patterson, C. (1973) Sufficient Conditions for Convergence in the Finite Element Method for any Solution of Finite Energy. In: Mathematics of Finite Elements and Applications, Ed. J.R. Whiteman, Academic Press, London.

Patterson, C and Sheikh, M.A. (In print) Regular Boundary Integral Equations for Fluid Flow. In: Numerical Methods in Laminar and Turbulent Flow, Eds. C. Taylor and B.A. Schrefler, Pineridge Press, Swansea.

Saada, A.S. (1974) Elasticity: Theory and Applications, Pergamon Press, New York.

Sneddon, I.N. (1951) Fourier Transforms, Mcgraw Hill, New Jersey.

THE DISPLACEMENT DISCONTINUITY METHOD IN THREE DIMENSIONS

W. Scott Dunbar[1] and Don L. Anderson[2]

ABSTRACT

The displacement discontinuity method is a type of boundary integral method wherein the fundamental solution used is the static displacement due to a displacement discontinuity on a finite segment in an infinite or semi-infinite medium. The extension of the method to three-dimensional problems is presented together with some examples of its application. It is shown that the use of the Volterra integral to derive the fundamental solution not only clearly identifies the method as a standard boundary integral procedure with a different influence function, but allows the development of several useful extensions of the method in a straightforward manner, and also provides a natural method of linking the method to a finite element computer code.

INTRODUCTION

The displacement discontinuity method (DDM) is a type of boundary integral method wherein the fundamental solution used is the displacement due to a constant displacement discontinuity on a finite segment in an infinite or semi-infinite elastic medium. Boundary value problems in mechanics are solved by distributing a series of such displacement discontinuity segments over a boundary on which the displacement or stress is known. The solution is found by adjusting the magnitude of each displacement discontinuity to match the boundary conditions.

The DDM has been developed in two dimensions by Crouch [1973, 1976a,b]. Although it is generally applicable to most elastic boundary value problems, the method has mainly been applied to geomechanical problems such as the analysis of stresses and displacements near underground excavations. In this regard, one of the principal advantages of the method is its ability to model cracks and faults in a realistic fashion. A

[1]Senior Research Engineer, Weidlinger Associates, Menlo Park,CA
[2]Professor, Department of Civil Engineering, University of British Columbia, Vancouver, Canada

numerical advantage of the method is that there is no singularity in displacement at the point of application of the boundary force or stress. Rather, because of the assumption of a constant displacement discontinuity, stress singularities occur at the edges of the boundary segments.

This paper discusses the development of the DDM in three dimensions. The present formulation is for infinite media, but it will be shown that the extension to semi-infinite media is relatively easy. The primary motivation for this development is the potential application of a three-dimensional formulation to problems in mining engineering.

DERIVATION OF FUNDAMENTAL SOLUTIONS

The fundamental solutions required for the DDM are the displacements due to a constant displacement discontinuity over an arbitrarily-oriented segment in a three-dimensional elastic medium. The displacement discontinuities may be normal to the plane (mode 1) or parallel to the plane (modes 2 and 3).

Many authors have derived such equations. Much of this work has appeared in the geophysical literature [Steketee, 1958; Chinnery, 1961; Maruyama, 1964, 1966; Press, 1965; Mansinha and Smylie, 1971] and relates to faulting by shear failure. By means of a Galerkin vector fomulation Rongved [1957] and Rongved and Frasier [1958] derived the displacements due to constant displacement discontinuities as an arbitrarily-oriented rectangular segment in infinite and semi-infinite media. Berry and Sales [1962] derived the displacements in infinite and semi-infinite transversely isotropic media for a constant mode 1 discontinuity over a rectangular segment parallel to one axis. Chen [1964] completed the derivation for mode 2 and 3 discontinuities.

For reasons which will become apparent later, the derivation of the displacements due to an arbitrarily-oriented rectangular segment in infinite media will be done by means of the Volterra integral [Hirth and Lothe, 1968, p. 93] rather than a Galerkin vector formulation. The coordinate system used is shown in Fig. 1. The origin of the local coordinate system $l = (l_1, l_2, l_3)$ is at the center of the rectangular segment. The source coordinates c_2 and c_3 are coincident with l_2 and l_3 ; $c_1 = 0$. The dimensions of the segment are $-a \leq c_2 \leq a$, $-b \leq c_3 \leq b$. The displacements u' in the local coordinate system l are given by

$$u_i'(l) = \int_{-a}^{a} \int_{-b}^{b} d_j(c) \Big\{ \lambda U_i^{m,m}(l,c)\delta_{jk}$$

$$+ \mu\left[U_i^{j,k}(1,c) + U_i^{k,j}(1,c)\right]\Big\} n_k(c)dc_2dc_3 \qquad (1)$$

where

$d_j(c)$ = the displacement discontinuity in the $1_j = c_j$ direction at c corresponding to a mode j displacement discontuity;

$n_k(c)$ = the kth component of the unit normal vector to the segment at c;

$U_i^j(1,c)$ = the displacement in the 1_i direction due to a force in the 1_j direction at c; i.e., the Green's function of the medium; and

λ,μ = Lame's constant and the shear modulus, respectively.

The notation $U_i^{j,k}$ means the derivative of U_i^j with respect to c_k. Expressions for U may be found in Love [1944, p. 185], Mindlin and Cheng [1950] and Press [1965]. The displacement discontinuity d_j is reckoned positive if it acts in the positive c_j direction on the side of the segment with the positive normal $n(c)$. For the planar segment $n(c) = n = (1,0,0)$.

To derive the fundamental equations for the DDM, the displacement discontinuity d is assumed constant over the segment. The remaining part of the integrand in eq. (1) may then be integrated analytically without too much algebraic difficulty. The evaluation of the resulting function at the four corners of the segment was done within a computer program. The displacements in the local coordinate system may be transformed to the global system g by means of the matrix equation

$$u = C^t u'$$

where u is the vector of displacements in the global coordinate system and C is the transformation matrix which transforms a vector in the global coordinate system to one in the local coordinate system. The superscript t denotes the matrix transpose. Thus, the displacements at any point due to a displacement discontinuity d on a segment may be written

$$u = Ud = \sum_{j=1}^{3} U_{ij} d_j \qquad (2)$$

where U_{ij} is the displacement in the g_i direction due to the mode j displacement discontinuity d_j on the segment, i.e., an influence coefficient.

By differentiation, the strains may be derived in the local coordinate system and, by substitution into Hooke's law, the stresses. If S' is the stress tensor in the local coordinate system, the stress tensor, S, in the global coordinate system is given by

$$S = C^t S' C$$

Given S and the normal to any segment, the mutually orthogonal tractions, t, on the segment due to a displacement discontinuity, d, on another segment may be computed by the product

$$t = Td = \sum_{j=1}^{3} T_{ij} d_j \qquad (3)$$

where T_{ij} is the traction in the l_i direction due to the mode j displacement discontinuity d_j on the segment.

For stress boundary conditions, the DDM is formulated as follows: N displacement discontinuity segments are distributed over a boundary on which the stress is known. The traction vector $t(i)$ at the centroid of the ith boundary segment due to a discontinuity $d(j)$ at the jth boundary segment is given by the superposition

$$t(i) = \sum_{j=1}^{N} T(i,j) d(j)$$

where $T(i,j)$ is the 3x3 matrix relating tractions at the ith boundary segment to displacement discontinuities at the jth segment. By writing a similar equation for $1 \leq 1 \leq N$, a system of equations results which may be solved for $d(j)$, $1 \leq j \leq N$. Given the latter, the displacements and stresses may be computed anywhere in the medium using eqs. (2) and (3).

Displacement and mixed boundary conditions may also be handled by selecting the appropriate parts of eqs. (2) and (3) when constructing the system of equations. The details may be found in Crouch [1976a,b]. The problems dealt with in the examples have stress boundary conditions only.

DEVELOPMENT OF COMPUTER PROGRAM

Based on the latter formulation, a computer program entitled DDM3 was written in FORTRAN to solve three-dimensional stress boundary value problems. The central element of this program is 27 function subroutines representing the nine displacement derivatives for each of the three modes of displacement discontinuity. Each routine is the FORTRAN rendition of the result of integrating eq. (1) for a particular i and j and differentiating with respect to a local coordinate l_k. The displacement derivative is computed by evaluating the function at each of the four corners of the segment. Each function was tested by numerical differentiation of the appropriate one of the nine displacement functions. The displacement functions were tested by numerical integration of eq. (1).

The remainder of the program consists of subroutines to compute the matrix of influence coefficients, an equation solving algorithm, and several matrix and vector multiplication routines. A flow diagram is presented in Fig. 2. The program is not in production form. However, none of the examples presented in the next section required more than five seconds of execution time on the CDC 7600. The bulk of the computational effort lies in the calculation and decomposition of the matrix of influence coefficients.

One problem which immediately presented itself was the large size of the system of equations to be solved. Since there are three degrees of freedom on each segment, a system of 3N equations results. The system is typically fully populated and unsymmetric. If the number of segments, N, approaches 40, computer storage may be a problem and one must resort to out of core equation solvers for such systems. One such algorithm was developed by Hofmeister [1978] and was implemented in the program.

Other devices to reduce the size of the system of equations were also implemented. For some problems, there exist planes of symmetry. On either side of a plane of symmetry, the shear displacement discontinuities will be equal in magnitude but opposite in sign [Crouch, 1976b]. The normal displacement discontinuities will be equal. Given the geometry and location of the segments on one side of a plane of symmetry, the effect of the segments on the other side may be incorporated by superposition after multiplication by the appropriate sign. Thus, the size of the system of equations may be reduced by a factor of 2 for each plane of symmetry.

Depending on the boundary conditions, the size of the system of equations may be reduced. For example, purely hydrostatic stress boundary conditions do not induce shear displacement discontinuities. The members of the system of equations corresponding to such discontinuities may thus be eliminated, resulting in a significant saving of storage. This is particularly relevant to deep mining problems where the boundary conditions are almost hydrostatic.

By far, the most difficult problem is the visualization of the three-dimensional model and the definition of its location and geometry. The ideal solution would be to link the problem to an interactive graphics system where a light pen would describe the geometry on a screen. A possibly less expensive alternative would be to design a simple physical modeling system with more angular flexibility than TinkerToy or Meccano.

EXAMPLES

The computer program was extensively tested and checked against known analytical solutions for three- and two-dimensional problems.

The simplest available three-dimensional problem, which is amenable to an approximate discretization into rectangular segments, is that of a penny-shaped crack opened by a constant pressure parallel to the crack axis. The discretization is shown in Fig. 3a. Nineteen segments were used to model one-quarter of the crack. The remainder of the crack was modeled by symmetry conditions about the axes $g_2 = 0$ and $g_3 = 0$. Since the boundary condition is a normal pressure, there is only a mode 1 displacement discontinuity at the crack. The other two modes are zero. Consequently, the order of the system of equations was reduced to 19 instead of 19x3 = 57.

The radial and axial stresses along the g_1 axis are shown in Fig. 4a. They are compared with the analytical solution given by Sneddon [1946]. For this particular discretization, the agreement near the crack surface is not particularly good. As pointed out by Crouch [1976a, p. 307] in relation to another problem, this is due to the assumption of a constant displacement discontinuity over the segments. At larger distances from the crack, the agreement is within 1-2%, which is more than adequate for most geomechanical applications.

The two-dimensional problem of a circular hole in an infinite medium under a uniaxial stress at infinity was also analyzed by means of the three-dimensional DDM. The discretization is shown in Fig. 3b. In this case, the length of the segments in the g_3 direction was made very long in order to simulate the two-dimensional plane strain conditions. Fifty segments were used to discretize one-quarter of the hole, the remainder being modeled by symmetry conditions about the axes $g_1 = 0$ and $g_2 = 0$. To create a free surface at the boundary, the boundary tractions were set equal to the negative of the tractions applied at infinity. Since a plane strain condition is simulated, there are no mode 3 displacement discontinuities. The order of the system of equations was therefore reduced to 100 from 150, and could therefore be solved in core.

The numerical results along the g_1 axis are shown in Fig. 4b where they are compared with the analytical solution given in Jaeger and Cook [1976, p. 249]. The agreement in this case is excellent, even near the boundary of the hole. This latter problem was used as a more stringent test of the coordinate transformations used to calculate the stresses and tractions.

As an example of the type of geomechanical problem which may be solved by the three-dimensional DDM, the mining layout shown in Fig. 5 was modeled. Two ore seams, known as the East and West limbs, are to be mined by a series of 60 by 80 metre panels, each separated by a 20 metre pillar. For reasons of economic access to each limb, it is desirable to locate the mineshaft at the location shown. The problem is to determine values of displacements due to the mining operation along the shaft for use in design of the shaft.

Each panel was modeled by 32 segments as shown in Fig. 5. The initial stress on each panel was assumed to be hydrostatic and related to depth by

$$s_{11}(g_2) = s_{22}(g_2) = s_{33}(g_2) = 26g_2 \text{ kPa}$$

The Young's modulus and Poisson's ratio of the medium were assumed to be 60 GPa and 0.25, respectively. The coordinates of the points a,b,c,d and e shown in Fig. 5 are sufficient to define the location and geometry of the limbs and shaft. The mining operation was modeled by applying the negative of the initial (hydrostatic) tractions to each segment. This produced a free surface at the panels. Each of the panels shown in Fig. 5 was mined individually and the resulting displacements along the shaft were stored on a disk file. The displacements due to any sequence of panel mining were then calculated by superposition.

The horizontal (u_1 and u_3) displacements along the shaft are shown in Figs. 6a and 6b. After mining the panels shown on the East limb, the u_1 displacements are very small and positive. However, they become larger and change sign once the panels shown on the West limb are mined. The u_3 displacements are greater in magnitude than the u_1 displacements after mining the East limb and do not change appreciably after mining the West limb. This may be expected considering the geometry and sequence of mining. The vertical (u_2) displacements were relatively unaffected by the mining sequence.

One could continue mining the panels in any order and compute the resulting displacements. However, the displacements are nearly negligible and the induced stresses may be of more importance to design.

FUTURE DEVELOPMENTS

The use of the Volterra integral readily allows a number of developments of the theory which would be useful in applications. These are discussed below.

1. Extension to semi-infinite media

To derive the fundamental equations for the displacements and stresses in a semi-infinite medium, the Green's functions for such a medium would be substituted into eq. (1) and integrated, assuming a constant displacement discontinuity, as before. The resulting expresssions, which are additive to those of the infinite medium, could be incorporated into the function subroutines of the computer program in a straightforward manner.

2. Integration over general quadrilaterals

With rectangular segments, it is difficult to model surfaces whose boundaries are not at right angles to one another. An example is a boundary segment whose edge lies along the line bd in Fig. 5. To model such a boundary, it would be necessary to integrate eq. (1) over a general quadrilateral or a triangle.

By means of suitable transformations, a general quadrilateral may be transformed into a square region over which the integration may be performed. Consider the quadrilateral shown in Fig. 7 whose area is denoted R'. The transformation to R' from the standard region R is

$$c_2 = \sum_{i=1}^{4} h_i(x_2,x_3)c_{2i} = p(x_2,x_3) \qquad c_3 = \sum_{i=1}^{4} h_i(x_2,x_3)c_{3i} = q(x_2,x_3) \tag{4}$$

where c_{2i} and c_{3i} are the local coordinates of the corners of the quadrilateral and the functions h_i are given by

$$h_1 = \tfrac{1}{4}(1+x_2)\ (1+x_3) \qquad h_2 = \tfrac{1}{4}(1-x_2)\ (1+x_3)$$

$$h_3 = \tfrac{1}{4}(1-x_2)\ (1-x_3) \qquad h_4 = \tfrac{1}{4}(1+x_2)\ (1-x_3)$$

[Bathe and Wilson, 1976, p. 131]. An integral of a function f over R' is then given by

$$\iint_{R'} f(c_2,c_3)dc_2dc_3 = \iint_{R} f\left[p(x_2,x_3),\ q(x_2,x_3)\right]\ J(x_2,x_3)\ dx_2dx_3$$

where $J(x_2,x_3)$ is the Jacobian of the transformation (4). Substitution of the above transformation in eq. (1) results in a very complicated integral which must be evaluated numerically. However, the use of Gaussian integration of order two along the x_2 and x_3 axes requires a total of four function evaluations, which is the same as that required to evaluate the displacement or its derivative due to a rectangular segment. Consequently, the time required to evaluate a displacement or its derivative due to a general quadrilateral segment should not be much greater than that required for a rectangular segment. An experiment was carried out to test this and the times required were nearly equal. Order 2 Gaussian integration is probably not sufficient for complete accuracy, considering the order of the function to be evaluated (which is approximately six), but is likely to be sufficient for geomechanical problems. This is currently under study.

3. Representation of the displacement discontinuity on segment

In some applications, such as those involving fracture, it may be desirable to represent the displacement discontinuity, d , by some function other than a constant. A suitable representation would avoid stress singularities at the edges of a segment. If the coordinate transformation functions, h , used in the last section are also used to interpolate d in terms of its values at the corners of a segment when integrating eq. (1), then a linear isoparametric boundary segment would result. Compatibility at the edges between segments would be guaranteed by such a formulation [Bathe and Wilson, 1976, p. 144]. However, it is possible that the amount of numerical integration required may be prohibitively expensive. Lachat and Watson [1976] discussed a successful implementation of such a procedure in a conventional boundary integral code.

4. Linkage with finite elements

In the finite element analysis of geomechanical problems, it is often necessary to model fractures or, more commonly, boundaries at infinity. Conversely, there may be portions of a model which are not amenable to analysis by the DDM because of inhomogeneities or inelastic behavior. The ideal solution to this problem is to find a method of coupling a DDM analysis to a finite element analysis. A method of accomplishing this was outlined by Zienkiewicz [1977] and Zienkiewicz et al [1977] in terms of conventional boundary integral procedures. Their objectives were to derive a symmetric stiffness matrix of a boundary-type element which could be assembled into the conventional finite element stiffness matrix in a standard fashion. The following outlines a very similar development for the DDM.

Referring to Fig. 8, the potential energy of the region H may be written

$$\frac{1}{2}\int_B u^t t \; dB \;-\; \int_B \bar{u}^t t \; dB \qquad (5)$$

where u and t are the displacements due to the displacement discontinuity segments along the boundary B and $\bar{u}$ are the displacements along B due to the finite element mesh. The latter displacements are given in terms of the nodal displacements w by the finite element shape function N :

$$\bar{u} = Nw$$

Substituting this and eqs. (2) and (3) into eq. (5) and minimizing with respect to d results in

$$\frac{1}{2}\left[\int_B U^t T \, dB + \int_B T^t U \, dB\right] d - \left[\int_B T^t N \, dB\right] w$$

or

$$K_d d + K_{dw} w = 0 \qquad (6)$$

Minimizing with respect to w results in

$$K_{dw}^t \, d = 0 \qquad (7)$$

Eliminating d between eqs. (6) and (7) yields the required stiffness matrix for the region H :

$$K_H = - K_{dw}^t \, K_d^{-1} \, K_{dw}$$

which is symmetric and can be directly assembled into the finite element stiffness matrix. However, notice that as the matrix K_{dw} is calculated by integration along the boundary, contributions from each boundary node will be introduced. This means that the bandwidth of K_H will generally be large and will increase the bandwidth of the finite element stiffness matrix.

To ensure continuity along the interface B , the interpolation of the vector d of displacement discontinuities should be the same as that used in the adjacent finite elements. If the finite elements are of the isoparametric type, the isoparametric boundary segment discussed in the last section would be a natural choice of interpolation and would likely lead to a compact computer code as well as a versatile method of modeling different boundary shapes.

A program to implement the above element in a three-dimensional finite element code is currently being developed.

CONCLUSIONS

The development of the DDM for the analysis of three-dimensional elastic media is relatively straightforward if the Volterra integral is used to derive the fundamental solution. The use of the Volterra integral permits the identification of the DDM as a boundary integral procedure with a different influence function. It also allows:

1) a relatively easier means of extending the technique to semi-infinite media,

2) more flexible representation of boundary surfaces in three dimensions by means of integration over general quadrilaterals, and

3) a natural linkage with finite element analysis.

A working computer code has been developed and tested. Two major difficulties in the computer implementation of the technique are the large size of the system of equations to be solved and the definition of the location and geometry of three-dimensional models. The large size of the system of equations can be decreased by either the use of symmetry or the elimination of non-existent modes of displacement discontinuity, or both. Alternatively, the equations can be solved by an out of core solution algorithm. The ability to visualize and prepare the input data for three-dimensional models would be greatly enhanced by the use of an interactive graphics system.

ACKNOWLEDGMENTS

The authors would like to thank Dr. Ken Matthews and Mr. Byron Stewart of Golder Associates in Vancouver, B.C., for suggesting the mining problem shown in this paper. The first author would like to gratefully acknowledge the assistance provided by Dr. Tom Hughes of Stanford University with regard to the problem of linking the DDM with finite elements.

REFERENCES

BATHE, K.J. and E.L. WILSON [1976] Numerical Methods in Finite Element Analysis, Prentice-Hall, Inc., Englewood Cliffs, NJ.

BERRY, D.S. and T.W. SALES [1962] An Elastic Treatment of Ground Movement due to Mining - III. Three Dimensional Problem, Transversely Isotropic Ground, J. Mech. Phys. Solids, 10, 73-83.

CHEN, W.T. [1964] Displacement Discontinuity Over a Transversely Isotropic Halfspace, IBM J. Res. and Dev., 8, 435-442.

CHINNERY, M.A. [1961] Deformation of the Ground Around Surface Faults, Bull. Seis. Soc. Amer., 51, 355-372.

CROUCH, S.L. [1973] Two Dimensional Analysis of Near Surface, Single Seam Extraction, Int. J. Rock Mech. Min. Sci. & Geomech. Abstr., 10, 85-96.

__________ [1976a] Solution of Plane Elasticity Problems by the Displacement Discontinuity Method, Int. J. Num. Meth. Eng., 10, 301-343.

__________ [1976b] Analysis of Stresses and Displacements Around Underground Excavations: An Application of the Displacement Discontinuity Method, University of Minnesota Geomechanics Report, Dept. of Civil and Mineral Engineering, University of Minnesota.

HIRTH, J.P. and J. LOTHE [1968] Theory of Dislocations, McGraw-Hill, Inc., New York.

HOFMEISTER, L.D. [1978] An Out of Core Equation Solver for Full, Unsymmetric Matrices, Int. J. Num. Meth. Eng., 12, 721-731.

JAEGER, J.C. and N.G.W. COOK [1976] Fundamental of Rock Mechanics, John Wiley & Sons, Inc., New York.

LACHAT, J.C. and J.O. WATSON [1976] Effective Numerical Treatment of Boundary Integral Equations: A Formulation for Three Dimensional Elasto-Statics, Int. J. Num. Meth. Eng., 10, 991-1006.

LOVE, A.E.H. [1944] A Treatise on the Mathematical Theory of Elasticity, Dover Publications, New York.

MANSINHA, L. and D.E. SMYLIE [1971] The Displacement Fields of Inclined Faults, Bull. Seis. Soc. Amer., 61, 1433-1440.

MARUYAMA, T. [1964] Statical Elastic Dislocations in an Infinite and Semi-Infinite Medium, Bull. Earthq. Res. Inst., 42, 289-368.

__________ [1966] On Two Dimensional Elastic Dislocations in an Infinite and Semi-Infinite Medium, Bull. Earthq. Res. Inst., 44, 811-871.

MINDLIN, R. and D. CHENG [1950] Nuclei of Strain in the Semi-Infinite Solid, J. Appl. Phys., 21, 926-930.

PRESS, F. [1965] Displacements, Strains and Tilts at Teleseismic Distances, J. Geophys. Res., 70, 2395-2412.

RONGVED, L. [1957] Dislocation Over a Bounded Plane Area in an Infinite Solid, J. Appl. Mech., 79, 252-254.

__________ and J.T. FRASIER [1958] Displacement Discontinuity in the Elastic Half-Space, J. Appl. Mech., 25, 125-128.

SNEDDON, I.N. [1946] The Distribution of Stress in the Neighbourhood of a Crack in an Elastic Solid, Proc. Roy. Soc. A., 187, 229-260.

STEKETEE, J.A. [1958] Some Geophysical Applications of the Elasticity Theory of Dislocations, Can. J. Phys., 36, 1158-1198.

ZIENKIEWICZ, O.C. [1977] The Finite Element Method, 3rd ed., McGraw-Hill, New York.

________, D.W. KELLY and P. BETTESS [1977] The Coupling of the Finite Element Method and Boundary Solution Procedures, Int. J. Num. Meth. Eng., 11, 355-375.

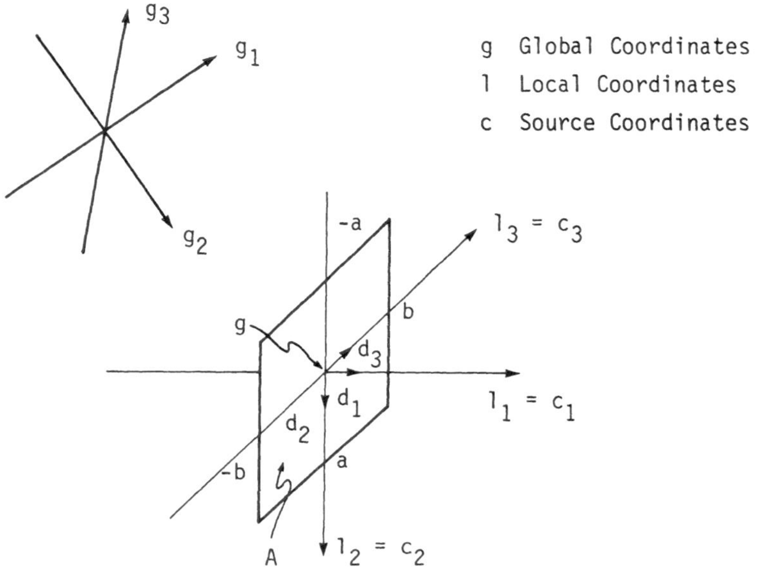

Figure 1. Global and local coordinate systems of dislocation segment

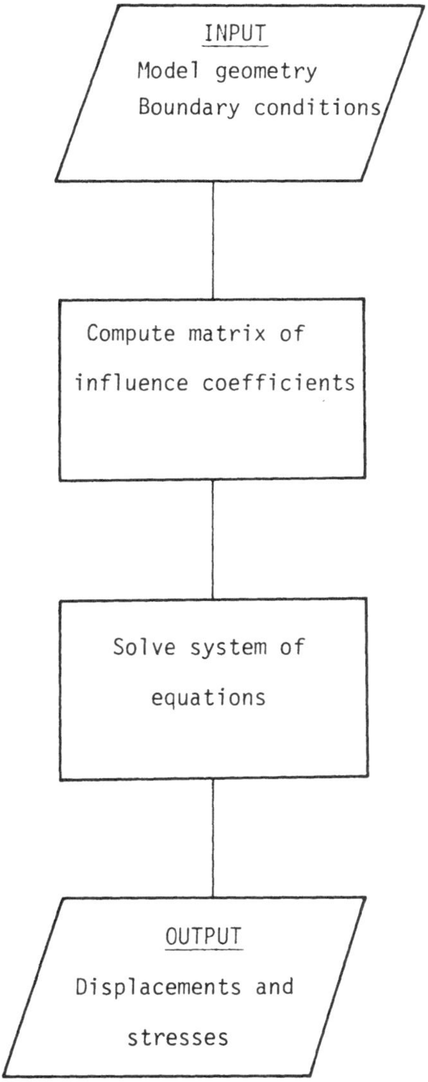

Figure 2. Organization of program DDM3

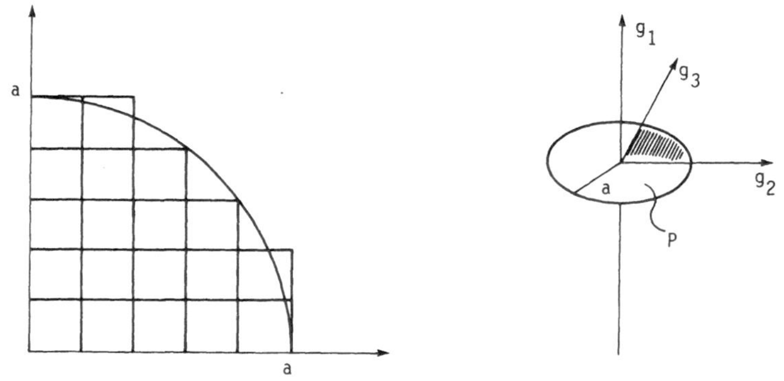

Figure 3a. Discretization of penny-shaped crack

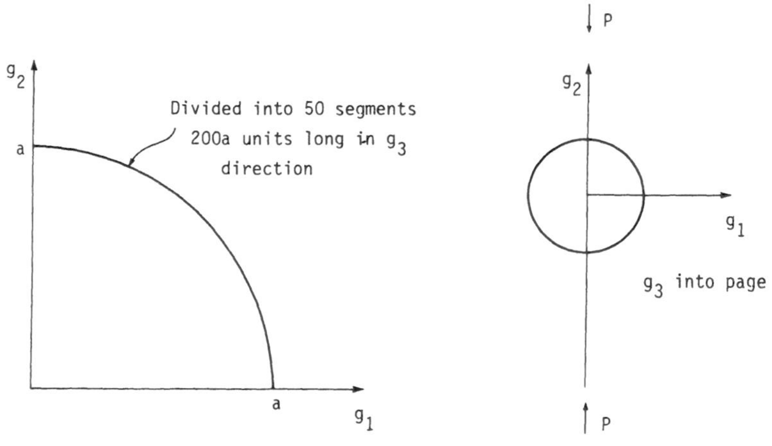

Figure 3b. Discretization of circular hole in infinite medium two-dimensional

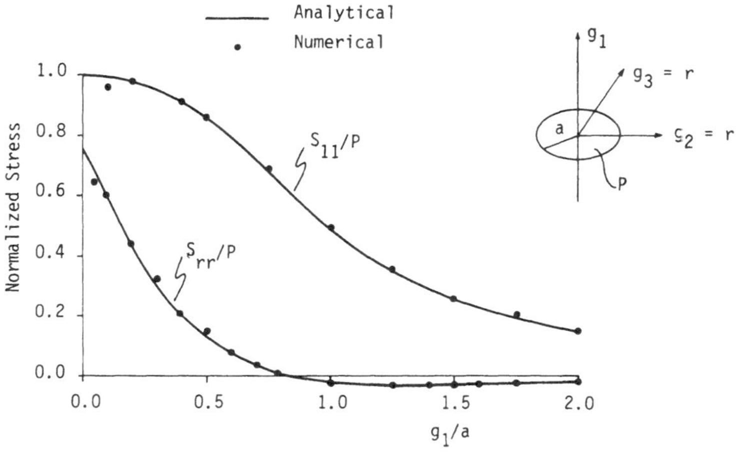

Figure 4a. Analysis of penny shaped crack

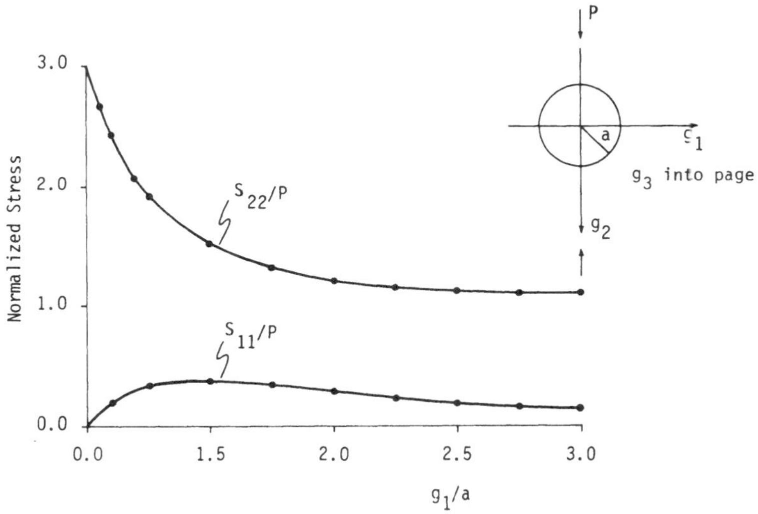

Figure 4b. Two-dimensionalized analysis of circular hole in infinite media

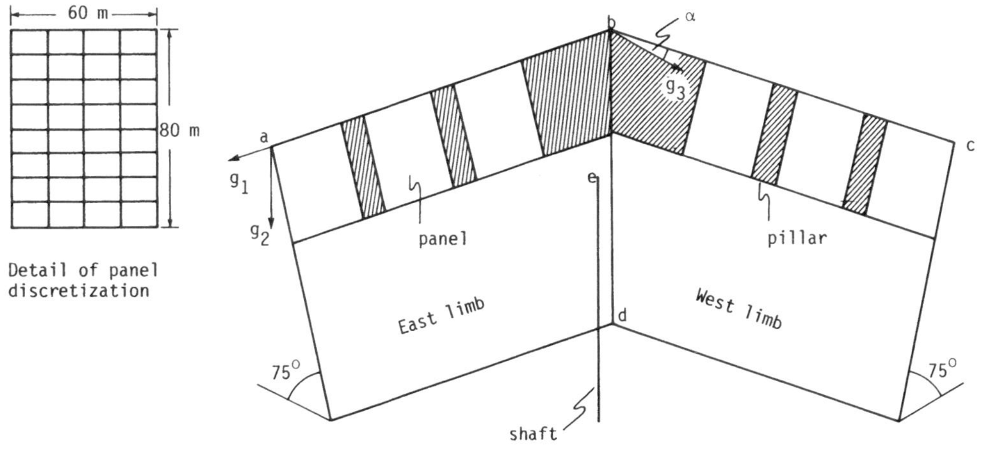

a: (300, 1000, 0)

b: (0, 1000, 0)

c: $(-200\tan 15^{\circ},\ 1000{+}200\tan^2 15^{\circ},\ 300\cos\alpha)$ $\quad \alpha = \sin^{-1}\left(\frac{2\tan 15^{\circ}}{3}\right)$

d: $(200\tan 15^{\circ},\ 1200,\ 200\tan 15^{\circ})$

e: (150, 1000, 150)

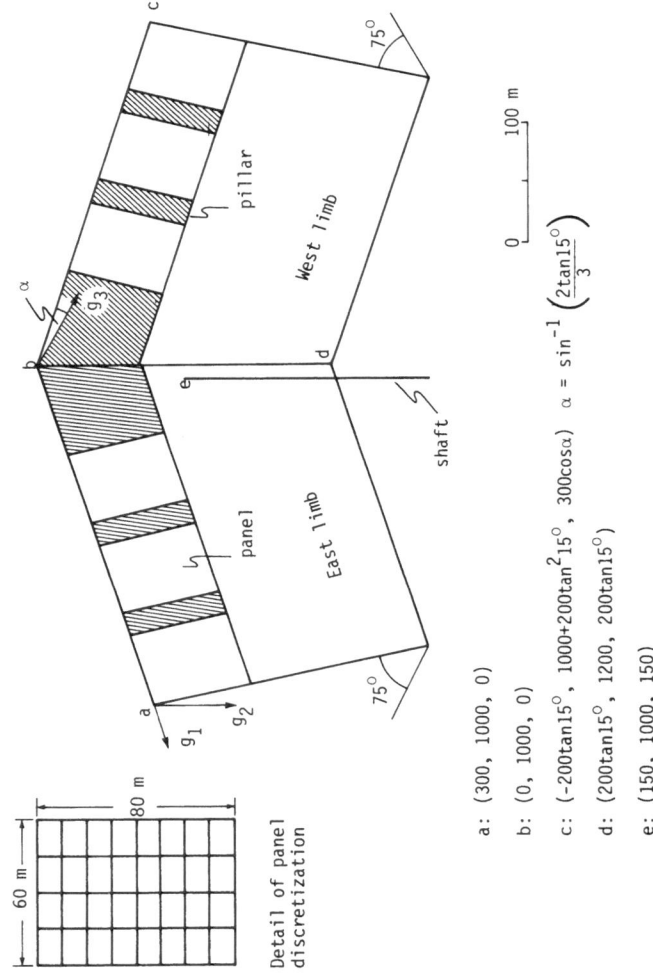

Figure 5. Orebody configuration and mining plan

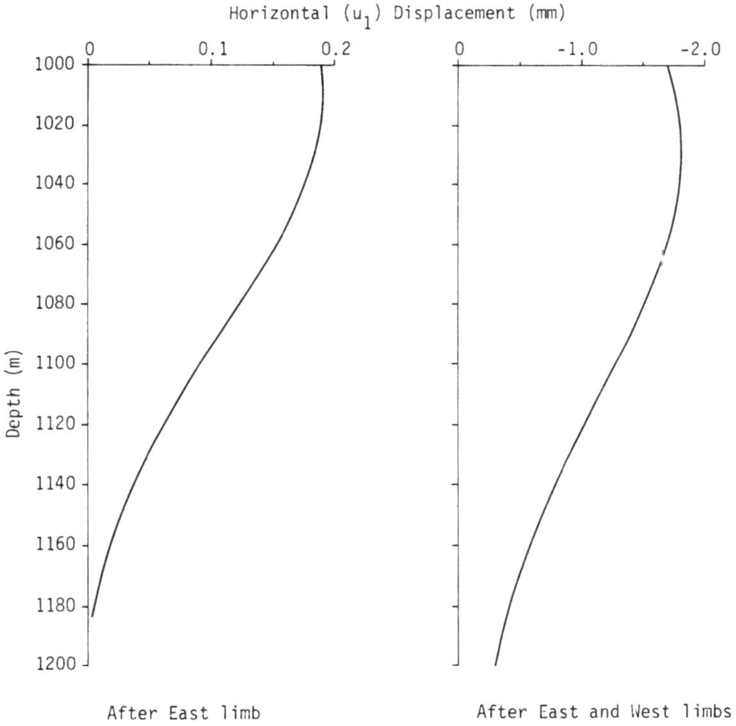

Figure 6a. Horizontal (u_1) displacements of shaft due to mining sequence shown in Figure 5

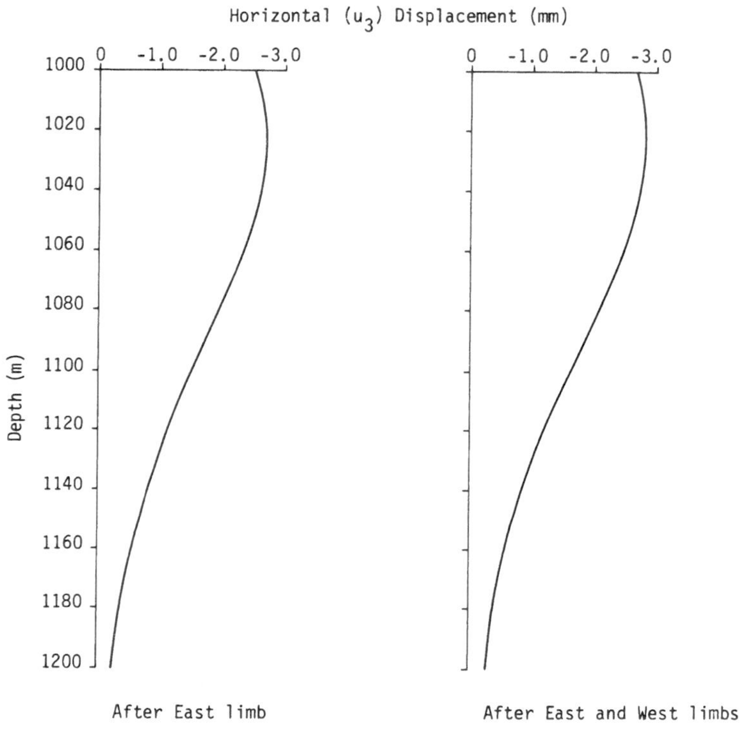

Figure 6b. Horizontal (u_3) displacements of shaft due to mining sequence shown in Figure 5

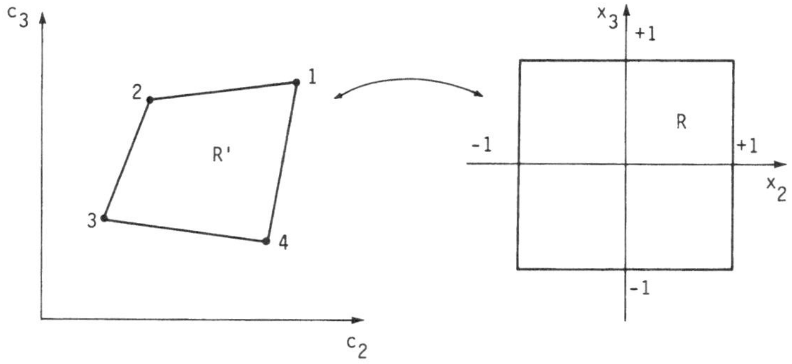

Figure 7. Transformation of general quadrilateral to a standard region

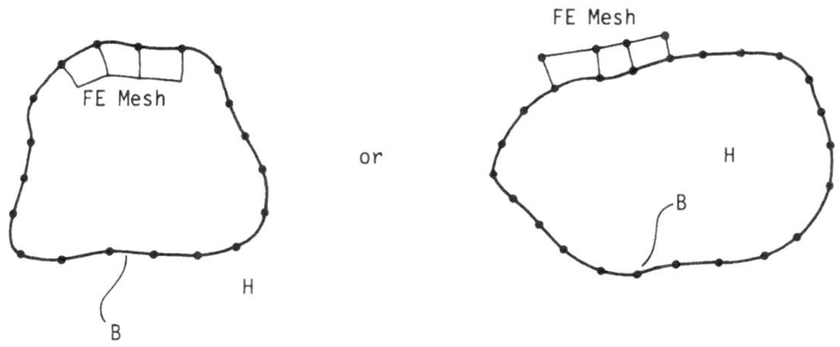

Figure 8. Linkage of DDM with finite element method

A UNIFIED SECOND ORDER BOUNDARY ELEMENT METHOD FOR STRUCTURES ANALYSIS

Michel Dubois
Research Engineer

CETIM - France

ABSTRACT

The classical modelizations used in structures analysis fail for singular situations giving high gradients. All numerical methods lead then to inaccurate or instable results, due to the geometrical and/or loading singularities.

High gradients zones (even very localized) are precisely damaging areas, especially responsible for fatigue collapses. Therefore, it seems particularly interesting to introduce a better modelization in order to eliminate the mentioned difficulties, and insuring by the same way a complete compatibility in complex structures analysis.

From the mathematical point of view, we are lead to construct a second order modelization with free rotations : kinematics of continua are defined by the first gradient of displacements (classical strains) and, further, by second order terms, homogeneous to second gradient components. In addition to classical (symmetrical) stress, we have to take into account skew-symmetric components and couple-stress which are second order terms. The second order quantities are negligible compared to the first order ones in low gradients zones, but they cannot be omitted in singular high gradients zones.

The main advantage of the new modelization is the existence of a common variables system used for all kinds of structures: three dimensional thick structures, shells, beams .

From the numerical point of view, the existence of a common variational principle gives after discretization a common system of algebraic equations, which is automatically reduced in function of geometrical data and in function of a test related to the mesh fineness, according to singular zones.

For the isotropic elastostatic case presented above, the four additional elasticity coefficients are defined in terms of the two classical ones and in function of a mesh size parameter, in order to recover the classical formulation in regular zones.

By using distribution's theory, the field equations are transformed into two integral equations relating displacements, rotations, stress and couple-stress. Taking limits at collocation points on these relations, we obtain the boundary integral equations to be numerically solved.

The integral equations kernels are the elementary solutions corresponding to unit concentrated forces and couples. These elementary solutions are calculated by FOURIER's transform and HORMANDER's method. The numerical treatment is then the same as used for standard B.E. programs of CETIM.

NOTATION

Let V be a (normed) vectorial space of $\vec{v}$ admissible elements defined on a S-structure /1/ . We introduce a bilinear symmetrical positive definite and continuous application of $V \times V$ in $\mathbb{R}$. If $\vec{v}_1$, $\vec{v}_2 \in V$, if V_I is a definition subdomain of V with boundary S_I (I = 3, 2, 1 according to the subdomain is three, two or one dimensional), we designate as $< \vec{v}_1 , \vec{v}_2 >_{V_I}$ and $< \vec{v}_1 , \vec{v}_2 >_{S_I}$ the real numbers that the bilinear application associates to $\vec{v}_1$, $\vec{v}_2$ on V_I and on S_I.

If I = 3, V_3 is a volumic subdomain and S_3 is the boundary surface of V_3 ; the bilinear application is then defined by :

$$< \vec{v}_1, \vec{v}_2 >_{V_3} = \int_{V_3} \vec{v}_1 \cdot \vec{v}_2 \, dV_3, \quad < \vec{v}_1 , \vec{v}_2 >_{S_3} = \int_{S_3} \vec{v}_1 \cdot \vec{v}_2 dS_3 \tag{1}$$

where $\vec{v}_1$, $\vec{v}_2$ is the scalar product, dV_3 is the differential elementary volume, dS_3 is the differential elementary area.

If I = 2, V_2 is the reference surface of a shell and S_2 is the boundary curve of V_2 ; the bilinear application is then defined by :

$$< \vec{v}_1, \vec{v}_2 >_{V_2} = \int_{V_2} \vec{v}_1 \cdot \vec{v}_2 dV_2, \quad < \vec{v}_1, \vec{v}_2 >_{S_2} = \int_{S_2} \vec{v}_1 \cdot \vec{v}_2 \, dS_2 \tag{2}$$

where dV_2 is the differential elementary area on the reference surface and dS_2 is the differential elementary arc on the boundary curve.

If I = 1, V_1 is the reference fibre of a beam, and S_1 represents the boundary points of V_1 ; the bilinear application is then defined by :

$$< \vec{v}_1, \vec{v}_2 >_{V_1} = \int_{V_1} \vec{v}_1 \cdot \vec{v}_2 dV_1, \quad < \vec{v}_1, \vec{v}_2 >_{S_1} = \vec{v}_1 \cdot \vec{v}_2 \Big|_{S_1} \tag{3}$$

where dV_1 is the differential elementary arc and $\vec{v}_1 \cdot \vec{v}_2 \Big|_{S_1}$ designates the value of the scalar product to be taken at boundary points S_1 .

Assume S'_I and S''_I be the complementary parts of S_I where kinematical conditions and efforts are respectively prescribed.

We use a curvilinear system of coordinates ζ^i (i = 1,2,3) which are intrinsic coordinates for the numerical discretization. We note $\vec{x}$ the position vector on V_I with regard to a global cartesian reference frame and $\vec{g}_i$ are the basis-vectors of the local natural reference frame:

$$\vec{g}_i = \partial_i \vec{x} \tag{4}$$

In a general way, ∂_i is the derivative with respect to ζ_i, so that :

$$\partial_i \vec{v} = v^j \mid_i \vec{g}_j \tag{5}$$

where $v^j\mid i$ is the covariant derivate. At last, $\vec{n}$ will be the exterior normal on the boundary S_I of V_I.

GENERALIZED MODELIZATIONS

The kinematical quantities are defined on each substructure V_I by displacements ${}^I\vec{u}$ and free rotations ${}^I\vec{\theta}$ as by generalized strains ${}^I\vec{\varepsilon}_i$ and ${}^I\vec{k}_i$.

The efforts are defined by stresses and couple stresses ${}^I\vec{\sigma}^i$ and ${}^I\vec{\mu}^i$. Prescribed efforts are forces and moments densities ${}^I\underline{\vec{f}}$ and ${}^I\underline{\vec{c}}$ given on V_I as forces and couples ${}^I\underline{\vec{\sigma}}$ and ${}^I\underline{\vec{\mu}}$ given on S''_I

- If I = 3, the above fields are volumic distributions on V_3. Indexes i take then values 1,2,3. The {} symmetrical components of ${}^3\vec{\varepsilon}_i$ in the natural reference frame reduce to classical strain ε_{ij} :

$$\{ {}^3\varepsilon_{ij} \} = \varepsilon_{ij} \quad \text{with } i, j = 1,2,3 \tag{6}$$

The corresponding symmetrical components of ${}^3\vec{\sigma}^i$ reduce to classical stress σ^{ij} :

$$\{ {}^3\sigma^{ij} \} = \sigma^{ij} \quad \text{with } i, j = 1,2,3 \tag{7}$$

But the generalized modelization involves skew-symmetrical components of ${}^3\vec{\varepsilon}_i$, ${}^3\vec{\sigma}^i$ and components of ${}^3\vec{k}_i$, ${}^3\vec{\mu}^i$ which do not appear in classical theory.

- If I = 2, the fields are surface distributions on reference surfaces V_2 of shells. V_2 being defined by $\zeta^3 = 0$ indexes i take values $\alpha = 1,2$. The $^2\vec{\varepsilon}_\alpha$ are membrane strains $\vec{E}\alpha$ and $^2\vec{k}\alpha$ are strains $\vec{K}_\alpha$ due to shells curvatures changes :

$$^2\vec{\varepsilon}_\alpha = \vec{E}_\alpha \qquad ^2\vec{k}_\alpha = \vec{K}_\alpha \qquad \alpha = 1,2 \tag{8}$$

Components of $^2\vec{\sigma}^\alpha$ along $\vec{g}_\beta = \vec{g}_\beta\ (\zeta^1, \zeta^2, \zeta^3 = 0)$ with $\beta = 1,2$ are membrane efforts $N^{\alpha\beta}$; components of $^2\vec{\sigma}^\alpha$ along $\vec{g}_3$ orthogonal to $\vec{g}_1$, $\vec{g}_2$ represent transversal sharing forces $N^{\alpha 3}$:

$$^2\sigma^{\alpha\beta} = N^{\alpha\beta}, \qquad ^2\sigma^{\alpha 3} = N^{\alpha 3} \quad \text{with } \alpha,\beta = 1,2 \tag{9}$$

Components of $^2\vec{\mu}^\alpha$ along $\vec{g}_\beta$ are bending moments $M^{\alpha\beta}$ and components of $^2\vec{\mu}^\alpha$ along $\vec{g}_3$ represent twisting moments $M^{\alpha 3}$:

$$^2\mu^{\alpha\beta} = M^{\alpha\beta} \qquad ^2\mu^{\alpha 3} = M^{\alpha 3} \quad \text{with } \alpha,\beta = 1,2 \tag{10}$$

- If I = 1, the fields are linear distributions on reference fibres V_1 of beams. Indexes i have then the single value 1, where we define V_1 by $\zeta^2 = \zeta^3 = 0$ and get $\vec{g}_1 = \vec{g}_2\ (\zeta^1, \zeta^2 = \zeta^3 = 0)$ with $\vec{g}_2$, $\vec{g}_3$ orthogonal to $\vec{g}_1$.

The $^1\vec{\varepsilon}_1$ and $^1\vec{k}_1$ are then reduction elements $\vec{\mathcal{E}}$ and $\vec{\mathcal{K}}$ of strain torsor at ζ^1 :

$$^1\vec{\varepsilon}_1 = \vec{\mathcal{E}} \quad , \quad ^1\vec{k}_1 = \vec{\mathcal{K}} \tag{11}$$

while $^1\vec{\sigma}^1$ and $^1\vec{\mu}^1$ are reduction elements $\vec{N}$ and $\vec{M}$ of contact effort torsor at ζ^1 :

$$^1\vec{\sigma}^1 = N \qquad , \qquad ^1\vec{\mu}^1 = \vec{M} \tag{12}$$

Vector $(^1\vec{\sigma}^1 \cdot \vec{g}_1)\ \vec{g}_1$ is a traction or compression at ζ^1 according to $^1\vec{\sigma}^1.\vec{g}_1$ is positive or negative ;

$^1\vec{\sigma}^1 - (^1\vec{\sigma}^1.\vec{g}_1)\ \vec{g}_1$ is the shearing force ;

$(^1\vec{\mu}^1 .\vec{g}_1)\ \vec{g}_1$ is the bending moment and $^1\vec{\mu}^1 - (^1\vec{\mu}^1 .\vec{g}_1)\vec{g}_1$ is the twisting moment.

BASIC EQUATIONS

The generalized strains have covariant components ${}^{I}\varepsilon_{ij}$ ${}^{I}k_{ij}$ in the natural reference frame :

$$ {}^{I}\varepsilon_{ij} = {}^{I}\vec{\varepsilon}_{i} \cdot \vec{g}_{j} \qquad {}^{I}k_{ij} = {}^{I}\vec{k}_{i} \cdot \vec{g}_{j} \tag{13} $$

where the generalized strain vectors are defined by compatibility equations :

$$ {}^{I}\vec{\varepsilon}_{i} = \partial_{i}{}^{I}\vec{u} + \vec{g}_{i} \times {}^{I}\vec{\theta} \quad , \quad {}^{I}\vec{k}_{i} = \partial_{i}{}^{I}\vec{\theta} \tag{14} $$

with x being the vectorial cross product. Equation(14)are symbolically written :

$$ ({}^{I}\vec{\varepsilon}_{i} , {}^{I}\vec{k}_{i}) = C \, ({}^{I}\vec{u}, {}^{I}\vec{\theta}) \tag{15} $$

where C is the linear first order differential operator defined by (14).

Kinematical boundary conditions are :

$$ {}^{I}\vec{u} = {}^{I}\underline{\vec{u}} \quad , \quad {}^{I}\vec{\theta} = {}^{I}\underline{\vec{\theta}} \quad \text{on } S'_{I} \tag{16} $$

where ${}^{I}\underline{\vec{u}}$ and ${}^{I}\underline{\vec{\theta}}$ are prescribed distributions on parts S'_{I} of S_{I} .

Generalized efforts have contravariant components ${}^{I}\sigma^{ij}$, ${}^{I}\mu^{ij}$ in the natural reference frame :

$$ {}^{I}\vec{\sigma} = {}^{I}\vec{\sigma}^{i} n_{i} = {}^{I}\sigma^{ij} n_{i} \vec{g}_{j} = {}^{I}\sigma^{j}_{n} \vec{g}_{j} \; , $$

$$ {}^{I}\vec{\mu} = {}^{I}\vec{\mu}^{i} n_{i} = {}^{I}\mu^{ij} n_{i} \vec{g}_{j} = {}^{I}\mu^{j}_{n} \vec{g}_{j} \tag{17} $$

Transformation of ${}^{I}\vec{\sigma}^{i}$, ${}^{I}\vec{\mu}^{i}$ into ${}^{I}\vec{\sigma}_{n}$, ${}^{I}\vec{\mu}_{n}$ is symbolically written :

$$ ({}^{I}\vec{\sigma}_{n} , {}^{I}\vec{\mu}_{n}) = N \, ({}^{I}\vec{\sigma}^{i} , {}^{I}\vec{\mu}^{i}) \tag{18} $$

Vectors ${}^{I}\vec{\sigma}^{i}$ and ${}^{I}\vec{\mu}^{i}$ verify equilibrium equations for the quasistatic case here to be considered :

$$ \partial_{i}{}^{I}\vec{\sigma}^{i} + {}^{I}\underline{\vec{f}} = 0 \quad , \quad \vec{g}_{i} \times {}^{I}\vec{\sigma}^{i} + \partial_{i}{}^{I}\vec{\mu}^{i} + {}^{I}\underline{\vec{c}} = 0 \tag{19} $$

which are symbolically written :

$$D\left({}^{I}\vec{\sigma}^{\,i},\ {}^{I}\vec{\mu}^{\,i} \right) = -\left(\underline{{}^{I}\vec{f}},\ \underline{{}^{I}\vec{c}} \right) \qquad (20)$$

Boundary conditions on efforts are

$$^{I}\vec{\sigma} = \underline{^{I}\vec{\sigma}},\quad ^{I}\vec{\mu} = \underline{^{I}\vec{\mu}} \quad \text{on } S''_{I} \qquad (21)$$

where $\underline{^{I}\vec{\sigma}}$ and $\underline{^{I}\vec{\mu}}$ are prescribed distributions on parts S_I'' of S_I. At last, we have to define constitutive relations for the linear elastic case to be considered:

$$^{I}\sigma^{ij} = \frac{\partial W_I}{\partial\, ^{I}\varepsilon_{ij}}, \qquad ^{I}\mu^{ij} = \frac{\partial W_I}{\partial\, ^{I}k_{ij}} \qquad (22)$$

which are symbolically written :

$$\left(^{I}\vec{\sigma}^{\,i},\ ^{I}\vec{\mu}^{\,i} \right) = R\left(^{I}\vec{\varepsilon}_i,\ ^{I}\vec{k}_i \right) \qquad (23)$$

Strain energy densities W_I are quadratic positive definite forms of principal invariants of :
$^{I}\varepsilon_{ij}$, $^{I}k_{ij}$, with GRI coefficients :

$$GR3 = (\lambda_3, \mu_3, \rho_3, \phi_3, \psi_3, \chi_3),\ GR2 = (\lambda_2, \mu_2, \rho_2, \nu_2, \phi_2, \psi_2, \chi_2, \tau_2)$$
$$GR1 = (\lambda_1, \mu_1, \rho_1, \phi_1, \psi_1, \chi_1) \qquad (24)$$

The generalized modelizations defined by the above equations are called IDGM according to whether $I = 3, 2, 1$. The generating model is 3DGM while 2DGM and 1DGM may be considered as simplified representations.

The mesh size parameter is defined as e/e_o where e is a typical measure of local mesh size whereas e_o is a prescribed (adjustable data) critical dimension.

If $e \geqslant e_o$, GR3 reduce to the classical R3 coefficients so that 3DGM reduces to classical 3DM elasticity.

If $e < e_o$, the generating model coefficients are obtained as functions of R3 and e/e_o :

- If $e \to e_o$, $GR3 \to R3 = (\lambda, \mu)$
- If $e < e_o$, $GR3 = GR3(\lambda, \mu, e/e_o)$ (25)

The GR2 and GR1 are then obtained from GR3 by the same way as in classical 2MD shells and 1MD beams theories [2],[3].

FUNCTIONAL FORMULATION

Transformation of basic equations into integral relations is performed by a former presented method /4/, /5/, making use of distributions theory /8/. In a first step, functions equations are transformed into linear continuous functionals equations. The $^I\vec{u}$, $^I\vec{\theta}$ are prolonged by zero if $\vec{x} \notin V_I$ by using the characteristic function :

$$H(V_I) = \begin{cases} 1 & \text{if } \vec{x} \in V_I \\ 0 & \text{if } \vec{x} \notin V_I \end{cases} \tag{26}$$

and we apply in distribution sense at $(H(V_I)\,^I\vec{u}, H(V_I)\,^I\vec{\theta})$ the composed second order linear differential operator :

$$L = D \circ R \circ C \tag{27}$$

where D, R, C, are respectively defined by (20), (23),(15).

We use distributions derivation formulas like :

$$\partial_i (H(V_I)\ \vec{v}) = H(V_I)\,\partial_i \vec{v} - n_i \vec{v}\,\delta(S_I) \tag{28}$$

and :

$$\partial_j \partial_i (H(V_I)\,\vec{v}) = H(V_I)\ \partial_j\ \partial_i \vec{v} - \partial_j (n_i \vec{v}\,\delta(S_I)) - n_j\ \partial_i \vec{v}\,\delta(S_I) \tag{29}$$

where $n_i \vec{v}\,\delta(S_I)$ is the distribution on S_I defined on each admissible $\vec{w}$ by :

$$< n_i \vec{v}\,\delta(S_I), \vec{w} > \; = \; < n_i \vec{v}, \vec{w} >_{S_I} \tag{30}$$

whose value is given by (1), (2) or (3) according to I = 3, 2 or 1.

As result we have the functional equation :

$$L(H(V_I)\,^I\vec{u}, H(V_I)\,^I\vec{\theta}) = H(V_I) L(^I\vec{u}, {}^I\vec{\theta}) - T(^I\vec{u}, {}^I\vec{\theta})\,\delta(S_I) - {}^tT((^I\vec{u}, {}^I\vec{\theta})\,\delta(S_I)) \tag{31}$$

where T is the composed first order linear operator :

$$T = N \circ R \circ C \tag{32}$$

with N defined by (18), and where ^tT is the transposed operator in the sense of partial differential equations theory.

Making use of (16), (19) and (21), we obtain finally for(31):

$$L(H(V_I)\,^I\vec{u}, H(V_I)\,^I\vec{\theta}) = -H(V_I)(^I\underline{\vec{f}}, {}^I\underline{\vec{c}}) - (^I\vec{\sigma}_n, {}^I\vec{\mu}_n)\,\delta(S_I) - {}^tT((^I\vec{u}, {}^I\vec{\theta})\,\delta(S_I)) \tag{33}$$

which is the functional formulation of equations (13)to (23).

INTEGRAL RELATIONS

Let ${}^{I}E = ({}^{I}U, {}^{I}H)$ be the tempered elementary solution of L : The (6 x 6) ${}^{I}U$ and ${}^{I}H$ matrix has components

$$ {}^{I}U_{ij} = {}^{I}\vec{U}_i \cdot \vec{g}_j \quad , \quad {}^{I}H_{ij} = {}^{I}\vec{H}_i \cdot \vec{g}_j \tag{34} $$

where ${}^{I}\vec{U}_i$ and ${}^{I}\vec{H}$ are respectively displacements and rotations corresponding to concentrated forces and couples at origin in direction $\vec{g}_i$: ${}^{I}\vec{U}_i$ and ${}^{I}\vec{H}_i$ are tempered solutions of :

$$ L\,({}^{I}\vec{U}_i, {}^{I}\vec{H}_i) = -\,(\vec{g}_i\,{}^{I}\delta(0),\ \vec{g}_i\,{}^{I}\delta(0)) \tag{35} $$

with ${}^{I}\delta(0)$ being the I dimensional DIRAC distribution. The elementary solution ${}^{I}E$ is obtained by HORMANDER'S method [7]

By convolution product of ${}^{I}E$ with (33), we obtain then

$$ \begin{aligned} (H(V_I)\,{}^{I}\vec{u},\ H(V_I)\,{}^{I}\vec{\theta})(\vec{x}) = {} & <({}^{I}U, {}^{I}H)(\vec{x}-\vec{\xi}), ({}^{I}\vec{\underline{f}}, {}^{I}\vec{\underline{c}})(\vec{\xi})>_{V_I} \\ & + <({}^{I}U, {}^{I}H)(\vec{x}-\vec{\xi}), ({}^{I}\vec{\sigma}_n, {}^{I}\vec{\mu}_n)(\vec{\xi})>_{S_I} \\ & - <(T({}^{I}U, {}^{I}H))(\vec{x}-\vec{\xi}), ({}^{I}\vec{u}, {}^{I}\vec{\theta})(\vec{\xi})>_{S_I} \end{aligned} \tag{36} $$

which are integral relations available for any $\vec{x} \in V_I$ (I = 3,2,1), $\vec{\xi}$ being an integration variable.

BOUNDARY INTEGRAL EQUATIONS

Transformation of (36) into boundary integral equations is performed by taking the limit on (36) of $\vec{x} \in V_I$ to $\vec{x} \in S_I$

We obtain then integral equations :

$$ \begin{aligned} \alpha(\vec{x})\,({}^{I}\vec{u}, {}^{I}\vec{\theta})(\vec{x}) = {} & <({}^{I}U, {}^{I}H)(\vec{x}-\vec{\xi}), ({}^{I}\vec{\underline{f}}, {}^{I}\vec{\underline{c}})(\vec{\xi})>_{V_I} \\ & + <({}^{I}U, {}^{I}H)(\vec{x}-\vec{\xi}), ({}^{I}\vec{\sigma}_n, {}^{I}\vec{\mu}_n)(\vec{\xi})>_{S_I} \\ & - <(T({}^{I}U, {}^{I}H))(\vec{x}-\vec{\xi}), ({}^{I}\vec{u}, {}^{I}\vec{\theta})(\vec{\xi})>_{S_I} \end{aligned} \tag{37} $$

with $\vec{x} \in S_I$ and where $\alpha(\vec{x})$ is a coefficients matrix depending on the local geometry of S_I.

The numerical treatment of (37) is quite the same as the one used in standard programs of CETIM [6] for classical elasticity. Boundaries S_I are discretized by finite elements S_I^e, the unknowns :

${}^{I}\vec{u}$, ${}^{I}\vec{\theta}$, ${}^{I}\vec{\sigma}_n$, ${}^{I}\vec{\mu}_n$ are isoparametrically interpolated by classical shape functions, and (37) is solved at collocation points $\vec{x}_c$.

The obtained algebraic system has coefficients which are integrals on S_I^e of $({}^IU, {}^IH)$ and $T({}^IU, {}^IH)$. These integrals are computed by ordinary GAUSS quadrature rules if $\vec{\xi} \notin S_I^e$ containing $\vec{x}_c$. When $\vec{\xi} \in S_I^e$ containing $\vec{x}_c$, the most singular coefficients (arising from $T({}^IU, {}^IH)$) are obtained by the unicity condition :

$$\alpha\,({}^I\vec{u}, {}^I\vec{\theta}) + \langle T({}^IU, {}^IH), ({}^I\vec{u}, {}^I\vec{\theta}) \rangle_{S_I} = 0$$

with ${}^I\vec{u} = \vec{a} + \vec{b} \times \vec{x}$, ${}^I\vec{\theta} = \vec{b}$ where $\vec{a}$, $\vec{b}$ are arbitrary constant vectors.

REFERENCES

/1/ Dubois, M. (1978) Méthodes numériques de calcul des structures. Part. 2. N.T.I. - Calcul, CETIM, Senlis, France.

/2/ Dubois, M. (1980) Bases d'un système compatible de calcul des structures. Second International Symposium on Innovative Numerical Analysis in Applied Engineering Science, Montréal, Canada.

/3/ Dubois, M. (1981) Etude d'un système compatible de calcul des structures. Recherche sous contrat D.G.R.S.T. n° 80.E.0885, France.

/4/ Dubois, M.,Lachat, J.C., (1972) The Integral formulation of boundary value problems. Variational Methods in Engineering, University of Southampton.

/5/ Dubois, M. , Lachat, J.C., Watson, J.O. (1973) Le Calcul des Structures par la méthode des équations intégrales. C.R. quatrième Congrès Canadien de Mécanique Appliquée, Montréal, Canada.

/6/ Dubois, M. , Lange, D., Serres, D., Chaudouet, A., Boissenot, J.M., Loubignac, G., Percie du Sert, B.(1978) Application de la méthode des équations intégrales à la mécanique. N.T - Calcul, CETIM, Senlis, France.

/7/ Hörnander, L. (1965) Lineinîe diferentialnîe operatorî s ciastîmi proizvednîmi, Moskva, C.C.C.P.

/8/ Schwartz, L. (1966) Théorie des distributions. Hermann, Paris, France.

METHOD OF BOUNDARY INTEGRAL EQUATIONS FOR ANALYSIS OF THREE DIMENSIONAL CRACK PROBLEMS

J. Balaš and J. Sládek

Institute of Construction and Architecture, Slovak Academy of Sciences, 885 46 Bratislava, CSSR

1. INTRODUCTION

The solution of spatial problems for finite region with cracks often encounters difficulties which results in the fact that there is a incomparably smaller number of solved spatial problems for crack than that of plane problems. There are almost no exact solutions of this kind at all. Those prevailing are numerical solutions mostly using the finite elements method (FEM). Among the earlier papers these by Tracey [1] and Benzley [2] should be mentioned. The calculations made by this method are rather inefficient and require large computers. It has been shown that the method of boundary integral equations (BIE) is more efficient for these problems not claiming such storage capacities as the finite element method. The BIE method is very well suited to the solution of three-dimensional stress concentration problems reducing the latter to boundary solutions, i.e. requiring only the elements on the boundary to be defined. In contradistinction to them, the finite elements require many interior nodes and approximations involving discontinuity of stresses. The fact that the BIE method allows direct solutions for boundary displacements with no modelling of the internal stresses and stress intensity factor analysis, requiring only crack opening displacements, implies that there is no need to determine the stress inside the body, near the crack tip [3].

In spite of the foregoing advantages, there will be a large number of surface elements where more complicated problems are treated. For this reason, the problem solution is split in two auxiliary problems. In the first part, a finite region without crack will be solved by the BIE method. In the second part (using the so-called perturbance method) the solution will be concerned with the state of stress and strain of a crack in infinite space. Using the Schwarz algorithm we obtain then the final solution.

2. BOUNDARY-INTEGRAL EQUATION ANALYSIS

The theoretical deduction of integral equations being known [4], only the fundamental equations and their numerical solutionare presented. For the deduction of the integral equations we use the reciprocity theorem (the Green formula in the theory of potentials, the Betti's theorem concerning reciprocity of virtual displacements) as well as the corresponding singular solution of differential equations the so called Kelvin problem of elasticity. The combination of both these partial solutions is well-known from the classical theory of elasticity as the Somigliano's identity. The above mentioned method includes only physical quantities i.e. loads and displacements on the surface of the region and which permits also a stable numerical solution for the unsmooth geometries (corners and edges) . The studies region may be simply or multiply connected this feature by no means complicating the numerical procedure.

The set of integral equations for the solution of elasto-static problems due to surface loads relates the boundary values of tractions $t_i(x)$ and displacements $u_i(x)$.

$$C_{ij}(x)u_j(x) + \int_S T_{ij}(x,y)u_j(y)ds = \tag{1}$$

$$= \int_S U_{ij}(x,y)t_j(y)ds$$

where C_{ij} is a constant, in the case of continuous tangent plane at x, $C_{ij}(x) = \delta_{ij}/2$. The kernel functions $T_{ij}(x,y)$ and $U_{ij}(x,y)$ are the tractions and displacements in the x_i direction at $y(y_1, y_2, y_3)$ due to orthogonal unit loads in the x_j directions at $x(x_1, x_2, x_3)$.

$$U_{ij}(x,y) = \frac{1+\nu}{8\pi(1-\nu)Er}\left\{(3-4\nu)\delta_{ij} + r_{,i}\, r_{,j}\right\} \tag{2}$$

$$T_{ij}(x,y) = -\frac{1}{8\pi(1-\nu)r^2}\left\{(1-2\nu)\frac{\partial r}{\partial n}\delta_{ij} + \right. \tag{3}$$

$$\left. + 3\frac{\partial r}{\partial n} r_{,i}\, r_{,j} + (1-2\nu)\,(n_i\, r_{,i} - n_j\, r_{,j})\right\}$$

where

$$r_{,i} = \frac{\partial r}{\partial y_i} = \frac{y_i - x_i}{r} \quad ,$$

ν and E are elastic constants, r is the distance between points x and y, δ_{ij} is the delta-function, u_i is a component of the unit outward-normal of the surface. The repeated index in a product is to be understood as a summation over all the indices.

The given boundary conditions are the tractions t_j on a part of the boundary and the displacements u_j and/or the combination of both t_j and u_j on the remainder of the boundary. The solution of BIE [4] gives the other unknown parts of the functions t_j and u_j over all the surface. However, the BIE are exactly solvable only for a few types of regions and must, therefore, be solved numerically. In the last case we use isoparametrical shape functions for the representation of both the geometry and the boundary data [5]. The approach is known from the FEM.

The Cartesian coordinates of an arbitrary point of an element are expressed in terms of nodal coordinates y_i^e and shape functions $N^e(\xi)$ of local coordinates ξ.

$$y_i(\xi) = N^e(\xi)\, y_i^e \qquad (4)$$

where

$$N^1(\xi) = \frac{1}{4}\,(\xi_1+1)\,(\xi_2+1)\,(\xi_1+\xi_2-1)$$

$$N^5(\xi) = \frac{1}{2}\,(\xi_1+1)\,(1-\xi_2^2) \quad \text{etc. } [6]$$

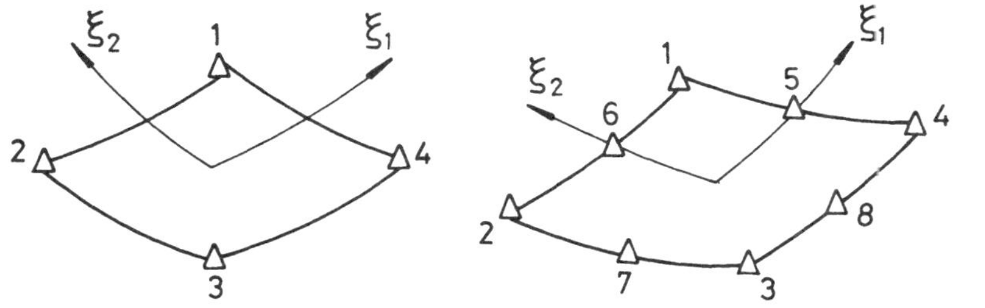

Fig. 1. Surface elements: a) linear approximation
b) quadratic approxim.

By differentiating equation (4) with respect to ξ_1 we obtain the tangent vectors to the coordinate curves by means of which the Jacobean and the normal to the element surface is obtained.

Similarly, the displacements $u_i(x)$ and the tractions $t_i(x)$ on the surface are expressed in terms of their values at the nodal points and the shape functions.

$$\Phi(\xi) = M^b(\xi)\,\Phi^b \tag{5}$$

where $M^b(\xi)$ stand for linear approximation

$$M^1(\xi) = \frac{1}{4}(\xi_1 + 1)(\xi_2 + 1)$$

$$M^2(\xi) = \frac{1}{4}(\xi_2 + 1)(1 - \xi_1)$$

$$M^3(\xi) = \frac{1}{4}(1 - \xi_1)(1 - \xi_2)$$

$$M^4(\xi) = \frac{1}{4}(1 + \xi_1)(1 - \xi_2)$$

For quadratic approximation the following formula is true

$$M^b(\xi) = N^b(\xi) \qquad (6)$$

The points x, at which the integral equations are expressed are internal points of the surface element not identical with the nodal points in terms of which the boundary coordinates, the displacements and/or the tractions are expressed. Generally the number of the equations after the discretization of BIE and the number of unknown values of the nodal displacements and/or tractions are different, i.e. in the matrix form

$$\left[T_{ij}\right]\left\{u_j\right\} - \left[U_{ij}\right]\left\{t_j\right\} = \left\{0\right\} \quad \begin{array}{l} i = 1,2\ldots n \\ j = 1,2\ldots p \end{array} \qquad (7)$$

where $\left[T_{ij}\right]$ and $\left[U_{ij}\right]$ are matrices of the coefficients determined from the numerical solution of the BIE (1), $\left\{u_j\right\}$ and $\left\{t_j\right\}$ are the displacement vector and the traction vector, respectively, of the nodal points.

In the form (6) the problem is not solvable, it is, however, possible to minimize the error function by the least squares method [7]

$$\sum_{i=1}^{n} \left\{\varepsilon_i\right\}^2 + \sum_{i=1}^{n} \left(\left[T_{ij}\right]\left\{u_j\right\} - \left[U_{ij}\right]\left\{t_j\right\}\right)^2 = \min \qquad (8)$$

The resulting system of equations is given in the form

$$\left[T_{ih}\right]^T \left[T_{ij}\right] \left\{u_j\right\} = \left[T_{ih}\right]^T \left[U_{ij}\right] \left[t_j\right] \qquad (9)$$

The matrix of this system of equations (n x n) is symmetric, fully populated. The non-singularity condition of the system is $n \geq p$ except the static determination, i.e. the number of equations in (7) should be greater or equal to the number of the unknown parameters.

To obtain the matrix coefficients T_{ij}, U_{ij} in (9) we solve numerically the integrals of the product of the kernel functions T_{ij} (3) and U_{ij} (2) and the shape functions $N_u(\xi)$ of the displacements and $N_t(\xi)$ of the tractions with the use of the Gaussian quadrature formulas.

$$\int_{-1}^{1}\int_{-1}^{1} f(\eta_1, \eta_2)\, d\eta_1\, d\eta_2 = \tag{10}$$

$$\sum_{i=1}^{n_1}\sum_{j=1}^{n_2} A_i^{(n_1)} A_j^{(n_2)} f\left(\eta_i^{(n_1)}, \eta_j^{(n_2)}\right) + \varepsilon ,$$

where ε is the value of the error and n_1, n_2 are the orders of the used quadrature formula.

In the case of singular elements the transformation of the local coordinates to local polar coordinates ς, φ having their origin at the singularity point [7] leads to a reduction of the order of the singularity by one degree. It means that in this transformation the integrals with a weak singularity are reduced to integrals over smooth functions. The strong singularity $(1/r^2)$ cannot be eliminated from the integrals. For their evaluation, however, the solution of the following homogeneous integral equation may be used

$$C_{ij}(x)\, u_j(x) + \int_S T_{ij}(x,y)\, u_j(y)\, ds = 0 \tag{11}$$

The solution of this equation yields the function

$$u_i(x) = a_i + \epsilon_{ijk}\, x_j\, b_k , \tag{12}$$

the physical interpretation of which are rigid displacements and rotations of the body; a_i, b_k (i,k= 1, 2, 3) are arbitrary constants, ϵ_{ijk} is an anti-symmetric unit tensor.

From equations (11) and (12) by the choice of constant $a_j = u_j(x)$ there follows the relation

$$C_{ij}(x)\, u_j(x) + \int_{S_o} T_{ij}(x,y)\, u_j(x)\, ds = \tag{13}$$

$$= - \int_{S-S_o} T_{ij}(y,x)\, u_j(x)\, ds$$

and when this is used we receive

$$C_{ij}(x)\, u_j(x) + \int_{S_o} T_{ij}(x,y)\, u_j(x)\, ds = \tag{14}$$

$$= - \int_{S-S_o} T_{ij}(x,y)\, u_j(x)\, ds - \int_{S_o} T_{ij}(x,y) \left[u_j(y) - u_j(x) \right] ds$$

where S_o denotes the area of the element with a singularity point.

The first of the right hand side integrals in equation (14) is obtained as a sum of integrals over all the elements but for the singularity (hence they need not be evaluated once more) and the second one contains at most a weak singularity and may, therefore, be evaluated numerically in terms of the local polar coordinates.

The stresses in the interior of the studies region are obtained by differentiating u at the internal points [8]

$$\sigma_{ij}(x) = \int_S D_{kij}(x,y)\, t_k(y)\, ds - \tag{15}$$

$$- \int_S S_{kij}(x,y)\, u_k(y)\, ds$$

where

$$D_{kij} = \frac{1}{r^2}\left\{ (1-2\nu)\left[\delta_{ki}\, r_{,j} + \delta_{kj}\, r_{,i} - \delta_{ij}\, r_{,k}\right] + \right.$$

$$\left. + 3r_{,i}\, r_{,j}\, r_{,k}\right\} \frac{1}{8\pi(1-\nu)}$$

$$S_{kij} = \frac{2\mu}{r^3}\left\{ 3\,\frac{\partial r}{\partial n}\left[(1-2\nu)\,\delta_{ij}\, r_{,k} + \right.\right.$$

$$\left. + \nu(\delta_{ik}\, r_{,j} + \delta_{jk}\, r_{,i}) - 5r_{,i}\, r_{,j}\, r_{,k}\right]$$

$$+ 3\nu\,(n_i\, r_{,j}\, r_{,k} + n_j\, r_{,i}\, r_{,k}) + (1-2\nu)$$

$$(3n_k\, r_{,i}\, r_{,j} + n_j\, \delta_{ik} + n_i\, \delta_{jk}) -$$

$$\left. - (1-4\nu)\, n_k\, \delta_{ij}\right\} \frac{1}{8\pi(1-\nu)}$$

3. CRACK OF ARBITRARY CONTOUR IN INFINITE SPACE

In this part of the paper we shall discuss the perturbation method of solution departing from the work of Weaver [9] who has derived the system of integral equations for unknown dislocations of a spatial crack of arbitrary contour in an infinite space. All the other quantities describing the state of stress and strain near the crack may be expressed in terms of unknown dislocations will be calculated after the transformation of the system of integral equations to a system of linear equations.

We shall study an infinite elastic space using Cartesian coordinates x_i. According to the preceding paragraph, the j-th stress component due to a unit force in an i-th direction at point ζ_k is expressed in terms of equation (3)

$$T_{ij}(x,\zeta) = -\frac{K}{r^2}\left[\frac{\partial r}{\partial n}\left(\delta_{ij} + \frac{3r_{,i}\,r_{,j}}{1-2\nu}\right) - \right. \tag{16}$$

$$\left. - n_j\, r_{,i} + n_i\, r_{,j}\right]$$

where

$$K = \frac{1-2\nu}{8\pi(1-\nu)}$$

We shall assume that region D consists of two parts mutually complementary to the infinite space.

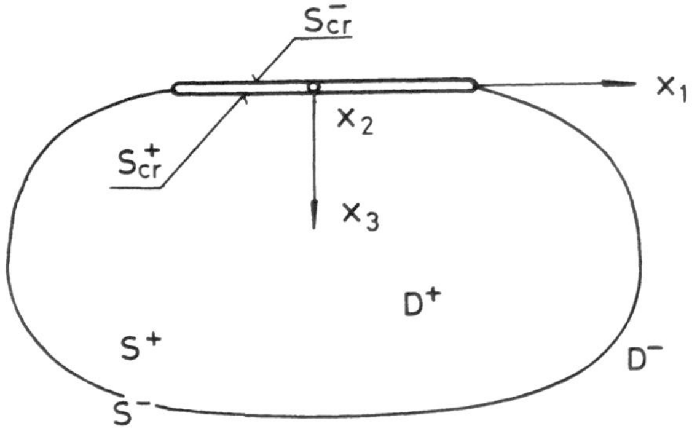

Fig. 2. Crack of arbitrary contour in infinite space

The internal region is infinite and denoted by D^+. The external D^- is its complement. The common surface of D^+ and D^- is denoted by $S^\pm$. The plane part $S^\pm$ (crack) is denoted by $S_{cr}^\pm$.

According to Kupradze [10] it holds for $x \in D^+$

$$u_i(x) = \int_{S^+} t_j(y)\, U_{ij}(y,x)\, ds - \tag{17}$$

$$- \int_{S^+} u_j(y)\, T_{ij}(y,x)\, ds$$

$$0 = \int_{S^-} t_j(y)\, U_{ij}(y,x)\, ds - \int_{S^-} u_j(y)\, T_{ij}(y,x)\, ds \tag{18}$$

Let us assume the crack loads to be symmetric with respect to $x_3 = 0$ i.e. $t_j^+ = -t_j^-$. Similarly $T_{ij}^+ = -T_{ij}^-$ and $U_{ij}^+ = U_{ij}^-$. We obtain by summing equations (17) and (18)

$$u_i(x) = - \int_{S_{cr}^+} \Delta u_j(y)\, T_{ij}^+(y,x)\, ds \tag{19}$$

where

$$\Delta u_j = u_j^+ - u_j^-$$

The stress components are obtained from equation (19)

$$\sigma_{\alpha 3} = \mu (u^+_{\alpha,3} + u^+_{3,\alpha}) \qquad \text{for } \alpha = 1,2 \tag{20}$$
$$\sigma_{33} = 2\mu \left[\nu u_{\alpha,\alpha} + (1-\nu)\, u^+_{3,3} \right] \frac{1}{1-2\nu}$$

and by integration by parts we get

$$u_{i,\alpha} = - \int_{S_{cr}^+} \Delta u_j(y) \frac{\partial T_{ij}^+(y,x)}{\partial x_\alpha}\, ds =$$

$$= \int_{S_{cr}^+} \Delta u_j(y) \frac{\partial T_{ij}^+(y,x)}{\partial y_\alpha}\, ds =$$

$$= - \int_{S_{cr}^+} \Delta u_{j,\alpha}\, T_{ij}^+(y,x)\, ds$$

Because on the crack there is a vector of normal
$n_i^+ = -\delta_{i3}$

$$T_{ij} = -\frac{K}{r^2}\left[\frac{\zeta_3}{r}\left(\delta_{ij} + \frac{3r_{,i}\, r_{,j}}{1-2\nu}\right) + \delta_{j3}\, r_{,i} - \delta_{i3} r_{,j}\right]$$

Hence, form the preceding relationships and from equation (20) we receive

$$\sigma_{\alpha 3} = \frac{E\,(1-\nu)}{16\pi(1-\nu^2)} \int_{S_{cr}^+} \left[\delta_{\beta\alpha}\, r_{,\gamma} - \delta_{\alpha\gamma} r_{,\beta} + \frac{3r_{,\alpha}\, r_{,\beta}\, r_{,\gamma}}{1-2\nu}\right] \frac{\Delta u_{\beta,\gamma}}{r^2}\, \partial S_{cr} \tag{21}$$

$$\sigma_{33} = \frac{E}{8\pi(1-\nu^2)} \int_{S_{cr}^+} \frac{r_{,\alpha}\, \Delta u_{3,\alpha}}{r^2}\, \partial S_{cr}$$

for $\alpha, \beta, \gamma = 1, 2$

The integrals in equation (21) are understood in the sense of Cauchy's principal values. Now, we can take up the numerical solution of the system of integral equations of first order (21) where displacements on the surface of the crack to which the known loads σ_{33} and/or $\sigma_{\alpha 3}$ are applied. The crack surface is approximated by means of surface elements. Then, the Cartesian coordinates of any point of the element are expressed in terms of coordinates of nodal points and shape functions in terms of local coordinates. For this purpose, we can use immediately the elements shown in Fig. 1 with the corresponding shape functions. As it will be shown later for penny-shaped cracks or elliptical cracks it is advantageous to use a transition element. Such an element permits matching of one element with two elements side-by-side [1]. For the sake of simplicity, a formulation of the transition element with the introduction of transition intermediate nodes is presented here. However, the method is general and can be easily applied to other and higher-order elements.

In Fig. 3 the edge AB of element I has a discontinuity in the displacement slope at node C. The variation of any displacement u along edge AB can be expressed as

$$u = \frac{1}{2}\,(1-\xi)\;u^A + \frac{1}{2}\,(1+\xi)\;u^B + (1-|\xi|)\,\Delta u^C \tag{22}$$

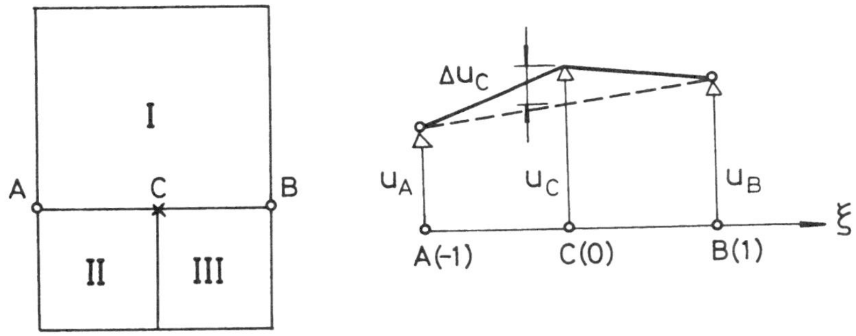

Fig. 3. The transition quadrilateral element and displacement field

If we express Δu_C in terms of displacement u at the nodal points we obtain from equation (22)

$$u = \frac{1}{2}\,(|\xi|-\xi)u_A + \frac{1}{2}\,(|\xi|+\xi)u_B + (1-|\xi|)u_C \tag{23}$$

The procedure used to obtain the shape functions for the transition elements may be summarized as follows: Write the shape functions for the corner nodes as if there were no intermediate transition node. Write the shape function for the intermediate node treating the displacement at the intermediate node as a departure Δu_C from the linear displacement variation defined by the displacements at the corner nodes. Modify the shape function at the corner nodes in such a way that the displacement at the intermediate node becomes the total displacement at the node u_C. Note that the shape function at the intermediate node does not change if $N_6 = N_7 = N_8 = 0$

$$N_1 = \frac{1}{4}(1-\xi)(1-\eta) - \frac{1}{4}(1-|\xi|)(1-\eta)$$

$$N_2 = \frac{1}{4}(1-\xi)(1+\eta)$$

$$N_3 = \frac{1}{4}(1+\xi)(1-\eta) - \frac{1}{4}(1-|\xi|)(1-\eta)$$

$$N_4 = \frac{1}{4}(1+\xi)(1+\eta)$$

$$N_5 = \frac{1}{2}(1-|\xi|)(1-\eta)$$

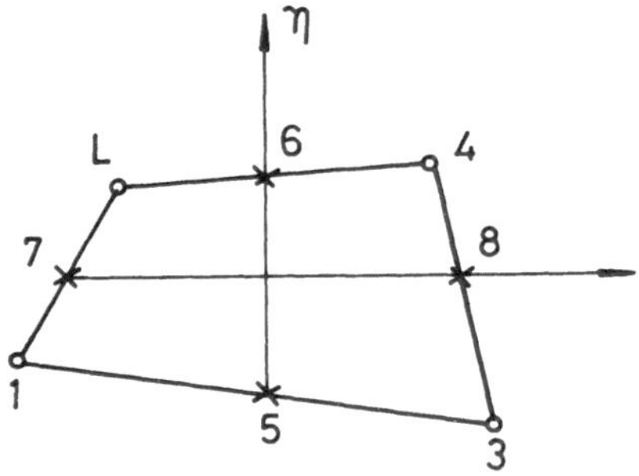

Fig. 4. The transition quadrilateral element and the shape function

In the present case the integrand of equation (10) is discontinuous at $\xi = 0$ and the integral is split into two continuous integrals.

The gradient of the shape functions can now be written as

$$\begin{pmatrix} \frac{\partial N_i}{\partial x} \\ \frac{\partial N_i}{\partial y} \end{pmatrix} = \left[J\right]^{-1} \begin{pmatrix} \frac{\partial N_i}{\partial \xi} \\ \frac{\partial N_i}{\partial \eta} \end{pmatrix} \qquad (24)$$

where

$$\left[J\right] = \begin{pmatrix} \frac{\partial x}{\partial \xi} & \frac{\partial y}{\partial \xi} \\ \frac{\partial x}{\partial \eta} & \frac{\partial y}{\partial \eta} \end{pmatrix}$$

By substiting (24) into integral equation (21) and after performing the corresponding numerical integrations we get a system of linear equations for the unknown displacements in the crack.

The strong singularity integrals will be calculated by erecting a normal at each nodal point on which the auxiliary points P_j^l $(l = 1,2,3,4)$ lie for which the integral equation (21) will be written.

The subintegral expressions at these points will be regular and the value of the integral for point P_j^o lying on the crack surface (nodal point) will be found by extrapolation of integrals for the auxiliary points (12). A beam including an eccentrically situated crack of rectangular shape with respect to the neutral axis is subjected to a bending moment. The stresses at the place of the crack are then

$$p(x_1) = p_o + p_1 \frac{x_1}{a}$$

$$p_o = \frac{Mh_1}{I}\left(1 - \frac{h}{h_1}\right)$$

$$p_1 = M\,\frac{a}{I}$$

I - moment of inertia of the cross-section area of the beam

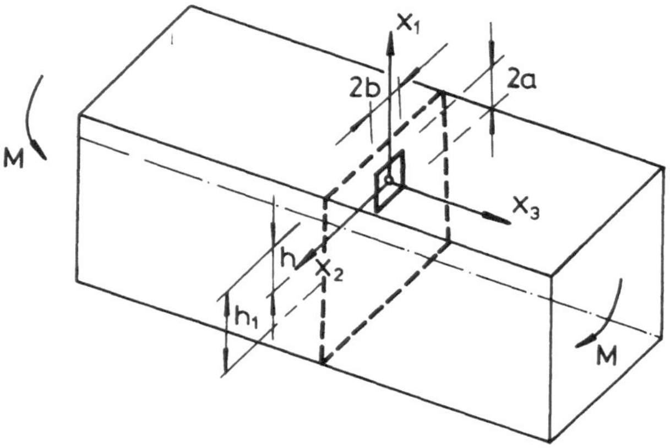

Fig. 5. Crack in a beam subjected to a bending moment

The stress intensity factor K_I is related to the displacements near the edge of the crack. Kassir and Sih have shown that the two-dimensional equations for the plane strain crack hold for the three-dimensional smoothed crack, too [13].

Thus for the opening mode

$$u_3 = \frac{K_I}{2\mu}\left(\frac{r}{2\pi}\right)^{1/2} \sin\frac{\varphi}{2}\left[\,4(1-\nu)-2\cos^2\frac{\varphi}{2}\,\right] \qquad (25)$$

where (r,φ) is a local polar coordinate system in a plane perpendicular to the edge.

Then a stress intensity factor is computed from equation (25) for $\varphi = \pi$ by means of the extrapolation method.

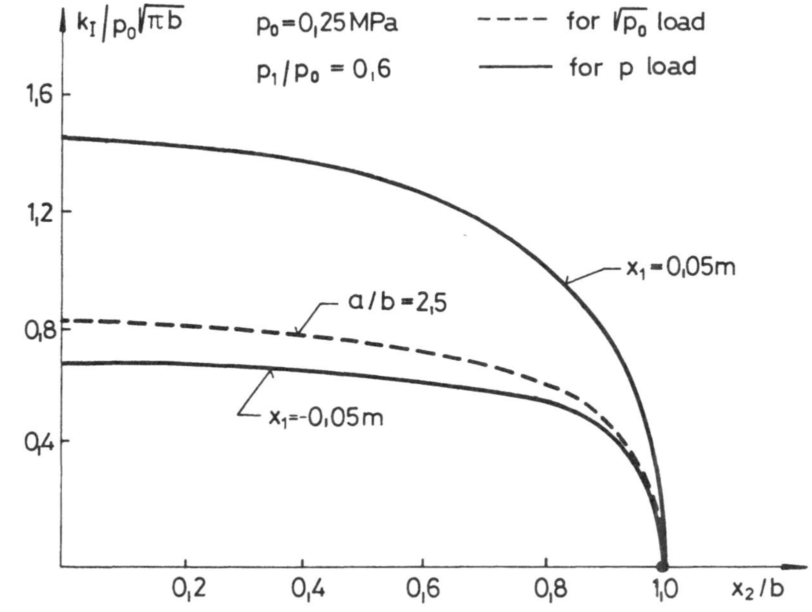

Fig. 6. The stress intensity factor on the shorter crack sides

It is noted that at point $x_2 = b$ $K_I \neq 0$ because at this point the contourof the crack is not smooth and the stress singularity is not equal to 1/2. Further details of the singularity calculation at this point may be found in [14].

4. INFLUENCE OF FREE EDGE ON VARIATION OF STRESS INTENSITY FACTOR

The free edge near the crack engenders an

increase of the local stress due to the crack. It means that the stress intensity factor will be greater than in the case of crack in an infinite space. The solution of such problems is of a great practical importance owing to the fact that cracks occur very often near free edges or on their surface.

Smith [15] studied a semi-penny-shaped surface crack in a half-space by combining the Levy's solution of a half-space with the solution for a penny-shaped crack in an infinite space. The mentioned method cannot be used for finite regions only for semi-infinite ones. More general numerical methods lay high claims on the computers. In order not to exaggerate the claims on the computer the problem of the state of stress for a crack in a finite region is split into two problems described in the preceding paragraphs.

To illustrate the suggested method, let us study a tensioned beam including a penny-shaped crack and an elliptical crack in its interior.

As the first step let us calculate the stresses at the place of the crack for a tensioned beam. Hence, these stresses, only with opposite sign, will be used as loads of the crack in the infinite space and according to the perturbation solution we shall calculate the stresses on the beam surface. These stresses with the opposite sign will be loading the beam and again the stresses at the place of the crack will be calculated. This procedure will be repeated several times. The number of iterations used will depend on how far the crack is from the external boundary of the finite region.

For the given region it is sufficient to calculate only once the left side of the system of linear equations for different loads. Using this as basis we can calculate integrals with a strong singularity. Once we know the displacements for the given geometry under a certain load (e.g. constant) we can calculate the diagonal coefficients of the left side matrix.

We know analytical solutions for the penny-shaped crack [12] and the elliptical crack [13] in an infinite space. The calculation of displacements for constant loads on the basis of an analytical solution is rather simple. It is not so simple in the case of a general load. For this reason we calculate first the diagonal coefficients of the left side matrix for a constant load and then for the known coefficients of the matrix the displacements due to the general load will be calculated. The advantage of this procedure consists in eliminating the sources of error consisting in the calculations on singular elements.

The stress intensity factor for an elliptic crack with ratio $a/b = 2,5$ and the free edge influence ratio $b/h = 0,75$ is normed by stress-intensity factor for a penny-shaped crack for an infinite space loaded by constant pressure σ_o.

The BIE method has been found to offer an efficient and accurate tool of fracture mechanics analysis using a numerical solution of a set of boundary constraint equations.

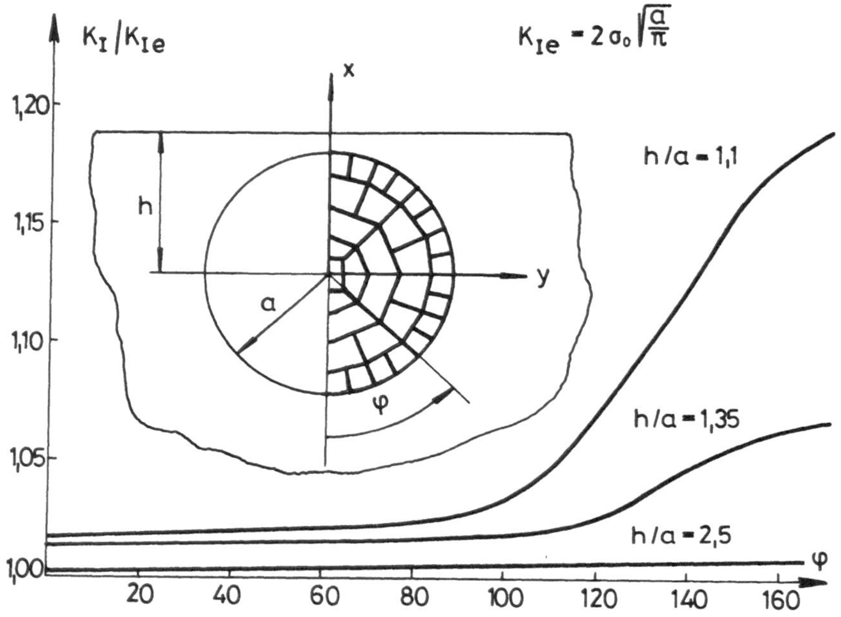

Fig. 7. Influence of free edge on stress-intensity factor for penny-shaped crack

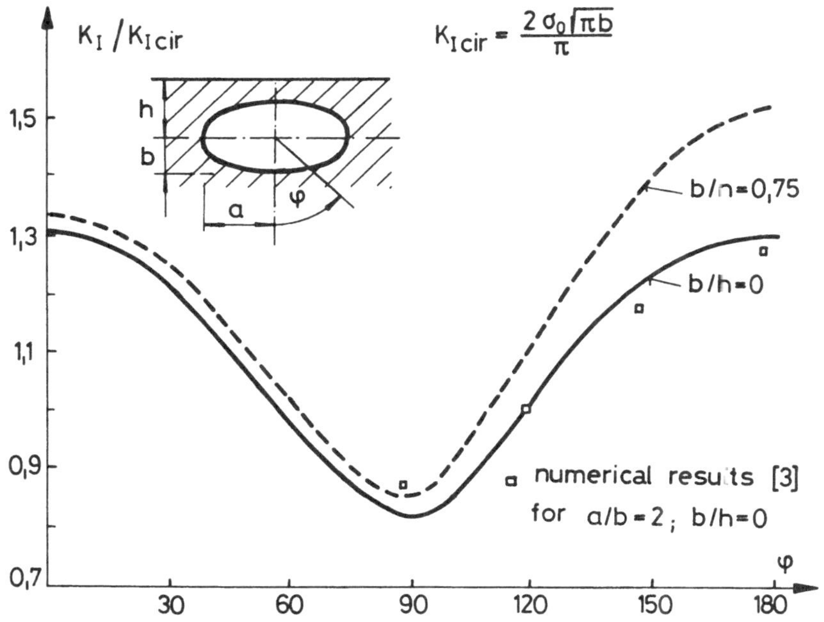

Fig. 8. Influence of free edge on stress intensity factor K_I for elliptical crack

References:

[1] D.M. Tracey: 3-D elastic singularity element for evaluation of K along an arbitrary crack front; Int. J. Fracture 9, 1973, 340-343

[2] S.E. Benzley: Representation of singularities with isoparametric finite elements, Int. J. Num. Meth. Eng. 8, 1974, 537-545

[3] T.A. Cruse, G.J. Meyers: Three dimensional fracture mechanics analysis, Journal of the structural division, Proceedings of the American Society of Civil Engineers, 103, 1977, 309-320

[4] T.A. Cruse: Application of the boundary integral equation method to three dimensional stress analysis, Computers and Structures, 3, 1973, 509-527

[5] J.C. Lachat, J.O. Watson: Effective numerical treatment of boundary integral equations: A formulation for three dimensional elastostatics, Int. J. Num. Meth. in Engng. 10, 1976, 991-1005

[6] O.C. Zienkiewicz: The Finite Method in Engineering Science, Mc Graw-Hill, London, 1971

[7] V. Kompiš: Boundary integral equation method for three dimensional elastostatic problems and a formulation of the problem for large displacements, Proc. of 14-th Jugosl. kongres racion.: primenjene mechanike C, 1978, 113-120

[8] C.A. Brebbia, R. Nakaguma: Boundary elements in stress analysis, Journal of the Engineering mechanics division, Proceedings of the American society of Civil Engineers, 105 No EM1, 1979, 55-69

[9] J. Weaver: Three dimensional crack analysis, Int. J. Solids and Structures, 13, 1977, 321-330

[10] V.D. Kupradze: Metody potenciala v teorii uprugosti, Moskva, Fiz.-mat. izd., 1963

[11] A.K. Gupta: A finite element for transition from a fine to a coarse grid, Int. J. Num. Meth. in Eng., 12, 1978, 35-45

[12] J. Sládek: Numerical methods in Fracture mechanics, Thesis, Slovak Academy of Science, Bratislava, 1980 (in Slovak)

[13] M.K. Kassir, G.C. Sih: Three dimensional stress distribution around an elliptical crack under arbitrary loadings, Journal of Appl. Mech., Trans. ASME ser. E, 33, 1966, 601-611

[14] V.L. Rvačev, V.S. Procenko: Kontaktnyje zadači teorii uprugosti deja neklasičeskich oblastej, Naukova dumka 1977, Kijev

[15] F.W. Smith et. al.: Stress intensity factors for penny-shaped crack, Part I Semi-infinite solid, Journal of Appl. Mech., Trans. ASME ser. E, 34, 1967, 953-959

CYCLIC SYMMETRY AND SLIDING BETWEEN STRUCTURES BY THE BOUNDARY INTEGRAL EQUATION METHOD

Anne CHAUDOUET

CETIM - France

INTRODUCTION

Seven years ago, CETIM developed a computer program based on the Boundary Integral Equation method to analyse three dimensional structures under mechanical loading. At that time, the boundary conditions that could be taken into account by the program were :

- given displacements on nodes, lines or elements,
- given tensions on lines or elements, and concentrated forces,
- gravitation and centrifugal forces

  The symmetries with respect to the surfaces x=o, y=o and z=o could also be taken into account, and a subregioning process allowed to analyse bodies made of different materials, or to analyse contact between structures when the contact zone was known and when the contact was perfect.

The advantages of this method are well known :

- great saving of time in input and output operations since only the surface of the structure has to be represented, and results are given only on the surface and in areas described by the user,
- great saving in computing time since unknowns are only on the surface thus leading to smaller systems than the ones that are generated by the use of other methods such as Finite Difference or Finite Element ones.

This is the reason why, since 1975, all the analyses of three dimensional compact bodies have been performed by using this program. |1|, |2|, |12|.

Then, in order to be able to take into account thermal loading the program was developed in such a way that elastic or steady state thermal analysis could be performed by the same program. The results of the steady state thermal analysis being then used as data for elastic analysis under thermal loading. |3|, |5|, |6|.

In this paper, we present two of the latest models developed at CETIM on its program based on the Boundary Integral Equation Method to analyse three dimensional structures : cyclic symmetry and perfect sliding between structures.

SHORT REVIEW OF THE METHOD

Any linear partial differential equation with constant coefficients can be transformed into a Boundary Integral Equation.

Once the fundamental solution is known, by applying a reciprocity theorem, a relation is obtained between the quantity that appears as unknown in the linear partial differential equation and other quantities on the surface. Then, by a limiting process a relation between unknowns that lie only on the surface is obtained (fig.1) |7|.

Two examples are given below, the first one concerns elasticity problems whose Governing Equation is the Navier's Equation, (fig.2), the second one concerns steady state heat transfer problems whose governing equation is the Laplace Equation (fig.3).

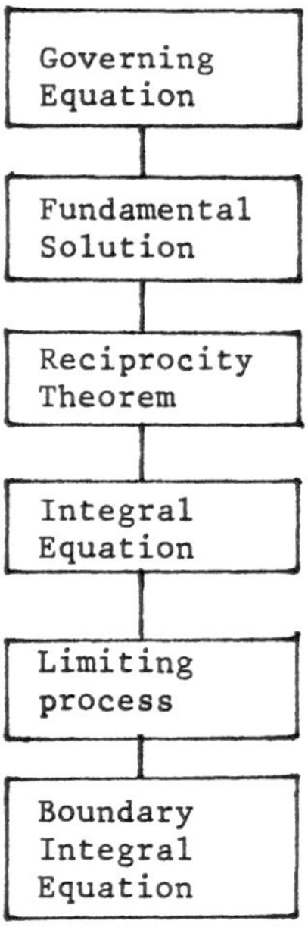

General process to obtain BIE
fig.1

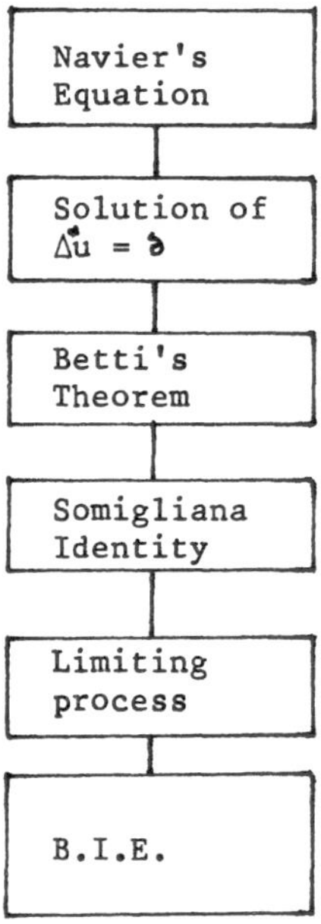

Process applied to Elasticity
fig.2

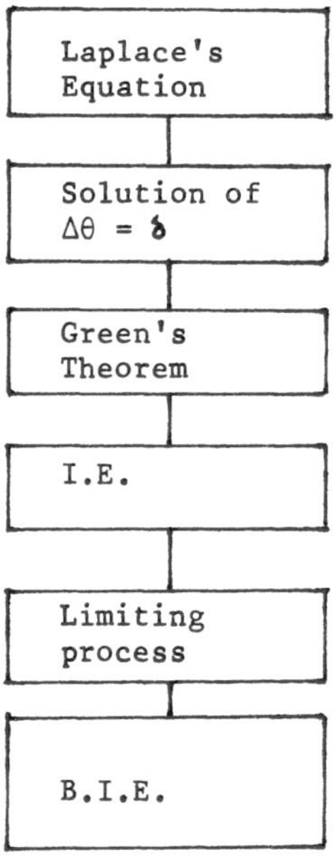

Process applied to steady state heat transfer problems
fig.3

with : δ : Dirac function (fig.1 and 3)
u : displacement (fig.2)
θ : temperature (fig.3)

The final equations obtained by this process are :

- For Navier's Equation :

$$C_{ij}(x)\, u_j(x) + \int_S T_{ij}(x,y) u_j(y) ds_y = \int_S U_{ij}(x,y) t_j(y) ds_y \qquad (1)$$

with : $i = 1,2,3$; $j = 1,2,3$
t = tension

- For Laplace's Equation :

$$C(x)\theta(x) + \int_S T(x,y)\theta(y) ds_y = \int_S U(x,y)\, \psi(y) ds_y \qquad (2)$$

with : Ψ = Flux

In those equations, the Kernels T_{ij} and T are varying as $1/R^2$ and the Kernels U_{ij} and U are varying as $1/R$ in the three dimensional case (fig.4).

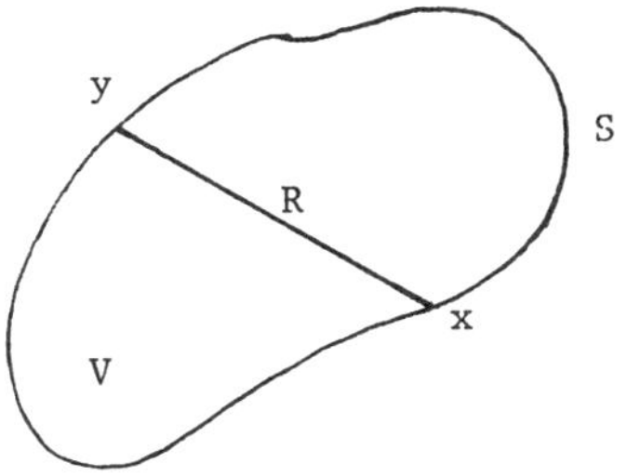

fig.4

|4|,|9|,|10|,|11|,|13|, |14|

To solve a particular problem : three kinds of data are needed :

- the geometry,
- the characteristics of the material
- the boundary conditions (fig.5 and 6)

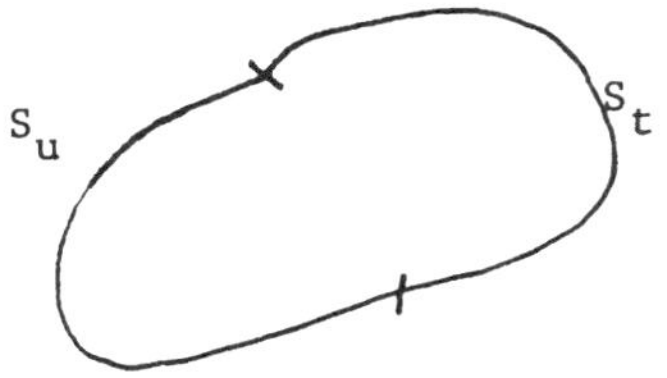

$u=\overline{u}$ on S_u , $t=\overline{t}$ on S_t

Elastic boundary Conditions

fig.5

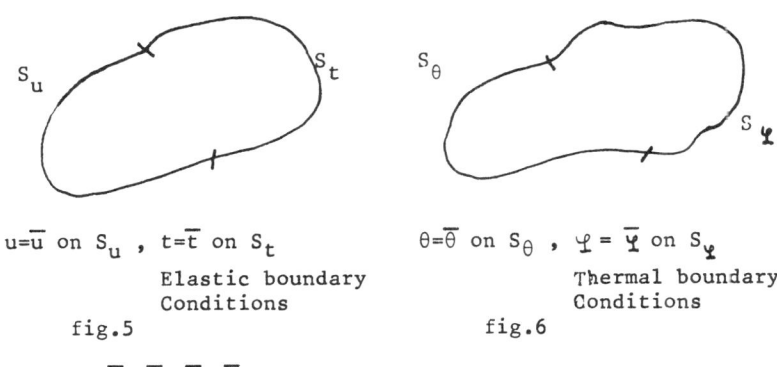

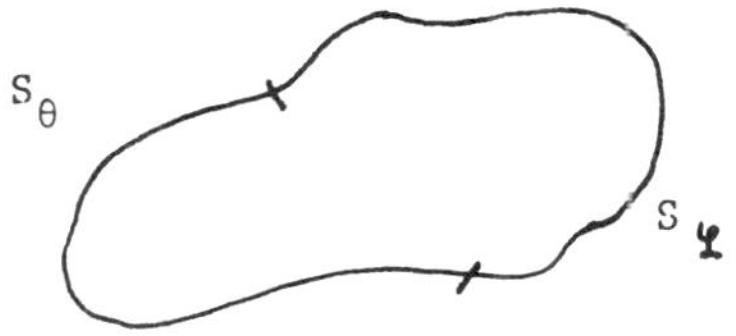

$\theta=\overline{\theta}$ on S_θ , $\Psi = \overline{\Psi}$ on S_Ψ

Thermal boundary Conditions

fig.6

with : $\overline{u}$, $\overline{t}$, $\overline{\theta}$, $\overline{\Psi}$ = Known quantities

The geometry is represented by 8-nodes quadrangular elements and by 6-nodes triangular elements, (fig.7). On each element the unknown may vary linearly or quadratically |15|.

triangular element

quadrangular element

fig.7

In the quadratic case, equations (1) or (2) are written at each node of the elements describing the structure, where the unknowns are located. In the linear case, this is restricted to corner nodes. The number of equations written at each node and the number of unknowns at those nodes is 3 in the elastic case, and 1 in the thermal case.

When using a subregioning process, (fig.8) equations (1) or (2) are written for every subregion, with, on the interface : |8|

- $u_A = u_B \quad (\theta_A = \theta_B)$

- $t_A = -t_B \quad (\varphi_A = -\varphi_B)$

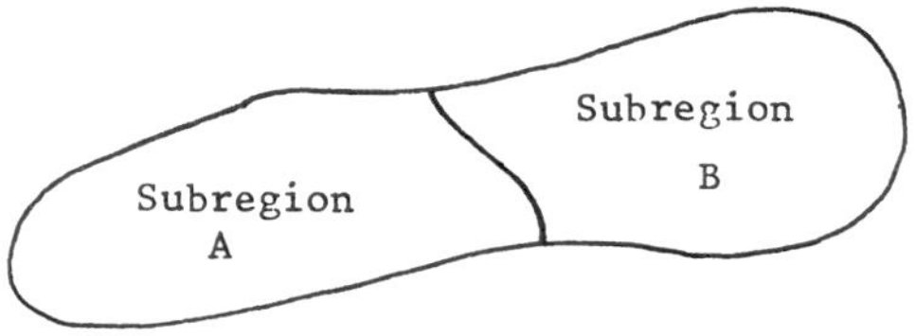

fig.8

Finally, in the elastic case, equations (1) are written in the direction of given displacements at each node.

CYCLIC SYMMETRY

When analysing a body whose geometry is the same from one sector to another around a given axis, and when the loading is radial, it is possible to restrain the analysis of the behaviour of the total structure to the one of only one sector by imposing a normal displacement equal to zero on every element of the radial surfaces representing the sector. But when the loading is not radial (for example, torsion problem), it is no longer possible to study the behaviour of a sector alone with the following boundary conditions : given tension or given displacement on the surface, because on surfaces "inside the volume" (fig.9), tension and displacements are both unknowns.

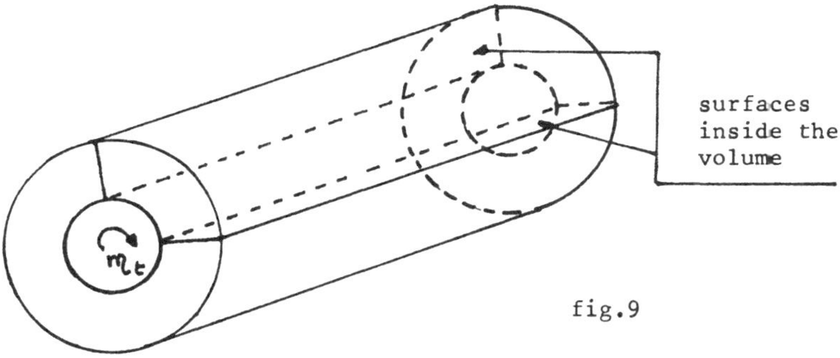

fig.9

Representing the whole structure would lead to a very big linear system of equations. If the boundary conditions (tension and displacement) are the same from one sector to another, by representing each sector by a subregion, it would be possible to save computing time when building the linear system of equations by writing that the matrix corresponding to subregion I is the same as the one corresponding to the first subregion provided all the subregions are the same and the unknowns and equations at each corresponding node of every subregion are the same up to a rotation. But during the reduction of the system, the matrix would become fully populated and computing cost would remain high. A more powerful way to proceed is presented here.

Let consider the structure of figure 9.

Only one sector is modelized (fig.10).

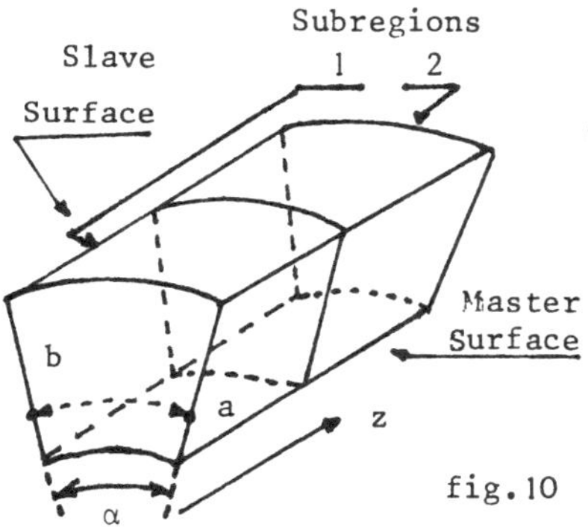

fig.10

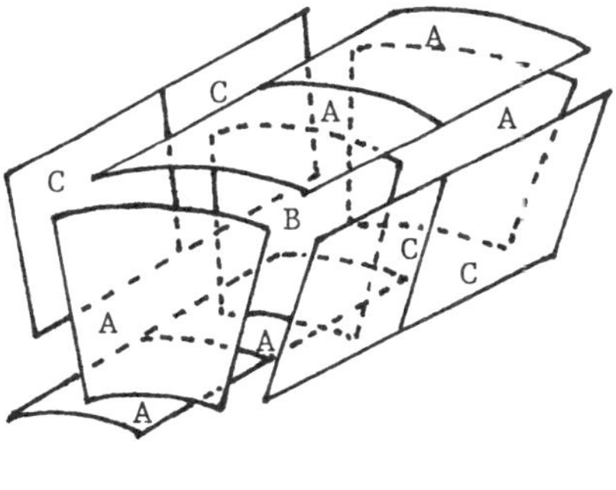

fig.11

On surfaces (A), which correspond to real boundary surfaces of the body, one group of equations is written at each node, the unknown being either the displacement or the tension.

On surface (B), which corresponds to the interface between subregion 1 and subregion 2 both displacement and tension are unknown. On points on this surface a group of equations is written considering the point as belonging to subregion 1, and another group of equations is written considering the point as belonging to subregion 2, the number of equations written at each node being then equal to the number of unknowns.

On surface (C), both tension and displacement are unknown, but only one group of equations is written at each node. In case of cyclic symmetry, the following relations between two corresponding points (fig.10) may be written.

- $u_a = \tilde{u}_b$ (3)

- $t_a = \tilde{t}_b$ (4)

with u_a and t_a = displacement and tension at point a in a given basis

and $\tilde{u}_b$ and $\tilde{t}_b$ = displacement and tension at point b in a basis deduced from the basis at point a by a rotation of angle α around the z axis.

Using these relations, there is an equal number of unknowns and equations :

- number of unknowns : 6 at point a
  6 at point b
- number of equations: 3 equations (1) at point a
  3 equations (1) at point b
  3 relations (3)
  3 relations (4)

Relations (3) and (4) can be taken into account in two ways :

- they can either be added to the initial system of equations in which the unknowns are u_a, u_b, t_a, t_b on the cyclic surfaces

- or they can be taken into account implicitly in the system of equations by writing that unknowns at point a (point belonging to the "master surface") are displacement and that unknowns at point b (point belonging to the "slave surface") are tensions ; the unknowns being expressed in the local basis of the master surface. If we consider only the displacement, in the first case, for any equation there is matrix term U_{ij} for the unknown u_a expressed in the bases B_a , and there is a matrix term U_{ij} for the unknown u_b expressed in the bases B_b . In the second case, for any equation the matrix term is $U_{uj} + \tilde{U}_{ij}$ for the unknown u_a, U_{ij} being expressed in the basis B_a and $\tilde{U}_{ij}$ being expressed in the basis B_b deduced from B_a by a rotation of α around the z axis.

The second method leads to a smaller system of equations than the first one and it is this method that was chosen to be included in the program. Furthermore, if the corresponding points are in the same subregions, the structure of the matrix remains the same as the one without cyclic symmetry and so computing time spent to solve the linear system of equations does not increase. Finally, a special treatment must be performed for points on the z-axis. For these points, only 3 equations (for the elastic case) are written at each node (the "master" node and the "slave" one being the same), and there should be 6 unknowns. But on the z-axis we have :

- $u_r = 0$
- $u_\theta = 0$ (5)
- $\sigma_{\theta z} = 0$ or $t_z = 0$ on the "cyclic surfaces"

The only unknowns that remain are then : u_z - t_θ - t_r with t_θ and t_r : tensions on the cyclic surface, or t_z - t_θ - t_z when the displacement along the z-axis is known, t_z being a tension on an element perpendicular to the z-axis.

For the thermal case, the analysis is performed in almost the same way as the one described above. The differences are :

- no rotation is taken into account (the equations and unknowns being scalars)
- the condition on the z-axis which eliminates one unknown is : no flux on the cyclic surface on this axis.

Industrial example : Analysis of a head of Jack

Consider the structure shown in figure 12.

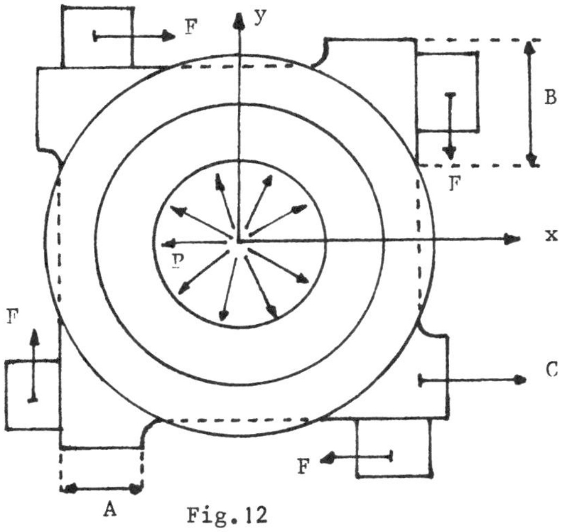

Fig. 12

It is visible that, from a geometrical point of view, it was possible to restrict the analysis of the structure to the analysis of a sector of 90 degrees. As the loading of the structure was not radial, it was not possible to give boundary conditions on the surfaces x = o and y = o.
But as the loading was the same on each pawn, the analysis of the structure could be restricted to the analysis of only one sector by using cyclic symmetry.

The structure that was modelled is shown in figure 13.

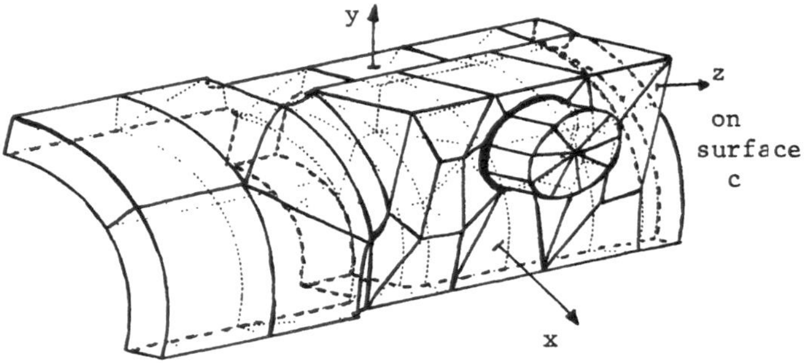

fig.13

The mesh was composed of 2 subregions, 116 elements, 319 nodes leading to a system of about 1200 unknowns.

As far as the meshing of the cyclic surfaces is concerned, only the surface x = o was meshed ; the meshing of the surface y = o being made automatically by the program in order to have the most precise correspondence possible between unknowns on these surfaces.

Zero normal displacement was given on elements corresponding to contact zones between the head of Jack and the exterior body (fig.13).

Fives loadcases were taken into account, each one correspon-ding to a different position of the head of Jack with respect to the exterior body.

Each loadcase was composed of :

- Internal pressure,
- Pressure on some elements of the cylindrical surface of the pawn, this pressure (zone and intensity) was determined earlier by a 2D calculation by the Finite Element Method taking into account contact with friction phenomena.

- Traction on the same elements to take friction into account,
- Pressure on some elements on surfaces A and B in order to have no momentum applied to the structure (those elements were corresponding also to contact zones).

The conclusion of the study was that the stresses were too important at the basis of the pawn for the force that had to be transmitted. In order to come back to a reasonable level of stresses in this area it was decided to increase the diameter of the pawn.

The computation time needed for this analysis was 210 seconds (CP) on a CDC 7600.

SLIDING ON INTERFACE

When there is contact between two bodies, three kinds of contact may exist :

- perfect,
- contact with friction,
- perfect sliding.

When the contact zone is known and when the contact is perfect, the global structure may be analysed by writing conditions of continuity on the displacement field and of equilibrium on the tension field, in the contact zone.

When the contact zone is known and when the kind of contact is perfect sliding, the conditions are (fig.14).

- In the normal direction : $u_{na} = u_{nb}$ and $t_{na} = - t_{nb}$
- In the tangential directions : $t_{ta} = 0$ and $t_{tb} = 0$

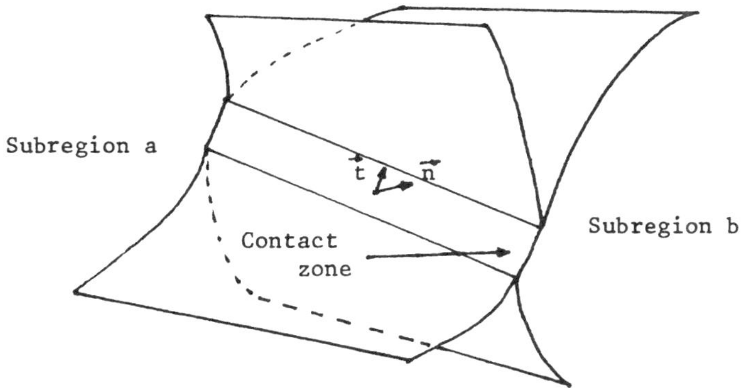

fig.14

The six unknowns at each node on the interface between subregion a and subregion b, in the contact zone are then

u_n, t_n, u_{t1a}, u_{t2a}, u_{t1b}, u_{t2b}

with n : normal direction, and t_1 and t_2 two tangential directions.

When the displacement is known in a tangential direction then the corresponding unknown is the tension on the surface in that direction.

Finally, when the surface of the contact zone is not smooth (fig.15), then for nodes on line L sliding can occur only in the direction d the unknowns are then :

- u_{da}, u_{db} = displacement in the d direction in subregion a and b
- u_{n_1}, t_{n_1} = displacement and tension in the normal direction of element 1
- u_{n_2}, t_{n_2} = displacement and tension in the normal direction of element 2

with t_{n_1} : unknown tension on element 1 and t_{n_2} : unknown tension on element 2 (t_{n_1} on element 2 and t_{n_2} on element 1 beeing equal to zero).

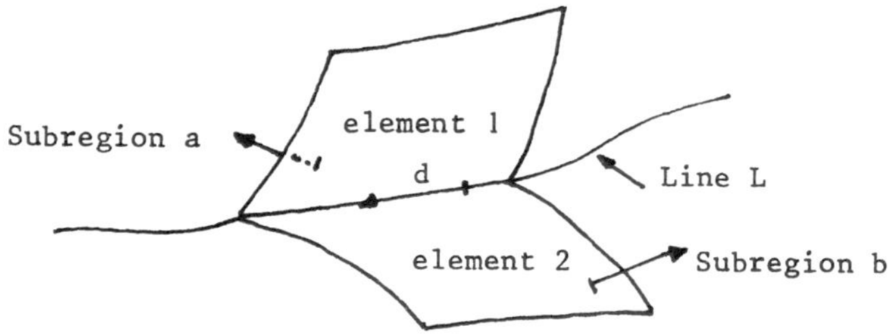

fig.15

<u>Theoretical example : cylindrical shaft sliding into a hollow cylinder</u>

To test the algorithm, we have compared numerical results to analytical ones. Condsidering a shaft sliding into a hollow cylinder (fig.16) the upper surface of the shaft being loaded with a pressure p ; the analytical solution is

- In the shaft :

$$u_r\ (r,z) = -\frac{\nu pr}{E}\left[1 - \frac{1-\nu}{2}\ \frac{R_2^2-R_1^2}{R_2^2}\right]$$

$$u_z\ (r,z) = \frac{pz}{E}\left[1 - \nu^2\ \frac{R_2^2-R_1^2}{R_2^2}\right]$$

$$\sigma_{rr}(r,z) = \sigma_{\theta\theta}(r,z) = \frac{\nu}{2}\ p\ \frac{R_2^2-R_1^2}{R_2^2}$$

$$\sigma_{zz}(r,z) = p$$

$$\sigma_{rz} = \sigma_{\theta z} = \sigma_{r\theta} = 0 \quad ; \quad u_\theta = 0$$

- In the hollow cylinder :

$$u_r(r,z) = -\frac{\nu p}{2E}\ \frac{r}{R_2^2}\left[(1-\nu)R_1^2 + (1+\nu)\frac{R_2^2-R_1^2}{r^2}\right]$$

$$u_z(r,z) = \frac{\nu^2 p}{E}\ \frac{R_1^2}{R_2^2}\ z$$

$$\sigma_{rr}(r,z) = -\frac{\nu}{2}\ p\ R_1^2\left[\frac{1}{R_2^2} - \frac{1}{r^2}\right]$$

$$\sigma_{\theta\theta}(r,z) = -\frac{\nu}{2}\ p\ R_1^2\left[\frac{1}{R_2^2} + \frac{1}{r^2}\right]$$

$$\sigma_{rz} = \sigma_{\theta z} = \sigma_{r\theta} = \sigma_{zz} = 0 \quad ; \quad u_\theta = 0$$

with E : Young Modulus and ν : Poisson Ratio

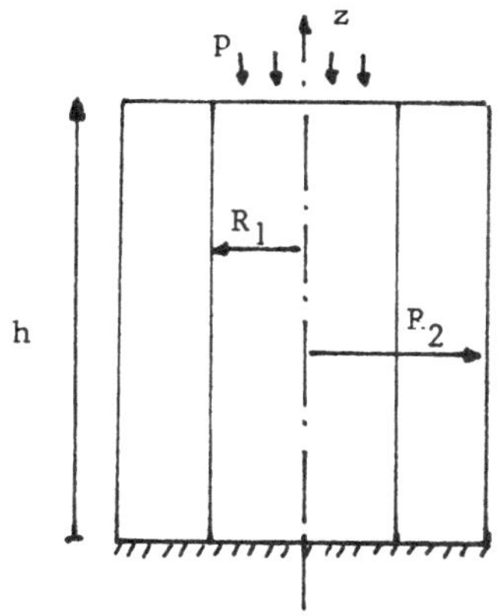

fig.16

The test was done with : $h = 10$ mm
$R_1 = 10$ mm
$R_2 = 20$ mm
$p = -30$ daN/mm^2
$E = 21000$ daN/mm^2
$\nu = .3$

The modelized structure is shown in figure 17.

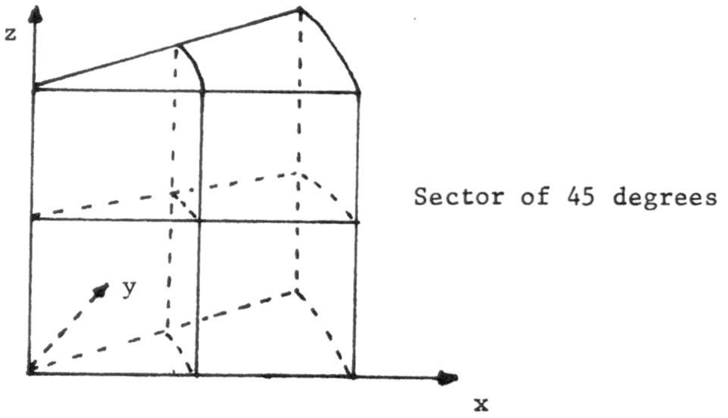

fig.17

The elements on surfaces $\theta = 0$ and $\theta = 45°$ had zero normal displacement. The comparison between analytical and numerical results is given in table 1 for the displacements and in table 2 for the stresses.

On those tables, we can see that the difference between theoretical and numerical results is usually under 10 % of the theoretical result. Furthermore, the biggest differences can be noticed for midside nodes, the numerical results at corner nodes beeing usually very near the theoretical ones.

	Shaft			Hollow cylinder		
	Theoretical Result mm	Numerical result		Theoretical Result mm	Numerical result	
		Min	Max		Min	Max
$u_r(r=5mm)$	$1.58\ 10^{-3}$	$1.53\ 10^{-3}$	$1.55\ 10^{-3}$			
$u_r(r=10mm)$	$3.16\ 10^{-3}$	$3.13\ 10^{-3}$	$3.14\ 10^{-3}$	$3.16\ 10^{-3}$	$3.13\ 10^{-3}$	$3.14\ 10^{-3}$
$u_r(r=15mm)$				$2.42\ 10^{-3}$	$2.39\ 10^{-3}$	$2.40\ 10^{-3}$
$u_r(r=20mm)$				$2.14\ 10^{-3}$	$2.11\ 10^{-3}$	$2.13\ 10^{-3}$
$u_z(z=2.5mm)$	$-3.33\ 10^{-3}$	$-3.31\ 10^{-3}$	$-3.33\ 10^{-3}$	$-8.03\ 10^{-5}$	$-7.97\ 10^{-5}$	$-8.41\ 10^{-5}$
$u_z(z=5mm)$	$-6.66\ 10^{-3}$	$-6.63\ 10^{-3}$	$-6.69\ 10^{-3}$	$-1.61\ 10^{-4}$	$-1.55\ 10^{-4}$	$-1.62\ 10^{-4}$
$u_z(z=7.5mm)$	$-9.99\ 10^{-3}$	$-9.37\ 10^{-3}$	$-1.00\ 10^{-2}$	$-2.41\ 10^{-4}$	$-2.27\ 10^{-4}$	$-2.35\ 10^{-4}$
$u_z(z=10mm)$	$-1.33\ 10^{-2}$	$-1.33\ 10^{-2}$	$-1.35\ 10^{-2}$	$-3.21\ 10^{-4}$	$-2.99\ 10^{-4}$	$-3.25\ 10^{-4}$

Table 1

	Shaft (*)			Hollow cylinder		
	Theoretical Result daN/mm^2	Numerical result		Theoretical Result dan/mm^2	Numerical result	
		Min	Max		Min	Max
σ_r	-3.375	-3.19	-3.47			
σ_θ	-3.375	-3.13	-3.63			
σ_r	- 30	-29.7	- 30			
σ_r(r=10mm)				-3.375	-3.15	-3.47
σ_θ(r=10mm)				5.625	5.53	5.72
σ_r(r=15mm)				-.875	-1.21	-1.28
σ_θ(r=15mm)				3.125	3.07	3.12
σ_r(r=20mm)				0	10^{-9}	10^{-1}
σ_θ(r=20mm)				2.250	2.19	2.25
σ_z				0	10^{-3}	10^{-1}

* Results on the axis of the shaft are not taken into account.

Table 2

CONCLUSION

These two last developments enable us either to analyse the behaviour of structures at lower cost (cyclic symmetry) or to treat new kinds of problems (sliding on interface). When analysing contact problems when the contact zone is not known, it is obvious that the method described above cannot be used right away. In that case an incremental method should be used to determine the contact zone. But since up to now, all inquiries for which sliding had be taken into account were restricted to cases where the contact zone was known, we have decided not to include the complete algorithm in our program. We consider it more important to give new possibilities to our program such as those to take into account elastic support in elastic analysis, or thermal radiations in thermal analysis.

BIBLIOGRAPHY

|1| Boissenot, J.M., Lachat, J.C. and Watson, J.O. (1974) Etude par équations intégrales d'une éprouvette C.T. 15. Revue de Physique Appliquée, Vol.9, pp. 611-615.

|2| Boissenot, J.M., Gazagne, L. and Lange, D. (Mai 1977) Some industrial applications of the boundary integral technique in the field of 3-D elastostatics. International symposium on innovative numerical analysis in applied engineering science Versailles.

|3| Chaudouet, A. (Juin 1980). Les Equations Intégrales de Frontière : outil d'analyse mécanique de pièces industrielles. 2ème colloque international sur les développements nouveaux dans les méthodes numériques de l'ingénieur - Montréal.

|4| Chaudouet, A. and Loubignac, G. (July 1979). Analysis of bodies made of incompressible material under thermal loading by the boundary integral equation method. Numerical Method in Thermal Problems, Swansea.

|5| Chaudouet, A. and Loubignac, G. Boundary Integral Equations used to solve thermoelastic problems application to standard and incompressible materials. Numerical Methods in thermal problems - Ed. John Wiley & Sons Ltd.

|6| Chaudouet. A, Loubignac, G. and Serres, D.(Octobre 80) Utilisation de la méthode des Equations Intégrales pour l'analyse élastique de pièces tridimensionnelles soumises à des chargements thermiques. Congrès AFIAP - Paris.

|7| Dubois, M. and Lachat, J.C. (1972). The integral formulation of boundary value problems. Variational Methods in Engineering. University of Southampton.

|8| Lachat, J.C. (1975). A further development of the boundary integral technique for elastostatics, Ph. D. thesis. University Southampton.

|9| Lachat, J.C. and Watson, J.O. (March 1977). Progress in the use of boundary integral equations illustrated by examples. Computer methods in applied mechanics and engineering. Volume 10 n° 3.

|10| Lachat, J.C. and Watson, J.O. (1975). A second generation boundary integral equation program for three dimensional elastic analysis, (Appl. Mech. Div. ASME 11, New-York.00-00.

BIBLIOGRAPHY

|11| Lachat, J.C. and Watson, J.O. (1976). Effective numerical treatment of boundary integral equation Int. J. Numer. Metho. Eng. 00.

|12| Lange, D. (13 Janv. 1978). Fracture analysis using the boundary integral equation method. Com. Internation. Conf. Num. Methods Fracture Mech., Swansea, 9.

|13| Rizzo, F.J. and Shippy, D.J. (1977). An advanced Boundary Integral Equation method for 3D thermoelasticity. Int. J. Num. Meth. Eng. Vol. 10 p. 1753.

|14| Shaw, R.P. (1974). An integral equation approach to diffusion. Int. J. Heat Transfer. Vol. 17 p. 693.

|15| Zienkiewicz, O.Z. (1978). The finite element method in engineering science. Third edition Mc Graw Hill, London.

ON BOUNDARY INTEGRAL EQUATIONS FOR CIRCULAR CYLINDRICAL SHELLS

H. Antes

Ruhr-University, Bochum, West-Germany

ABSTRACT

The so-called direct approach establishes boundary integral equations with the aid of Somigliana's boundary identity. This identity is based on the Betti's reciprocal work theorem and the so-called fundamental solutions of the basic equations. For elasticity problems in general these singular solutions are due to point loads.

In this work for the linear theory of thin shells a new reciprocal theorem is deduced. There the interaction energy of two elastic states is expressed by derivatives of the displacements and the stress functions.

The knowledge of stress functions, which solve the equilibrium equations for the cases of point loads directed along the three coordinate lines (Antes 1980), gives the possibility to derive from this new reciprocal theorem boundary integral formulations for the solution as well of the geometrical as of the statical boundary value problem. Here this derivation is realized for the case of circular cylindrical shells.

BASIC EQUATIONS

Let Ω be the simply-connected interior of the middle surface of a shell and Γ its boundary. In an operator-matrix formulation the equilibrium conditions for the stress resultant $\underset{\sim}{\sigma}$, the vector of the symmetric membrane forces $\bar{n}^{\alpha\beta}$ and bending moments $m^{(\alpha\beta)}$ are (Antes 1980)

$$\underline{D}^T \cdot (\underset{\sim}{\sigma} + \underset{\sim}{\sigma}^{K*}) = \underline{D}^T \cdot \underset{\sim}{\sigma} + \underset{\sim}{p}^* = \underset{\sim}{0} \qquad (1)$$

in the interior Ω, and

$$\underline{V}^T \cdot \underline{R}_S \cdot (\underset{\sim}{\sigma} + \underset{\sim}{\sigma}^K) - \underset{\sim}{p}^* = \underset{\sim}{0} \tag{2}$$

on the boundary Γ_p, the part of Γ where static terms are prescribed. Note that on the boundary Γ only four quantities can be given independently. These shall be indicated by asterisks.

These and the following equations are written using vector and matrix notations to get as short and clearly arranged expressions as possible. The explicit forms are given for the operator-matrices in an appendix and for the vector quantities in the table.

The relations between the displacement components u_α and the stretching and bending strains $\alpha_{(\alpha\beta)}$ and $\beta_{(\alpha\beta)}$, respectively, are almost identical

$$\underline{D} \cdot \underline{N} \cdot \underset{\sim}{u} = \underset{\sim}{\varepsilon} + \underset{\sim}{\varepsilon}^{Q*} . \tag{3}$$

The so-called extra-strain $\underset{\sim}{\varepsilon}^Q$ is produced by an inhomogeneity, an incompatibility $\underset{\sim}{\eta}^*$ of the compatibility equations:

$$\underline{D}^T \cdot \underline{B} \cdot (\underset{\sim}{\varepsilon} + \underset{\sim}{\varepsilon}^{Q*}) = \underline{D}^T \cdot \underline{B} \cdot \underset{\sim}{\varepsilon} + \underset{\sim}{\eta}^* = \underset{\sim}{0} . \tag{4}$$

The equations (4) are identically satisfied by each $\underset{\sim}{u}$ if the relations (3) are used. Besides, $\underset{\sim}{u}$ must fulfil on Γ_u the prescribed geometric conditions

$$\underline{R}_G \cdot \underset{\sim}{u} = \underset{\sim}{g}^* . \tag{5}$$

A conclusion from analogy gives

$$\underset{\sim}{\sigma} + \underset{\sim}{\sigma}^{K*} = -\underline{B} \cdot \underline{D} \cdot \underline{N} \cdot \underset{\sim}{\psi} = \underline{B}^T \cdot \underline{D} \cdot \underline{N} \cdot \underset{\sim}{\psi} , \tag{6}$$

a rule to determine stresses out of functions $\underset{\sim}{\psi}$, the so-called stressfunctions $\underset{\sim}{\psi}$, which solve the homogeneous equilibrium equations (1) identically.

These differential equation systems (1) and (4) with their pertinent rules (5) and (3), respectively, to gain homogeneous solutions, are connected by the constitutive equations

$$\underset{\sim}{\varepsilon} = \underline{K} \cdot \underset{\sim}{\varepsilon} , \tag{7}$$

$$\underset{\sim}{\sigma} = \underline{K}^{-1} \cdot \underset{\sim}{\varepsilon} . \tag{8}$$

It is easy to show that it is possible to describe the whole problem by only one equation system.
If we use the relation (3), which solves the equations (4) identically, and the constitutive equations (8) we can substitute the stress $\underset{\sim}{\sigma}$ in the equation system (1)

$$\underline{D}^T \cdot \underline{K}^{-1} \cdot (\underline{D} \cdot \underline{N} \cdot \underset{\sim}{u} - \underset{\sim}{\varepsilon}^{Q*}) + \underset{\sim}{p}^* = \underset{\sim}{O} \quad . \tag{9}$$

This equation (9), the reduction of the two systems (1) and (4), can be called the Navier's equation of the linear shell theory.

Alternatively we can use the relation (6) to fulfil the equilibrium equations (1). Then, expressing the strain $\underset{\sim}{\varepsilon}$ with the help of the constitutive equation (7) and this relation (6) by stress functions $\underset{\sim}{\psi}$, the only equation system to solve will be

$$\underline{D}^T \cdot \underline{B} \cdot \underline{K} \cdot (\underline{B}^T \cdot \underline{D} \cdot \underline{N} \cdot \underset{\sim}{\psi} - \underset{\sim}{\sigma}^{K*}) + \underset{\sim}{\eta}^* = \underset{\sim}{O} \quad . \tag{10}$$

In three-dimensional elasticity theory an equivalent system is called Beltrami equation.

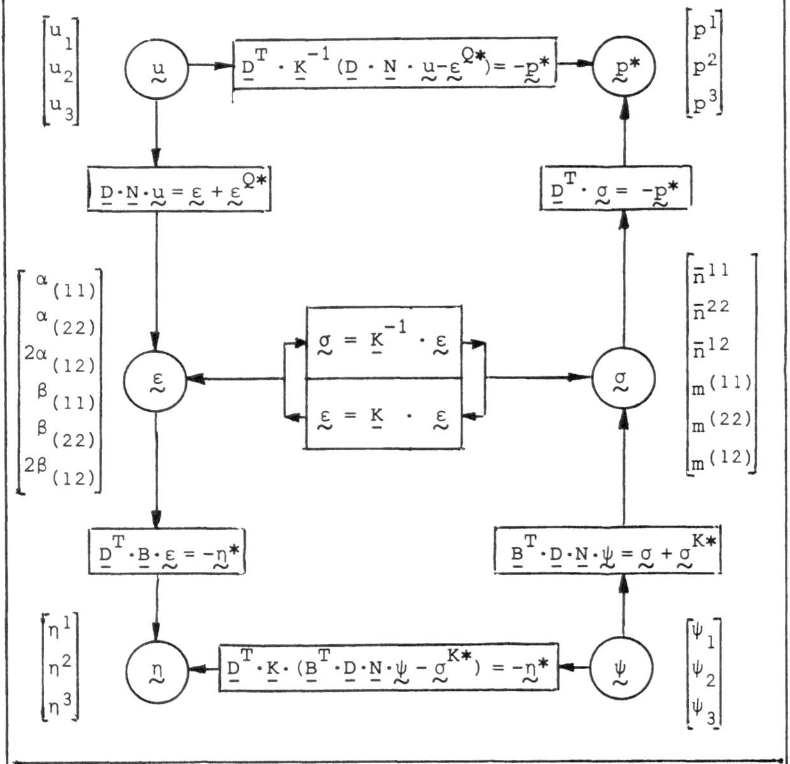

Table : Schematic presentation of the basic equations in a linear theory of thin shells

This table gives a review about connections between these basic equations, similar to the diagrams given from Tonti (1972) for a large class of physical theories.

RECIPROCAL THEOREM

The foundation of the above-mentioned direct approach to integral equations is Betti's reciprocal theorem combining displacements and internal as well as external loadings of two different elastic states. The original basis is the equality of the double internal deformation energy and the external work (Brebbia 1978).

For shells this can be expressed in the vector-matrix notation as follows

$$\int_\Omega \underset{\sim}{\varepsilon}^T \cdot \underset{\sim}{\sigma}\, d\Omega = \int_\Omega \underset{\sim}{u}^T \cdot \underset{\sim}{p}^* \, d\Omega + \int_\Gamma (\underline{R}_G \cdot \underset{\sim}{u})^T \cdot \underline{V}^T \cdot (\underline{R}_S \cdot \underset{\sim}{\sigma})\, d\Gamma \,. \tag{11}$$

Introducing so-called modified strains

$$\underset{\sim}{E} = \underline{B} \cdot \underset{\sim}{\varepsilon} \tag{12}$$

and modified stresses (Antes 1980)

$$\underset{\sim}{\Sigma} = (\underline{B}^T)^{-1} \cdot \underset{\sim}{\sigma} \quad \text{or} \quad \underset{\sim}{\sigma} = \underline{B}^T \cdot \Sigma \quad , \tag{13}$$

the double internal deformation energy may be rewritten as

$$\int_\Omega \underset{\sim}{\varepsilon}^T \cdot \underset{\sim}{\sigma}\, d\Omega = \int_\Omega \underset{\sim}{\varepsilon}^T \cdot \underline{B}^T \cdot \underset{\sim}{\Sigma}\, d\Omega = \int_\Omega (\underline{B} \cdot \underset{\sim}{\varepsilon})^T \cdot \underset{\sim}{\Sigma}\, d\Omega = \int_\Omega \underset{\sim}{E}^T \cdot \underset{\sim}{\Sigma}\, d\Omega \;. \tag{14}$$

Using the relation (6)

$$\underset{\sim}{\sigma} + \underset{\sim}{\sigma}^{K*} = \underline{B}^T \cdot (\underset{\sim}{\Sigma} + \underset{\sim}{\Sigma}^{K*}) = \underline{B}^T \cdot \underline{D} \cdot \underline{N} \cdot \underset{\sim}{\psi} \tag{15}$$

it is possible to transform the surface integral (14) by Stokes' theorem (Antes 1979)

$$\begin{aligned} \int_\Omega \underset{\sim}{E}^T \cdot \underset{\sim}{\Sigma}\, d\Omega &= \int_\Omega \underset{\sim}{E}^T \cdot (\underline{D} \cdot \underline{N} \cdot \underset{\sim}{\psi} - \underset{\sim}{\Sigma}^{K*})\, d\Omega = \\ &= \int_\Omega [(\underline{D}^T \cdot \underset{\sim}{E})^T \cdot \underset{\sim}{\psi} - \underset{\sim}{E}^T \cdot \underset{\sim}{\Sigma}^{K*}]\, d\Omega + \\ &+ \int_\Gamma (\underline{R}_S \cdot \underset{\sim}{E})^T \cdot \underline{V} \cdot (\underline{R}_G \cdot \underset{\sim}{\psi})\, d\Gamma \,. \end{aligned} \tag{16}$$

If we have problems without incompatibilities, that is, with homogeneous compatibility equations (4)

$$\underline{D}^T \cdot \underline{B} \cdot \underset{\sim}{\varepsilon} = \underline{D}^T \cdot \underset{\sim}{E} = \underset{\sim}{O} \quad \text{in } \underset{\sim}{\Omega} \quad , \tag{17}$$

the equation (16) will be reduced to

$$\int_\Omega \underset{\sim}{E}^T \cdot (\underset{\sim}{\Sigma} + \underset{\sim}{\Sigma}^K)\,d\Omega = \int_\Gamma (\underline{R}_S \cdot \underset{\sim}{E})^T \cdot \underline{V} \cdot (\underline{R}_G \cdot \underset{\sim}{\psi})\,d\Gamma \;. \tag{18}$$

If we consider two different states the interaction energy is given equally by

$$\int_\Omega \overset{1}{\underset{\sim}{\varepsilon}}{}^T \cdot \overset{2}{\underset{\sim}{\sigma}}\,d\Omega = \int_\Omega \overset{1}{\underset{\sim}{E}}{}^T \cdot \overset{2}{\underset{\sim}{\Sigma}}\,d\Omega =$$

$$= \int_\Gamma (\underline{R}_S \cdot \overset{1}{\underset{\sim}{E}})^T \cdot \underline{V} \cdot (\underline{R}_G \cdot \overset{2}{\underset{\sim}{\psi}})\,d\Gamma - \int_\Omega \overset{1}{\underset{\sim}{E}}{}^T \cdot \overset{2}{\underset{\sim}{\Sigma}}{}^{K*}\,d\Omega \;. \tag{19}$$

Since the constitutive equations (7) and (8) are uniquely invertible the equation (19) can be written as the following reciprocal theorem:

$$\int_\Gamma (\underline{R}_S \cdot \overset{1}{\underset{\sim}{E}})^T \cdot \underline{V} \cdot (\underline{R}_G \cdot \overset{2}{\underset{\sim}{\psi}})\,d\Gamma - \int_\Omega \overset{1}{\underset{\sim}{E}}{}^T \cdot \overset{2}{\underset{\sim}{\Sigma}}{}^{K*}\,d\Omega$$

$$= \int_\Gamma (\underline{R}_S \cdot \overset{2}{\underset{\sim}{E}})^T \cdot \underline{V} \cdot (\underline{R}_G \cdot \overset{1}{\underset{\sim}{\psi}})\,d\Gamma - \int_\Omega \overset{2}{\underset{\sim}{E}}{}^T \cdot \overset{1}{\underset{\sim}{\Sigma}}{}^{K*}\,d\Omega \;. \tag{20}$$

Here only modified strains $\underset{\sim}{E}$ and stress functions $\underset{\sim}{\psi}$ are used to express the unknown quantities. This new formulation of the reciprocal theorem can be combined with the well known Betti's reciprocal theorem, for shells in the here introduced vector-matrix notation given by

$$\int_\Gamma (\underline{R}_G \cdot \overset{1}{\underset{\sim}{u}})^T \cdot \underline{V}^T \cdot (\underline{R}_S \cdot \overset{2}{\underset{\sim}{\sigma}})\,d\Gamma + \int_\Omega \overset{1}{\underset{\sim}{u}}{}^T \cdot \overset{2}{\underset{\sim}{p}}{}^*\,d\Omega$$

$$= \int_\Gamma (\underline{R}_G \cdot \overset{2}{\underset{\sim}{u}})^T \cdot \underline{V}^T \cdot (\underline{R}_S \cdot \overset{1}{\underset{\sim}{\sigma}})\,d\Gamma + \int_\Omega \overset{2}{\underset{\sim}{u}}{}^T \cdot \overset{1}{\underset{\sim}{p}}{}^*\,d\Omega \;. \tag{21}$$

If we apply the equation (11) the interaction energy formulated in equation (19) can be given in the following well known form

$$\int_\Omega \overset{1}{\underset{\sim}{\varepsilon}}{}^T \cdot \overset{2}{\underset{\sim}{\sigma}}\,d\Omega = \int_\Gamma (\underline{R}_G \cdot \overset{1}{\underset{\sim}{u}})^T \cdot \underline{V}^T \cdot (\underline{R}_S \cdot \overset{2}{\underset{\sim}{\sigma}})\,d\Gamma + \int_\Omega \overset{1}{\underset{\sim}{u}}{}^T \cdot \overset{2}{\underset{\sim}{p}}{}^*\,d\Omega \;. \tag{22}$$

This yields immediately a mixed formulation of the reciprocal theorem:

$$\int_\Gamma (\underline{R}_G \cdot \overset{1}{\underset{\sim}{u}})^T \cdot \underline{V}^T \cdot (\underline{R}_S \cdot \overset{2}{\underset{\sim}{\sigma}})\,d\Gamma + \int_\Omega \overset{1}{\underset{\sim}{u}}{}^T \cdot \overset{2}{\underset{\sim}{p}}{}^*\,d\Omega$$

$$= \int_\Gamma (\underline{R}_S \cdot \overset{2}{\underset{\sim}{E}})^T \cdot \underline{V} \cdot (\underline{R}_G \cdot \overset{1}{\underset{\sim}{\psi}})\,d\Gamma - \int_\Omega \overset{2}{\underset{\sim}{E}}{}^T \cdot \overset{1}{\underset{\sim}{\Sigma}}{}^{K*}\,d\Omega \;. \tag{23}$$

BOUNDARY INTEGRAL EQUATION

The above derived reciprocal theorem (23) can be the basis of a boundary integral equation for the unknown state "1". With it the necessary assumptions are weaker than for Betti's reciprocal theorem. It is only to postulate from the singular state "2", the fundamental state caused by singular unit loads in the point $\underset{\sim}{x}_p$

$$\overset{2}{\underset{\sim}{p}}{}^{*} \mathrel{\hat{=}} \delta(\underset{\sim}{x} - \underset{\sim}{x}_p)\underset{\sim}{e}_k \ , \qquad k = 1,2,3 \ , \tag{24}$$

that the modified strains $\overset{2}{\underset{\sim}{E}}_k$ and the stress $\overset{2}{\underset{\sim}{\sigma}}_k$ are known in the interior Ω and on the whole boundary Γ. This is much less than the knowledge of the displacement field $\overset{2}{\underset{\sim}{u}}$.

For each unit load in the point $\underset{\sim}{x}_p$ we get from equation (23) an integral representation of the displacement component $\overset{1}{u}_k$ in this point $\underset{\sim}{x}_p$

$$\begin{aligned} \overset{1}{u}_k(\underset{\sim}{x}_p) = \int_\Gamma [& (\underline{R}_G \cdot \overset{1}{\underset{\sim}{\psi}}(\underset{\sim}{y}))^T \cdot \underline{V}^T \cdot (\underline{R}_S \cdot \overset{2}{\underset{\sim}{E}}_k(\underset{\sim}{x}_p,\underset{\sim}{y})) - \\ & -(\underline{R}_G \cdot \overset{1}{\underset{\sim}{u}}(\underset{\sim}{y}))^T \cdot \underline{V}^T \cdot (\underline{R}_S \cdot \overset{2}{\underset{\sim}{\sigma}}_k(\underset{\sim}{x}_p,\underset{\sim}{y}))]d\Gamma_y - \\ & -\int_\Omega \overset{2}{\underset{\sim}{E}}{}_k^T(\underset{\sim}{x}_p,\underset{\sim}{x}) \cdot \overset{1}{\underset{\sim}{\Sigma}}{}^{K*}(x)d\Omega_x . \end{aligned} \tag{25}$$

Collecting the three vectors $\underset{\sim}{E}_k$ and $\underset{\sim}{\sigma}_k$, respectively, in matrices

$$\underline{G}^*(\underset{\sim}{x}_p,\underset{\sim}{y}) = [\underset{\sim}{E}_1(\underset{\sim}{x}_p,\underset{\sim}{y}),\ \underset{\sim}{E}_2(\underset{\sim}{x}_p,\underset{\sim}{y}),\ \underset{\sim}{E}_3(\underset{\sim}{x}_p,\underset{\sim}{y})] \ , \tag{26}$$

$$\underline{S}^*(\underset{\sim}{x}_p,\underset{\sim}{y}) = [\underset{\sim}{\sigma}_1(\underset{\sim}{x}_p,\underset{\sim}{y}),\ \underset{\sim}{\sigma}_2(\underset{\sim}{x}_p,\underset{\sim}{y}),\ \underset{\sim}{\sigma}_3(\underset{\sim}{x}_p,\underset{\sim}{y})] \ , \tag{27}$$

the three equations (25) can be combined to one vector integral equation

$$\begin{aligned} \underset{\sim}{u}^T(\underset{\sim}{x}_p) = \int_\Gamma [& (\underline{R}_G \cdot \underset{\sim}{\psi}(\underset{\sim}{y}))^T \cdot \underline{V}^T \cdot (\underline{R}_S \cdot \underline{G}^*(\underset{\sim}{x}_p,\underset{\sim}{y})) - \\ & -(\underline{R}_G \cdot \underset{\sim}{u}(\underset{\sim}{y}))^T \cdot \underline{V}^T \cdot (\underline{R}_S \cdot \underline{S}^*(\underset{\sim}{x}_p,\underset{\sim}{y}))]d\Gamma_y - \\ & -\int_\Omega \underline{G}^*(\underset{\sim}{x}_p,\underset{\sim}{x}) \cdot \underset{\sim}{\Sigma}^{K*}(\underset{\sim}{x})d\Omega_x \ . \end{aligned} \tag{28}$$

This equation states an integral relation between the unknown displacement field in the interior and the unknown boundary reactions.

If the boundary integral disappears, for example, because all boundary values of the fundamental state are zero - such a fundamental state is often called Green's state -, the displacement field $\underset{\sim}{u}$ can be calculated directly by equation (28).

But in general it is necessary to determine the unknown boundary reactions $\underline{R}_G \cdot \underset{\sim}{\psi}$ on Γ_p and/or $\underline{R}_G \cdot \underset{\sim}{u}$ on Γ_u. A boundary integral equation can solve this problem. For that purpose in the equation (28) the singular point $\underset{\sim}{x}_p$ is to move towards the boundary Γ: $\underset{\sim}{x}_p \in \Omega \to \bar{\underset{\sim}{y}} \in \Gamma$.

Dependent on the discontinuity of $\underline{R}_S \cdot \underline{S}^*(\bar{\underset{\sim}{y}},\underset{\sim}{y})$ and $\underline{R}_S \cdot \underline{G}^*(\bar{\underset{\sim}{y}},\underset{\sim}{y})$ for $\underset{\sim}{y} \to \bar{\underset{\sim}{y}}$ and on the smoothness of the boundary Γ (Hartmann 1980) the limiting values of the boundary integrals yield

$$\Big(\int_\Gamma - \Big|\int_\Gamma\Big|\Big)\Big[(\underline{R}_G \cdot \underset{\sim}{u}(\underset{\sim}{y}))^T \cdot \underline{V}^T \cdot (\underline{R}_S \cdot \underline{S}^*(\bar{\underset{\sim}{y}},\underset{\sim}{y})) -$$

$$-(\underline{R}_G \cdot \underset{\sim}{\psi}(\underset{\sim}{y}))^T \cdot \underline{V}^T \cdot (\underline{R}_S \cdot \underline{G}^*(\bar{\underset{\sim}{y}},\underset{\sim}{y}))\Big]d\Gamma_y =$$

$$= \underline{C}_1(\bar{\underset{\sim}{y}}) \cdot (\underline{R}_G \cdot \underset{\sim}{u}(\bar{\underset{\sim}{y}})) - \underline{C}_2(\bar{\underset{\sim}{y}}) \cdot (\underline{R}_G \cdot \underset{\sim}{\psi}(\bar{\underset{\sim}{y}})) \quad . \quad (29)$$

Here $|\int|$ denotes the Cauchy principal value of the boundary integral. The characteristic matrices $\underline{C}_1(\bar{\underset{\sim}{y}})$ and $\underline{C}_2(\bar{\underset{\sim}{y}})$ are calculated as the limiting values (Hartmann 1980)

$$\underline{C}_1(\bar{\underset{\sim}{y}}) = \lim_{\rho \to 0} \int_{\Gamma_\rho} \underline{V}^T \cdot (\underline{R}_S \cdot \underline{S}^*(\bar{\underset{\sim}{y}},\underset{\sim}{y} = \bar{\underset{\sim}{y}} + \rho\underset{\sim}{w}))d\Gamma_y \quad , \quad (30)$$

$$\underline{C}_2(\bar{\underset{\sim}{y}}) = \lim_{\rho \to 0} \int_{\Gamma_\rho} \underline{V}^T \cdot (\underline{R}_S \cdot \underline{G}^*(\bar{\underset{\sim}{y}},\underset{\sim}{y} = \bar{\underset{\sim}{y}} + \rho\underset{\sim}{w}))d\Gamma_y \quad . \quad (31)$$

With these limiting values and the relation (29) we get from the integral relations (28) the following three boundary integral equations, here given as one vector equation

$$\underset{\sim}{u}(\bar{\underset{\sim}{y}}) + \underline{C}_1(\bar{\underset{\sim}{y}}) \cdot (\underline{R}_G \cdot \underset{\sim}{u}(\bar{\underset{\sim}{y}})) + \Big|\int_\Gamma\Big|(R_G \cdot \underset{\sim}{u}(\underset{\sim}{y}))^T \cdot \underline{V}^T \cdot (\underline{R}_S \cdot \underline{S}^*(\bar{\underset{\sim}{y}},\underset{\sim}{y}))d\Gamma_y =$$

$$= \underline{C}_2(\bar{\underset{\sim}{y}}) \cdot (\underline{R}_G \cdot \underset{\sim}{\psi}(\bar{\underset{\sim}{y}})) + \Big|\int_\Gamma\Big|(\underline{R}_G \cdot \underset{\sim}{\psi}(\underset{\sim}{y})^T \cdot \underline{V}^T \cdot (\underline{R}_S \cdot \underline{G}\ (\bar{\underset{\sim}{y}},\underset{\sim}{y}))d\Gamma_y -$$

$$-\int_\Omega \underline{G}^*(\bar{\underset{\sim}{y}},\underset{\sim}{x}) \cdot \underset{\sim}{\Sigma}^{K*}(\underset{\sim}{x})d\Omega_x \quad . \quad (32)$$

In the linear shell theory at any point of the boundary four quantities are given and four reactions are unknown. The essential geometric components are u_1, u_2, u_3 and $\partial u_3/\partial\nu$. They determine uniquely the boundary values $\underline{R}_G \cdot \underset{\sim}{u}$. Formally the same is valid for the static boundary values $\underline{R}_G \cdot \underset{\sim}{\psi}$. Thus it is necessary to obtain a fourth equation.

For that purpose we form the directional derivative of the third component equation of the integral relation (25) in a fixed direction $\underset{\sim}{z}$, with respect to the coordinates of $\underset{\sim}{x}_p$ (Bezine 1978):

$$\frac{\partial u_3}{\partial z}(\underset{\sim}{x}_p) = \int_\Gamma [(\underline{R}_G \cdot \underset{\sim}{\psi}(\underset{\sim}{y}))^T \cdot \underline{V}^T \cdot (\underline{R}_S \cdot \frac{\partial \underset{\sim}{E}_3^*}{\partial z}(\underset{\sim}{x}_p, \underset{\sim}{y}))$$

$$-(\underline{R}_G \cdot \underset{\sim}{u}(\underset{\sim}{y}))^T \cdot \underline{V}^T \cdot (\underline{R}_S \cdot \frac{\partial \underset{\sim}{\sigma}_3^*}{\partial z}(\underset{\sim}{x}_p, \underset{\sim}{y}))] d\Gamma_y$$

$$-\int_\Omega \frac{\partial \underset{\sim}{E}_3}{\partial z}(\underset{\sim}{x}_p, \underset{\sim}{x}) \cdot \Sigma^{K*}(\underset{\sim}{x})\, d\Omega_x \quad . \tag{33}$$

When $\underset{\sim}{x}_p$ tends to a point $\bar{\underset{\sim}{y}}$ of Γ we put $\underset{\sim}{z} = \bar{\underset{\sim}{\nu}}$ where $\bar{\underset{\sim}{\nu}}$ is the unit outward normal vector at the point $\bar{\underset{\sim}{y}}$. Analogous to the above given derivations limiting values are calculated. The result, the fourth boundary integral equation

$$\frac{\partial u_3}{\partial \nu}(\bar{\underset{\sim}{y}}) + C_3(\bar{\underset{\sim}{y}})\underline{R}_G \cdot \underset{\sim}{u}(\bar{\underset{\sim}{y}}) + |\int_\Gamma| (\underline{R}_G \cdot \underset{\sim}{u}(\underset{\sim}{y}))^T \cdot \underline{V}^T \cdot (\underline{R}_S \cdot \frac{\partial \underset{\sim}{\sigma}_3^*}{\partial \nu}(\bar{\underset{\sim}{y}}, \underset{\sim}{y}) d\Gamma_y =$$

$$= C_4(\bar{\underset{\sim}{y}})\underline{R}_G \cdot \underset{\sim}{\psi}(\bar{\underset{\sim}{y}}) + |\int_\Gamma| (\underline{R}_G \cdot \underset{\sim}{\psi}(\underset{\sim}{y}))^T \cdot \underline{V}^T \cdot (\underline{R}_S \cdot \frac{\partial \underset{\sim}{E}_3^*}{\partial \nu}(\bar{\underset{\sim}{y}}, \underset{\sim}{y}) d\Gamma_y$$

$$-\int_\Omega \frac{\partial \underset{\sim}{E}_3}{\partial \nu}(\bar{\underset{\sim}{y}}, \underset{\sim}{x}) \cdot \underset{\sim}{\Sigma}^{K*}(\underset{\sim}{x}) d\Omega_x \tag{34}$$

can contain kernels which behave like $1/r^2$. In this case we revert to the integral relation (33) and integrate by parts with respect to the curvilinear abscissa s tangential to the boundary Γ(Bezine 1978). This procedure gives new kernels like $1/r$. Thus these integrals can be interpreted as Cauchy principal values.

Note that, in general, fundamental state descriptions, where the modified strains $\underset{\sim}{E}_k$ are not derived from the pertinent displacement field $\underset{\sim}{u}_k$, will not satisfy the equation (17). Then the corresponding integral term (see equation (16)) must be supplemented in the reciprocal theorems (20) and (23), in the integral relations (25) and (28), and moreover in the boundary integral equations (32) and (34).

FUNDAMENTAL STATES FOR CIRCULAR CYLINDRICAL SHELLS

If the above given integral formulation of boundary value problems is used, at least the modified strains $\underset{\sim}{E}_k$ and the stresses $\underset{\sim}{\sigma}_k$ must be known. It is not necessary to have the total description of the fundamental state. Both these quantities are determined if singular stress functions $\underset{\sim}{\psi}_k$ have been found which solve the equilibrium equations (1) for the cases of unit point-loads.

It is known (Chernyshev 1963) that the behaviour of shells in the neighbourhood of singular load points is almost like the behaviour of plates and disks, respectively. Thus the stress functions ψ_α of the plate and the Airy's stress function ψ_3 of the disk must be appropriate starting-points.

For circular cylindrical shells the basic equations given before are simplified because coordinate lines can be chosen such that the metric tensor $a_{\alpha\beta}$ is the unit tensor $\delta_{\alpha\beta}$ and the only non-zero component of the curvature $b_{\alpha\beta}$ is $b_{22} = 1/R$.

The necessary condition for the fundamental state "2"

$$\int_\Omega \overset{1}{\underset{\sim}{\varepsilon}}{}^T \cdot \overset{2}{\underset{\sim}{\sigma}} \, d\Omega = \overset{1}{u}_k(\underset{\sim}{x}_p) + \int_\Gamma (\underline{R}_G \cdot \overset{1}{\underset{\sim}{u}})^T \cdot \underline{V}^T \cdot (\underline{R}_S \cdot \overset{2}{\underset{\sim}{\sigma}}) \, d\Gamma \tag{35}$$

becomes substituting the stress function $\overset{2}{\psi}$ by equation (6) and the displacement field $\underset{\sim}{u}$ by the relation (3)

$$\int_\Omega (\underline{D} \cdot \underline{N} \cdot \overset{1}{\underset{\sim}{u}})^T \cdot (\underline{B}^T \cdot \underline{D} \cdot \underline{N} \cdot \overset{2}{\underset{\sim}{\psi}}) \, d \;= \overset{1}{u}_k(\underset{\sim}{x}_p) +$$

$$+ \int_\Gamma (\underline{R}_G \cdot \overset{1}{\underset{\sim}{u}})^T \cdot \underline{V}^T \cdot (\underline{R}_S \cdot \overset{2}{\underset{\sim}{\sigma}}) \, d\Gamma \,. \tag{36}$$

Considering the special case of a circular cylindrical shell and assuming homogeneous boundary conditions this equation (36) is explicitly (Antes 1980):

$$\int_\Omega [u_{1,1}(\psi_{3,22} + \frac{1}{R}\psi_{2,2}) - u_{1,2}(\psi_{3,12} + \frac{1}{R}\psi_{2,1})$$

$$+ u_{2,2}(\psi_{3,11} - \frac{1}{R}\psi_{1,1}) - u_{2,1}(\psi_{3,12} - \frac{1}{R}\psi_{1,2})$$

$$+ \frac{1}{R}(\psi_3 u_{3,11} - \psi_{3,11} u_3) + (\psi_{1,2} u_{3,21} - \psi_{1,1} u_{3,22})$$

$$+ (\psi_{2,2} u_{3,21} - \psi_{2,2} u_{3,11})] d\Omega = u_k(\underset{\sim}{x}_p) \;. \tag{37}$$

For singular normal point-load $p_{3E}\delta(\underset{\sim}{x} - \underset{\sim}{x}_p)\underset{\sim}{e}_3$ the corresponding stress functions of a thin plate can be used. We assume free coefficients in order to have the possibility to consider all shell influences ($x = x_1 - x_{1p}$, $y = x_2 - x_{2p}$)

$$\psi_1^{(3)} = a_1 y \cdot \arctan\frac{x}{y} + b_1 \cdot x \ln r^2 + c_1 x \;, \tag{38}$$

$$\psi_2^{(3)} = a_2 x \cdot \arctan\frac{x}{y} + b_2 \cdot y \ln r^2 + c_2 y \;. \tag{39}$$

The condition (37) determines $a_2 = 1/2\pi$ and requires

$$\psi_3^{(3)} \hat{=} -\frac{1}{R} \int \psi_2^{(3)}(x,y)\,dy =$$

$$= -\frac{1}{R}\,[a_2\, xy \cdot \arctan\frac{x}{y} + (b_{31}x^2 + b_{32}y^2)\ln r^2\,] \,. \qquad (40)$$

Analysing the equilibrium of a small neighbourhood of the singular point $\underset{\sim}{x}_p$ (Antes 1976), for example in the direction $\underset{\sim}{e}_1$

$$\int_{y=-b}^{a} [\bar{n}^{11}(a,y) - \bar{n}^{11}(-a,y)]dy + \int_{x=-a}^{a} [\bar{n}^{12}(x,b) - \bar{n}^{12}(x,-b)]dx =$$

$$= \int_{-b}^{b} [\frac{1}{R}(\psi_{2,2}(a,y) - \psi_{2,2}(-a,y)) + (\psi_{3,22}(a,y) - \psi_{3,22}(-a,y))]dy +$$

$$+ \int_{-a}^{a} [\frac{1}{R}(\psi_{2,1}(x,-b) - \psi_{2,1}(x,b)) + (\psi_{3,12}(x,-b) - \psi_{3,12}(x,b))]dx = 0, \qquad (41)$$

we get $a_1 = -a_2$. The other coefficients are not determined by these statical considerations. Thus the behaviour of the corresponding strains must be studied. In order to get real boundary integral equations, the compatibility equations (17) have to be satisfied in their homogeneous form

$$\underline{D}^T \cdot \underset{\sim}{E} = \underline{D}^T \cdot \underline{B} \cdot \underset{\sim}{\varepsilon} = 0 \quad . \qquad (42)$$

For circular cylindrical shells these conditions are explicitly

$$\frac{1}{R}\alpha_{(22),1} + \beta_{(22),1} - \beta_{(12),2} = 0 \quad , \qquad (43)$$

$$\frac{1}{R}(\alpha_{(11),2} + 2\alpha_{(12),1}) + \beta_{(11),2} - \beta_{(12),1} = 0 \quad , \qquad (44)$$

$$-\alpha_{(11),22} - \alpha_{(22),11} + 2\alpha_{(12),12} + \frac{1}{R}\beta_{(11)} = 0 \quad . \qquad (45)$$

The free coefficients of the stress functions should be chosen in such a way that at least no essential singular terms rest in $\underline{D}^T \cdot \underset{\sim}{E}$. If this is not possible completely, the remaining terms $\underset{\sim}{\eta}^*$ can be considered as coming from an extra-strain $\underset{\sim}{\varepsilon}^{Q*}$

$$\underline{D}^T \cdot \underset{\sim}{E} = \underset{\sim}{\eta}^* = \underline{D}^T \cdot \underline{B} \cdot \underset{\sim}{\varepsilon}^{Q*} = \underline{D}^T \cdot \underset{\sim}{E}^{Q*} \quad . \qquad (46)$$

Then, integrating by parts these terms we get

$$\int\limits_{\Omega}(\underline{D}^{T}\cdot\underset{\sim}{E}^{Q*})\cdot\underset{\sim}{\psi}\,d\Omega = \int\limits_{\Omega}\underset{\sim}{E}^{Q*T}\cdot(\underset{\sim}{\Sigma}+\underset{\sim}{\Sigma}^{K*})\,d\Omega - \int\limits_{\Gamma}(\underline{R}_{S}\cdot\underset{\sim}{E}^{Q*})^{T}\cdot\underline{V}\cdot(\underline{R}_{G}\cdot\underset{\sim}{\psi})\,d\Gamma. \tag{47}$$

For this case instead of the reciprocal theorem (23) the following formulation is to be used as the basis for the integral equations:

$$\int\limits_{\Gamma}(\underline{R}_{G}\cdot\overset{1}{\underset{\sim}{u}})^{T}\cdot\underline{V}^{T}\cdot(\underline{R}_{S}\cdot\overset{2}{\underset{\sim}{\sigma}})\,d\Gamma + \int\limits_{\Omega}\overset{1}{\underset{\sim}{u}}{}^{T}\cdot\overset{2}{\underset{\sim}{p}}{}^{*}\,d\Omega =$$

$$= \int\limits_{\Gamma}(\underline{R}_{S}\cdot(\overset{2}{\underset{\sim}{E}}-\underset{\sim}{E}^{Q*}))^{T}\cdot V\cdot(\underline{R}_{G}\cdot\overset{1}{\underset{\sim}{\psi}})\,d\Gamma + \int\limits_{\Omega}[\underset{\sim}{E}^{Q*T}\cdot\overset{1}{\underset{\sim}{\Sigma}} + (\underset{\sim}{E}^{Q*}-\overset{2}{\underset{\sim}{E}})^{T}\cdot\overset{1}{\underset{\sim}{\Sigma}}{}^{K*}]\,d\Omega \quad . \tag{48}$$

Here we have found the following three stress functions as the best possible ones, provided we use the approximations (38), (39) and (40):

$$\psi_1^{(3)} = \frac{1}{2\pi}\left[-y\cdot\arctan\frac{x}{y} - \frac{(1+\nu)}{4}\,x\cdot\ln r^2 + \frac{(1-\nu)}{4}x\right], \tag{49}$$

$$\psi_2^{(3)} = \frac{1}{2\pi}\left[\ x\cdot\ \arctan\frac{x}{y} - \frac{(1+\nu)}{4}\,y\cdot\ln r^2 + \frac{(1-\nu)}{4}y\right], \tag{50}$$

$$\psi_3^{(3)} = -\frac{1}{R}\frac{1}{2\pi}\left[xy\cdot\arctan\frac{x}{y} + \frac{(3-\nu)}{16}\,(x^2-y^2)\ln r^2\right] . \tag{51}$$

By adequate considerations and calculations the stress functions $\underset{\sim}{\psi}^{(1)}$ and $\underset{\sim}{\psi}^{(2)}$ for the unit-point loads in the direction of the other two coordinate lines can be found too.

CONCLUSIONS

In this paper new formulations of reciprocal theorems for the linear theory of thin non-shallow shells are given. These can be used, as has also been shown, to formulate boundary integral equation if the fundamental states have been found. The research has not yet been finished. Thus it was not possible to check the practicability by numerical examples.

APPENDIX

The operator matrices, which are used to get short expressions for the basic equations of the linear shell theory, are given explicitly in the following. There D_α signify the covariant differentiation $|_\alpha$ along the coordinate lines θ^α. As usual $a_{\alpha\beta}$ and b_α^β mean the covariant metric tensor and the curvature tensor, respectively. ν_α denotes the components of the outward unit normal vector.

$$
\underline{D}^T = \begin{bmatrix}
D_1 & 0 & D_2 & \begin{matrix}-2b_1^1 D_1\\ -b_1^1|_1\end{matrix} & \begin{matrix}-2b_2^1 D_2\\ -b_2^1|_2\end{matrix} & \begin{matrix}-2b_1^1 D_2\\ -2b_1^1|_2\end{matrix} & -2b_2^1 D_1 \\[2ex]
0 & D_2 & D_1 & \begin{matrix}-2b_1^2 D_1\\ -b_1^2|_1\end{matrix} & \begin{matrix}-2b_2^2 D_2\\ -b_2^2|_2\end{matrix} & \begin{matrix}-2b_1^2 D_2\\ -2b_2^2|_1\end{matrix} & -2b_2^2 D_1 \\[2ex]
b_{11} & b_{22} & 2b_{12} & \begin{matrix}-b_{1\lambda}b_1^\lambda\\ +D_{11}\end{matrix} & \begin{matrix}-b_{2\lambda}b_2^\lambda\\ +D_{22}\end{matrix} & \begin{matrix}-2b_{1\lambda}b_2^\lambda\\ +2D_{12}\end{matrix} &
\end{bmatrix}
$$

$$
\underline{V}^T = \begin{bmatrix}
\nu_1 & 0 & \nu_2 & -\nu_1 b_1^1 & -\nu_2 b_2^1 & \begin{matrix}-\nu_1 b_2^1\\ -\nu_2 b_1^1\end{matrix} & 0 & 0 \\[2ex]
0 & \nu_2 & \nu_1 & -\nu_1 b_1^2 & -\nu_2 b_2^2 & \begin{matrix}-\nu_1 b_2^2\\ -\nu_2 b_1^2\end{matrix} & 0 & 0 \\[2ex]
0 & 0 & 0 & 0 & 0 & 0 & \nu_1 & \nu_2 \\[2ex]
0 & 0 & 0 & \nu_1 & 0 & \nu_2 & 0 & 0 \\[2ex]
0 & 0 & 0 & 0 & \nu_2 & \nu_1 & 0 & 0
\end{bmatrix}
$$

$$
\underline{K} = \begin{bmatrix} \underline{F} & \underline{0} \\ \underline{0} & \frac{12}{t^2}\,\underline{F} \end{bmatrix}, \qquad
\underline{K}^{-1} = \begin{bmatrix} \underline{F}^{-1} & \underline{0} \\ \underline{0} & \frac{t^2}{12}\,\underline{F}^{-1} \end{bmatrix},
$$

$$
F = \frac{1}{Et}\begin{bmatrix} a_{11}^2 & \begin{matrix}(1+\nu)a_{12}^2 \\ -\, a_{11}a_{22}\end{matrix} & 2a_{12}a_{11} \\ \text{symm.} & a_{22}^2 & 2a_{12}a_{22} \\ & & \begin{matrix}2(1+\nu)a_{11}a_{22} \\ +2(1-\nu)a_{12}^2\end{matrix} \end{bmatrix},
$$

$$
\underline{F}^{-1} = \frac{Et}{1-\nu^2}\begin{bmatrix} (a^{11})^2 & \begin{matrix}(1-\nu)(a^{12})^2 \\ +\nu a^{11}a^{22}\end{matrix} & a^{12}a^{11} \\ \text{symm.} & (a^{22})^2 & a^{12}a^{22} \\ & & \begin{matrix}\frac{1-\nu}{2}\, a^{11}a^{22} \\ +\frac{1+\nu}{2}\,(a^{12})^2\end{matrix} \end{bmatrix},
$$

$$
\underline{B}^T = \frac{1}{a}\begin{bmatrix} 0 & b_1^1-b_2^2 & -b_2^1 & 0 & -1 & 0 \\ b_2^2-b_1^1 & 0 & -b_1^2 & -1 & 0 & 0 \\ b_2^1 & b_1^2 & 0 & 0 & 0 & \frac{1}{2} \\ 0 & 1 & 0 & 0 & 0 & 0 \\ 1 & 0 & 0 & 0 & 0 & 0 \\ 0 & 0 & -\frac{1}{2} & 0 & 0 & 0 \end{bmatrix},
$$

$$
a = a_{11}a_{22} - a_{12}^2 ,
$$

$$
\underline{N} = \begin{bmatrix} 1 & 0 & 0 \\ 0 & 1 & 0 \\ 0 & 0 & -1 \end{bmatrix} .
$$

$$\underline{R}_S := \begin{pmatrix} 1 & 0 & 0 & 0 & 0 & 0 \\ 0 & 1 & 0 & 0 & 0 & 0 \\ 0 & 0 & 1 & 0 & 0 & 0 \\ 0 & 0 & 0 & 1 & 0 & 0 \\ 0 & 0 & 0 & 0 & 1 & 0 \\ 0 & 0 & 0 & D_1 & 0 & D_2 \\ 0 & 0 & 0 & 0 & D_2 & D_1 \end{pmatrix}$$

$$\underline{R}_G := \begin{pmatrix} 1 & 0 & 0 \\ 0 & 1 & 0 \\ 0 & 0 & 1 \\ -b_1^1 & -b_1^2 & -D_1 \\ -b_2^1 & -b_2^2 & -D_2 \end{pmatrix}$$

REFERENCES

Antes, H. (1976) Über singuläre Lastfälle in einer linearen Schalentheorie und ihre finite Behandlung. Ingenieur-Archiv 45, 99 - 114.

Antes, H. (1979)On Dual-Complementary Variational Principles Generated from Dual Functionals in Linear Shell Theory. Acta Mechanica, 33, 55 - 67 (German, English summary).

Antes, H. (1980) Über Fehler und Möglichkeiten ihrer Abschätzung bei numerischen Berechnungen von Schalentragwerken (Habilitation thesis). Mitteilungen Institut für Mechanik 19, Ruhr-Universität Bochum, W.-Germany.

Bezine, G. (1978) Boundary Integral Formulation for Plate Flexure with Arbitrary Boundary Conditions.Mech. Res. Comm. 5, 197 - 206.

Brebbia, C. A. (1978) The Boundary Element Method for Engineers. Pentech Press, Plymouth, London.

Chernyshev, G. N. (1963) On the action of concentrated forces and moments on an elastic thin shell of arbitrary shape. J. Appl. Math.Mech. 27, 172 - 184.

Hartmann, F. (1980) Elastische Potentiale in Gebieten mit Ecken. Dissertation, University Dortmund, W.-Germany.

Tonti, W. (1972) On the Mathematical Structure of a Large Class of Physical Theories. Rend. Accad. Naz. Lincei. Class. Sci. fis. mat. nat. 52, 48 - 56.

THE BOUNDARY ELEMENT METHOD APPLIED TO TWO-DIMENSIONAL CONTACT PROBLEMS WITH FRICTION

Torbjörn Andersson

Linköping Institute of Technology, Dept of Mech Eng,
S-581 83 Linköping, Sweden

INTRODUCTION

Contact problems and the study of load transfer in mechanical assemblages are of great importance in mechanical engineering. For nearly one century, since H Hertz (1881) published his famous work on normal contact between elastic bodies, much research has been performed in this area, both theoretical and experimental work. An interesting survey of the mechanics of contact between solid bodies is given by Kalker (1977). He gives an account for the classical formulation of the contact problem as well as for the variational one. The latter has been used especially for numerical calculations, for instance with the finite element method, (FEM). In most contact problems the contact area is a function of the external forces. When friction has to be taken into account, the whole load history has to be followed. Thus when using numerical methods, contact problems have to be solved by iteration and in the frictional case also with incremental technique. A lot of computer time has to be spent which makes it important for the system matrix to be small. When sliding occurs in the contact zone, the normal and tangential forces have to be coupled with a friction parameter. The possibility of coupling the normal- and tangential tractions in the matrix system is an advantage for the Boundary Element Method (BEM) over the displacement FEM. This makes it fruitful to use BEM for solving contact problems with friction.

This paper describes a technique for applying BEM to two-dimensional contact problems with friction. The pressure and shear stress distribution between two elastic bodies in contact and the resulting stress distribution for the case of no body forces are calculated. By use of a standard boundary element program the relations between increments for displacements and tractions at the boundaries of the bodies are determined. Considering the friction properties inside the contact area, the two systems of equations are linked together and solved numerically. To get the correct contact area consistent with the external forces the load is applied in increments corresponding to the element model of the contact. In each increment the correct adhesion and slide zone are calculated with iterative technique. The solution from the contact program can be treated by a standard BEM-program for calculating the stresses and strains of the two bodies.

Applications are made to some practical problems and the BEM solutions are compared to solutions from other methods.

The influence of body forces are easy to take into consideration. The general equations can be extended to three-dimensional problems.

BEM-FORMULATION OF THE TWO-DIMENSIONAL CONTACT PROBLEM WITH FRICTION

Basic relationship

Consider two linearly elastic bodies Ω^A and Ω^B bounded by cylindrical surfaces. The traces of the boundary surfaces in the (x_1,x_2)-plane are the curves Γ^A and Γ^B respectively. The geometry and all variable quantities are independent of the x_3-coordinate. The problem is thus considered as a plane problem in the (x_1,x_2)-system, Figure 1.

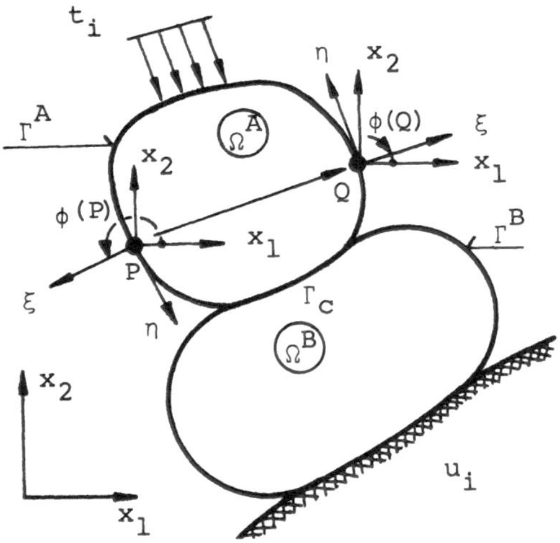

Figure 1 Definition of the problem. Two bodies in contact along the boundary Γ_c

The tractions acting on the boundaries are denoted t_i and the displacement by u_i. The contribution due to body forces is neglected in the following presentation. On the boundary Γ_c

the two bodies are in contact with each other. The contact region can be separated into two parts. On one part Γ_{cs} sliding occurs i.e. the surfaces in contact move relative to each other. This part is called the slip region. On the other part Γ_{ca}, called the adhesion region, no sliding occurs. Thus

$$\Gamma^A_c = \Gamma^A_{ca} + \Gamma^A_{cs}$$

$$\Gamma^B_c = \Gamma^B_{ca} + \Gamma^B_{cs}$$

Γ^A_c and Γ^B_c are the contact boundaries in the undeformed state. According to small displacement theory, $\Gamma_c \approx \Gamma^A_c \approx \Gamma^B_c$.

The relation between increments of boundary values for displacements and tractions in the local coordinate system, defined in Figure 1, may be written Andersson (1980)

$$\int_{\Gamma^k} H^k_{ij}(P,Q)\Delta u^k_j(Q)d\Gamma(Q) = \int_{\Gamma^k} U^k_{ij}(P,Q)\Delta t^k_j(Q)dT(Q),\ k=A,B \qquad (1)$$

From the previous load state a normal gap between the contact boundaries can be calculated, Figure 2. This gap is dencted $u^{o,n}_1$ where n is the load step under consideration.

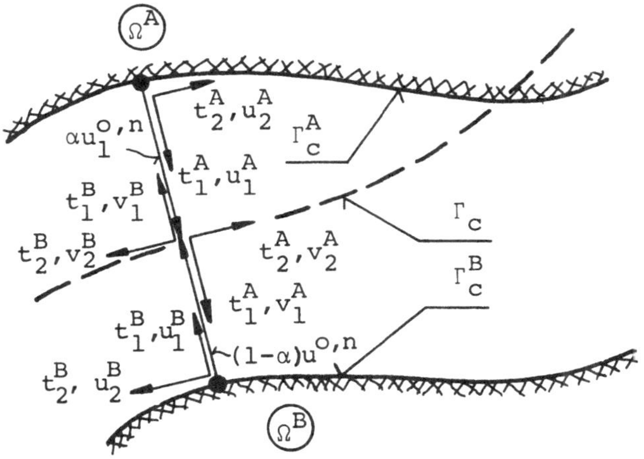

Figure 2 Definition of local coordinate system, the gap $u^{o,n}_1$ and the parameter α

Four new displacement components v^k_i (k = A,B) in the local coordinate system can now be defined.

$$\Delta v_1^A = \Delta u_1^A - \alpha u_1^{o,n} \qquad \Delta v_2^A = \Delta u_2^A \tag{2a}$$

$$\Delta v_1^B = \Delta u_1^B - (1-\alpha) u_1^{o,n} \qquad \Delta v_2^B = \Delta u_2^B \tag{2b}$$

α is a function of the location of the contact boundary and is defined in Figure 2.

In the slip region the relation between the normal- and tangential traction is defined by the coefficient of friction μ.

$$t_2 = \pm \mu \cdot t_1 \tag{3}$$

The sign in (3) depends on the relative displacements so that the tangential force works against the relative displacement and energy is dissipated.

The friction coefficient μ is assumed to depend on the total effective slip v_e, Fredriksson (1976)

$$\mu(v_e) = \mu_m \{1 - (1- \frac{\mu_i}{\mu_m}) \cdot e^{-hv_e} \} \tag{4a}$$

$$v_e = \Sigma |\Delta v_e| , \quad \Delta v_e = \Delta u_2^A + \Delta u_2^B \tag{4b}$$

The parameter h is the hardening coefficient. In the case when the initial friction μ_i and the limit friction μ_m are equal the ideal Coulomb slip is assumed, Figure 3.

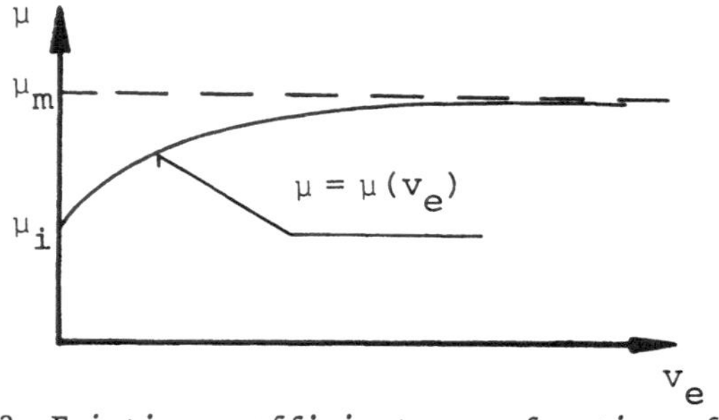

Figure 3 Friction coefficient as a function of effective slip

When the friction coefficient is a function of the slip, it is also a function of the load. Hence, μ is continuously changing during the loading. In the BEM model the load is changed in finite increment. The new μ values for each element pair are then computed after each load step. If the "exact" μ value has to be used in each load step, iterations should be performed to get those values.

Since this iterative technique needs a lot of computer time the μ values are computed approximately from the previous load step without iterations. An effect of this technique is that an element which changes the contact status from adhesion to slip or slip to slip with changed μ value has a residual tangential traction

$$\Delta t^{\varepsilon,n} = \pm \mu^{n-1} t_1^{n-1} - t_2^{n-1} \tag{5}$$

which in the slip to slip case also can be stated, Figure 4

$$\Delta t^{\varepsilon,n} = \pm (\mu^{n-1} - \mu^{n-2}) t_1^{n-1} \tag{6}$$

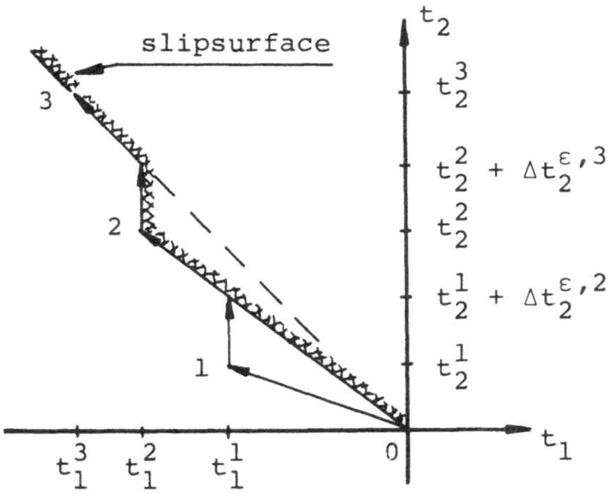

Figure 4 Contact conditions for a pair of contact elements

First load step 0-1: adhesion
Second -"- 1-2: slip
Third -"- 2-3: slip

In slip state the increment in tangential traction is

$$\Delta t_2^n = \pm \mu^{n-1} \Delta t_1^n + \Delta t_2^{\varepsilon,n} \tag{7}$$

This ensures that after each load step the condition for slip, $t_2 = \pm \mu t_1$ is valid, see Figure 4.

The contact conditions are then stated

$$\Gamma_c : \Delta v_1^A + \Delta v_1^B = 0, \quad \Delta t_1^A - \Delta t_1^B = 0, \quad t_1^A < 0, \quad t_1^B < 0 \tag{8a}$$

$$\Gamma_{ca}: \Delta v_2^A + \Delta v_2^B = 0, \quad \Delta t_2^A - \Delta t_2^B = 0 \tag{8b}$$

$$\Gamma_{cs}: \Delta t_2^A - \Delta t_2^B = 0, \quad \Delta t_2^k = \pm\mu\Delta t_1^k + \Delta t_2^{\varepsilon,k} \quad (k = A,B) \tag{8c}$$

The contact variables for body B in equation (1) are now eliminated by use of the contact conditions, equation (8). If these contact variables are written without superscript A the following system of integral equations is obtained

$$\left\{\begin{aligned}
&\int_{\Gamma^A-\Gamma_c^A} T_{ij}^A \,\Delta u_j^A d\Gamma + \int_{\Gamma_{ca}^A} T_{ij}^A \Delta v_j d\Gamma + \int_{\Gamma_{cs}^A} (T_{i1}^A \Delta v_1 + T_{i2}^A \,\Delta v_2^A) d\Gamma = \\
&= \int_{\Gamma^A-\Gamma_c^A} U_{ij}^A \,\Delta t_j^A d\Gamma + \int_{\Gamma_{ca}^A} U_{ij}^A \Delta t_j d\Gamma + \int_{\Gamma_{cs}^A} (U_{i1}^A \pm \mu\, U_{i2}^A) \Delta t_1 d\Gamma + \\
&- \int_{\Gamma_c^A} \alpha T_{ij}^A \, u_1^{o,n} \, d\Gamma + \int_{T_{cs}^A} U_{i2}^A \,\, \Delta t_2^{\varepsilon,n} d\Gamma \qquad (9a) \\
&\int_{\Gamma^B-\Gamma_c^B} T_{ij}^B \,\Delta u_j^B d\Gamma + \int_{\Gamma_{ca}^B} -T_{ij}^B \,\Delta v_j d\Gamma + \int_{\Gamma_{cs}^B} (-T_{i1}^B \,\Delta v_1 + T_{i2}^B \,\Delta v_2^B) d\Gamma = \\
&= \int_{\Gamma^B-\Gamma_c^B} U_{ij}^B \,\Delta t_j d\Gamma + \int_{\Gamma_{ca}^B} U_{ij}^B \,\Delta t_j d\Gamma + \int_{\Gamma_{cs}^B} (U_{i1}^B \pm \mu U_{i2}^B) \,\Delta t_1 d\Gamma \;\; + \\
&- \int_{\Gamma_c^B} (1-\alpha) T_{i1}^B \, u_1^{o,n} \, d\Gamma + \int_{\Gamma_{cs}^B} U_{i2}^B \,\Delta t_2^{\varepsilon,n} d\Gamma \qquad (9b)
\end{aligned}\right.$$

In equations (9) there are for every point on the boundary one pair of variables

$$\Gamma^k - \Gamma_c^k : \; (\Delta u_i^k, \Delta t_i^k) \tag{10a}$$

$$\Gamma_{ca}^k : \; (\Delta v_i, \Delta t_i) \tag{10b}$$

$$\Gamma_{cs}^k : \; (\Delta v_1, \Delta t_1), \; (\Delta v_2^k, \pm \mu\Delta t_1) \tag{10c}$$

Outside the contact region there is one relation for one unknown quantity and inside the contact region there are two relations for two unknown quantities.

The problem is now solvable since, for every boundary point, there is one relation for every unknown quantity.

The parameter α in equations (9) represent the correct contact boundary and can be calculated if Δv_1 are set to zero ($u_1^{o,n} \neq 0$). The solution is however performed practically by giving trail values to α, for example 0.0, 0.5 or 1.0 and solving for the misfit Δv_1 in the contact area.

The problem is linear if the contact-, adhesion- and slip areas are assumed to be constant during the load increment. Hence, the principal of superposition is valid and the problem can be separated into two parts. One part depends on the two last integrals in (9a) and (9b) and the other on the load increment. The last two integrals represent the work needed to get the new contact area forced together and the residual tangential traction to complete the total traction to perform $t_2 = \pm \mu t_1$. With this separation the solution can be done for a load increment up to the total load and scaled afterwards. In the discrete case the scaling is made up to the point where two new elements are contacting each other or where an element pair gets positive normal traction, Figure 5.

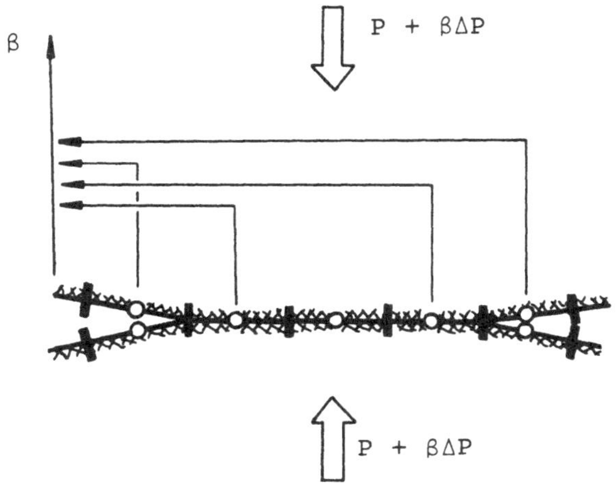

Figure 5 Scaling of the load increment. The scaling parameter β is calculated from the conditions for positive normal traction inside the contact region or geometrical incompatibility outside the contact region

Until now it has been assumed that the adhesion and slip zones are the correct ones and that the tangential traction in the slip zone has the right sign. One way to find the correct solution in each increment is to get a solution for an assumed contact condition that is not violating itself. In the assumed adhesion zone the total tangential traction has to satisfy $|t_2| < \mu|t_1|$. In the slip zone the friction force should work against the relative displacement increments. The condition sign $(t_2) \neq$ sign $(\Delta u_2^A + \Delta u_2^B)$ has thus to be fulfilled.

If the contact conditions are not violated the load can be incremented and the displacement and tractions succesively calculated as

$$t^n = \Delta t^{\varepsilon,n} + \beta \Delta t^n + t^{n-1} \tag{11a}$$

$$u^n = \Delta u^{\varepsilon,n} + \beta \Delta u^n + u^{n-1} \tag{11b}$$

In equation (11) subscript i is omitted.

Boundary element discretization

For the numerical solution of equations (9) the integrals are divided into a sum of integrals over line segments or elements of the boundary. Over each element the geometry, displacements and tractions are assumed to vary according to interpolation functions X_m, Φ_m and ψ_m respectively. The interpolation functions give the variation over the element in terms of values in discrete points or nodes. If $\bar{x}_{im}$, $\Delta\bar{u}_{im}$ and $\Delta\bar{t}_{im}$ are the nodal values for the element the variation can be written

$$x_i = X_m \bar{x}_{im}, \quad \Delta u_i = \Phi_m \Delta\bar{u}_{im}, \quad \Delta t_i = \psi_m \Delta\bar{t}_{im} \tag{12}$$

From equations (2) and (12) it can be concluded that Δv_i and $u_i^{o,n}$ are interpolated by Φ_m.

The interpolation functions are now substituted into equations (9). Integration is performed over the boundary elements. If two variables from different elements act in the same node the results have to be added (assembled). The total number of nodes is N on Γ^A and M on Γ^B. The summation is performed over $N-N_c$ and $M-M_c$ nodes outside the contact area and over N_c and M_c nodes inside the contact area. Note that $N_c = M_c$ and that these nodes have to correspond to each other.

It is now more convenient to work with matrix notations. If the point P successively coincide with every N+M node 2 x (N+M) relations can be established. Collecting the contact variables in the two left hand side vectors the whole system of equations can be written in matrix notation as

$$A\,[x, \beta y] = B\,[c, \beta d] \tag{13}$$

or written out in extensio

$$\left[\begin{array}{ccccc|ccccc} T^A & T^A_{1a} & T^A_{1s} & T^A_{2a} & T^A_{2s} & 0 & -U^A_{1a} & (-U_{1s} \pm \mu U^A_{2s}) & -U^A_{2a} & 0 \\ \hline 0 & T^B_{1a} & T^B_{1s} & T^B_{2a} & 0 & T^B & -U^B_{1a} & (-U^B_{1s} \pm \mu U^B_{2s}) & -U^B_{2a} & T^B_{2s} \end{array}\right] \left[\begin{array}{c|c} \Delta u^A & \beta u^A \\ \Delta v_{1a} & \beta v_{1a} \\ \Delta v_{1s} & \beta v_{1s} \\ \Delta v_{2a} & \beta v_{2a} \\ \Delta v^A_{2s} & \beta v^A_{2s} \\ \hline \Delta u^B & \beta u^B \\ \Delta t_{1a} & \beta t_{1a} \\ \Delta t_{1s} & \beta t_{1s} \\ \Delta t_{2a} & \beta t_{2a} \\ \Delta v^B_{2s} & \beta v^B_{2s} \end{array}\right] =$$

$$= \begin{bmatrix} U^A & 0 & -T^A_1 & U^A_{2s} & 0 \\ 0 & U^B & 0 & U^B_{2s} & -T^B_1 \end{bmatrix} \begin{bmatrix} 0 & \beta t^A \\ 0 & \beta t^B \\ \alpha u_1^{o,n} & 0 \\ \Delta t_2^{\varepsilon,n} & 0 \\ (1-\alpha) u_1^{o,n} & 0 \end{bmatrix} \tag{14}$$

These are two linear systems of equations, each one with 2 x (M+N) relations for 2 x (M+N) unknown nodal quantities. Thus it is solvable if rigid body motion is suppressed.

NUMERICAL SOLUTION AND THE COMPUTER PROGRAM SYSTEM

Forming and solving the system of equations

The numerical solution technique is described for two-dimensional contact problems with friction. The boundary element used is the constant displacement and traction element with one node at the center of the element, Figure 6.

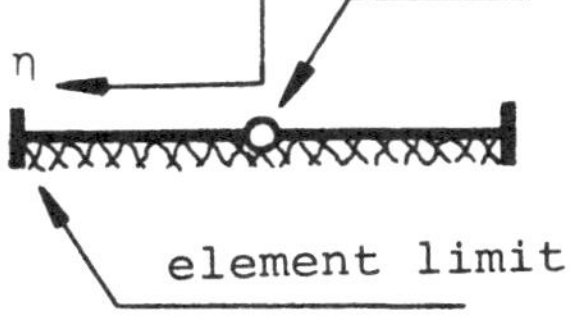

Figure 6 Boundary element used

Even if the formulation is made for a special element it can easily be extended to elements with shape functions of higher order and also to three-dimensional contact problems. The fact that using shape functions with higher order than linear may imply geometrical incompatibilities near the contact boundaries. The same effect is found in FEM-fracture mechanics, Bäcklund and Aronsson (1978). This limits the advantage of such elements. The two bodies under consideration are divided into boundary elements in the usual way, see for example Brebbia (1978). In the expected contact area the pair of contact elements opposite to each other should be modeled in the same length.

With a standard boundary element program the relation of displacement versus traction in the (x_1,x_2)-system can be formed for the two bodies A and B. For each pair of contact elements we define local coordinate systems (ξ,η). The ξ-axis is determined from the normal vectors of the two elements, $\hat{n}_A$ and $\hat{n}_B$ respectively, (see Figure 7).

$$\hat{n}_\xi^A = \frac{\hat{n}^A - \hat{n}^B}{|\hat{n}^A - \hat{n}^B|} = -\hat{n}_\xi^B \tag{15}$$

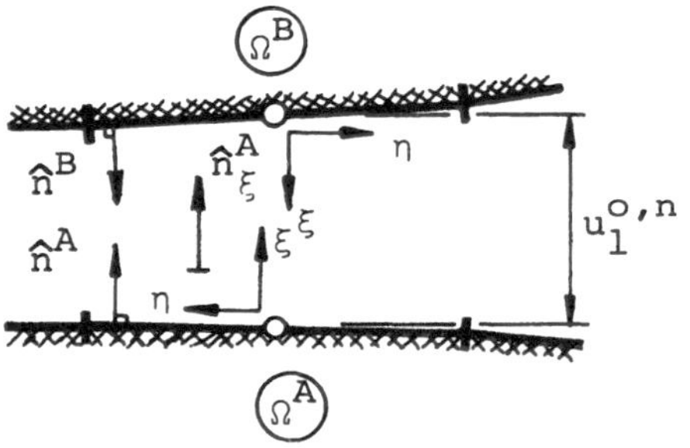

Figure 7 Definition of local coordinate systems for a pair of contact elements

The initial gap $\Delta u_\xi^{o,n}$ between the elements referred to the (ξ,η)-system is calculated from the vectors between the element nodes in the last increment.

$$u_\xi^{o,n} = (\bar{r}_{A-B} - \bar{u}^A - \bar{u}^B) \cdot \hat{n}_\xi^A \tag{16}$$

$\bar{r}_{A-B}$ is the vector between the nodes in undeformed state. The variables which belong to the contact area are transformed to the (ξ,η)-system. The two matrix systems are then linked together in accordance with the contact conditions, equation (8), to from the equation (14). Outside the contact area traction or displacement are known. The variables can be rearranged so that the unknown variables are placed in the left hand side vectors and the known quantities in the right hand side vectors. The right hand side vectors which arise after matrix multiplication are called $[b_1, \beta b_2]$ and the unknown vectors are called $[y_1, \beta y_2]$. Hence, if the system matrix is called A, the system of equations simply becomes

$$A\,[y_1, \beta y_2] = [b_1, \beta b_2] \tag{17}$$

If the rigid body motion is suppressed, the equation (17) can be solved for $[y_1, \beta y_2]$.

Scaling of the contact solution

By use of the equations (2) and (8) all increments in the boundary conditions are known for the two bodies. The total displacement and traction are calculated from equations (11). For each contact pair, β can be calculated to give zero contact pressure

$$\beta = -\frac{\Delta t^{\varepsilon,n} + t^{n-1}}{\Delta t^{n}}, \quad \Delta t^{n} \neq 0 \tag{18}$$

For the element pairs in close vicinity to the contact zone, β are calculated for zero gap, i.e. these new elements are contacting each other.

$$\beta = \frac{\left[\bar{r}_{A-B} - \sum_{k=A,B} (\Delta u^{\varepsilon,n} + \Delta u^{n-1})_k\right] \cdot \hat{n}^A_\xi}{(\Delta u^{n,A} + u^{n,B}) \cdot \hat{n}^A_\xi} \tag{19}$$

Depending on whether the load is increased or decreased the smallest of the β-values from equation (18) or the largest β-value from equation (19) are selected and the total state is calculated for that β-value.

Validity of assumed contact conditions

After scaling the solution the assumed condition for adhesion, slip and sign for the frictional tractions has to be verified. The condition for the adhesion area is

$$|t_2| \leq \mu |t_1| \tag{20}$$

and for the slip area where energy dissipates

$$\operatorname{sign}(t_2) \neq \operatorname{sign}(\Delta u_1^A + \Delta u_2^B) \tag{21}$$

If these conditions are satisfied the correct contact solution is found. If not, new contact conditions have to be chosen and the solution procedure repeated once again. After some iterations the correct contact solution is found for the load state and the problem is thereby solved. It remains to be proved that the solution found with this technique is unique.

Stresses and strains inside the two bodies
The BEM solution for each increment found from the friction program is written to disc store. From this solution stresses and strains inside and on the boundary of the two bodies may be calculated using an ordinary BEM-program, see for example Brebbia (1978) or Andersson (1979).

Computer program
A computer program system for solving elastostatic contact problems with friction using the Boundary Element Method is developed. The program system is built up of three programs and data are transformed between the program by use of disc store, Figure 8. The programs are written for two-dimensional plane strain or plane stress problems and linear elastic material properties. The shape functions for describing displacement and traction variation over each element are assumed to be constant.

The programs are:

- ISOTROPIC — A conventional BEM-program
- GENERATOR — A computer program for generating the T- and U-matrices
- FRICTION — A computer program for solving two-dimensional elastostatic contact problems with friction using incremental and iterative technique.

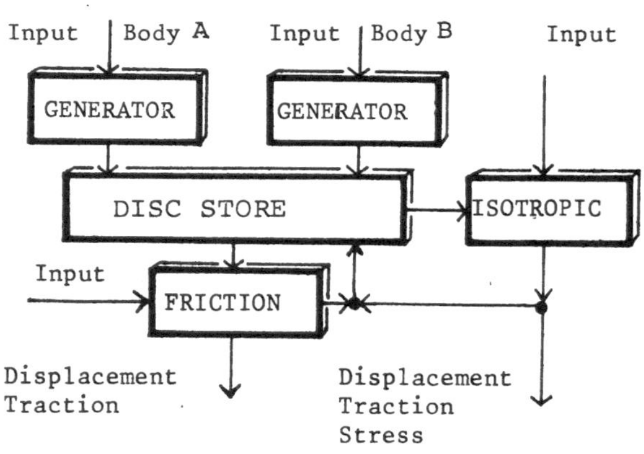

Figure 8 Configuration of the computer program system for solution of elastostatic contact problems with friction

APPLICATION

In order to examine the accuracy of the described technique three problems have been solved. The first two solutions are compared to analytical solutions for the frictionless case. The third problem is a typical engineering problem where no other solutions are available.

Problem 1

An elastic roller on an elastic foundation is considered. The BEM model of the two bodies is shown in Figure 9.

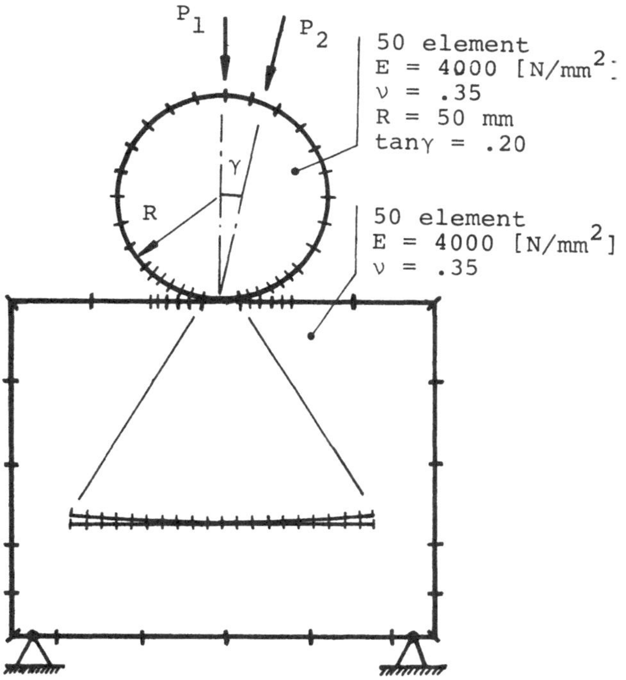

Figure 9 Elastic roller on an elastic foundation, BEM models

The non-linear behavior between contact length and applied force P_1 is compared to Hertz solution in Figure 10. The contact pressure and shear stress distribution for two different load cases with varying μ-values are shown in Figure 11. The actual parameters for the BEM solution are compared to Hertz solution in Table 1. For the BEM solution of the frictionless case, see Andersson(1980). The contact pressure seems to be very little influenced by the friction properties.

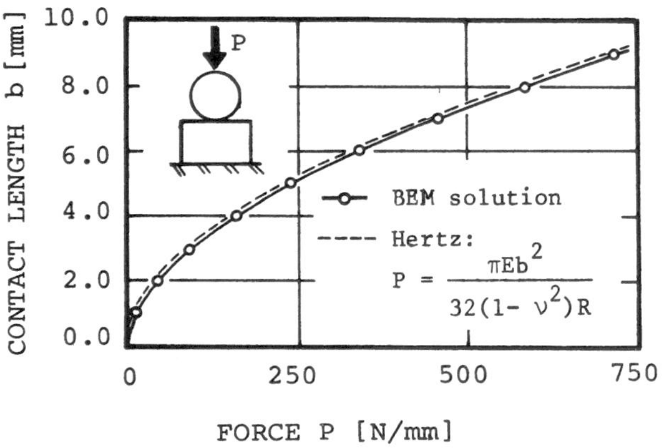

Figure 10 Contact length versus applied force. Comparison between BEM and Hertz solutions

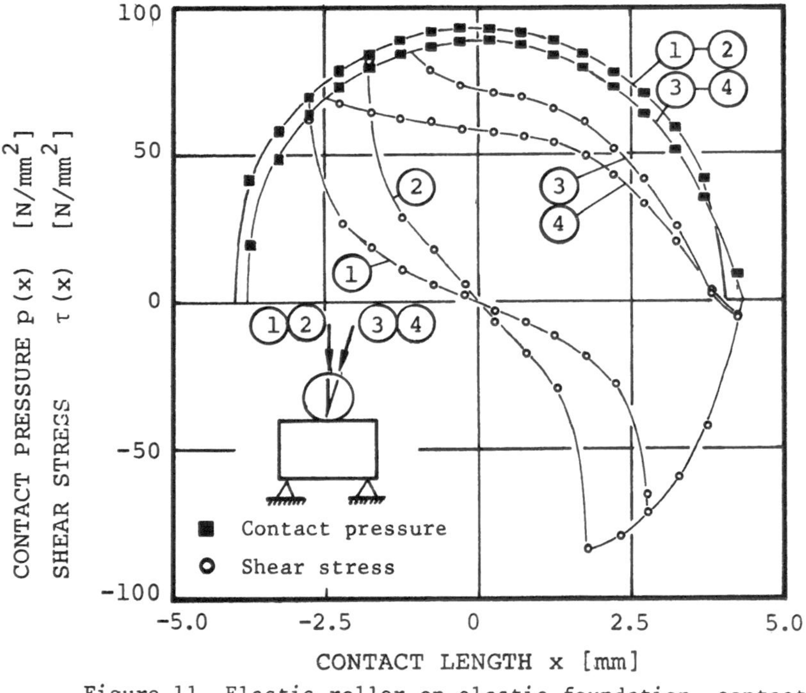

Figure 11 Elastic roller on elastic foundation, contact pressure and shear stress distribution on the foundation

Load case	①	②	③	④	
load step	8	8	9	9	Nr
μ	.05	.01	.25	.30	
P_1	591	591	0	0	N/mm
P_2	0	0	549	549	"
p_{max}	95	95	90	90	N/mm^2
p_{max} Hertz	93	93	88	88	"
b	8.0	8.0	8.1	8.1	mm
b_{Hertz}	8.1	8.1	7.8	7.8	"

Table 1 Hertz problem with friction. Comparison of the BEM and Hertz solutions (Hertz: μ = 0)

Problem 2

A circular disc is placed in a cylindrical cutout of an infinite plate. The radii of the hole of the plate and the disc are approximately the same. In the BEM calculations the clearance is assumed to be $\Delta R/R$ = 0.003. The BEM model is shown in Figure 12.

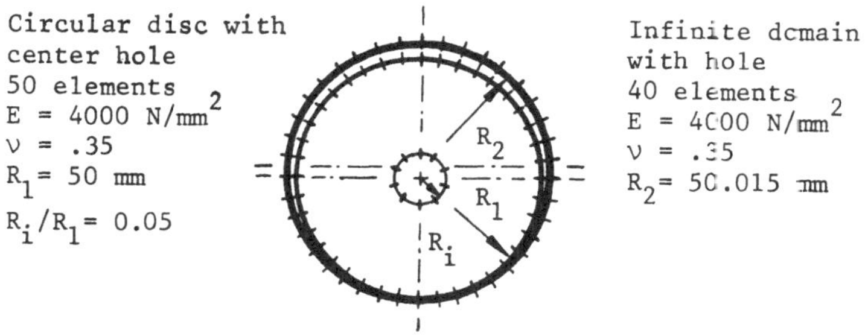

Figure 12 Circular disc in a circular hole of an infinite domain, BEM models

In the center of the disc there has to be a hole where the load is applied. The disc is loaded by a force Q and a moment M. The contact pressure and shear stress distribution for three different cases with varying μ-value and moment are shown in Figure 13.

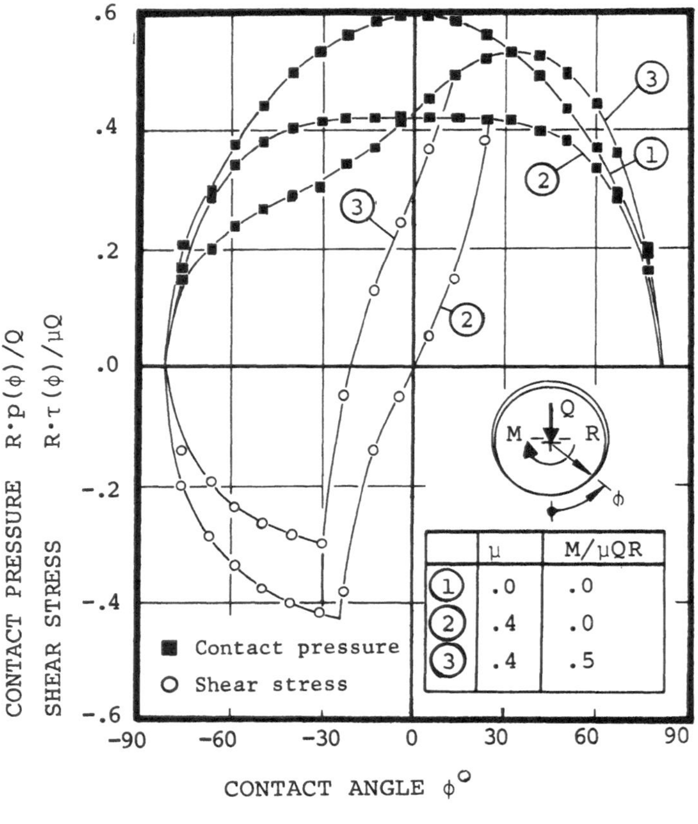

Figure 13 Contact pressure and shear stress distribution for the circular disc

The analytical solution without friction is given by Persson (1964) but also calculated with BEM by Andersson(1980). The BEM solution for $\mu = 0$ is in good accordance with those solutions. The friction has in this problem a clear influence on the maximum contact pressure since the frictional forces take a part of the load.

<u>Problem 3</u>
Load transfer in mechanical assemblage is of practical interest

since high stress gradients often occur and the stiffness properties often determine the load transfer. A problem where a part κ of the load P is transferredbetween the sheets by a steel rivet is shown in Figure 14.

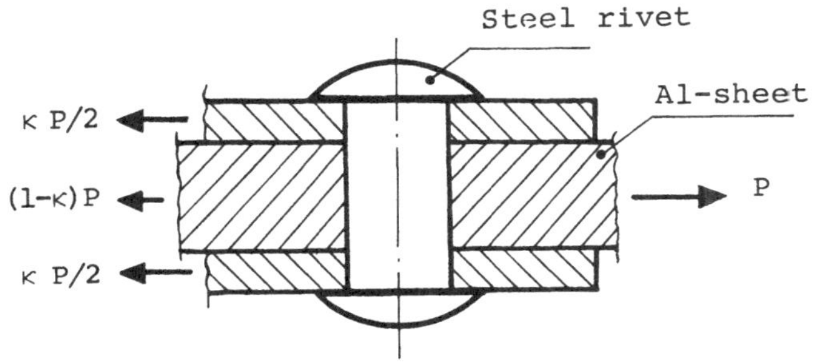

Figure 14 Load transfer in a mechanical assemblage

The middle sheet is assumed to be made of aluminium. No bending effects in the rivet or the sheet are considered. In Figure 15 the BEM model of the problem is shown.

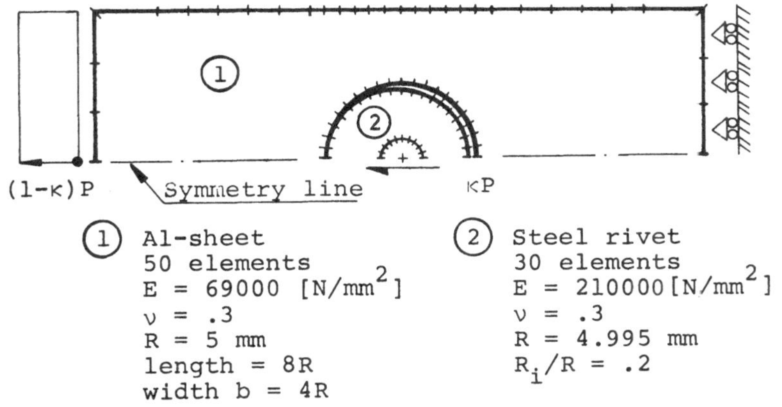

Figure 15 Steel rivet in an aluminium sheet, BEM model

The contact pressure and shear stress distribution for two μ-values and different load transfer κ are shown in Figure 16.

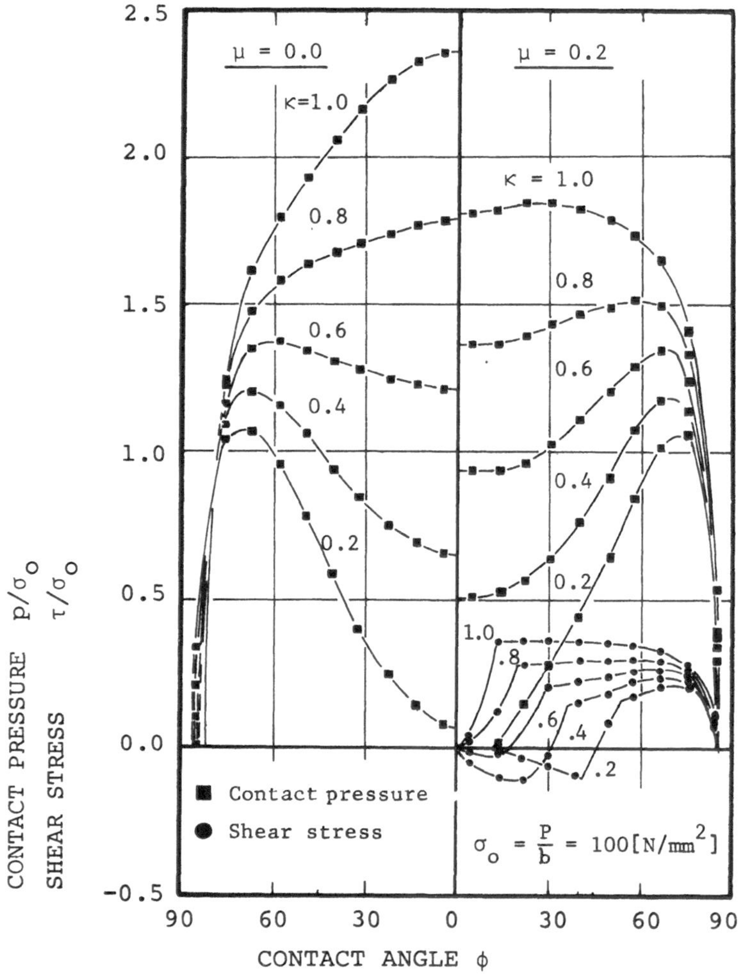

Figure 16 Contact pressure and shear stress distribution for different values of the load transfer

The friction properties influence the contact pressure in accordance with the load transfer by the friction forces. In all cases the clearance is assumed to be $\Delta R/R = 0.001$ and the contact area starts growing from $\phi = 0$.

CONCLUSIONS

In the paper two boundary integral equations for elastic contact including friction between two-dimensional plane structures have been presented. The initial normal gap in the contact zone can be treated in both of the equations using a parameter α. The influence of applied load as well as the initial gap and residue tractions are treated as two different loads resulting in two linear equation systems. The solution is afterwards found by superposition and a scaling technique determined by the element model of the contact region.

The discretization procedure with a simple boundary element and the numerical solution gives good agreement to analytical solutions. The choice of α in the numerical solution has no influence in the presented problems. In the case of an illconditioned matrix system for one of the bodies it might, however, have a numerical effect. The parameter α could then be used to avoid illconditioning.

The incremental and iterative procedure works well for the presented problems. For a more general contact problem where the contact, slip and adhesion zones are changing discontinuously,it would be necessary to restart from a previous load step with an intelligent guess.

REFERENCES

Andersson, T. (1979) The Boundary Element Method with Application to Contact Mechanics, Linköping Institute of Technology, Department of Mechanical Engineering, Report R-135, Linköping. (In Swedish)

Andersson, T. (1979) MARCUS*ISOTROPIC, TA*GENERATOR and TA*CONTACT. Computer package for Analysis of Elastostatic, Frictionless Contacts Between Plane Structures - Users Manual, Linköping Institute of Technology, Department of Mechanical Engineering, Report R-136, Linköping, (In Swedish)

Andersson, T. (1980) "The Boundary Element Method Applied to Two-dimensional Contact Problems". Proceedings of the Second International Seminar on Recent Advances in Boundary Element Methods, Southampton March 1980, Ed. C.A. Brebbia, CML Publications

Argyris, J. H., et al. (1977) ASKA User's Reference Manual, ISD Report No. 73, Rev. D, Stuttgart

Brebbia, C. A. (1978) The Boundary Element Method for Engineers, Pentech Press, London

Bäcklund, J. and Aronsson, C.G. (1978) Effects of Geometrical Incompatibilities on Stress Intensity Factors Calculated by the Finite Element Method. Proc. First International Conference on Numerical Methods in Fracture Mechanics (eds A. R. Luxmore and D. R. J. Owen), Swansea 9-13 Jan., 1979, pp 281-291

Cruse, T. A. (1977) Mathematical Foundation of the Boundary Integral Equation Method in Solid Mechanics, AFOSR-TR-77-1002, Pratt and Whitney Aircraft Group, United Technologies Corporation, East Hartford, Connecticut 06108

Fredriksson, B. (1976) Finite Element Solution of Surface Nonlinearities in Structural Mechanics, Computers & Structures, 6: 281-290

Kalker, J. J. (1977) A Survey of the Mechanics of Contact Between Solid Bodies. ZAMM75, T3-T17

Persson, B. G. A. (1964) On the Stress Distribution of Cylindrical Elastic Bodies in Contact. Diss. Chalmers Inst. of Technology, Gothenburg, Sweden

Rizzo, F. J. (1967) An Integral Equation Approach to Boundary Value Problems of Classical Elastostatics. Q. Appl. Math. 25, pp 83-95

QUASISTATIC INDENTATION OF A RUBBER COVERED ROLL BY A RIGID ROLL - THE BOUNDARY ELEMENT SOLUTION

R. C. Batra

Department of Engineering Mechanics, Univ. of Missouri-Rolla

ABSTRACT

The linear elastic problem involving the indentation of a compressible rubberlike layer bonded to a rigid cylinder and indented by another rigid cylinder is analyzed by the boundary element method. For the same contact width, the pressure at the contact surface is found to depend noticeably upon the thickness of the layer and the Poisson ratio of the material of the layer. Results computed and presented graphically include the pressure distribution over the contact surface, the shape of the indented surface and the stress distribution at the bond surface.

INTRODUCTION

Traction in vehicles, nip action in cylindrical rolls in the paper-making process and in the textile industry, and friction drives are examples of the kind of problem studied herein. Each of these problems involves indentation, by a steel or granite cylinder, of a rubberlike layer bonded to a cylindrical core made also of steel or granite. Hertz's (1881) solution for determining stresses in two bodies in contact is applicable when both bodies are homogeneous and linear elastic and the dimensions of the contact region are small compared with the dimensions of the bodies. The contact problem involving two rollers, one of which is assumed perfectly hard, in contact with their axes parallel is a special case of Hertz's problem and has been investigated by Thomas and Hoersch (1930) on the assumption that plane strain state of deformation prevails. Solutions of the problem when the deformable roller is not homogeneous but consists of a thin elastic layer bonded to a hard supporting core has been given by Hannah (1951), and Hahn and Levinson (1974). Whereas Hannah's approach involves the numerical solution of an integral equation, Hahn and Levinson use an Airy stress function and obtain the solution in the form of a double series one of which converges rather slowly.

In the aforementioned studies the rubberlike layer is taken to be linear elastic and the deformations involved are presumed to be infinitesimal so that the linear theory applies. Also for the case of infinitesimal deformations, Batra et al. (1976) solved the problem by taking the rubberlike layer to be made of a thermorheologically simple material. In order to explore the effects of material and geometric nonlinearities, Batra (1980, 1981) recently studied the problem in which the rubberlike layer is modeled as a homogeneous and incompressible or nearly incompressible Mooney-Rivlin material. We note that Batra used the finite element method to solve the problem.

Experimental work involving varying thicknesses of the rubberlike layer and different combinations of the diameters of mating cylinders has been carried out by Spengos (1965). However, the only material property listed for the rubber is the durometer hardness. For the linear elastic problem involving incompressible material, the durometer hardness is sufficient to find Young's modulus and hence solve the problem. Since the length to diameter ratio for the rolls used by Spengos is of the order of one, the assumption of plane strain state of deformation made in the work referenced above is not valid. For such geometrical configurations one needs to solve the three dimensional problem.

It seems that the Boundary Element Method has an advantage over the Finite Element Method for three dimensional problems since at least the data preparation for the former method is easier. With the ultimate aim of solving the three dimensional problem, in this paper, we solve the two dimensional (plane strain) problem by the Boundary Element Method. Results obtained from the use of the boundary element method are found to compare very favorably with those obtained by Hahn and Levinson (1974).

Formulation of the problem

The system to be studied and the location of rectangular Cartesian axes are shown in Fig. 1. Since the length of rolls perpendicular to the plane of paper is large as compared to their diameters we assume that plane strain state of deformation prevails. Also the deformations of the rubberlike layer are presumed to be infinitesimal so that a linear theory applies. The rubberlike layer is assumed to be made of a homogeneous elastic material. In the absence of friction at the contact surface, the deformations of the rubberlike layer are symmetrical about the line joining the centers of two rolls.

Neglecting the effect of body forces such as gravity, equations governing the quasistatic deformations of the rubber-

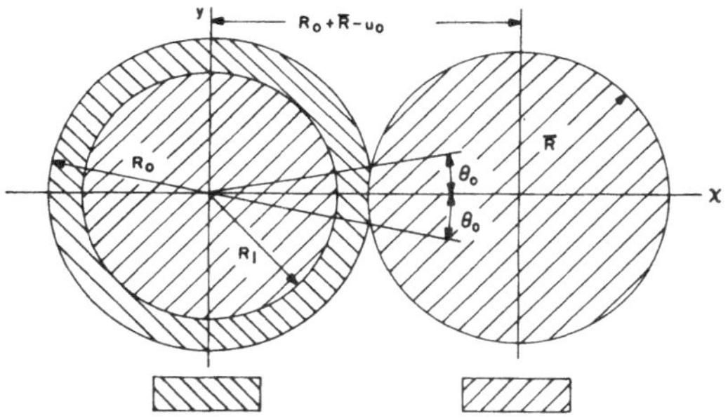

Fig. 1 System to be studied

like layer are

$$\rho = \rho_o (1 - u_{i,i}), \tag{1}$$

$$(L_{ijk\ell}\, u_{k,\ell})_{,j} = 0, \qquad (i,j = 1,2) \tag{2}$$

$$L_{ijk\ell} = \frac{E\nu}{(1+\nu)(1-2\nu)} \delta_{ij}\, \delta_{k\ell} + \frac{E}{2(1+\nu)} (\delta_{ik}\, \delta_{j\ell} + \delta_{i\ell}\, \delta_{jk}). \tag{3}$$

The pertinent boundary conditions are:

at the inner surface S_1 $(x_i x_i = R_1)$, $\quad u_i = o$, (4)

at the outer surface S_2 $(x_i x_i = R_0)$;

$$e_i \sigma_{ij} n_j = 0, \tag{5}$$

$$n_i \sigma_{ij} n_j = 0, \qquad x_1 \geq b \tag{6}$$

$$n_i \sigma_{ij} n_j = p(x_1, x_2), \quad x_2 < b, \tag{7}$$

$$\lim p(x_1, x_2) = 0, \quad x_2 \to b, \tag{8}$$

and at the plane surface S_3 $(x_2 = 0)$;

$$u_2 = 0, \tag{9}$$

$$\sigma_{12} = 0, \tag{10}$$

where

$$\sigma_{ij} = L_{ijk\ell} u_{k,\ell}. \tag{11}$$

In these equations, ρ is the mass density of the material particle in the deformed state, ρ_0 is its mass density in the undeformed state, $\underset{\sim}{u}$ is the displacement of a material particle, E and ν are respectively Young's modulus and Poisson's ratio for the material of the rubberlike layer, a comma followed by an index i indicates partial differentiation with respect to x_i, the summation convention is used, $\underset{\sim}{e}$ is a unit vector tangent to the surface, $\underset{\sim}{n}$ is an outward unit normal to the surface, δ_{ij} is the Kronecker delta, σ_{ij} is the Cauchy stress tensor and the function p gives the pressure distribution at the contact surface. The condition (8) implies that the normal pressure at the boundary of the contact surface vanishes. This ensures that a contact problem rather than a punch problem is being solved. We note that the semicontact width b and the pressure p (x_1,x_2) at the contact surface are not known a priori but are to be determined as a part of the solution. As has been employed by Hahn and Levinson (1974), the boundary condition (7) can be replaced by a displacement type boundary condition. We find it convenient to use the boundary condition (7) rather than the equivalent displacement type boundary condition.

BOUNDARY ELEMENT FORMULATION OF THE PROBLEM

Taking the inner product of both sides of Equation (2) with w_i, integrating the resulting equation over the domain Ω consisting of the region occupied by the layer above the plane $x_2 = 0$ and using the divergence theorem, we obtain

$$\int_{\Omega} (L_{ijk\ell}\, w_{i,j})_{,\ell}\, u_k \, d\Omega = \int_{\partial\Omega} (g_k u_k - f_i w_i)\, ds \tag{12}$$

where

$$f_i = L_{ijk\ell}\, u_{k,\ell}\, n_j, \tag{13}$$

$$g_k = L_{ijk\ell}\, w_{i,j}\, n_\ell. \tag{14}$$

Details of deriving Equation (12) from Equation (2) are given by Brebbia (1978). We now choose w_i to be a solution of

$$(L_{ijk\ell}\, w_{i,j})_{,\ell} - \Delta_k^{(m)} = 0 \tag{15}$$

where $\Delta_k^{(m)}$ is the Dirac delta function and represents a unit load at point m in the k direction. With this choice of w_i, Equation (12) becomes

$$u_i^{(m)} + \int_{\partial\Omega} f_i\, w_i\, ds = \int_{\partial\Omega} g_k u_k\, ds \tag{16}$$

Note that w_i and g_i are displacements and tractions due to a unit concentrated load at the point m in the k direction. Considering unit forces acting in the three directions, Equation (14) can be written as

$$u_i^{(m)} + \int_{\partial\Omega} W_{i\ell}\, f_\ell\, ds = \int_{\partial\Omega} G_{ik}\, u_k\, ds \tag{17}$$

where $W_{i\ell}$ and $G_{i\ell}$ represent the displacements and tractions in the ℓ direction due to a unit force acting in the i direction. Equation (17) is valid for the particular point m where these forces are applied. Expressions for $W_{i\ell}$ and $G_{i\ell}$ are given in Brebbia's book (1978). In Equation (17) the integrations are over the boundary of the domain. At points of the boundary, either surface tractions f_i or displacements u_i or a suitable combination of the two are known. Equation (17) enables us to solve for the unknown surface tractions or unknown surface displacements at each point of the boundary.

COMPUTATION AND DISCUSSION OF RESULTS

In order to solve the problem by the boundary element method, the boundary of the rubberlike layer in the first quadrant is divided into a large number of segments as shown in Fig. 2. Displacements and surface tractions within each segment (boundary element) are assumed to be constant. The mesh is finer within approximately twice the semicontact width. As has been shown in earlier studies (Batra, 1980, 1981), the boundary conditions on the vertical plane have essentially no effect on the pressure profile at the contact surface and the deformations of the rubberlike layer within the vicinity of the contact region. In the results presented herein the vertical plane is taken to be traction free. The computer program given in Brebbia's (1978) book for constant boundary elements has been modified to solve the present contact problem.

Half nip width b and the form of the pressure function p are presumed. With this additional data the problem is well defined and can be solved. Having solved the problem a check is made to insure that the deformed surface in the assumed contact zone matches, within a prescribed tolerance, with the circular profile of the indentor and that the nodal point just outside the contact area has not penetrated into the indentor. If the second condition is not satisfied implying thereby that the nodal point just outside the presumed contact width has penetrated into the indentor, either the value of b is increased or the total load is decreased. However, if the second condition is satisfied and the first is not, the values of p at various nodal points are adjusted so that the deformed shape of the assumed contact area conforms to the circular profile of the indentor. The deformed surface of the roll cover is taken to match with the profile of the indentor if the distance of each nodal point on the contact surface from the indentor is less than 1.5 percent of the indentation u_0 (see Fig. 1). Usually, with a little experience one can make pretty good estimates of b and p (x_1,x_2) so that the entire process converges in four or five iterations. The total load P is obtained by integrating p over the contact area.

Test cases To ensure that the discretization of the boundary was adequate, the contact problem with $R_0 = \bar{R} = 454.7$ mm, $R_1 = 441$ mm, $b = 22.9$ mm, $E = 8.96 \times 10^5$ N/m^2, and $\nu = .3$ was solved. The pressure on the contact surface matched exactly with that obtained by Hahn and Levinson (1974). Stresses on the bond surface could not be compared since such information is not given in Hahn and Levinson's work for $\nu = .3$ and our formulation is not valid for $\nu = .5$.

Next we solved the problem for a different set of material and geometric parameters, namely that $R_0 = 61$ mm, $\bar{R} = 76.2$ mm, $R_1 = 46.5$ mm, $b = 12.7$ mm, $\nu = .45$ and $E = 1 \times 10^6$ N/m^2. Batra (1981) has solved this problem by the finite element method. The pressure profile at the contact surface came out to be virtually the same by the two methods. For this case we plot, in Fig. 3, the variation of the normal stress σ_{rr} and the shear stress $\sigma_{r\theta}$ at the bond surface with the angular distance from the center line of rollers. These stresses decay to essentially 0 as θ/θ_0 approaches 3. Thus the assumption that the vertical plane is traction free is verified to be true. In this and all other figures $\theta_0 = b/R_0$.

Effect of thickness of the layer Keeping $R_0 = 61$ mm, $\bar{R} = 76.2$ mm, $b = 12.7$ mm, $\nu = .3$ and $E = 1 \times 10^6$ N/m^2 fixed, the value of R_1 was varied. The total load required and the resultant indentation for various thicknesses of the layer are listed in Table 1. As expected, the load required to keep the contact width constant increases as the thickness of the layer decreases. For thinner layer the indentation is less even though the load is more. However, the indentation/thickness increases with the decrease in the thickness of the layer.

Thickness/R_0	Load P/(Gb)	Indentation/b	$\frac{\text{Indentation}}{\text{Thickness}}$
.2343	2.162	.2784	.2486
.2092	2.182	.2550	.2550
.1883	2.216	.2368	.2631
.1674	2.272	.2190	.2738
.1454	2.303	.1974	.2820
.1172	2.633	.1802	.3218

Table 1. Load and indentation for different thicknesses of the layer

Fig. 4 depicts the pressure profile at the contact surface for two different thicknesses of the layer. Note the change in the shape of the curve near the ends of the contact width as the thickness of the layer is reduced. With the decrease in the thickness of the layer it seems that the pressure at the center increases more than the decrease in the pressure near the ends so that the total load increases. Such a change in the curvature of the curve near the ends of the contact width is also apparent in Meijer's (1968) work which studied the indentation, by a rigid cylinder, of a layer fixed to a rigid plane surface.

In Fig. 5 is shown the deformed surface of the rubberlike layer for two different thicknesses of the layer. Note that the scales along the vertical and horizontal axes are different. Therefore, the undeformed position of the rubberlike layer plots as an ellipse rather than as a circle. It seems that the change in the curvature of the deformed surface near the ends of the contact zone is heavily dependent upon the thickness of the rubberlike layer.

Effect of Poisson's ratio To investigate the effect of Poisson's ratio on the load required to cause the same contact width the values of R_0 = 61 mm, $\bar{R}$ = 76.2 mm, R_1 = 46.5 mm, b = 12.7 mm, and the shear modulus G = 3.45 x 10^5 N/m^2 are kept fixed. The total load required and the resulting indentation for various values of ν are listed in Table 2. These data show that

Poisson's Ratio	Load/(Gb)	Indentation/b
.41	1.878	.2850
.42	1.951	.2844
.43	2.030	.2840
.45	2.162	.2784
.47	2.154	.2599
.48	2.152	.2506
.49	2.147	.2414

Table 2. Load and indentation for various values of Poisson's ratio

the total load increases with the increase in the value of Poisson's ratio upto a certain value of ν. As ν is increased beyond .45, the total load changes little but the pressure distribution on the contact surface differs in that it is more at the center and less near the ends. This becomes clear from Fig. 6 in which we plot the pressure distribution at the contact surface for two different values of ν.

Remarks

We should emphasize that the results presented above and the conclusions drawn are applicable only to the geometric configuration studied herein. For other values of geometric parameters, one will expect similar qualitative and not necessarily quantitative results.

For the present problem the CPU time and the core requirement for the finite element solution and the boundary element solution were virtually the same. In the boundary element solution of the problem the grid used had 10 elements across the thickness of the layer and 35 elements across the circumference. The boundary element solution perhaps gave slightly better values of tractions at the bond surface. The grid data in both cases were generated internally in the computer program and required the same effort.

Even though it is usually said that the finite element solution generates a lot of unneeded information about stresses and strains within a body, to us, it seems to be an advantage especially in a problem like the one studied here. In such a problem one does not know a priori where the maximum principal stress or the maximum principal strain will occur. It is more expensive to compute stresses and strains at internal points from the boundary element solution than it is from the finite element solution.

REFERENCES

Batra, R. C., Levinson, M and Betz, E. (1976) Rubber Covered Rolls--The Thermoviscoelastic Problem. A Finite Element Solution, Int. J. Num. Meth. Engng. 10: 767-785.

Batra, R. C. (1980) Rubber Covered Rolls--The Nonlinear Elastic Problem, Trans. ASME - J. Appl. Mechs., 47: 82-86.

Batra, R. C. (1981) Quasistatic Indentation of a Rubber Covered Roll by a Rigid Roll, Int. J. Num. Meth. Engng. 16: (to appear).

Brebbia, C. A. (1978) the Boundary Element Method for Engineers, John Wiley & Sons, New York.

Hahn, H. T. and Levinson, M (1974) Indentation of an Elastic Layer Bonded to a Rigid Cylinder--I. Quasistatic Case Without Friction. II - Unidirectional Slipping with Coulomb Friction, 16: 489-5.

Hannah, Margaret (1951) Contact Stress and Deformation in a Thin Elastic Layer, Quart. Journ. Mech. and Applied Math. 4: 94-105.

Hertz, H. (1881) J. F. Math. 92: 156-171.

Meijers, P. (1968) The Contact Problem of a Rigid Cylinder on an Elastic Layer, Appl. Sci. Res. 18: 353-383.

Spengos, A. C. (1965) Experimental Investigation of Rolling Contact, Trans. ASME - J. Appl. Mechs., 32: 859-864.

Thomas, H. R. and Hoersch, V. A. (1930) University of Ill. Eng. Exp. Sta. Bull. 212.

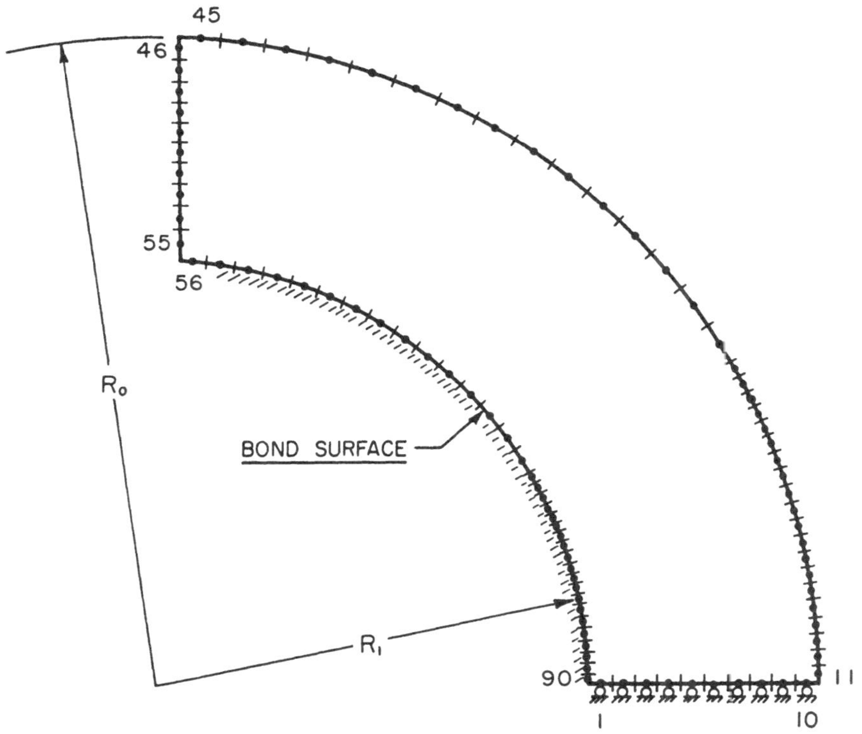

Fig. 2 Boundary element grid

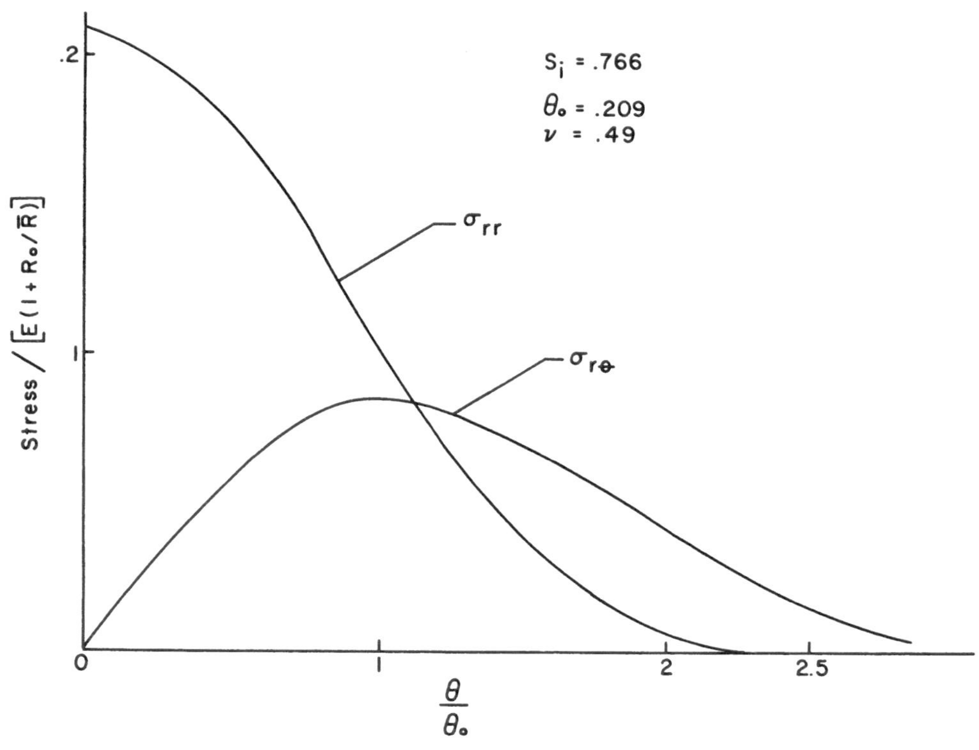

Fig. 3 Stresses at the bond surface

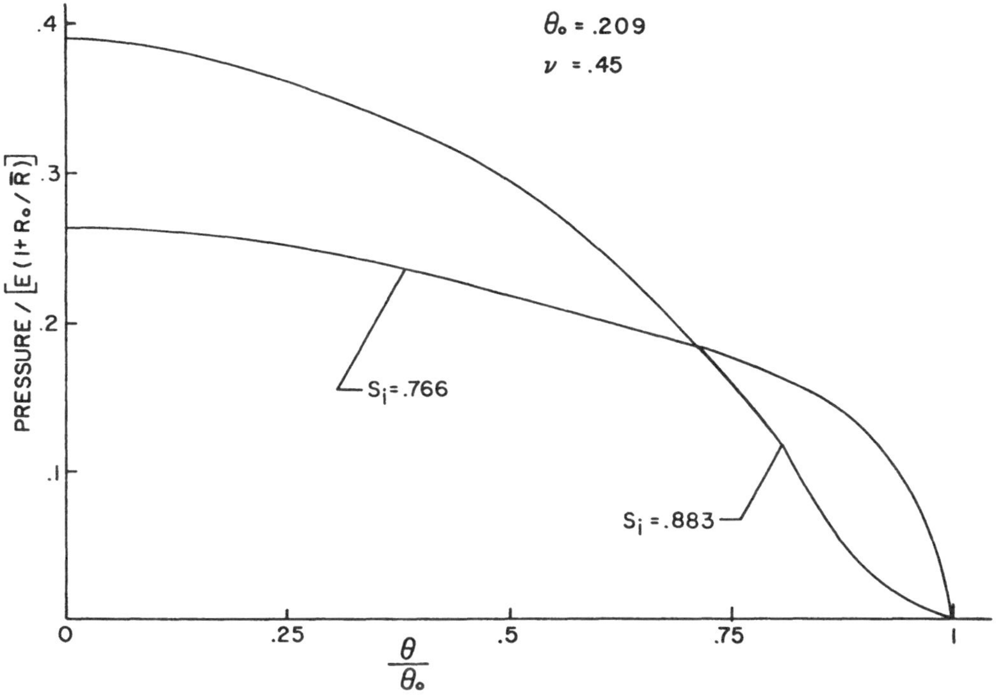

Fig. 4 Pressure distribution at the contact surface; effect of layer thickness

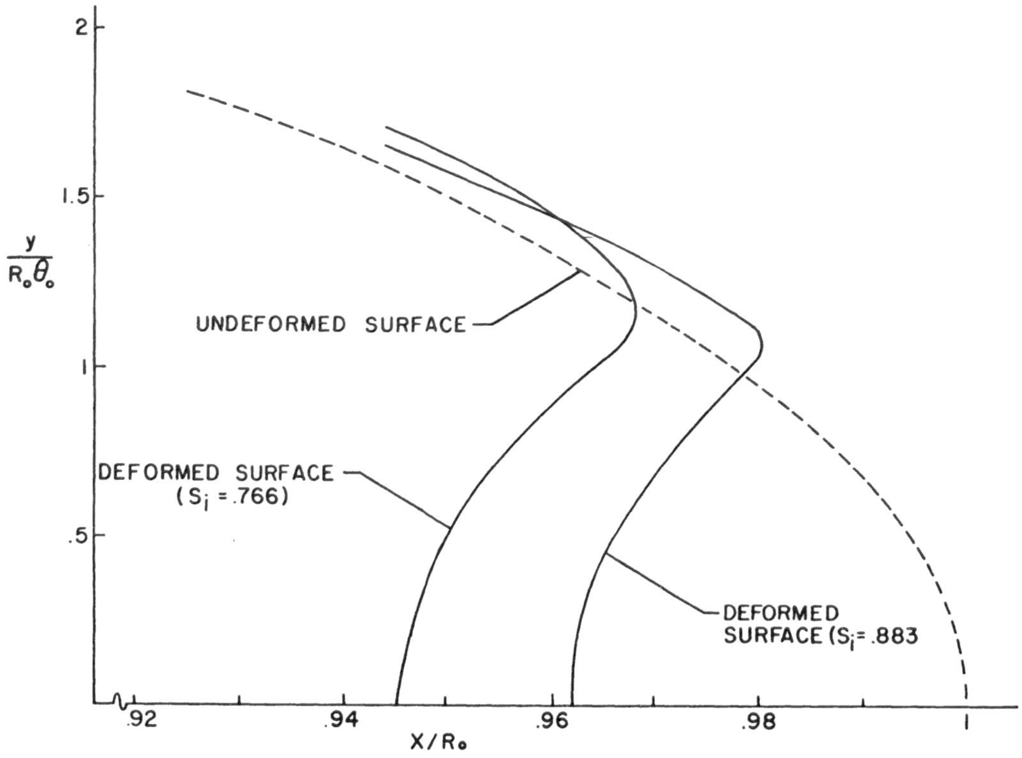

Fig. 5 Deformed surface of the rubberlike layer

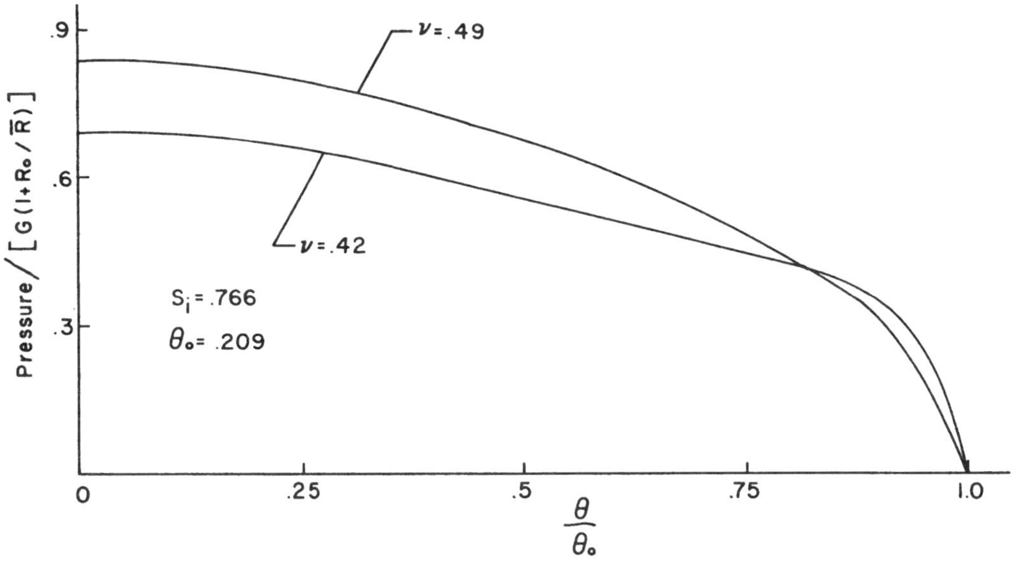

Fig. 6 Pressure distribution at the contact surface; effect of Poisson's ratio

AN IMPROVED BOUNDARY ELEMENT METHOD FOR PLATE VIBRATIONS

George K.K. Wong and James R. Hutchinson

Graduate Student, Mechanical Engineering Department, Stanford University
Professor, Civil Engineering, University of California, Davis

ABSTRACT

The boundary element method is used to determine the natural frequencies and mode shapes of thin elastic plates of uniform thickness with arbitrary boundary conditions and arbitrarily shaped edges. The boundary element formulation leads to a pair of integral equations that contain the boundary condition variables of displacement, normal slope, bending moment, and effective shear, providing an effective means of solving plate problems with arbitrary boundary conditions. The boundary elements are taken as straight lines or as circular arcs, and unknown boundary functions are assumed to vary linearly along the elements.

INTRODUCTION

In a 1979 paper by the authors [1], a formulation was presented by which the natural frequencies of thin uniform elastic plates could be found. The formulation was limited to simply supported and clamped edges only, and was also limited to straight line segments. The numerical results from that research, however, were very encouraging and this present research is an attempt to remove many of the limitations from the previous solution method. The problem has, therefore, been reformulated for boundary condition variables of displacement, normal slope, bending moment, effective shear, and twisting moment discontinuity at corners. Also, in addition to straight line segments, one has the option of using circular arc elements.

The formulation of integral equations for static loads on plates was presented by Bézine and Gamby [2] in 1978, and independently by Stern [3] in 1979. The present formulation for the vibration problem closely follows Stern's development. Previous research employing boundary elements for plate vibration include a recent paper by Bézine [4], in which the problem is treated by a mixed boundary element-finite element approach. In a 1974 paper by Vivoli and Filippi [5], the vibration problem is formulated in a similar way to that used in this paper, but there are differences in both the formulation and application of the integral equations.

At the present time, the computer program which we have developed is not yet fully operational, but preliminary results indicate that accurate frequencies can be found with a small number of elements.

FORMULATION

The governing partial differential equation for the free vibration of a homogenous, isotropic, linear elastic thin plate of uniform thickness is

$$D\nabla^4\bar{w} + \rho\frac{\partial^2\bar{w}}{\partial t^2} = 0 \quad \text{for } \bar{w} \text{ in } \Omega \tag{1}$$

where ∇^4 is the biharmonic operator, D is the flexural stiffness, $\bar{w}$ is the vertical displacement, ρ is the mass per unit area, t is the time and Ω is the area of the plate bounded by curve Γ.

The time dependence of Equation (1) is removed by assuming that $\bar{w}$ varies sinusoidally with time; thus

$$\bar{w}(x, y, t) = w(x, y) \sin(\omega t) \tag{2}$$

Substituting Equation (2) into Equation (1), the final time-independent differential equation is

$$\nabla^4 w - \lambda^4 w = 0 \qquad \text{for } w \text{ in } \Omega \tag{3}$$

where λ is the frequency parameter defined as

$$\lambda^4 = \omega^2\rho/D \tag{4}$$

and ω is the circular frequency of the plate. Equation (3) is the governing differential equation for the natural frequencies and mode shapes of the vibrating plate. Let E be a singular solution that satisfies Equation (3) for any point inside of Ω. Hence,

$$\nabla^4 E - \lambda^4 E = 0 \qquad \text{for } E \text{ in } \Omega \tag{5}$$

Multiplying Equations (3) and (5) by E and w respectively, then subtracting and integrating over the region Ω gives

$$\iint_{\Omega}(E\nabla^4 w - w\nabla^4 E)\,dA = 0 \tag{6}$$

Equation (3) can be reduced to a boundary integral by making use of the Rayleigh-Green indentity with K corners [2,3] as:

$$\iint_{\Omega}(E\nabla^4 w - w\nabla^4 E)\,dA = \frac{1}{D}\int_{\Gamma}\{V_n(E)w - M_n(E)\frac{dw}{dn} + \frac{dE}{dn}M_n(w) - E\,V_n(w)\}ds$$

$$- \sum_{k=1}^{K} < M_t(w) > E - < M_t(E) > w = 0 \qquad (7)$$

where $V_n(w)$, $M_n(w)$, $M_t(w)$, $\frac{dw}{dn}$ and w are the effective shear, normal bending moment, twisting moment, normal slope and displacement respectively, and $<\cdot>$ represents the discontinuity jump in the direction of increasing arc length.

Using polar coordinates (r,θ) and letting β be the angle from the radial direction to the outer normal yields

$$M_n(w) = \frac{D}{2}\left\{-(1+\nu)\nabla^2 w + (1-\nu)\left[\left(\frac{1}{r}\frac{\partial w}{\partial r} + \frac{1}{r^2}\frac{\partial^2 w}{\partial \theta^2} - \frac{\partial^2 w}{\partial r^2}\right)\cos 2\beta - 2\frac{\partial}{\partial r}\left(\frac{1}{r}\frac{\partial w}{\partial \theta}\right)\sin 2\beta\right]\right\} \qquad (8\text{-a})$$

$$M_t(w) = -\frac{D(1-\nu)}{2}\left\{\left(\frac{1}{r}\frac{\partial w}{\partial r} + \frac{1}{r^2}\frac{\partial^2 w}{\partial \theta^2} - \frac{\partial^2 w}{\partial r^2}\right)\sin 2\beta + 2\frac{\partial}{\partial r}\left(\frac{1}{r}\frac{\partial w}{\partial \theta}\right)\cos 2\beta\right\} \qquad (8\text{-b})$$

$$V_n(w) = \left[-D\frac{\partial}{\partial r}\nabla^2 w + \frac{1}{r}\left(\frac{\partial}{\partial \theta}M_t(w) - \frac{\partial}{\partial \beta}M_t(w)\right]\cos\beta - \left[\frac{D}{r}\frac{\partial}{\partial \theta}\nabla^2 w + \frac{\partial}{\partial r}M_t(w)\right]\sin\beta + \frac{1}{R}\frac{\partial}{\partial \beta}M_t(w) \qquad (8\text{-c})$$

where R is the radius curvature at a regular boundary point with a sign convention that indicates a negative curvature when the center of curvature is on the outward normal, and ν is Poisson's ratio.

The integral equation (7) contains a singular point when r equals zero. In order to exclude this point a circular region bounded by curve Γ_ε with a radius ε is deleted from the region Ω. Hence, the integration is taken over two curves, Γ and Γ_ε, and letting the limit of ε approach zero. Capital letter P and small letter p denote the singular point inside Ω and on Γ respectively, while capital letter Q and small letter q denote any point on the curve Γ_ε and Γ respectively, as shown in Figure 1.

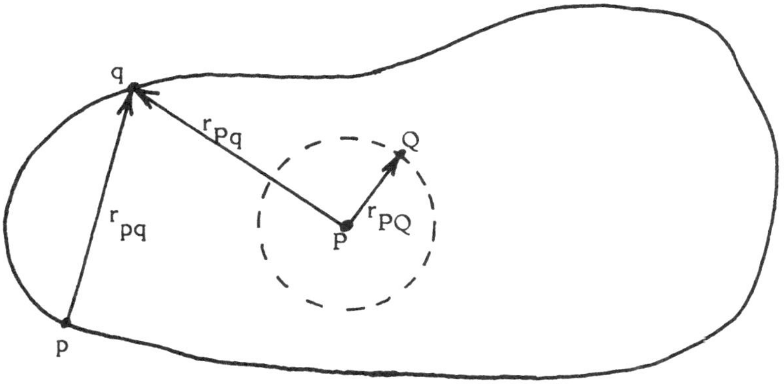

FIGURE 1

Hence, the integral equation (7) for any point P inside of Ω becomes

$$\frac{1}{D}\int_{\Gamma}\{V_n(E)\Big|_{Pq} w\Big|_q - M_n(E)\Big|_{Pq}\frac{dw}{dn}\Big|_q + \frac{dE}{dn}\Big|_{Pq} M_n(w)\Big|_q - E\Big|_{Pq} V_n(w)\Big|_q\}\; ds\Big|_q$$

$$+\ \Phi - \sum_{k=1}^{K} < M_t(w)\Big|_{q_k} > E\Big|_{Pq_k} - < M_t(E)\Big|_{Pq_k} > w\Big|_{q_k} = 0 \qquad (10\text{-a})$$

where

$$\Phi = \lim_{\varepsilon\to 0}\int_{\Gamma_\varepsilon}\frac{1}{D}\{V_n(E)\Big|_{Pq} w\Big|_q - M_n(E)\Big|_{Pq}\frac{dw}{dn}\Big|_q + \frac{dE}{dn}\Big|_{Pq} M_n(w)\Big|_q - E\Big|_{Pq} V_n(w)\Big|_q\} ds\Big|_q$$

(10-b)

By placing the origin of the polar coordinate at point P the singular function E that satisfies Equation (5) is

$$E(\lambda r) = i\ c\ J_o(\lambda r) + c\ Y_o(\lambda r) + d\ K_o(\lambda r) \qquad (11)$$

where J_o is the zero order Bessel Function of the first kind, Y_o is the zero order Bessel Function of the second kind, K_o is the zero order modified Bessel Function of the second kind, $i=\sqrt{-1}$ and

$$c = \frac{1}{8\lambda^2}\ , \qquad d = \frac{1}{4\lambda^2\pi} \qquad (12)$$

Making use of Equations (11), (8-a) and (8-c) and separating in real and imaginary parts gives

$$E^{i}(\lambda r) = i\ c\ J_o(\lambda r) \tag{13-a}$$

$$E^{R}(\lambda r) = c\ Y_o(\lambda r) + d\ K_o(\lambda r) \tag{13-b}$$

$$\frac{dE^{i}}{dn}(\lambda r) = N^{i} = -\ i\ \lambda c\ J_1\ (\lambda r)\ \cos\beta \tag{13-c}$$

$$\frac{dE^{R}}{dn}(\lambda r) = N^{R} = -\lambda[c\ Y_1\ (\lambda r) + d\ K_1\ (\lambda r)]\ \cos\beta \tag{13-d}$$

$$M_n(E)^{i} = M^{i} = i\ \{\ c\ \frac{D}{2}\ [(1+\nu) + (1-\nu)\ \cos 2\beta]\ \ \lambda^2\ J_o\ (\lambda r)$$
$$-\ c\ D\ \lambda\ (1-\nu)\ \frac{J_1(\lambda r)}{r}\ \cos 2\beta\ \} \tag{13-e}$$

$$M_n(E)^{R} = M^{R} = \frac{D}{2}\Big\{[(1+\nu) + (1-\nu)\ \cos 2\beta]\ \lambda^2\ [cY_o(\lambda r) - d\ k_o(\lambda r)]$$
$$-\ 2\ \lambda\ (1-\nu)\left[\frac{cY_1(\lambda r) + d\ K_1(\lambda r)}{r}\right]\cos 2\beta\Big\} \tag{13-f}$$

$$M_t(E)^{i} = T^{i} = -\ i\ c\ \frac{D\ (1-\nu)}{2}\left[\lambda^2\ J_o(\lambda r) - 2\lambda\ \frac{J_1(\lambda r)}{r}\right]\ \sin 2\beta \tag{13-g}$$

$$M_t(E)^{R} = T^{R} = -\ \frac{D\ (1-\nu)}{2}\Big[\lambda^2\ \left(cY_o(\lambda r) - d\ K_o(\lambda r)\right)$$
$$-\ \frac{2\lambda}{r}\left(cY_1(\lambda r) + d\ K_1(\lambda r)\right)\Big]\ \sin 2\beta \tag{13-h}$$

$$V_n(E)^{i} = V^{i} = -\ i\ c\ D\ \Big\{J_1(\lambda r)\ \Big[\lambda^3\ \cos\beta + \frac{\lambda^3(1-\nu)}{2}\ \sin 2\beta\ \sin\beta$$
$$+\ \frac{2\lambda(1-\nu)}{r}\left(\frac{\cos 3\beta}{r} - \frac{\cos 2\beta}{R}\right)\Big] + (1-\nu)\ \lambda^2\ J_o(\lambda r)\ \Big(\frac{\cos 2\beta}{R}$$
$$-\ \frac{\cos 3\beta}{r}\Big)\Big\} \tag{13-i}$$

$$V_n(E)^R = V^R = - D \lambda^3 \left(cY_1(\lambda r) - d\, K_1(\lambda r)\right) \cos\beta$$

$$+ D\,(1-\nu) \left[\frac{\lambda^2}{r} \left(cY_o(\lambda r) - d\, K_o(\lambda r)\right) - \frac{2\lambda}{r^2} \left(cY_1(\lambda r)\right.\right.$$

$$\left.\left. + d\, K_1(\lambda r)\right)\right] \cos 3\beta - D\,(1-\nu) \left[\frac{\lambda^2}{R} \left(cY_o(\lambda r) - d\, K_o(\lambda r)\right)\right.$$

$$\left. - \frac{2\lambda}{rR} \left(cY_1(\lambda r) + d\, K_1(\lambda r)\right)\right] \cos 2\beta - \frac{D\,(1-\nu)}{2} \lambda^3 \left[cY_1(\lambda r)\right.$$

$$\left. - d\, K_1(\lambda r)\right] \sin 2\beta \sin\beta \tag{13-j}$$

On the circular curve Γ_ε, the following observations can be made.

$$r_{PQ} = \varepsilon \quad ; \quad \beta = \pi \quad ; \quad R = -\varepsilon \tag{14}$$

Using these results, Eqs. (13) reduce to the following:

$$E^i\Big|_{PQ} = i\, c\, J_o\, (\lambda\varepsilon) \tag{15-a}$$

$$E^R\Big|_{PQ} = c\, Y_o(\lambda\varepsilon) + d\, K_o(\lambda\varepsilon) \tag{15-b}$$

$$N^i\Big|_{PQ} = i\, c\, \lambda\, J_1(\lambda\varepsilon) \tag{15-c}$$

$$N^R\Big|_{PQ} = \lambda\, [c\, Y_1(\lambda\varepsilon) + d\, K_1(\lambda\varepsilon)] \tag{15-d}$$

$$M^i\Big|_{PQ} = i\, c\, D \left[\lambda^2\, J_o(\lambda\varepsilon) - \lambda\,(1-\nu) \frac{J_1(\lambda\varepsilon)}{\varepsilon}\right] \tag{15-e}$$

$$M^R\Big|_{PQ} = D \left[\lambda^2\, c\, Y_o(\lambda\varepsilon) - d\, K_o(\lambda\varepsilon)\right.$$

$$\left. - \lambda\,(1-\nu) \left(\frac{c\, Y_1(\lambda\varepsilon) + d\, K_1(\lambda\varepsilon)}{\varepsilon}\right)\right] \tag{15-f}$$

$$T^i\Big|_{PQ} = T^R\Big|_{PQ} = 0 \tag{15-g}$$

$$V\Big|_{PQ}^{i} = i\ c\ D\ \lambda^3\ J_1(\lambda\varepsilon) \tag{15-h}$$

$$V\Big|_{PQ}^{R} = D\ \lambda^3 \left[c\ Y_1(\lambda\varepsilon) - d\ K_1(\lambda\varepsilon)\right] \tag{15-i}$$

Also, the limit of Eqs. (15) as ε approaches to zero is

$$\lim_{\varepsilon\to 0} E\Big|_{PQ}^{i} = ic \tag{16-a}$$

$$\lim_{\varepsilon\to 0} E\Big|_{PQ}^{R} = \lim_{\varepsilon\to 0} (\tfrac{2}{\pi}\ c-d)\ \log\lambda\varepsilon = 0 \tag{16-b}$$

$$\lim_{\varepsilon\to o} N\Big|_{PQ}^{i} = \lim_{\varepsilon\to 0} i\ \frac{c\lambda^2\varepsilon}{2} = 0 \tag{16-c}$$

$$\lim_{\varepsilon\to 0} N\Big|_{PQ}^{R} = \lim_{\varepsilon\to 0} -\ (\tfrac{2}{\pi}\ c-d)\ \frac{1}{\varepsilon} = 0 \tag{16-d}$$

$$\lim_{\varepsilon\to 0} M\Big|_{PQ}^{i} = \lim_{\varepsilon\to 0} icD\ [\lambda^2 - \lambda^2\ \frac{(1-\nu)}{2}] = icD\ \frac{(1+\nu)}{2}\ \lambda^2 \tag{16-e}$$

$$\lim_{\varepsilon\to 0} M\Big|_{PQ}^{R} = \lim_{\varepsilon\to 0} D\ \lambda^2\ (\tfrac{2}{\pi}\ c+d)\ \log\lambda\varepsilon \tag{16-f}$$

$$\lim_{\varepsilon\to 0} V\Big|_{PQ}^{i} = \lim_{\varepsilon\to 0} icD\ \frac{\lambda^4\varepsilon}{2} = 0 \tag{16-g}$$

$$\lim_{\varepsilon\to 0} V\Big|_{PQ}^{R} = \lim_{\varepsilon\to 0} -\ D\ \lambda^2\ (\tfrac{2}{\pi}\ c+d)\ \frac{1}{\varepsilon} \tag{16-h}$$

The results of Eqs. (16) are substituted into Eq. (10-b) and we have

$$\Phi = \lim_{\varepsilon\to 0}\int_0^{2\pi} \frac{1}{D}\left\{-icD\,\frac{(1+\nu)}{2}\,\frac{dw}{dn}\,\varepsilon d\theta\right\} + \lim_{\varepsilon\to 0}\int_0^{2\pi}\frac{1}{D}$$

$$\left\{-D\,\lambda^2\,(\tfrac{2}{\pi}\,c+d)\,\frac{1}{\varepsilon}\,w - D\,\lambda^2\,(\tfrac{2}{\pi}\,c+d)\,\log\lambda\varepsilon\,\frac{dw}{dn}\right\}\varepsilon\,d\,\theta$$

$$= -\,\lambda^2\,(\tfrac{2}{\pi}\,c+d)\,w\,(P)\,2\pi = -w\,(P) \tag{17}$$

Hence, a boundary integral equation valid for any point P inside of the region Ω is found. Writing the equation in both imaginary and real parts yields

$$\frac{1}{D}\int_r \{V^i\Big|_{Pq}\,w\Big|_q - M^i\Big|_{Pq}\,\frac{dw}{dn}\Big|_q + N^i\Big|_{Pq}\,M_n(w)\Big|_q - E^i\Big|_{Pq}\,V_n(w)\Big|_q\}\,ds\Big|_q$$

$$-\sum_{k=1}^{K}[\,<M_t(w)\Big|_{q_k}>\,E^i\Big|_{Pq_k} - <T^i\Big|_{Pq_k}>\,w\Big|_{q_k}] = 0 \tag{18-a}$$

and

$$\frac{1}{D}\int_r V^R\Big|_{Pq}\,w\Big|_q - M^R\Big|_{Pq}\,\frac{dw}{dn}\Big|_q + N^R\Big|_{Pq}\,M_n(w)\Big|_q - E^R\Big|_{Pq}\,V_n\,(w)\Big|_q\,ds\Big|_q$$

$$-\sum_{k=1}^{K}[\,<M_t(w)\Big|_{q_k}>\,E^R\Big|_{Pq_k} - <T^R\Big|_{Pq_k}>\,w\Big|_{q_k}] - w\Big|_P = 0 \tag{18-b}$$

where the kernals are defined by Eqs. (13-a to 13-j).

A second independent integral equation is obtained by taking a directional derivative of Eqs. (18) in a fixed direction ξ that makes an angle ψ with the x axis. A local ξ-η coordinate system is introduced at point P and the coordinates are rotated by an angle ψ with respect to the x axis. A new polar coordinate (r,ϕ) with respect to ξ-η is formed. Using this new polar coordinate system, the appropriate singular function that satisfies Eq. (5) is

$$E_\gamma(\lambda r) = i\,c\,J_1(\lambda r)\,\cos\phi + [cY_1(\lambda r) + d\,K_1(\lambda r)]\,\cos\phi \tag{19}$$

Again, making use of Eqs. (8a to 8c) and writing the kernals in imaginary and real parts gives

$$E^i_\gamma(\lambda r) = i\,c\,J_1(\lambda r)\,\cos\phi \tag{20-a}$$

$$E_\gamma^R(\lambda r) = [c\,Y_1(\lambda r) + d\,K_1(\lambda r)]\,\cos\phi \tag{20-b}$$

$$N_\gamma^i(\lambda r) = i\,c\,[\lambda\,J_o(\lambda r)\,\cos\phi\,\cos\beta - \frac{J_1(\lambda r)}{r}\cos(\phi+\beta)] \tag{20-c}$$

$$N_\gamma^R(\lambda r) = \lambda\Big(c\,Y_o(\lambda r) - d\,K_o(\lambda r)\Big)\,\cos\phi\,\cos\beta - \frac{\Big(c\,Y_1(\lambda r) + d\,K_1(\lambda r)\Big)}{r}\cos(\phi+\beta) \tag{20-d}$$

$$M_\gamma^i(\lambda r) = i\,c\,D\,\{J_1(\lambda r)\,[\frac{\lambda^2}{2}\cos\phi\,(1+\nu+(1-\nu)\cos 2\beta) - \frac{2(1-\nu)}{r^2}\cos(\phi-2\beta)] + (1-\nu)\,\lambda\,\frac{J_o(\lambda r)}{r}\cos(\phi-2\beta)\} \tag{20-e}$$

$$M_\gamma^R(\lambda r) = \frac{D\lambda^2}{2}\Big(c\,Y_1(\lambda r) - d\,K_1(\lambda r)\Big)\cos\phi\,[1+\nu+(1-\nu)\cos 2\beta] + D(1-\nu)\,[\frac{\lambda}{r}\Big(c\,Y_o(\lambda r) - d\,K_o(\lambda r)\Big) - \frac{2}{r^2}\Big(c\,Y_1(\lambda r) + d\,K_1(\lambda r)\Big)]\,\cos(\phi-2\beta) \tag{20-f}$$

$$T_\gamma^i(\lambda r) = -\,i\,c\,D(1-\nu)\,[J_1(\lambda r)\Big(\frac{\lambda^2}{2}\cos\phi\,\sin 2\beta + \frac{2}{r^2}\sin(\phi-2\beta)\Big) - \frac{J_o(\lambda r)}{r}\sin(\phi-2\beta)\,] \tag{20-g}$$

$$T_\gamma^R(\lambda r) = -\frac{D(1-\nu)}{2}\lambda^2\Big(c\,Y_1(\lambda r) - d\,K_1(\lambda r)\Big)\cos\phi\,\sin 2\beta + D(1-\nu)\,[\frac{\lambda}{r}\Big(c\,Y_o(\lambda r) - d\,K_o(\lambda r)\Big) - \frac{2}{r^2}\Big(c\,Y_1(\lambda r) + d\,K_1(\lambda r)\Big)]\,\sin(\phi-2\beta) \tag{20-h}$$

$$\begin{aligned} V_\gamma^i(\lambda r) = {} & i\,c\,D\,\{J_o(\lambda r)\,[\lambda^3\Big(\cos\phi\,\cos\beta + \frac{(1-\nu)}{2}\cos\phi\,\sin 2\beta\,\sin\beta\Big) \\ & + (1-\nu)\Big(\frac{3\lambda}{r^2}\cos(\phi-3\beta) - \frac{2\lambda}{rR}\cos(\phi-2\beta)\Big)] \\ & + J_1(\lambda r)\,[\frac{\lambda^2}{r}\Big(-\cos(\phi-\beta) + \frac{(1-\nu)}{2}(\cos(\phi-3\beta) + \sin(\phi-2\beta)\,\sin\beta \\ & + \cos 3\beta\,\cos\phi)\Big) - \frac{\lambda^2}{R}(1-\nu)\cos\phi\,\cos 2\beta - \frac{6(1-\nu)}{r^3}\cos(\phi-3\beta) \\ & + \frac{4}{r^2 R}(1-\nu)\cos(\phi-2\beta)]\} \end{aligned} \tag{20-i}$$

$$V_\gamma^R(\lambda r) = D\,\lambda^3 \left(c\,Y_o(\lambda r) + d\,K_o(\lambda r)\right) [\cos\phi\,\cos\beta$$

$$+ \frac{(1-\nu)}{2} \cos\phi\,\sin 2\beta\,\sin\beta] + \frac{D\lambda^2}{r} \left(c\,Y_1(\lambda r)\right.$$

$$\left. - d\,K_1(\lambda r)\right) [-\cos(\phi-\beta) + \frac{(1-\nu)}{2} (\cos(\phi-3\beta) + \sin(\phi-2\beta)\,\sin\beta$$

$$+ \cos 3\beta\,\cos\phi)] - \frac{D\lambda^2}{R} \left(c\,Y_1(\lambda r) - d\,K_1(\lambda r)\right) [(1-\nu)\,\cos\phi\,\cos 2\beta]$$

$$+ 3\,D\,(1-\nu)\,[\frac{\lambda}{r^2} \left(c\,Y_o(\lambda r) - d\,K_o(\lambda r)\right) - \frac{2}{r^3} \left(c\,Y_1(\lambda r)\right.$$

$$\left. + d\,K_1(\lambda r)\right)]\,\cos(\phi-3\beta) - 2\,D(1-\nu)\,[\frac{\lambda}{rR} \left(c\,Y_o(\lambda r) - d\,K_o(\lambda r)\right)$$

$$- \frac{2}{r^2 R} \left(c\,Y_1(\lambda r) + d\,K_1(\lambda r)\right)]\,\cos(\phi-2\beta) \tag{20-j}$$

Again, making use of the observations of Eq. (14) along Γ_ε, we have the following

$$E_\gamma^i\Big|_{PQ} = i\,c\,J_1(\lambda\varepsilon)\,\cos\phi \tag{21-a}$$

$$E_\gamma^R\Big|_{PQ} = [c\,Y_1(\lambda\varepsilon) + d\,K_1(\lambda\varepsilon)]\,\cos\phi \tag{21-b}$$

$$N_\gamma^i\Big|_{PQ} = -\,i\,c\,[\lambda\,J_o(\lambda\varepsilon) - \frac{J_1(\lambda\varepsilon)}{\varepsilon}]\,\cos\phi \tag{21-c}$$

$$N_\gamma^R\Big|_{PQ} = -\lambda\left(c\,Y_o(\lambda\varepsilon) - d\,K_o(\lambda\varepsilon)\right)\,\cos\phi + [\frac{c\,Y_1(\lambda\varepsilon) + d\,K_1(\lambda\varepsilon)}{\varepsilon}]\,\cos\phi \tag{21-d}$$

$$M_\gamma^i\Big|_{PQ} = i\,c\,D\,\{J_1(\lambda\varepsilon)\left(\lambda^2 - \frac{2(1-\nu)}{\varepsilon^2}\right) + (1-\nu)\,\lambda\,\frac{J_o(\lambda\varepsilon)}{\varepsilon}\}\cos\phi \tag{21-e}$$

$$M_{\gamma|_{PQ}}^{R} = D\,\lambda^2 \left(c\,Y_1(\lambda\varepsilon) - d\,K_1(\lambda\varepsilon)\right) \cos\phi + D\,(1-\nu)\,\left[\frac{\lambda}{\varepsilon}\left(c\,Y_o(\lambda\varepsilon) - d\,K_o(\lambda\varepsilon)\right) - \frac{2}{\varepsilon^2}\left(c\,Y_1(\lambda\varepsilon) + d\,K_1(\lambda\varepsilon)\right)\right] \cos\phi \qquad (21\text{-f})$$

$$T_{\gamma|_{PQ}}^{i} = -\,i\,c\,D\,(1-\nu)\,\{J_1(\lambda\varepsilon)\,\frac{2}{\varepsilon^2} - \lambda\,\frac{J_o(\lambda\varepsilon)}{\varepsilon}\}\,\sin\phi \qquad (21\text{-g})$$

$$T_{\gamma|_{PQ}}^{R} = D(1-\nu)\,\{\frac{\lambda}{\varepsilon}\left(c\,Y_o(\lambda\varepsilon) - d\,K_o(\lambda\varepsilon)\right) - \frac{2}{\varepsilon^2}\left(c\,Y_1(\lambda\varepsilon) + d\,K_1(\lambda\varepsilon)\right)\}\sin\phi \qquad (21\text{-h})$$

$$V_{\gamma|_{PQ}}^{i} = i\,c\,D\,\{-J_o(\lambda\varepsilon)\,[\lambda^3 + (1-\nu)\,\frac{\lambda}{\varepsilon^2}]\,\cos\phi + \frac{J_1(\lambda\varepsilon)}{\varepsilon}\,[\lambda^2 + \frac{2(1-\nu)}{\varepsilon^2}]\,\cos\phi\} \qquad (21\text{-i})$$

$$V_{\gamma|_{PQ}}^{R} = -\,D\,\lambda^3\,[c\,Y_o(\lambda\varepsilon) + d\,K_o(\lambda\varepsilon)]\,\cos\phi + D\,\frac{\lambda^2}{\varepsilon}\,[c\,Y_1(\lambda\varepsilon) - d\,K_1(\lambda\varepsilon)]\,\cos\phi - D\,(1-\nu)\,\left[\frac{\lambda}{\varepsilon^2}\left(c\,Y_o(\lambda\varepsilon) - d\,K_o(\lambda\varepsilon)\right) - \frac{2}{\varepsilon^3}\left(c\,Y_1(\lambda\varepsilon) + d\,K_1(\lambda\varepsilon)\right)\right]\,\cos\phi \qquad (21\text{-j})$$

Taking the limit of ε to zero, Equations (21) reduce to:

$$\lim_{\varepsilon\to 0} E_{\gamma|_{PQ}}^{i} = \lim_{\varepsilon\to 0} i\,c\,\frac{\lambda\varepsilon}{2}\,\cos\phi = 0 \qquad (22\text{-a})$$

$$\lim_{\varepsilon\to 0} E_{\gamma|_{PQ}}^{R} = \lim_{\varepsilon\to 0} -(\frac{2}{\pi}\,c - d)\,\frac{1}{\lambda\varepsilon}\,\cos\phi = 0 \qquad (22\text{-b})$$

$$\lim_{\varepsilon\to 0} N_{\gamma|_{PQ}}^{i} = \lim_{\varepsilon\to 0} -\,i\,c\,[\lambda - \frac{\lambda\varepsilon}{2\varepsilon}]\,\cos\phi = -\,\frac{ic\lambda}{2}\,\cos\phi \qquad (22\text{-c})$$

$$\lim_{\varepsilon\to 0} N_{\gamma|_{PQ}}^{R} = \lim_{\varepsilon\to 0} -\lambda\,(\frac{2}{\pi}\,c + d)\,\log\lambda\varepsilon\,\cos\phi - (\frac{2}{\pi}\,c - d)\,\frac{1}{\lambda\varepsilon^2}\,\cos\phi$$

$$= \lim_{\varepsilon\to 0} -\lambda\,(\frac{2}{\pi}\,c + d)\,\log\lambda\varepsilon\,\cos\phi \qquad (22\text{-d})$$

$$\lim_{\varepsilon\to 0} M^{i}_{\gamma}\Big|_{PQ} = \lim_{\varepsilon\to 0} \frac{\lambda^3\varepsilon}{2} = 0 \tag{22-e}$$

$$\lim_{\varepsilon\to 0} M^{R}_{\gamma}\Big|_{PQ} = \lim_{\varepsilon\to 0} [-\frac{D\lambda}{\varepsilon}(\frac{2}{\pi}c+d) + D(1-\nu)\lambda(\frac{2}{\pi}c+d)\frac{\log\lambda\varepsilon}{\varepsilon}]\cos\phi \tag{22-f}$$

$$\lim_{\varepsilon\to 0} T^{i}_{\gamma}\Big|_{PQ} = 0 \tag{22-g}$$

$$\lim_{\varepsilon\to 0} T^{R}_{\gamma}\Big|_{PQ} = \lim_{\varepsilon\to 0} \{D(1-\nu)\frac{\lambda}{\varepsilon}(\frac{2}{\pi}c+d)\log\lambda\varepsilon\}\sin\phi \tag{22-h}$$

$$\lim_{\varepsilon\to 0} V^{i}_{\gamma}\Big|_{PQ} = -\frac{\lambda^3}{2} \tag{22-i}$$

$$\lim_{\varepsilon\to 0} V^{R}_{\gamma}\Big|_{PQ} = \lim_{\varepsilon\to 0} - D\lambda^3(\frac{2}{\pi}c-d)\log\lambda\varepsilon\cos\phi$$

$$+ \lim_{\varepsilon\to 0} - D\lambda^2(\frac{2}{\pi}c+d)\frac{1}{\lambda\varepsilon^2}\cos\phi$$

$$- \lim_{\varepsilon\to 0} [\lambda(\frac{2}{\pi}c+d)\frac{\log\lambda\varepsilon}{\varepsilon^2} + (\frac{2}{\pi}c-d)\frac{2}{\varepsilon^4}]\cos\phi\, D(1-\nu) \tag{22-j}$$

Again, the results of Eqs. (22-a to 22-j) can be substituted in Eq. (10-6) in order to determine Φ. However, we see that the limit does not converge. This can be corrected by replacing w of Eqs. (10-a and 10-b) by $\hat{w} = w - w|_P$. Note, this assumption does not affect its derivatives, also note that along Γ_ε we have

$$\hat{w} = w\Big|_Q - w\Big|_P = \frac{dw}{dr}\Big|_Q (r_Q - r_P) + O(r^2) \tag{23-a}$$

Neglecting the quadratic term and noting that

$$r_Q - r_P = \varepsilon \quad \text{and} \quad \frac{dw}{dr}\Big|_Q = -\frac{dw}{dn}\Big|_Q \tag{23-b}$$

Equation (23-a) can, therefore, be rewritten as

$$\hat{w}\Big|_{PQ} = w\Big|_{Q} - w\Big|_{P} = -\varepsilon \frac{dw}{dn}\Big|_{Q} \qquad (24\text{-a})$$

Also, $\frac{dw}{dn}\Big|_P$ is

$$\frac{dw}{dn}\Big|_P = -\left(\frac{\partial w}{\partial \xi}\Big|_P \cos\phi + \frac{\partial w}{\partial \eta}\Big|_P \sin\phi\right) \qquad (24\text{-b})$$

When the results of Eqs. (24-a), (24-b) and (22-a) to (22-j) are substituted in Eq. (10-b), Φ is found to be

$$\Phi = \int_0^{2\pi} \lambda^2 \left(\frac{2}{\pi}c+d\right) \frac{\cos\phi}{\lambda} \frac{dw}{dn} d\phi = -\ \lambda\left(\frac{2}{\pi}c+d\right) \int_0^{2\pi} \left[\frac{\partial w}{\partial \xi}\Big|_P \cos^2\phi + \frac{\partial w}{\partial \eta}\Big|_P \cos\phi \sin\phi\right] d\phi = -\lambda \left(\frac{2}{\pi}c+d\right) \frac{\partial w}{\partial \xi}\Big|_P \pi = -\frac{1}{2\lambda} \frac{\partial w}{\partial \xi}\Big|_P \qquad (25)$$

Hence, the second boundary integral equation at point P is

$$\int_\Gamma \frac{1}{D} \left\{V_\gamma^i\Big|_{Pq} (w\big|_q - w\big|_q) - M_\gamma^i\Big|_{Pq} \frac{dw}{dn}\Big|_q + N_\gamma^i\Big|_{Pq} M_n(w)\big|_q - E_\gamma^i\Big|_{Pq} V_n(w)\big|_q\right\} ds\big|_q - \sum_{k=1}^{K} \left[< M_t(w)\big|_{P_k} > E_\gamma^i\Big|_{Pq_k} - < T_\gamma^i\Big|_{Pq_k} > (w\big|_{q_k} - w\big|_P)\right] = 0 \qquad (26\text{-a})$$

and

$$\int_\Gamma \frac{1}{D} \left\{V_\gamma^R\Big|_{Pq} (w\big|_q - w\big|_P) - M_\gamma^R\Big|_{Pq} \frac{dw}{dn}\Big|_q + N_\gamma^R\Big|_{Pq} M_n(w)\big|_q - E_\gamma^R\Big|_{Pq} V_n(w)\big|_q\right\} ds\big|_q - \sum_{k=1}^{K} \left[< M_t(w)\big|_{q_k} > E_\gamma^R\Big|_{Pq_k} - < T_\gamma^i\Big|_{Pq_k} > (w\big|_{q_k} - w\big|_P)\right] - \frac{1}{2\lambda} \frac{\partial w}{\partial \xi}\Big|_P = 0 \qquad (26\text{-b})$$

where the kernals are defined by Eqs. (20-a to 20-j).

The integral equations (18) and (26) are for a singular point P inside of the region Ω. The equations involve only the boundary unknowns of effective shear, bending moment, normal slope and displacement.

If the singular point p is on the boundary Γ instead of in the interior then the curve Γ_t would be around a portion of the arc, hence the integrations in Eqs. (17) and (25) would be over a semicircle for a regular boundary point, and over the interior angle for a corner point. The boundary integral equations valid for any point p on the boundary curve Γ thus become,

$$\frac{\alpha w}{2}\Big|_p = \frac{1}{D}\int_\Gamma \{V^R\big|_{pq}\, w\big|_q - M^R\big|_{pq}\, \frac{dw}{dn}\Big|_q + N^R\big|_{pq}\, M_n(w)\big|_q - E^R\big|_{pq}\, V_n(w)\big|_q\}ds\big|_q$$

$$- \frac{1}{D}\sum_{k=1}^{K} [\, \langle M_t(w)\big|_{q_k} \rangle\, E^R\big|_{pq_k} - \langle T^R\big|_{pq_k} \rangle\, w\big|_{q_k}\,] \qquad (27\text{-R})$$

$$K_\xi \frac{\partial w}{\partial \xi}\Big|_p + K_\eta \frac{\partial w}{\partial n}\Big|_p = \frac{1}{D}\int_\Gamma \{V^R_\gamma\big|_{pq}\, (w\big|_q - w\big|_p) - M^R_\gamma\big|_{pq}\, \frac{dw}{dn}\Big|_q + N^R_\gamma\big|_{pq}\, M_n(w)\big|_q$$

$$- E^R_\gamma\big|_{pq}\, V_n(w)\big|_q\, ds\big|_q \qquad (28\text{-R})$$

$$- \frac{1}{D}\sum_{k=1}^{K} [\, \langle M_t(w)\big|_{q_k} \rangle\, E^R_\gamma\big|_{pq_k} - \langle T^R_\gamma\big|_{pq_k} \rangle\, (w\big|_{q_k} - w\big|_{p_k})$$

$$0 = \int_\Gamma \{V^i\big|_{pq}\, w\big|_q - M^i\big|_{pq}\, \frac{dw}{dn}\Big|_q + N^i\big|_{pq}\, M_n(w)\big|_q - E^i\big|_{pq}\, V_n(w)\big|_q\}\, ds\big|_q$$

$$- \sum_{k=1}^{K} [\, \langle M_t(w)\big|_{q_k} \rangle\, E^i\big|_{pq_k} - \langle T^i\big|_{pq_k} \rangle\, (w\big|_{q_k})] \qquad (27\text{-I})$$

and

$$0 = \int_\Gamma \{ V^i_{\gamma}\big|_{pq} (w\big|_q - w\big|_p) - M^i_{\gamma}\big|_{pq} \frac{dw}{dn}\Big|_q + N^i_{\gamma}\big|_{pq} M_n(w)\big|_q - E^i_{\gamma}\big|_{pq} V_n n(w)\big|_q \} ds\big|_q$$

$$- \sum_{k=1}^{K} [\langle M_t(w)\big|_{q_k} \rangle E^i_{\gamma}\big|_{pq_k} - \langle T^i_{\gamma}\big|_{pq_k} \rangle (w\big|_{q_k} - w\big|_p) \qquad (28\text{-I})$$

where α is a coefficient which relates to the interior angle of a corner as:

$$\alpha = \frac{\text{interior angle}}{\pi}$$

if p is on a regular point (not a corner point) then α equals unity. The coefficient K_ξ and K_η of Eq. (2-R) are defined as

$$K_\xi = \frac{\alpha}{2\lambda} + \frac{\nu}{2\pi\lambda} [\sin 2\delta + \sin 2 (\alpha\pi - \delta)] \quad \text{and}$$

$$K_\eta = \frac{\nu}{4\pi\lambda} [\cos 2\delta - \cos 2 (\alpha\pi - \delta)]$$

where the angle δ is measured from the boundary to the ξ axis in a counterclockwise sense.

At a regular point (not a corner point) of the boundary Γ we have 4 independent variables (w, $\frac{dw}{dn}$, $M_n(w)$ and $V_n(w)$); however, on the corner point, we have a total of 8 independent variables since two distinct limiting values of $\frac{dw}{dn}$, $M_n(w)$ and $V_n(w)$ exist. The 8 variables are w, $\frac{dw}{dn}\big|_+$, $\frac{dw}{dn}\big|_-$, $M_n\big|_+$, $M_n\big|_-$, $V_n\big|_+$, $V_n\big|_-$, and $\langle M_t(w) \rangle$.

At a corner point, Eq. (28-R) can be rewritten in two independent representations.

1) let $\delta_1 = (2+\alpha)\frac{\pi}{2}$ (i.e. ξ bisects the exterior angle)

and Eq. (28-R) becomes

$$\frac{\alpha + \nu/\pi \sin \alpha\pi}{4 \pi \sin \left(\frac{\alpha\pi}{2}\right)} \left\{ \left.\frac{dw}{dn}\right|_{p+} + \left.\frac{dw}{dn}\right|_{p-} \right\} = \frac{1}{D}\int_{\Gamma}\{V_{\alpha}^{R} \ldots\} \, ds - \frac{1}{D} \sum[..] \qquad (28\text{-R-}1)$$

2) let $\delta_2 = \frac{(\alpha-1)\pi}{2}$ (i.e. $\perp$ to ξ_1)

and Eq. (28-R) is

$$\frac{\alpha - \nu/\pi \sin \alpha\pi}{4 \cos\lambda\left(\frac{\alpha\pi}{2}\right)} \left\{ \left.\frac{dw}{dn}\right|_{p+} - \left.\frac{dw}{dn}\right|_{p-} \right\} = \frac{1}{D} \int \{V_{\alpha}^{R} \ldots\} ds - \frac{1}{D} \sum_{k=1}^{k} [\ldots] \qquad (28\text{-R-}2)$$

NUMERICAL SOLUTION

For the numerical solution, it was decided to assume a linear variation of the unknown boundary variables along the arc length between nodes. Node points are chosen at each corner point, as well as at selected regular points along the boundary. It was further decided to consider only the real parts of Eqs. (27) and (28).

At a regular point, there are two known and two unknown boundary conditions and hence, two unknowns correspond to each regular point. At a corner point, five of the eight boundary point variables can be determined from considerations of the boundary conditions on each side of the corner, leaving three unknowns at each corner point.

The integral equations are applied at each node point on the boundary. At a regular point, Eqs. (27-R) and (28-R) are applied; whereas, at a corner point, Eqs. (27-R), (28-R-1) and (28-R-2) are applied. This leads to a homogeneous system of 2NR + 3NC simultaneous equations with 2NR + 3NC unknowns, where NR is the number of regular nodes and NC is the number of corner nodes.

For a given plate geometry and physical properties, the coefficients of the simultaneous equations are a function of the natural frequency alone. The solution process, then, consists of searching for values of the frequency which makes the determinant of the coefficients approach zero. On finding such a frequency, the relative values of the boundary point unknowns are evaluated. This eigenvector contains the information necessary to deduce the complete mode shape from Eqs. (18) and (26).

COMPUTER PROGRAM

In the present computer programs for the solution of this problem, only the real parts of Eqs. (27) and (28) are searched to find the roots. This procedure yields roughly the same number of false roots as true roots. If both the real and imaginary parts of Eqs. (27) and (28) are searched, then the true roots would be those in which both real and imaginary roots coincide. Searching both sets of equations, however, requires twice as much computer time. The true and false roots found from the real parts of Eqs. (27) and (28) are easily separated, if one looks at corresponding mode shapes.

The computer program is set up with automatic node generation along either straight line segments or circular arcs. At corner points the interior angle must also be input. The integration in the program is carried out by Gaussian quadrature, and the option is provided for using up to seven quadrature points.

At present, the program can only treat homogeneous boundary conditions. Indexing of the unknown function is performed by assigning an index number at each regular point for each of the two unknown functions and a zero for the two known (zero) functions. At corner points, two indices are assigned for each of the two unknown functions and an index for the twisting moment discontinuity is also required. Zeros are input for the known functions.

As of this writing, the computer program is not fully checked out, so we have omitted numerical results. Preliminary results for certain cases, however, indicate that the method is convergent and provides excellent results with a minimum number of boundary elements.

CONCLUSIONS

The method used in this paper provides a viable and useful means of determining frequencies and mode shapes of arbitrarily shaped thin uniform elastic plates. At this time, nothing can be said about the relative merits of this method as opposed to the finite element method or the mixed method proposed by Bézine (4). Since all of these methods produce acceptable results, the criteria for choice must be computer costs. Comparisons of these methods will be the subject of future research.

REFERENCES

1. Hutchinson, J. R. and G.K.K. Wong, "The Boundary Element Method For Plate Vibrations", Proceedings of the ASCE Seventh Conference on Electronic Computation, St. Louis, Missouri, pp. 297-311, August 1979.

2. Bézine, G. P., and D.A. Gamby, "A New Integral Equation Formulation For Plate Bending Problems", Recent Advances in Boundary Element Methods, edited by C. A. Brebbia, Pentech Press, London:Plymouth, pp. 327-342, July 1978.

3. Stern, M., "A General Boundary Integral Formulation for the Numerical Solution of Plate Bending Problems", International Journal of Solids and Structures, Vol. 15, pp. 769-782, 1979.

4. Bezine, G., "A Mixed Boundary Integral - Finite Element Approach to Plate Vibration Problems", Mechanics Research Communications, Vol. 7, pp. 141-150, 1980.

5. Vivoli, J. and P. Filippi, "Eigenfrequencies of Thin Plates and Layer Potentials", Journal of the Acoustical Society of America, Vol. 55, No. 3, pp. 562-567, March 1974.

Section III
Geomechanics

BOUNDARY ELEMENTS APPLIED TO SOIL-FOUNDATION INTERACTION

M. Ottenstreuer and G. Schmid

Ruhr-Universität Bochum, W-Germany

INTRODUCTION

The response of a structure to a given loading does not only depend on its geometric configuration but may also be influenced by the surrounding media. This holds particularly for dynamically loaded structures whose behaviour is strongly influenced by soil-structure interaction. In such cases the foundation environment can not be neglected.

An essential feature exhibited by soil is the "geometric damping", which is due to the fact that waves travel out to infinity. This results in energy dissipation even if the soil were assumed to consist of purely elastic material.

In early attempts to simulate the embedding of a structure in soil simple spring and dashpot elements, connected to the mathematical model of the structure were used.

A more refined method for the simulation of soil-structure interaction takes advantage of the possibilities offered by finite elements which allow the simultaneous discretization of the structure and the soil. The method is equally applicable to embedded foundations and inhomogeneous soil. However, it has serious disadvantages when applied to three dimensional problems since it requires extensive, complicated and expensive data management. Moreover geometric damping effects can only be represented approximately by semi-infinite elements (Waas 1972) which are limited in that the infinite extension can only be simulated in the horizontal direction.

Another possibility to calculate soil-structure interaction is the substructure-method which analyses the soil and the structure separately. Here the soil is assumed to be an elastic homogeneous halfspace and the influence of the foundation upon the soil is considered by using Green's functions.

This method, however, is not applicable to layered soils and to embedded foundations.

As an alternative we present the boundary element method which - together with the substructure technique - has several advantages in the treatment of soil-structure interaction problems. First it enables one to consider embedded foundations and foundations on a viscoelastic halfspace, and second, the coupling effects of several neighbouring foundations may be taken into account. Single foundations with, and without, embedment have been considered earlier by Dominguez (1978) for an elastic halfspace.

DYNAMIC STIFFNESS

The dynamic stiffness matrix of the foundations represents the complex, frequency dependent impedance function which characterizes the soil-structure interaction. Its derivation of the matrix is shown in the following.

The differential equations of motion are given by

$$(c_p^2 - c_s^2)\, u_{j,ji} + c_s^2\, u_{i,jj} + K_i = \frac{\partial^2 u_i}{\partial t^2} \tag{1}$$

where c_p is the velocity of dilatational waves,

c_s is the velocity of distortional waves,

u_i is the displacement component in the i-direction,

K_i is the body force component in the i-direction,

t is time.

The conditions along the boundary are

$$u_i = \bar{u}_i \quad , \tag{2 a}$$

$$\sigma_{ij}\, n_j = \bar{t}_i \quad , \tag{2 b}$$

where σ_{ij} is the stress tensor,

n_j is the component of the normal vector

and the initial conditions are taken as

$$u_i(t=o) = u_{io} \quad , \tag{3 a}$$

$$\frac{\partial u_i}{\partial t}(t=o) = \dot{u}_{io} . \tag{3 b}$$

Equations (1) through (3) may be transformed into the frequency domain by use of the Laplace transform yielding the differential equation

$$(c_p^2 - c_s^2)\, \tilde{u}_{j,ji} + c_s^2\, \tilde{u}_{i,jj} + \tilde{K}_i = s^2\, \tilde{u}_i - s u_{io} - \dot{u}_{io} . \tag{4}$$

Writing $s = i\omega$ and defining the body force to be

$$P_i = \tilde{K}_i + \dot{u}_{io} + i\omega\, u_{io} \tag{5}$$

the frequency dependent differential equation of motion takes the form

$$(c_p^2 - c_s^2)\,\tilde{u}_{j,ji} + c_s^2\,\tilde{u}_{i,jj} + \omega^2\,\tilde{u}_i + P_i = 0. \tag{6}$$

where ω is the circular frequency.
The transformed boundary conditions are

$$\tilde{u}_i = \bar{u}_i \;, \tag{7 a}$$

$$\tilde{\sigma}_{ij}\, n_j = T\tilde{u}_i = \bar{t}_i \tag{7 b}$$

where T is a differential operator.

The inverse transformation from the frequency into the time domain allows the calculation of the response of the system to any specified loading history.

The fundamental solution for Equation (6) as given by Cruse and Rizzo (1968) is an essential prerequisite for the application of the boundary element method. In this context it corresponds to the response of the system to an impulse in the form of a Dirac-delta function.

Writing Equation (6) as

$$(c_p^2 - c_s^2)\, u^*_{j,ji} + c_s^2\, u^*_{i,jj} + \omega^2 u_i + \delta_i = 0 \tag{8 a}$$

or in the abbreviated form

$$Lu + \delta = 0 \tag{8 b}$$

where L is a differential operator and δ is a point load, the fundamental solution of Equation (8) for the three-dimensional situation is

$$u^*_{ij} = \frac{1}{4\pi\,\rho c_s^2}\,(\psi\,\delta_{ij} - \chi\, r_{,i}\, r_{,j}) \;, \tag{9}$$

$$\chi = \left(-\frac{3c_s^2}{\omega^2 r^2} + \frac{3c_s}{i\omega r} + 1\right)\frac{e^{\frac{-i\omega r}{c_s}}}{r} - \frac{c_s^2}{c_p^2}\left(-\frac{3c_p^2}{\omega^2 r^2} + \frac{3c_p}{i\omega r} + 1\right)\frac{e^{\frac{-i\omega r}{c_p}}}{r} \;, \tag{10 a}$$

$$\psi = \left(-\frac{c_s^2}{\omega^2 r^2} + \frac{c_s}{i\omega r} + 1\right)\frac{e^{\frac{-i\omega r}{c_s}}}{r} - \frac{c_s^2}{c_p^2}\left(-\frac{c_p^2}{\omega^2 r^2} + \frac{c_p}{i\omega r}\right)\frac{e^{\frac{-i\omega r}{c_p}}}{r} \tag{10 b}$$

In Equations (9) and (10) r is the distance between the point load and a generic field point, and ρ is the mass density.

The application of the method of weighted residuals with u* as weighting function, yields the Betti-formula

$$\int_{B} (u^*Lu - uLu^*)\, d\Omega = \int_{\Gamma} (u^*Tu - uTu^*)\, ds \qquad (11)$$

where B is the domain,

Γ is the boundary,

dΩ is a volume element and

ds is an area element.

With the help of the fundamental solution the deformation of a point inside a domain can be expressed in terms of the boundary values (Brebbia, 1978):

$$u = \int_{\Gamma} (u^*Tu - uTu^*)\, ds. \qquad (12)$$

For a point on a smooth boundary the relation

$$\frac{1}{2}u = \int_{\Gamma} (u^*Tu - uTu^*)\, ds \qquad (13)$$

is found.

To obtain a numerical solution the boundary is divided into elements. In our case the tractions and the displacements are assumed to be constant within the elements. Thus, the integration can be replaced by a summation over all elements on the boundary, and Equation (13) becomes

$$\frac{1}{2}u = \sum_{i=1}^{n} \int_{\Gamma_i} u^* ds_i\, t^i - \sum_{i=1}^{n} \int_{\Gamma_i} t^*\, ds_i\, u^i. \qquad (14)$$

where t^i and u^i are constant for each element.
Equation (14) can also be written in matrix notation as follows

$$\frac{1}{2}\underline{u} = \underline{U}\,\underline{t} - \underline{T}\underline{u} \qquad (15)$$

where

$$\underline{U} = \int_{\Gamma} \underline{u}^*\, ds, \qquad (16\text{ a})$$

$$\underline{T} = \int_{\Gamma} \underline{t}^*\, ds \qquad (16\text{ b})$$

or in a more condensed form

$$\hat{\underline{T}}\,\underline{u} = \underline{U}\,\underline{t}, \qquad (17\text{ a})$$

$$\hat{\underline{T}} = \underline{T} + \frac{1}{2}\underline{I}. \qquad (17\text{ b})$$

Equation (17) allows us to establish the dynamic stiffness matrix of one or several foundations. The surface of the soil represents the boundary of the domain which is discretized by boundary elements. They may be divided into two parts: One part, with index i, stems from the foundation, the remaining part, with index a, from the traction-free surface. Then Equation (17) may be partitioned as follows:

$$\begin{bmatrix} \hat{\underline{T}}^{ii} & \hat{\underline{T}}^{ia} \\ \hat{\underline{T}}^{ai} & \hat{\underline{T}}^{aa} \end{bmatrix} \begin{bmatrix} \underline{u}^{i} \\ \underline{u}^{a} \end{bmatrix} = \begin{bmatrix} \underline{U}^{ii} & \underline{U}^{ia} \\ \underline{U}^{ai} & \underline{U}^{aa} \end{bmatrix} \begin{bmatrix} \underline{t}^{i} \\ \underline{t}^{a} \end{bmatrix} . \tag{18}$$

As all boundary elements belong to the same horizontal plane some of the coefficients become zero because at the surface

$$\frac{\partial r}{\partial n} = 0 , \tag{19 a}$$

$$r_{,3} = 0, \tag{19 b}$$

$$\delta_{12} = \delta_{21} = \delta_{13} = \delta_{31} = \delta_{23} = \delta_{32} = 0 . \tag{19 c}$$

Correspondingly some of the coefficients of the element-related submatrices in $\underline{T}$ vanish:

$$\begin{bmatrix} 0 & 0 & T_{13} \\ 0 & 0 & T_{23} \\ T_{31} & T_{32} & 0 \end{bmatrix} . \tag{20}$$

Force-displacement relation of the foundation

The six degrees of freedom of a rigid foundation consist of three translations $\underline{u}$ and three rotations $\underline{\psi}$. The kinematic relations between the rigid body degrees of freedom and the boundary element displacements are given by the following geometric constraints

$$u_{1i} = u_1 - \psi_3 \, r_{2i}, \tag{21 a}$$

$$u_{2i} = u_2 - \psi_3 \, r_{1i}, \tag{21 b}$$

$$u_{3i} = u_3 + \psi_1 \, r_{2i} - \psi_2 \, r_{1i} \tag{21 c}$$

which, written in matrix notation, read

$$\underline{u} = \underline{a} \, \underline{\tilde{u}} . \tag{22}$$

The corresponding forces, (Fig.1) acting in the direction of $\underline{u}$ and $\underline{\psi}$, can be obtained by integrating the tractions over the area of the boundary elements:

$$K_1 = \sum_i \int_{A_i} t_1 \, dA_i \quad , \quad M_i = \sum_i \int_{A_i} t_3 \, r_2 \, dA_i , \tag{23a,d}$$

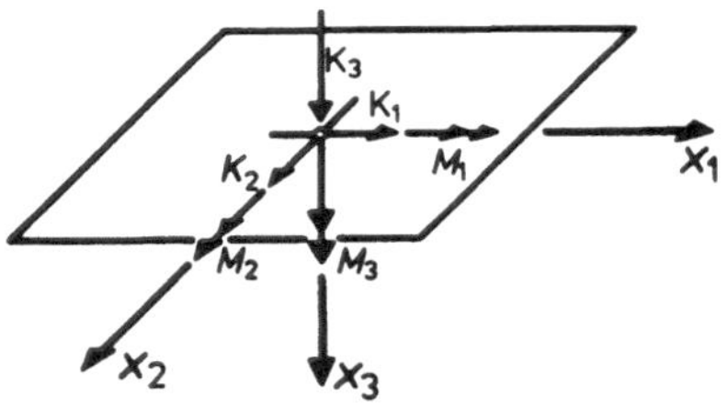

Figure 1 Forces on a single foundation

$$K_2 = \sum_i \int_{A_i} t_2 \, dA_i \quad , \quad M_2 = \sum_i \int_{A_i} t_3 \, r_1 \, dA_i \quad , \tag{23b,e}$$

$$K_3 = \sum_i \int_{A_i} t_3 \, dA_i \quad , \quad M_3 = \sum_i \int_{A_i} t_2 \, r_1 \, dA_i - \sum_i \int_{A_i} t_1 \, r_2 dA_i . \tag{23c,f}$$

In matrix notation Equation (23) reads:

$$\underline{P} = \underline{a}^T \, \underline{A} \, \underline{t} \tag{24}$$

where $\underline{A}$ is a diagonal matrix and stands for the area of the individual boundary elements of the foundation.

Depending on the particular assumptions concerning the motion, two stiffness relations may be derived.

Stiffness matrix with uncoupled boundary conditions (relaxed boundary conditions) If it is assumed that at the surface vertical loads result only in vertical displacements and horizontal loads produce only horizontal translations, then the coefficients in Equation (20) can be set equal to zero. Since the tractions outside the foundation are zero as well Equation (18) reduces to

$$\frac{1}{2} \begin{bmatrix} \underline{u}^i \\ \underline{u}^a \end{bmatrix} = \begin{bmatrix} \underline{U}^{ii} \\ \underline{U}^{ai} \end{bmatrix} \underline{t}^i \tag{25 a}$$

resulting in the decoupled system of equations

$$\frac{1}{2} \, \underline{u}^i = \underline{U}^{ii} \, \underline{t}^i \tag{25 b}$$

for the foundation. The surface deformation outside the foundation can be determined by

$$\frac{1}{2}\,\underline{u}^{a} = \underline{U}^{ai}\,\underline{t}^{i}. \tag{25 c}$$

Equation (25 b) shows that, within the above mentioned assumptions, the displacement-traction-relation of the foundation is not affected by the traction-free part of the surface. It is therefore sufficient to introduce boundary elements only at the location of the foundation itself. This simplification results in considerably reduced numerical effort.

The stiffness matrix can now be obtained by premultiplying Equation (25 b) with $(\underline{U}^{ii})^{-1}$ and using Equations (22) and (24). One obtains

$$\underline{K}\,\underline{\tilde{u}} = \underline{P} \tag{26}$$

where the complex dynamic stiffness is defined as

$$K = \frac{1}{2}\,\underline{a}^{T}\,\underline{A}\,(\underline{U}^{ii})^{-1}\,\underline{a}\,. \tag{27}$$

<u>Stiffness matrix with coupled boundary conditions</u> (non-relaxed boundary conditions) Here, tractions and displacements on the surface are not assumed to be decoupled. Thus, the traction-free surface outside the foundation has to be represented by boundary elements as well. It should be mentioned that for practical numerical calculation the boundary element discretization need not extend further than a few times the length of the foundation.

The stiffness is obtained by premultiplying Equation (18) by $\underline{A}\,\underline{U}^{-1}$ to give

$$\begin{bmatrix} \underline{Q}^{ii} & \underline{Q}^{ia} \\ \underline{Q}^{ai} & \underline{Q}^{aa} \end{bmatrix} \begin{bmatrix} \underline{u}^{i} \\ \underline{u}^{a} \end{bmatrix} = \underline{A} \begin{bmatrix} \underline{t}^{i} \\ \underline{0} \end{bmatrix} \tag{28}$$

where the definition

$$\underline{Q} = \underline{A}\,\underline{U}^{-1}\,\hat{\underline{T}} \tag{29}$$

has been introduced.

From Equation (28) $\underline{u}^{a}$ can be eliminated so that, with the assumption of rigid foundations, the relation between the foundation forces and displacements is obtained as

$$\underline{a}^{T}\,(\underline{Q}^{ii} - \underline{Q}^{ia}\,(\underline{Q}^{aa})^{-1}\,\underline{Q}^{ai})\,\underline{a}\,\underline{\tilde{u}} = \underline{a}^{T}\,\underline{A}\,\underline{t}\,. \tag{30}$$

The dynamic stiffness matrix is now given by

$$\underline{K} = \underline{a}^{T}\,(\underline{Q}^{ii} - \underline{Q}^{ia}\,(\underline{Q}^{aa})^{-1}\,\underline{Q}^{ai})\,\underline{a}\,. \tag{31}$$

Force-displacement relation of two foundations (Uncoupled boundary conditions)
To calculate the influence of a second foundation on the response of the first one, the stiffness matrix of two rigid foundations (Fig. 2) is derived.

Equation (17) is partitioned into three parts, where parts I and II correspond to foundations I and II and part "a" to the traction-free surface outside the foundations:

$$\begin{bmatrix} \hat{\underline{T}}^{I\,I} & \hat{\underline{T}}^{I\,II} & \hat{\underline{T}}^{I\,a} \\ \hat{\underline{T}}^{II\,I} & \hat{\underline{T}}^{II\,II} & \hat{\underline{T}}^{II\,a} \\ \hat{\underline{T}}^{a\,I} & \hat{\underline{T}}^{a\,II} & \hat{\underline{T}}^{aa} \end{bmatrix} \begin{bmatrix} \underline{u}^{I} \\ \underline{u}^{II} \\ \underline{u}^{a} \end{bmatrix} = \begin{bmatrix} \underline{U}^{I\,I} & \underline{U}^{I\,II} & \underline{U}^{I\,a} \\ \underline{U}^{II\,I} & \underline{U}^{II\,II} & \underline{U}^{II\,a} \\ \underline{U}^{a\,I} & \underline{U}^{a\,II} & \underline{U}^{aa} \end{bmatrix} \begin{bmatrix} \underline{t}^{I} \\ \underline{t}^{II} \\ \underline{t}^{a} \end{bmatrix} . \tag{32}$$

The assumption of uncoupled tractions and displacements yields three matrix equations

$$\hat{\underline{T}}^{I\,I} \; \underline{u}^{I} = \underline{U}^{I\,I} \; \underline{t}^{I} + \underline{U}^{I\,II} \; \underline{t}^{II} , \tag{33 a}$$

$$\hat{\underline{T}}^{II\,II} \; \underline{u}^{II} = \underline{U}^{II\,I} \; \underline{t}^{I} + \underline{U}^{II\,II} \; \underline{t}^{II} , \tag{33 b}$$

$$\hat{\underline{T}}^{aa} \; \underline{u}^{a} = \underline{U}^{a\,I} \; \underline{t}^{I} + \underline{U}^{a\,II} \; \underline{t}^{II} . \tag{33 c}$$

Again, the traction-free part of the surface has no influence

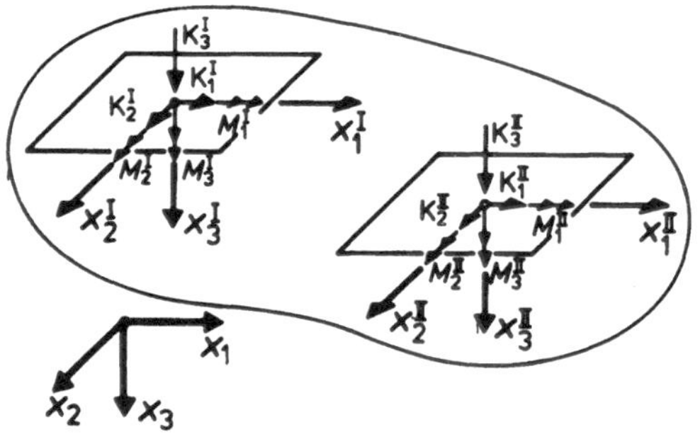

Figure 2 Forces on two foundations

on the stiffness of the foundations and need not be discretized by boundary elements.

The dynamic stiffness matrix for two rigid foundations is again given in the form

$$\frac{1}{2}\,\underline{a}^T\,\underline{A}\,\underline{U}^{-1}\,\underline{a} = \underline{K} \tag{34}$$

$\underline{a}$ now being the kinematic transformation matrix for two foundations.

NUMERICAL EXAMPLES

For numerical calculations the stiffness matrices are divided into real and imaginary parts which, mechanically speaking represent the spring and damping coefficients of the system

$$\underline{K} = \underline{K}^R + i\,\underline{K}^I\,. \tag{35}$$

The stiffness matrix for the static case $\underline{K}^o$ may be factored out, i.e.

$$\underline{K} = \underline{K}^o\,(\underline{K}^R + ia_o\,\underline{K}^I) \tag{36}$$

with
$$a_o = \frac{\omega \cdot b}{c_s} \tag{37}$$

where b is the half side-length of the foundation, and $\underline{K}^R$ and $\underline{K}^I$ are dimensionless stiffness coefficients. Inversion of $\underline{K}$ yields the flexibility matrix

$$\underline{F} = \underline{F}^R + i\,\underline{F}^I \tag{38}$$

again divided into real and imaginary part.
With shear modulus G and side-length b they can be written in dimensionless form:
for the displacements

$$\underline{f}^R = G \cdot b\,\underline{F}^R\,, \tag{39 a}$$

$$\underline{f}^I = G \cdot b\,\underline{F}^I \tag{39 b}$$

and for the rotations and the torsional degrees of freedom

$$\underline{f}^R = G \cdot b^3\,\underline{F}^R\,, \tag{39 c}$$

$$\underline{f}^I = G \cdot b^3\,\underline{F}^I\,. \tag{39 d}$$

Dynamic stiffness matrix for a single foundation

As an example for the numerical calculation of the stiffness matrix a quadratic foundation on a halfspace is chosen. In addition to geometric damping, material damping behaviour is taken into account by introducing a complex modulus of elasticity (Achenbach 1975)

$$\hat{E} = E\,(1+i\eta)\,. \tag{40}$$

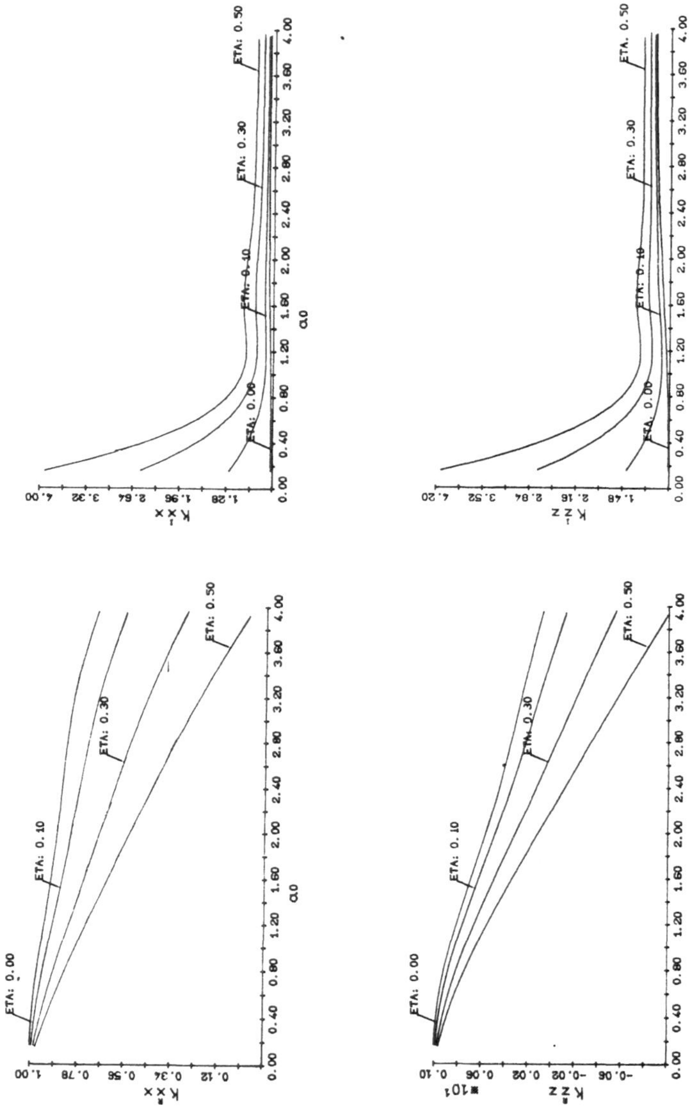

Figure 3   Real and imaginary part of the stiffness coefficients k_x and k_z; constant hysteretic damping

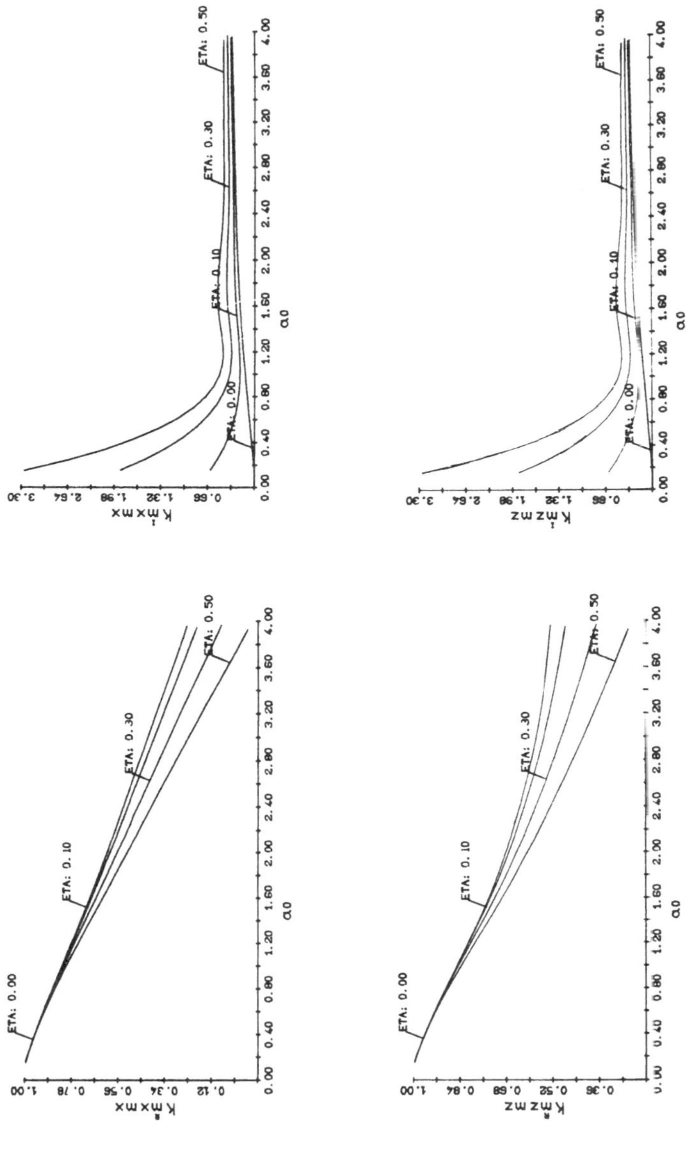

Figure 4 Real and imaginary part of the stiffness coefficients k_{mx} and k_{mz}; constant hysteretic damping

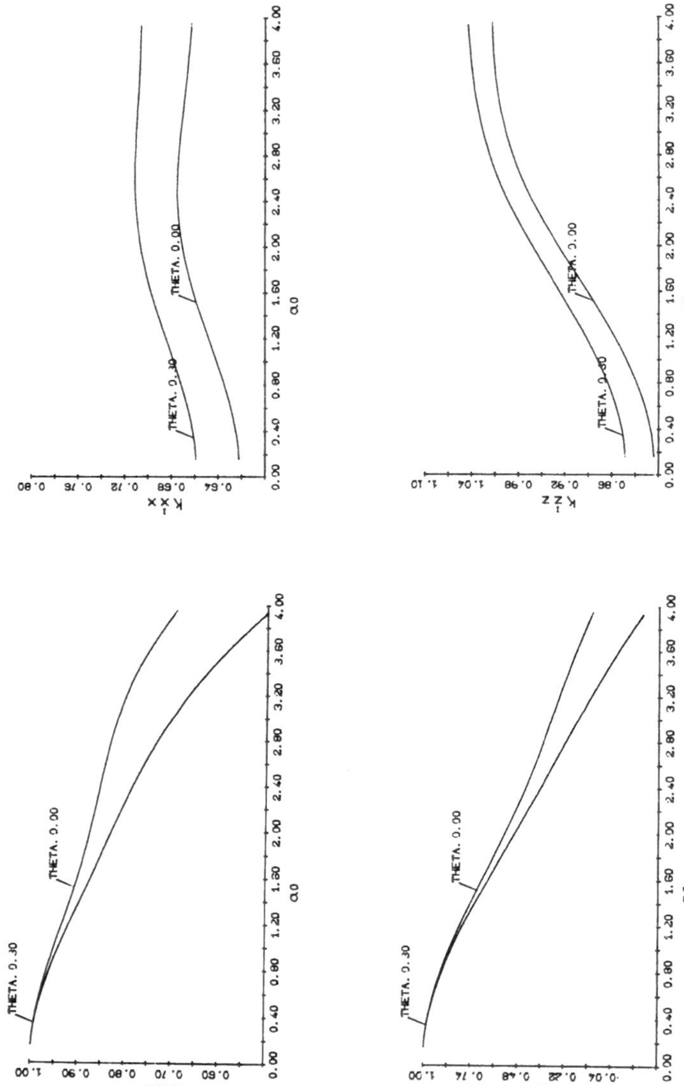

Figure 5 Real and imaginary part of the stiffness coefficients k_x and k_z; Voigt damping model

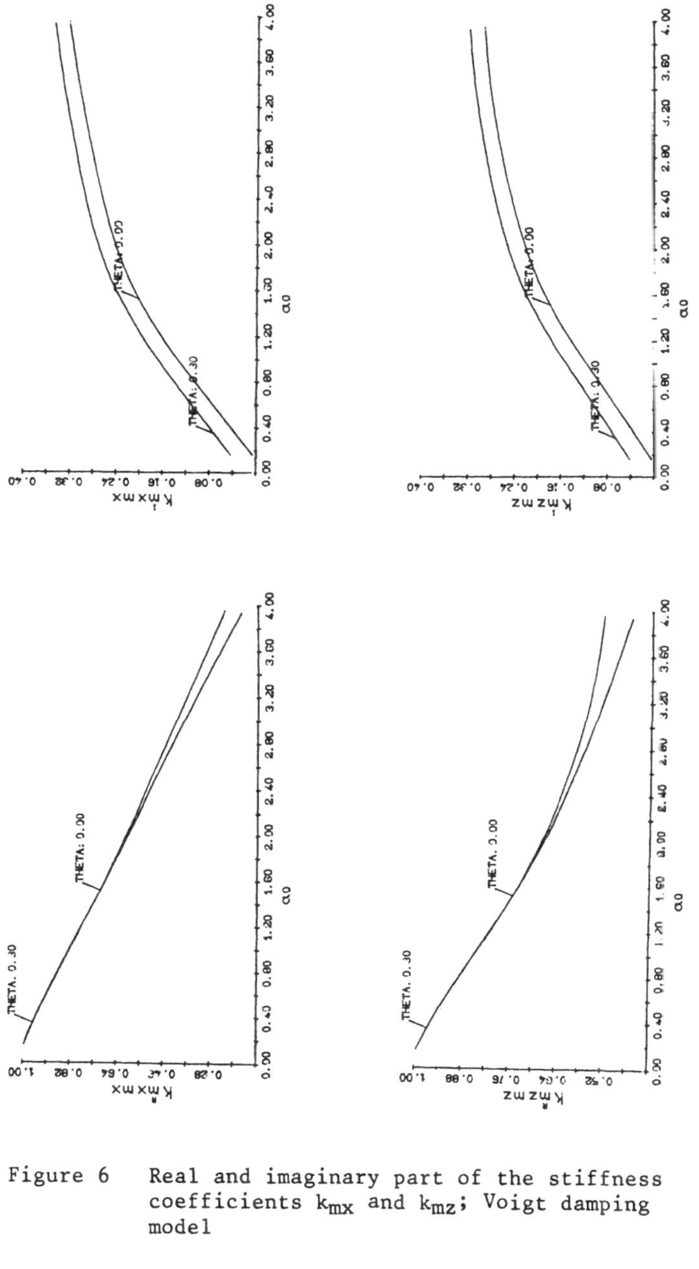

Figure 6 Real and imaginary part of the stiffness coefficients k_{mx} and k_{mz}; Voigt damping model

In Equation (40) η is the damping coefficient which depends on the chosen damping model.
E.g. for a "Voigt" model one has

$$\frac{\eta}{\omega} = \theta = \text{const.} \tag{41}$$

whereas for constant hysteretic damping η becomes

$$\eta = \omega \cdot \theta = \text{const.} \tag{42}$$

Figures 3 to 6 show the dependency of stiffness coefficients on the dimensionless frequency a_o for various values of the damping coefficients η and θ. It can be shown that the real parts of the dynamic stiffness matrix decrease with growing a_o which corresponds to a reduction of the stiffness of the spring in the model. Simultaneously damping increases in a way which depends on the chosen damping model.

It is important to point out the difference between the two models: The model with constant hysteretic damping shows a significant increase at small values of a_o and approaches the damping of the elastic-halfspace model at high frequencies. The "Voigt" model however shows a completely different behaviour in that material damping causes a nearly constant increase of damping over the whole frequency range and that it is almost independent of a_o. It could be shown (Ottenstreuer 1981) that these results are in good agreement with those found in the literature (Wong and Luco, 1976).

Dynamic stiffness matrix for two rigid foundations

As a second example the interaction between two rigid foundations (Fig. 7) is examined. Contrary to a single vibrating foundation even some offdiagonal coefficients of the stiffness matrix are present (Fig. 8). This demonstrates the coupling of the two foundations. Figures 9 to 10 show some of the stiffness

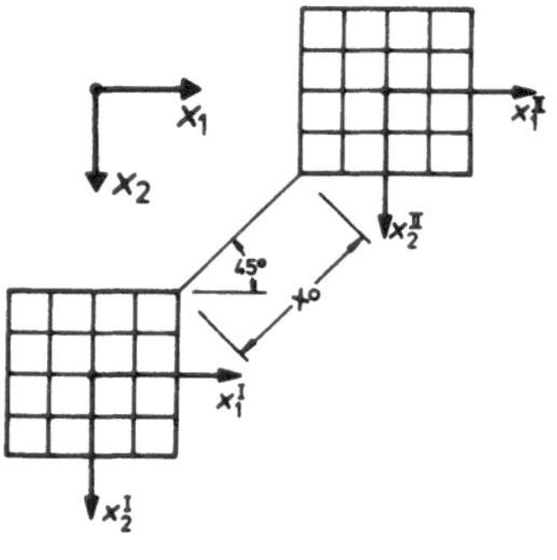

Figure 7 Two foundations with mesh

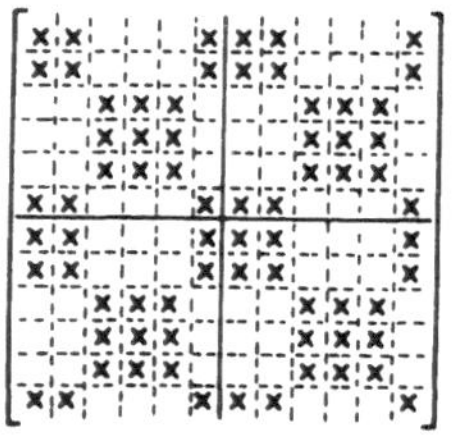

Figure 8 Non-zero stiffness coefficients

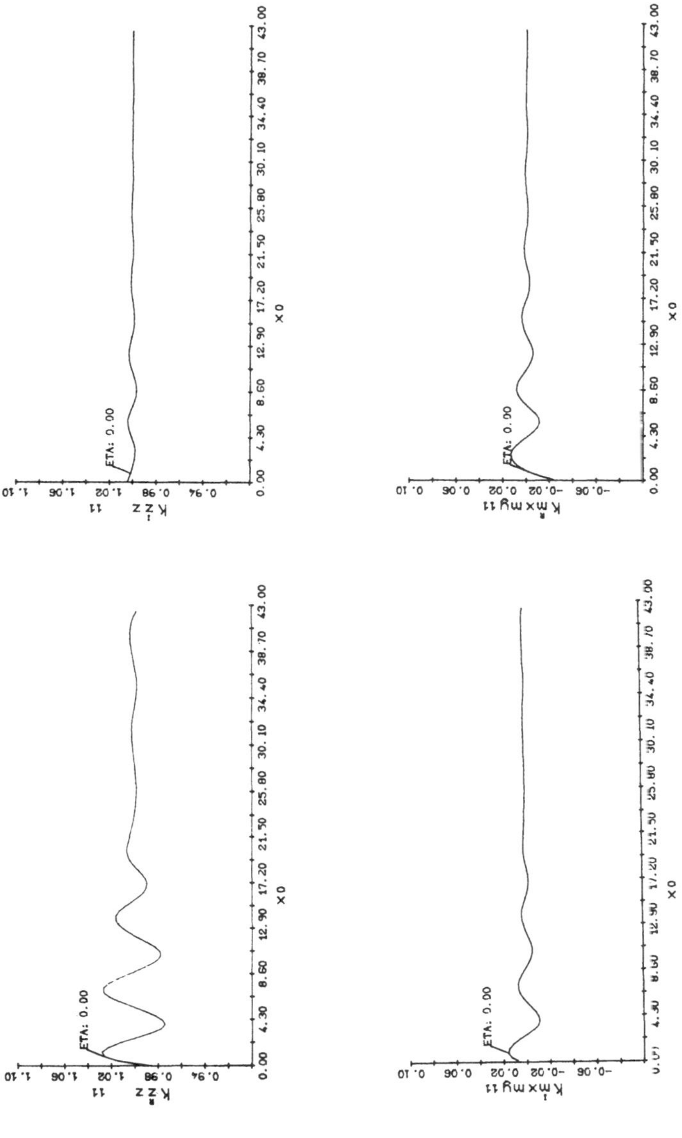

Figure 9 Real and imaginary part of the stiffness coefficients k_{zz11} and $k_{mx\ my\ 11}$

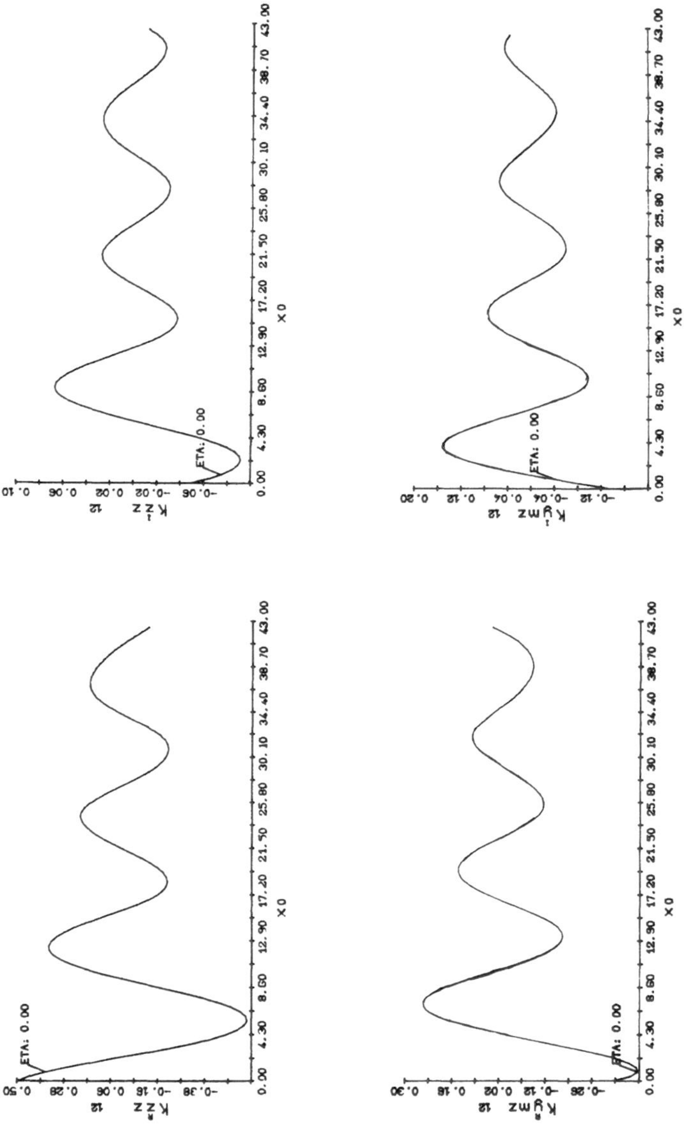

Figure 10 Real and imaginary part of the stiffness coefficients $k_{zz\ 12}$ and $k_{ymz\ 12}$ (cross coupling)

coefficients dependent on their relative distance. The influence of the offdiagonal terms becomes smaller with growing distance of the two foundations so that finally the influence of the coupling disappears. Thus, at large relative distance the two foundations behave in the same way as two single foundations.

CONCLUSION

From the results presented in this paper it can be concluded that the boundary element method-combined with the substructure technique often used in soil dynamics - is a suitable tool to simulate soil-structure interaction.

Although our results are given here only for rigid, flat foundations on homogeneous soil the method can easily be extended to embedded foundations on layered soil.

ACKNOWLEDGEMENT

The programming work was done by Margret Gibhardt and typing by Christa Hoogterp.

REFERENCES

Achenbach, J.D. "Wave Propagation in Elastic Solids" Amsterdam: North Publishing Company, 1975

Brebbia, C.A. "The Boundary Element Method for Engineers" London: Pentech Press, 1978

Cruse, T.A. and Rizzo, F.J. "A Direct Formulation and Numerical Solution of the General Transient Elastodynamic Problem I" Journal of Mathematical Analysis and Applications 22, 1968

Dominguez, J. "Dynamic Stiffness of Rectangular Foundations" Publication Nr. R 78-20, MIT, Department of Civil Eng., 1978

Ottenstreuer, M. "Das Verfahren der Randelemente - Ein Beitrag zur Darstellung der Wechselwirkung zwischen Bauwerk und Baugrund"
Dissertation, Ruhr-Universität Bochum, 1981

Waas, G. "Linear Two Dimensional Analysis in Soil Dynamics Problems in Semi-Infinite Layered Media"
Dissertation, University of California, Berkeley, Cal., 1971

Wong, H.L. and Luco, J.E. "Dynamic Response of Rigid Footings of Arbitrary Shape"
Earthquake Engineering and Structural Dynamics, 4 , 1976.

THE IMPLEMENTATION OF BOUNDARY ELEMENT CODES IN GEOTECHNICAL ENGINEERING

L. A. Wood

Lecturer, Department of Civil Engineering, Queen Mary College, London, U.K.

SYNOPSIS

The application of simple boundary element techniques in the field of geotechnical engineering is discussed. Particular emphasis is placed upon the coupling of standard finite elements to soil boundary elements in order to solve soil-structure interaction problems. Two computer programs which incorporate these techniques in the areas of foundation and pile/wall analysis and design are briefly described. In addition the use of these programs is demonstrated with respect to several case histories in which the agreement between predicted and observed behaviour is most encouraging.

INTRODUCTION

In the field of geotechnical engineering the semi-infinite nature of the soil domain may be efficiently modelled using boundary element techniques. Finite element programs which will handle the modelling of three-dimensional continua exist, but for semi-infinite domains the computation involved may become excessive. Furthermore, in the context of foundation design the three dimensional nature of the distribution of structural loads is such that the reduction of the problem to a two-dimensional axi-symmetric, plane stress or plane strain model, in order to facilitate economic analysis, is inappropriate. In order to exploit the potential offered by the boundary element approach two computer programs RAFTS and LAWPILE (Wood, 1978a and 1979a) have been written specifically for the analysis of foundations and, piles and walls subject to lateral loads. Both are based upon the coupling of standard structural finite elements with soil boundary elements.

The soil boundary elements are formulated on the basis of linear elastic behaviour but incorporate procedures to enable the effects of non-linear soil response to be assessed. Approximate extensions of the basic element formulation have been implemented in order to model layered (not neccessarily horizontal layers) and transversely isotropic continua. It is felt that these approximations based upon the solutions obtained by Mindlin (1936) and Gerrard and Harrison (1970) for homogeneous elastic continua, are acceptable in the context of soil where the assumption of elastic behaviour is itself questionable. Indeed the performance of the simple boundary element adopted has been shown to produce results which lie well within the normal tolerances associated with geotechnical design procedures, and suggest that the soil model is compatible with the level of soils information usually available.

The application of the simple boundary element is discussed below with respect to two general soil-structure interaction situations. Consideration is first given to predicting the behaviour of raft foundations using the program RAFTS and then to the use of a sister program LAWPILE for determining the behaviour of piles or walls subject to lateral forces. For both comparisons are shown between the actual recorded performance of structures and that predicted by back-analysis. The results are most encouraging and confirm the predictive power of this, albeit approximate, approach.

RAFTS

The development of the basic soil boundary element has taken place over a number of years and the acceptability of the approximate extension for layered elastic continua has been demonstrated by Wood(1977) with respect to a wide variety of inhomogeneous situations and elastic layers of finite extent. Subsequently the element has been enhanced (Hooper and Wood, 1977; Wood, 1978b) to include transversely isotropic behaviour. That is, a continuum within which the stiffness in the vertical direction differs from that in the horizontal plane. Thus, the elastic properties of the continuum are defined by five independent constants:

$$E_v, \quad E_h, \quad \nu_{vh}, \quad \nu_{hh}, \text{ and } \quad G_{vh};$$

where E and ν are Young's modulus and Poisson's ratio respectively, G is the shear modulus and suffixes v and h refer to the vertical and horizontal planes.

The use of a purely elastic analysis with regard to the design of raft foundations may give rise in some instances to somewhat unrealistic soil reactions under the edges of the foundation. This is particularly so in the case of stiff structures. Although, the adoption of a non-linear soil model (Wood and Larnach, 1975) produces a more realistic assessment of the raft behaviour, the increase in computational effort with respect to three-dimensional situations is rarely justified on economic grounds. Hence, the inclusion in RAFTS of a procedure which allows the imposition of an upper limit on the magnitude of the developed soil reactions in order to assess the effect of local shear failure of the soil. In addition similar routines are incorporated for the assessment of the effect of new construction on neighbouring structurures (Wood, 1980) and the effect of heave due to stress relief of the underlying strata.

Three examples are given below in order to demonstrate the flexibility of the use of RAFTS with respect to different soil conditions. In the first example consideration is given to the performance of a raft subjected to uplift forces due to swelling of the underlying clay strata. The other examples relate to two published case histories of rafts founded on sand and chalk strata respectively and illustrate the predictive power of the program.

Uplift

A 30m square, 2m thick raft resting on a clay soil 40m thick with:

$E_v = 10 + 1.0z$ MN/m^2, $E_h/E_v = 2.3$, $\nu_{vh} = 0.1$, $\nu_{hh} = -0.15$ and $G_{vh}/E_v = 0.661$;

where z is the depth below the underside of the raft, has been subjected to a uniform uplift pressure of 200kN/m^2. The raft material is of unit weight, 25kN/m^3 and elastic moduli E = 28GN/m^2, $\nu = 0.2$.

This example has been chosen to illustrate the ability of RAFTS to take account of loss of contact between the raft and the underlying soil. This point is clearly demonstrated in Fig. 1 where the stiff nature of the raft has caused it to lift away from the soil surface at the edge; and is shown in more detail in Fig. 2 where the contours of raft and soil movements are given.

Raft foundation on Sand

Recorded movements of the raft foundation for the reactors of Dungeness 'B' Nuclear Power Station, Kent have been presented by Dunn (1975). The 3.4m thick raft is founded at a depth of 9m below ground level on 30m of beach sand overlying mudstone. The layout of the reactor building is shown in Fig. 3.

In this case the mudstone has been taken as incompressible and the sand as exhibiting isotropic elastic properties;

E_v = 85MN/m^2 increasing linearly to 180MN/m^2, with $\nu = 0.1$

The datum for measured settlements was established after the construction of the raft itself and therefore corresponds to the imposition of the load from the superstructure of 500MN for the reactors and 115MN for the central fuel handling unit. In the analysis the raft concrete has been assigned a Young's modulus of 28GN/m^2 and Poisson's ratio of 0.15 and the raft

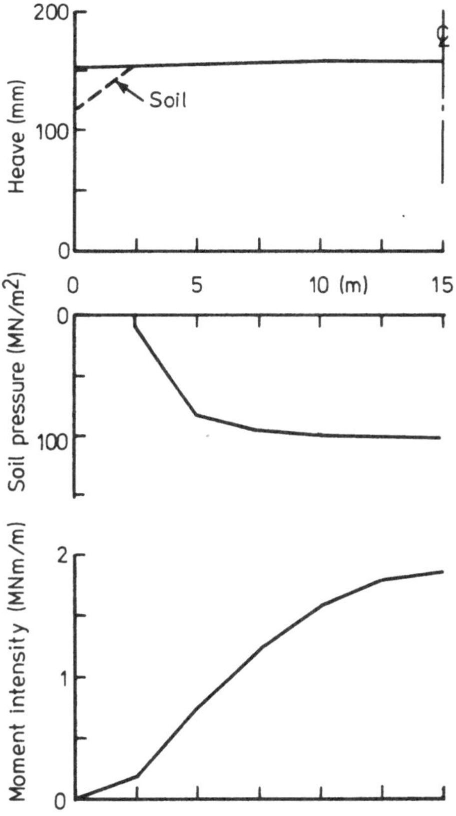

Figure 1 Computed results along centre-line

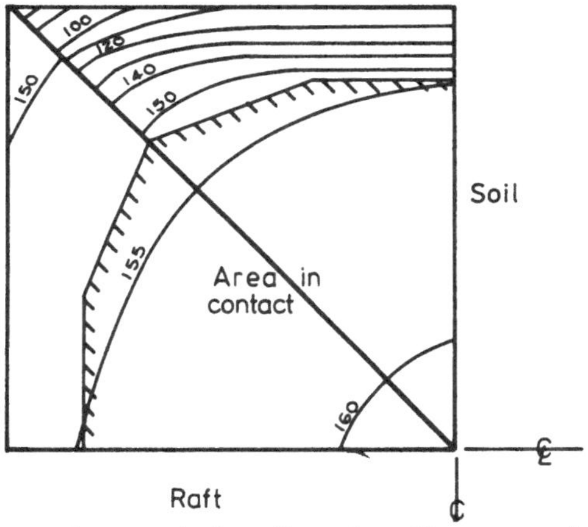

Figure 2 Contours of computed raft and soil movements (mm)

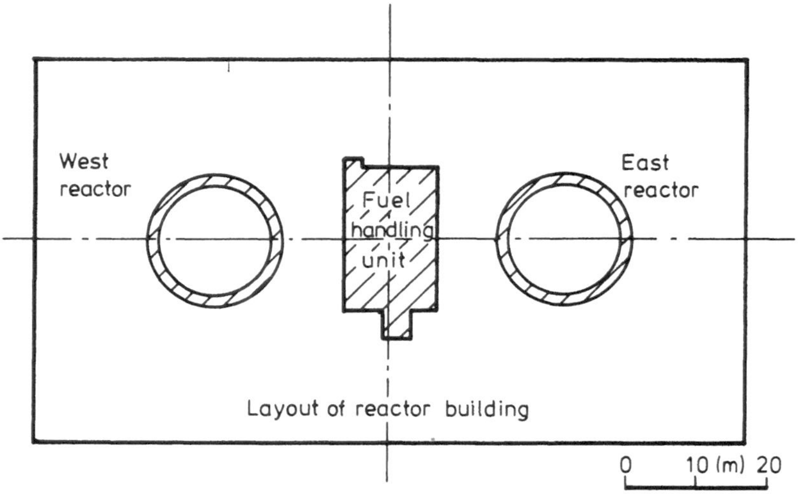

Fig. 3 Layout of reactor building

has been taken as symmetrical about both axes. Computed and measured settlements along the longitudinal axis of the raft are shown in Fig. 4; together with computed soil pressures and raft bending moments. It should be noted that the measured settlements also exhibited symmetry about the raft axes and that the agreement shown in Fig. 4 is very encouraging.

Silos on Chalk

Settlement observations for a complex of four grain silos supported on independent 23m diameter, 1.2m thick raft foundations founded on soft chalk have been reported by Burland and Davidson (1976). The layout of the complex, in which each silo weighs approximately 41MN and has a storage capacity of 120MN, is shown in Fig. 5.

The results obtained from plate bearing tests on the chalk are shown in Fig. 6 together with the assumed "elasto-plastic" curve with a Young's modulus of $120MN/m^2$, Poisson's ratio of 0.25 and a limiting soil stress of $0.7MN/m^2$. The chalk stratum has been taken as homogeneous and isotropic, and to be 60m thick. Comparison between computed and recorded settlements is somewhat complicated by the loading sequence adopted in practice and a more detailed explanation of the analysis is given in Wood (1979b). However, the results obtained for two of the silos are shown in Fig. 7 where computed settlements

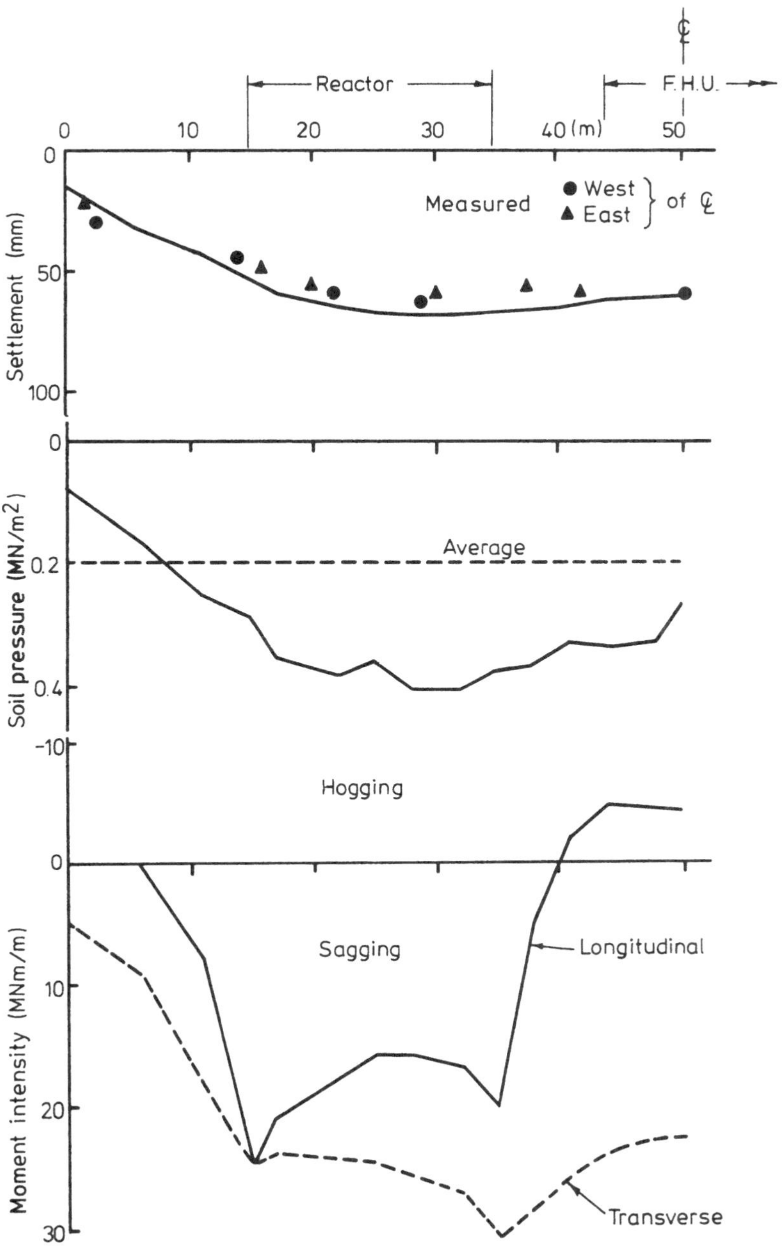

Figure 4 Computed results along longitudinal axis, Dungeness 'B' Nuclear Power Station

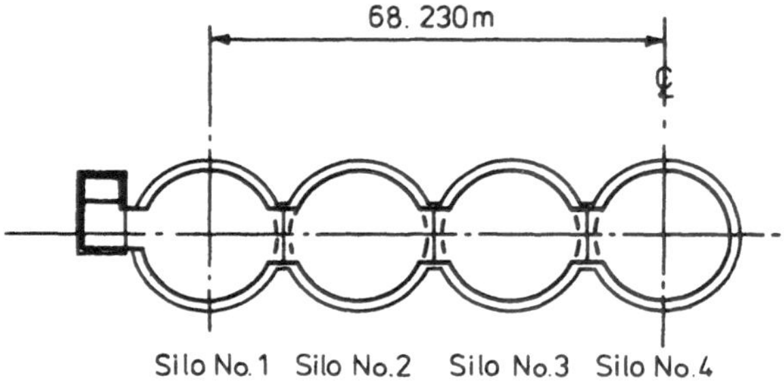

Figure 5 Plan of silo complex

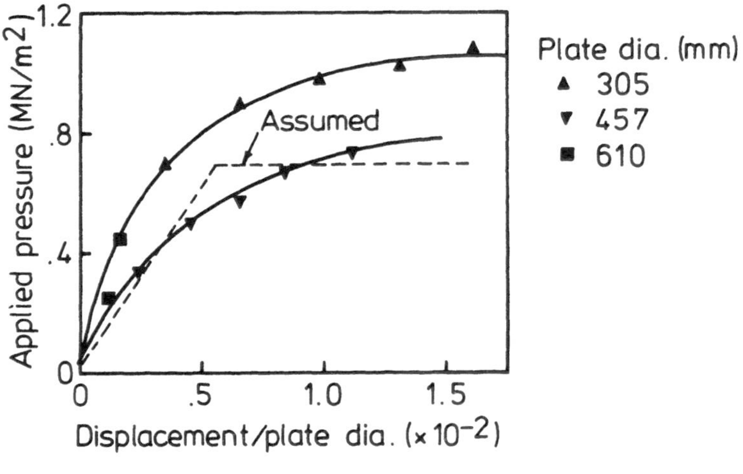

Figure 6 Results of plate loading tests

for both the elastic and for the "elasto-plastic" analysis are shown for comparison. Agreement is tolerably good given the complex behaviour of chalk, which exhibits creep characteristics, and is much improved when the "elasto-plastic" nature of the material is modelled by the imposition of an upper limit of $0.7 MN/m^2$ on the soil pressures.

LAWPILE

Many methods of analysis are available to the designer in order to quantify the effects of lateral loads on foundation members. These range from simple classical earth pressure theory through to elaborate non-linear finite element solu-

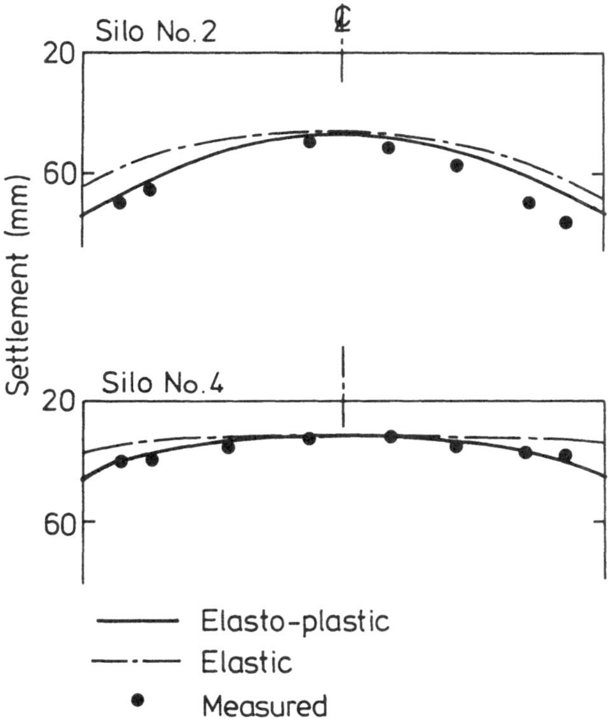

Figure 7 Computed and Observed settlement of silos

tions, with the middle ground occupied by methods based upon the theory of sub-grade reaction (also known as the beam on elastic foundation approach). The former ignore the interaction between the two basic components of the system, namely the soil and the structure, and the latter involve expensive data preparation and high solution costs, whilst the determination of equivalent spring stiffnesses with any degree of precision, as required in the sub-grade reaction methods, is a daunting task. LAWPILE has been developed in order to overcome the problems associated these approaches whilst still providing the designer with a knowledge of the overall stability of the structure and the magnitude of displacements and induced stress resultants.

The formulation of the simple boundary element is based upon the equations developed by Mindlin (1936) with an approximate extension to inhomogeneous soil desposits similar to that suggested by Poulos (1973). This approximation is analogous to that used in RAFTS and has been found to produce results of acceptable accracy for design purposes.

Hence, an initial elastic analysis is undertaken yielding the displacements, rotations, shears, induced moments and soil reactions. For an earth retaining structure such as a diaphragm wall the initial state of stress in the soil is taken as that corresponding to at rest, K_o conditions. Invariably the elastic analysis gives rise to some unrealistic soil reactions when consideration is given to the likely ultimate resistance that the soil is capable of sustaining.

In order to take account of local failure of the soil the computed elastic reactions are compared with limiting reactions obtained from the shear strength parameters of the soil. Where the elastic reactions exceed these, the soil reaction is maintained at the limit and a new solution of the displacements, etc., obtained. This procedure is repeated until all of the soil reactions lie within the limiting reaction envelope. In this manner the elasto-plastic behaviour of the soil is approximated to without incurring the high costs associated with more rigorous solutions.

For single piles or pile groups the limiting reactions are computed on the basis of a wedge type soil failure near the surface and flow around the pile at depth. In the case of a wall it is assumed that the soil pressures may not exceed the passive state or fall below the active condition.

Use of the program is illustrated below with respect to two examples, covering both of the main applications of the package.

Diaphragm Wall

In ordef to illustrate the use of the program in predicting the behaviour of a diaphragm wall consideration has been given to the case study reported by Burland and Hancock (1977) of the basement wall to the underground car park constructed at the Houses of Parliament, London. The wall has a nominal thickness of 1.0m and an over all height of approximately 28.0m, supporting retained soil to a height of about 17.0m The car park was constructed from the top, down, after the installation of the wall, by casting the uppermost floor slab and then excavating the material below down to the next slab level and so on. In this manner the floor slabs act as both temporary and permanent strutts to support the diaphragm wall. The soil profile and construction stages are shown in Fig. 8, together with the assumed soil properties used in the analysis.

The computed final wall displacements obtained from the analysis, taking account of the construction sequence, are given in Fig. 9, together with results obtained by other researchers employing more refined finite element methods. In

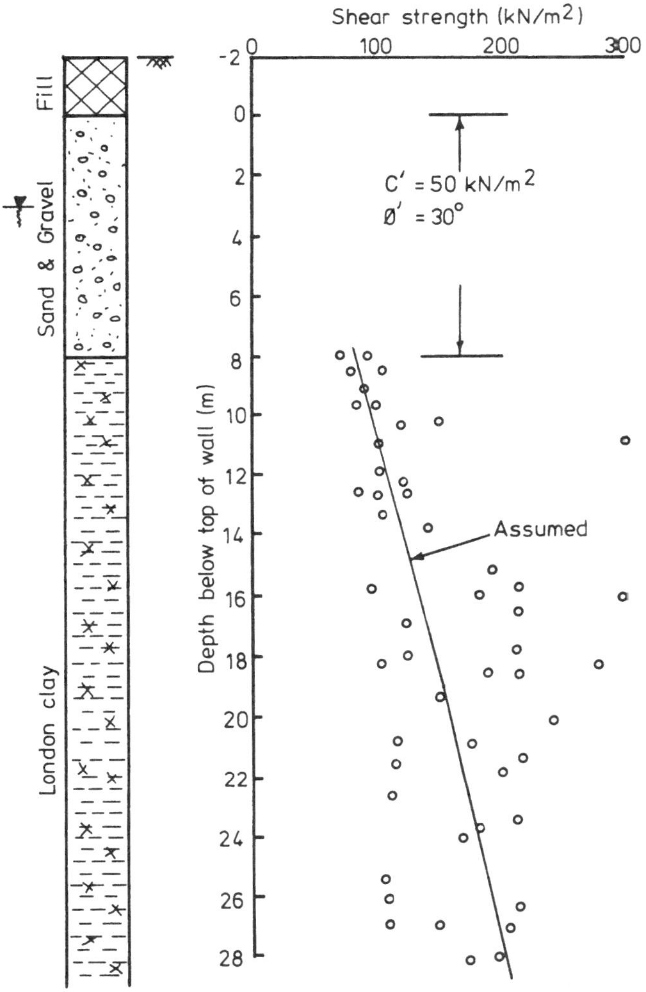

Figure 8 Soil Profile (diaphragm wall)

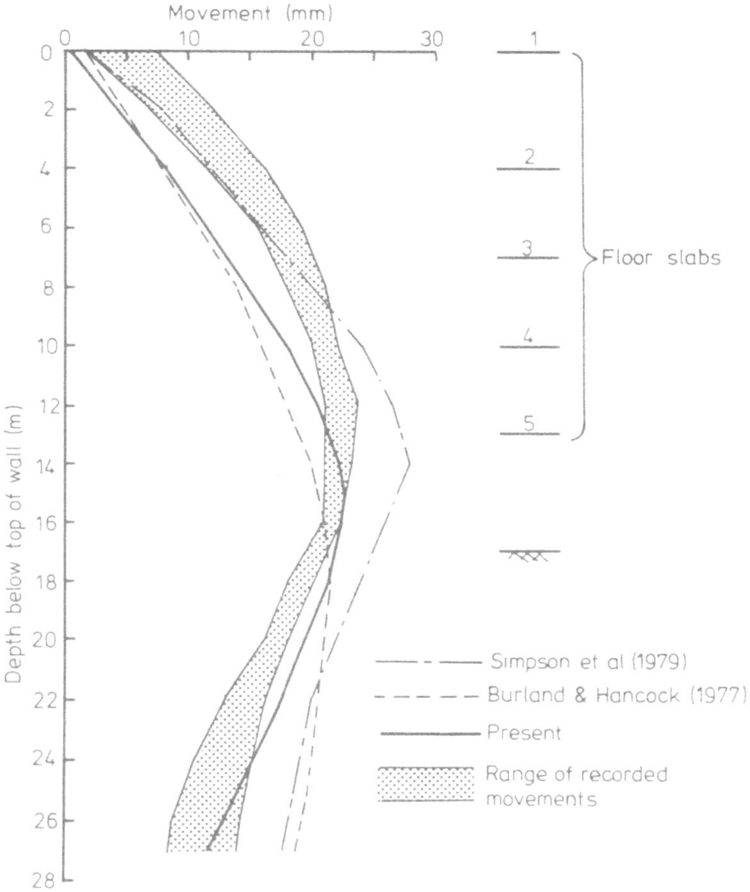

Figure 9 Computed and observed wall movements

addition the range of observed wall movements is shown for comparison. Fig. 10 shows the computed and observed maximum movements determined at each stage of construction as a percentage of the maximum final movement.

It is apparent that the agreement between the simple boundary element approach adopted herein and the more rigorous analyses is good and that the computed movements are in tolerable agreement with the observations.

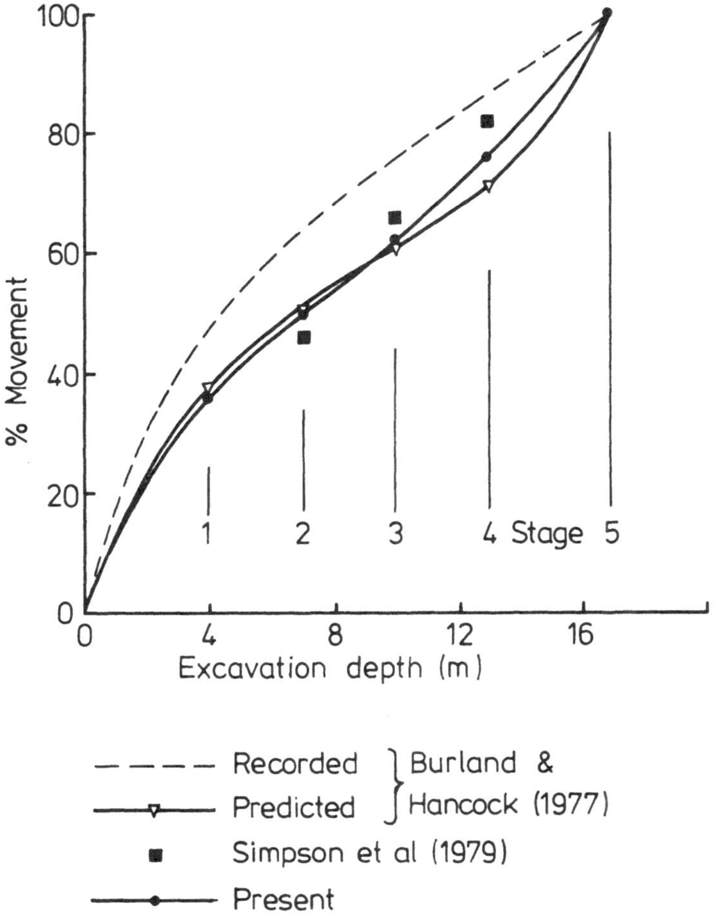

Figure 10 Computed and observed maximum wall, movements at each stage expressed as % of maximum movement at final stage

Sea-bed Anchor

As part of the development programme of a novel embedded anchor for marine applications (see, Harvey and Burley, 1977 and Nath et al. 1978) a series of lateral load tests have been conducted on a prototype anchor. The anchor has an unusual geometry enabling it to develop a large carrying capacity both with regard to vertical and horizontal applied forces; and is very efficient with respect to the load carried compared to its self-weight. The results of one such load test with the

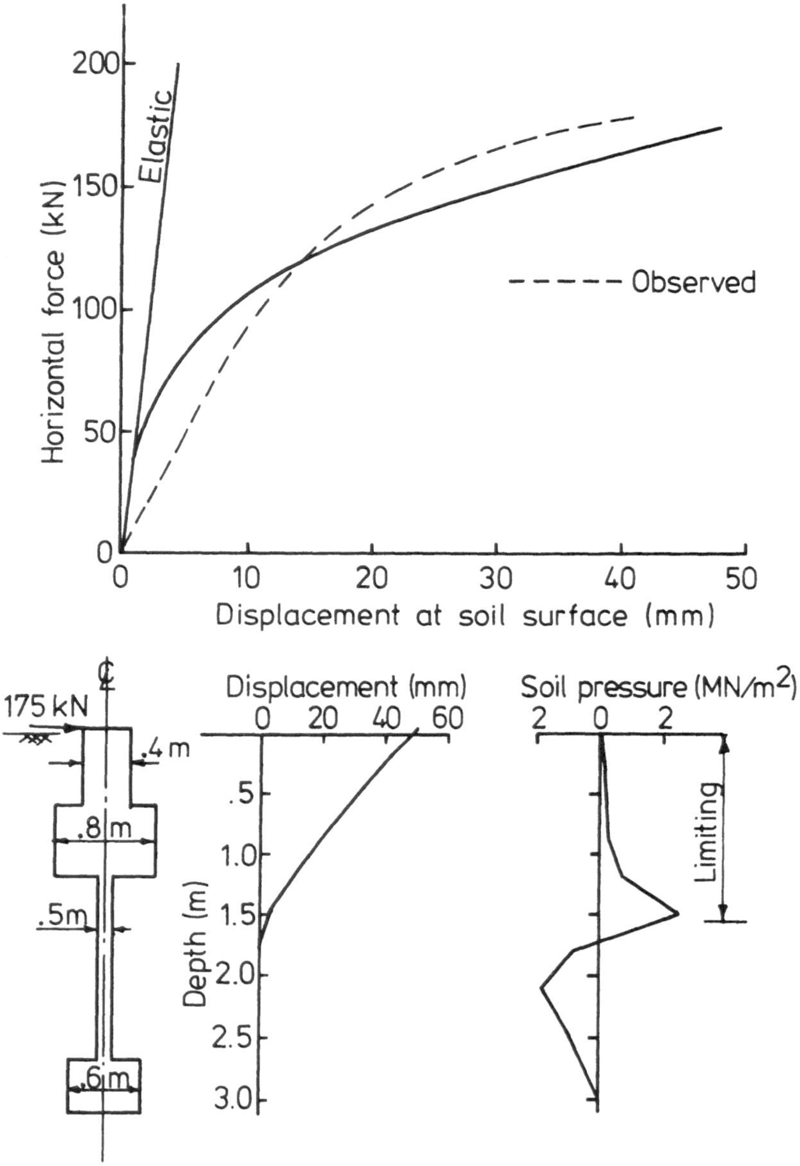

Figure 11 Sea-bed anchor

anchor embedded in a dense silty sand of unit weight $23kN/m^3$ with an angle of internal friction, $\phi = 39^0$ and $E = 50MN/m^2$ at the soil surface, increasing linearly to $\phi = 43^0$ and $E=200MN/m^2$ at a depth of 3.1m; with cohesion, $c = 5kN/m^2$ and $\nu = 0.1$; are shown in Fig. 11 together with the computer predictions. It should be noted that for a load of 175kN the computed soil pressures are limiting to a depth of 1.5m.

Bearing in mind the imponderables associated with a test of this type the agreement obtained between the computed and observed load-displacement curve is encouraging. The computed elastic curve is also shown for comparison and it is clear that the assumption of elastic behaviour would be valid for a maximum force approximately equal to one-quarter of the failure load only.

CONCLUSION

The two programs discussed illustrate the power of the boundary element approach in providing economic and realistic solutions to normal everyday geotechnical engineering problems. Although, the particular boundary element described above is based on elastic theory the various approximate extension that has been developed to take account of "elasto-plastic" soil response would appear to lead to reasonable agreement between computed and observed behaviour.

ACKNOWLEDGEMENTS

RAFTS and LAWPILE are available on the bureau network of United Computing Systems, whose help in the preparation of this paper is gratefully acknowledged.

REFERENCES

Burland, J.B. and Hancock, R.J.R. (1977) Underground car park at the House of Commons, London : Geotechnical aspects. The Structural Engineer, 55, 87.

Burland, J.B. and Davidson, W. (1976) A case study of cracking of columns supporting a silo due to differential foundation movement. Performance of Building Structures, Pentech Press, London.

Dunn, C.S. (1975) Settlement of a large raft foundation on sand. Settlement of Structures, Pentech Press, London.

Gerrard, C.M., and Harrison, W.J. (1970) Circular loads applied to a cross-anisotropic half space. CSIRO Aust. Div. Appl. Geomechanics Tech. Pap. No. 8.

Harvey, R.C. and Burley, E. (1977) A proposal for the development of a Sea-Bed anchor. Journal of the Society of Underwater Technology, 2, 9-11.

Hooper, J.A. and Wood, L.A. (1977) Comparative behaviour of raft and piled foundations. Proc. 9th Int. Conf. Soil Mechanics and Foundation Engineering, Tokyo.

Mindlin, R.A. (1936) Force at a point in the interior of a semi-infinite solid. Journal of Physics, 77.

Nath, B., Harvey, R.C., and Burley, E. (1978) The development of an embedded anchor to provide multi-directional restraint. Proc. Conf. on Offshore Structures, Oceanology Int. 78, Inst. Civ. Engrs., London.

Poulos, H.G. (1973) Load deflection prediction for laterally loaded piles. Aust. Geomechanics Journal, G3.

Simpson, B., O'Riordan, N.J and Croft D.D. (1979) A computer model for the analysis of ground movements in London Clay. Geotechnique, 29, 149.

Wood, L.A., and Larnach, W.J. (1975) The effects of soil-structure interaction on raft foundations. Settlement of Structures, Pentech Press, London.

Wood, L.A. (1977) The economic analysis of raft foundations. Int. Journal Num. Anal. Meth. Geomechanics, 1, 397-405.

Wood, L.A. (1978a) RAFTS - a program for the analysis of soil-structure interaction. Advances in Engineering Software, 1, 11-17.

Wood, L.A. (1978b) A simple boundary element approach to the prediction of the settlement of structures. Recent Advances in Boundary Element Methods. ed. C.A. Brebbia, Pentech Press, London.

Wood, L.A. (1979a) LAWPILE - a program for the analysis of laterally loaded pile groups and propped sheetpile and diaphragm walls. Advances in Engineering Software, 1, 173-179.

Wood, L.A. (1979b) A rational approach to the analysis of building structures taking full account of foundation movements. Engineering Software, ed. R.A. Adey, Pentech Press, London.

Wood, L.A. (1980) An unusual soil-structure interaction problem associated with the Thames Barrier Project. Proc. Instn. Civ. Engrs. Part 1, 68, 747-758.

BOUNDARY INTEGRAL METHOD FOR POROUS MEDIA

M. Predeleanu

Laboratoire de Mécanique et Technologie
Université Paris VI, E.N.S.E.T., Cachan (France)

INTRODUCTION

The physical non-homogeneity of some materials of technological interest such as soils, concrete, composite and metals has been modelled mainly by continuum theories applied to the bi-phasic media. The concept of porous medium is used at present not only in geomechanics but also in metal science to study the damage and rupture phenomena.

The fluid-saturated porous medium has received much attention due to the applications in hydraulics and mechanics of soils. The first model describing the uni-dimensional consolidation of soils was proposed by Terzaghi (1923), but Biot (1961) has developed a three-dimensional consolidation theory. Further generalizations of this theory have been made by its author and recently the non-linear effects have been included (1972).

The interest of the consolidation theory has increased because of satisfactory results obtained in the calculations of Venice subsidence (see, for instance Lewis and Schrefler (1978)) and of seismic response of Lower San Fernando Dam (see Zienkiewicz, et al. (1980)).

From the mathematical point of view, the consolidation Biot theory is governed by coupled equations system describing the interaction of fluid flow in a porous deformable skeleton. The linear rheological behaviour of solid phase and small strains kinematics leads to linear governing equations.

This paper is concerned with Biot consolidation theory, extended to a general class of viscoelastic bodies defined by Riemann-Stieltjes integral convolutions. Integral representations of the displacements and pore pressure are obtained so that they may be used for the formulation of the corresponding boundary integral equation method. The analogy with the coupled theory of the thermo-viscoelasticity - for which the present author has obtained some recent results (1979) - permits us to use the fundamental singular solutions in the theory of the porous media. This approach can be extended to the non-linear behaviour and coupled with other numerical methods.

GOVERNING EQUATIONS

Consider a porous deformable body, occupying a region B with a regular boundary ∂B, in the three-dimensional Euclidian space, referred to a fixed rectangular cartesian coordinate system ox_i, $i = 1, 2, 3$. By x_i will be noted the coordinates of the point $x \in B$.

Let t be the time variable, denoting by t the present time moment and by τ the values of t for $\tau < t$. We assume, for shortness, that every particle of the body is at rest for $\tau < 0$ and at $\tau = 0$, the body is in a free stress reference configuration.

The material under consideration is a bi-phasic continuum, composed of a compressible fluid which diffuses through a porous deformable solid. We are dealing with the average displacement of a volume element of material, sufficiently large with respect to the pore size. The displacement vector components of a fluid and solid particle are denoted respectively by U_i and u_i.

The small strain tensor ε_{ij} of the solid skeleton is defined by :

$$\varepsilon_{ij} = \frac{1}{2} (u_{i,j} + u_{j,i}) \tag{1}$$

The volumic dilatation of the fluid will be noted by $\Theta = U_{1,1}$ and the volumic dilatation of the solid skeleton with $\varepsilon = u_{1,1}$ where $u_{i,j} = \partial ui/\partial xj$.

Following the Biot theory, the measure of total stresses t_{ij} acting on the mixture is decomposed in "an effective " stress tensor σ_{ij} characterizing the solid skeleton and a spherical stress tensor $\sigma\delta_{ij}$ characterizing the fluid interaction . Thus

$$t_{ij} = \sigma_{ij} + \sigma\delta_{ij} \tag{2}$$

with $\sigma = - fp$, where f is the porosity of the material and p the fluid pressure.

Obviously to take into account a more general fluid influence on the solid phase, a general stress tensor must be considered. Various constitutive relations can be used for the theory of porous media. To include the visco-elastic effects of the mixture behaviour, we shall generalize the Biot theory by using the following linear-stress strain relations

$$\sigma_{ij} = 2N * d\,\varepsilon_{ij} + [A * d\varepsilon + Q * d\Theta]\,\delta_{ij} \qquad (3)$$

$$\sigma = Q * d\varepsilon + R * d\Theta \qquad (4)$$

where by $\Lambda * d\,\psi$ is noted the Stieltjes convolution of two real valued functions Λ and ψ defined by

$$(\Lambda * d\,\psi)\,(t) = \int_{-\infty}^{t} \Lambda(t-\tau)\,d\,\psi(t) \qquad (5)$$

under appropriate conditions formulated for instance by Gurtin and Sternberg (1962).

Equations (1) - (4) must be completed by the quasi-static stress field equations

$$(\sigma_{ij} + \sigma\delta_{ij})_{,j} + F_i = 0 \qquad (6)$$

and by the fluid flow equation of the Darcy type

$$\sigma_{,ii} = \beta\,(\dot{\Theta} - \dot{\varepsilon} + q) \qquad (7)$$

where F_i are body forces, β a filtration constant, q the external source of fluid $\dot{\Theta} = \partial\theta/\partial t$.

If it is assumed that the body is originally undisturbed, the initial conditions will be considered zero on $\bar{B} \times (-\infty, 0)$.

The stress boundary conditions are taken as

$$(\sigma_{ij} + \delta_{ij}\sigma)n_i = P_i(x,t), \quad x \in \partial B, \quad t > 0 \qquad (8)$$

or in the form :

$$\sigma_{ij}n_j = (1-f)P_i(x,t), \quad x \in \partial B, \quad t > 0 \qquad (9)$$

$$\sigma\,n_i = fP_i = g(x,t)n_i, \quad x \in \partial B, \quad t > 0 \qquad (10)$$

where n_i are the component of unit normal vector.

INTEGRAL REPRESENTATION

To deduce the integral representation of the displacement field u_i and fluid pressure p a generalized reciprocal theorem of the Betti type, deduced by Predeleanu (1968) will be used.

Consider two loading programs

$$J^{(\alpha)} = \{P_i^{(\alpha)},\ g^{(\alpha)},\ F_i^{(\alpha)},\ q^{(\alpha)}\}\alpha = 1,2 \qquad (11)$$

which induce two deformed configurations of the body B, defined by the fields

$$\phi^{(\alpha)} = \{u_i^{(\alpha)},\ U_i^{(\alpha)},\ \sigma_{ij}^{(\alpha)}, \sigma^{(\alpha)}\}, \alpha = 1,2 \qquad (12)$$

Reciprocal theorem

If the porous medium B is subjected to two loading programs $J^{(\alpha)}$, = 1,2, then governing fields of the corresponding deformed configurations verify the following reciprocity relation

$$L_{12}(J^1,\phi^2) \equiv \int_{\partial B} ds(x) \int_0^t P_i^{(1)}(x,t-\tau)\dot{u}_i^{(2)}(x,\tau)\, d\tau$$

$$+ \frac{1}{\beta}\int_{\partial B} \partial s(x) \int_0^t g^{(1)}(x,t-\tau)\sigma,_i^{(2)}(x,\tau) n_i d\tau$$

$$+ \int_B dv(x) \int_0^t F_i^{(1)}(x,t-\tau)\dot{u}_i^{(2)}(x,\tau) d\tau$$

$$+ \int_B dv(x) \int_0^t q^{(1)}(x,t-\tau)\sigma^{(2)}(x,\tau) d\tau =$$

$$= L_{21}(J^2,\ \phi^1) \qquad (13)$$

This theorem has been demonstrated by using the Laplace transform in respect of time variable t and it differs with other analogous theorems in the coupled theories by eliminating the volume integral terms concerning two unknown field variables. This reciprocity relation allows a more suitable use for the boundary integral approach. It is noted in passing that the dynamics effects can be taken into account and a similar theorem can be deduced as in coupled theory of thermoviscoelasticity (see Predeleanu 1979).

Integral representation of the Somigliana type

The use of the fundamental solutions of the governing equations and the reciprocity theorem permit us to deduce the integral representation of the Somigliana type for the displacement field and fluid pressure p.

Firstly, consider

$$\phi^k = \{a_i^k, V_i^k, A_{ij}^k,\ C^k\},\ k = 1,2,3, \qquad (14)$$

a fundamental solution of the governing equations, regular at infinity, with homogeneous initial condition corresponding to the loading

$$F_i(x,\zeta,t) = \delta(t)\delta(x-\zeta)\,\delta_{ki},\quad q(x,t) = 0 \qquad (15)$$

where δ is Dirac distribution.

By $a_i^k(x,\zeta,t)$ is noted the i-component of the solid displacement at x due to a concentrated load in the ox_k-direction, applied at a point ζ of the porous medium, that occupies the entire space ; V_i^k is the displacement of the fluid particle, A_{ij}^k, the total stress tensor and C^k, the corresponding solution for σ.

We obtain

$$u_k(\zeta,t)+ \int_{\partial B} ds(x) \int_0^t A_{ni}^k(x,\zeta,t-\tau)\dot{u}_i(x,\tau)\, d\tau + \frac{1}{\beta}\int_{\partial B} ds(x) \int_0^t C^k(x,\zeta,t-\tau)\sigma_{,i}(x,\tau)\, n_i d\tau = L(J,\phi^k) \qquad (16)$$

where

$$A_{ni}^k = A_{ij}^k n_j$$

$L(J,\phi^k)$ is a known function defined by Equation (13) where

$$J = \{ P_i,\ g,\ F_i,\ q \} \qquad (17)$$

and ϕ^k is given by Equation (14).
Secondly, consider

$$\phi = \{ b_i,\ W_i,\ B_{ij},\ e \} \qquad (18)$$

a fundamental solution of the governing equations regular at infinity, with homogeneous initial conditions, corresponding to the loading

$$F_i(x,t) = 0,\quad q(x,\zeta,t) = \delta(t)\,\delta(x-\zeta) \qquad (19)$$

From the reciprocity relation (13), we obtain :

$$\sigma(\zeta,t) + \int_{\partial B} ds(x) \int_0^t B_{ni}(x,\zeta,t-\tau)\, u_i(x,\tau)\, d\tau + \frac{1}{\beta}\int_{\partial B} ds(x) \int_0^t e(x,\zeta,t-\tau)\sigma_{,i}(x,\tau)\, n_i d\tau = L(J,\phi) \qquad (20)$$

where

$$B_{ni} = B_{ij} n_j .$$

J and ϕ are defined respectively by Equations (17) and (18). Equations (16) and (20) define a coupled linear integro-differential equations system for the unknown fields u_i and σ.

Analogous integral representations can be obtained for mixed boundary value problem.

The Laplace transform can be applied to eliminate the time dependence and reduce the viscoelastic problem to an associated elastic problem. Also, important simplifications of this system can be obtained if fundamental solutions of Green's type are used. In this case the homogeneous boundary condition is required for the singular solution of the governing equations

$$A_{ni}^{k} = B_{ni} = C^{k} = e = 0, \; x \in \partial B, \; t > 0 \tag{21}$$

From Equations (16) and (20) it is easy to see that the displacement field and the fluid pressure are directly determined in every interior point of the medium if the boundary and inside inputs are known :

$$\dot{u}_k(\zeta,t) = L(J,\phi^k), \zeta \in B, \; t > 0 \tag{22}$$

$$\sigma(\zeta,t) = L(J,\phi) \quad , \zeta \in B, \; t > 0 \tag{23}$$

It should be noted that, for appropriate definitions of the Green's functions, it is possible also to express the solution of the mixed boundary value problem exclusively in terms of prescribed functions by volume integrals and surface integrals.

Total volumic dilatation formulas

Let us define the total volumic dilatation of solid and fluid phase respectively by

$$D_1(t) = \int_V \varepsilon(x,t) \, dv(x) \tag{24}$$

$$D_2(t) = \int_V \Theta(x,t) \, dv(x) \tag{25}$$

By using the reciprocity theorem, the total volumic dilatations can be obtained in terms of boundary and inside prescribed data.

Thus, following the results obtained by Predeleanu and Nan (1968), we can write:

$$D_1(t) = \int_{\partial B} ds(x) \int_0^t m(t-\tau) P_i(x,\tau) \, x_i d\tau + \int_B dv(x) \int_0^t n(t-\tau) \, F_i(x,\tau) \, x_i d\tau + 3 \int_B dv(x) \int_0^t \left[m(t-\tau) + n(t-\tau)\right] \sigma(x,\tau) \, d\tau \tag{26}$$

$$D_2(t) = -\int_{\partial B} ds(x) \int_0^t n(t-\tau) P_i(x,\tau) \, x_i \, d\tau - \int_B dv(x) \int_0^t n(t-\tau) \, F_i(x,\tau) \, x_i \, d\tau + \tag{27}$$

$$+ 3 \int_B dv(x) \int_o^t [n(t-\tau) + r(t-\tau)] \sigma(x,\tau) d\tau$$

where the functions m, n and r are obtained by inverse Laplace transforms

$$m(t) = L^{-1} \frac{\bar{R}}{p(2\bar{N}\bar{R} + 3\bar{A}\bar{R} - 3\bar{Q}^2)}; n(t) = L^{-1} \frac{\bar{Q}}{p(2\bar{N}\bar{R} + 3\bar{A}\bar{R} - 3\bar{Q}^2)}$$

$$r(t) = L^{-1} \frac{2\bar{N} + 3\bar{A}}{3p(2\bar{N}\bar{R} + 3\bar{A}\bar{R} - 3\bar{Q}^2)} \tag{28}$$

by $\bar{f}(x,p)$ is noted the direct Laplace transform of a numerical function $f(x,t)$.

Obviously, the formulas (26) and (27) became very simple in the case of elastic behaviour of solid skeleton (see Predeleanu and Nan (1968).

BOUNDARY INTEGRAL EQUATION APPROACH

One of the most fruitful applications of the integral representations obtained in the previous section is the numerical approach by using the boundary integral equation method (BIEM). The BIEM has been utilized successfully in the non-coupled linearized theories of continua and that has been extended to some non-linear theories. In this section the BIEM is formulated for a coupled theory of fluid-saturated porous media and this method is proposed as an alternative to the finite element method (FEM) used for geotechnical calculations. It is proved by numerical applications that for infinite domains and also for the large domains the BIEM is more convenient that the FEM.

As is known, the boundary integral equations can be obtained from integral representation of Somigliana 's or Green's type if the point ζ inside B approaches an arbitrary point on the boundary ∂B .

For that let us note formally by integral symbol with an asterisk the following limit understood in the sense of its Cauchy principal value :

$$\lim_{\zeta \to y} \int_{\partial B} \Omega(x,\zeta,t)\, \omega(x,t) ds(x) = \overset{*}{\int_{\partial B}} \Omega(x,y,t)\, \omega(x,t) ds(x) \tag{29}$$

where $\zeta \in B$ and $y \in \partial B$.

Thus, from Equations (16) and (20), we obtain

$$\dot{u}_k(y,t) + \overset{*}{\int_{\partial B}} ds(x) \int_o^t A_{ni}^k(x,y,t-\tau)\dot{u}_i(x,\tau) d\tau -$$

$$\begin{aligned} &-\int^{*}_{\partial B} ds(x) \int_0^t a_i^k(x,y,t-\tau)\, t_{ni}(x,\tau)\, d\tau \;+ \\ &\frac{1}{\beta}\int^{*}_{\partial B} ds(x) \int_0^t C^k(x,y,t-\tau)\, \sigma_{,i}(x,\tau)\, n_i\, d\tau \;= \\ &\int_B dv(x) \int_0^t F_i(x,t-\tau)\, \dot{a}_i^{\,k}(x,y,\tau)\, d\tau \;+ \\ &\int_B dv(x) \int_0^t q(x,t-\tau)\, C^k(x,y,\tau)\, d\tau \;+ \\ &\frac{1}{\beta}\int^{*}_{\partial B} ds(x) \int_0^t g(x,t-\tau)\, C^k_{,i}(x,y,\tau)\, n_i\, d\tau \end{aligned} \tag{30}$$

$$\begin{aligned} &g(y,t) + \int^{*}_{\partial B} ds(x) \int_0^t B_{ni}(x,y,t-\tau)\, \dot{u}_i(x,\tau)\, d\tau \;- \\ &\int^{*}_{\partial B} ds(x) \int_0^t b_i(x,y,t-\tau)\, t_{ni}(x,\tau)\, d\tau \;+ \\ &\frac{1}{\beta}\int^{*}_{\partial B} ds(x) \int_0^t e(x,y,t-\tau)\, \sigma_{,i}(x,\tau)\, n_i\, d\tau \;= \\ &\int_B dv(x) \int_0^t F_i(x,t-\tau)\, \dot{b}_i(x,y,\tau)\, d\tau \;+ \\ &\int_B dv(x) \int_0^t q(x,t-\tau)\, e\,(x,t,\tau)\, d\tau \;+ \\ &\frac{1}{\beta}\int^{*}_{\partial B} ds(x) \int_0^t g(x,t-\tau)\, e_{,i}(x,y,\tau)\, n_i\, d\tau \end{aligned} \tag{31}$$

where $t_{ni} = t_{ij}\, n_j$

It is worth noting the following:

a. For the stress boundary value problem in which surface tractions and the fluid pressure are prescribed over the entire boundary ∂B for all time, the third terms of the Equations (30) and (31) are known. On the other hand for the mixed boundary value problem, these theorems contain the unknown fields on some parts of the boundary.

b. For certain singular Kernels Ω (like A_{ij} and B_{ij} for instance) and regularity conditions for the density ω and the boundary ∂B some additional terms (jumps) of type $\eta\omega(y,t)$ can be separated from the singular integral defined by Equations (30)-(31). The calculations of these limits are performed after eliminating the time convolution integrals by Laplace transformation. Otherwise, the regularity conditions must be imposed for the time and spatial integral operations.

c. The analogy between the consolidations theory of porous media and coupled theory of thermo-viscoelasticity allows the use of some fundamental solutions (see Nowacki (1962), Derski (1965)). If supplementary hypothesis concerning the non-coupled interaction between the two phases, is adopted, then important simplifications may be introduced. On the other hand, singular fundamental solutions known for the non-coupled problem can be used for coupled problems by using the method proposed by Ionescu-Cazimir (1964).

FINAL REMARKS

Another application field of the boundary solution approach is the coupling of BIEM with other analytical or numerical methods (the finite elements method or finite difference method, for instance). For that a natural or arbitrary boundary Γ can be introduced to separate the parts B_1 and B_2 of B. If in B_2, the BIEM is used, then the effects of B_2 on B_1 must be included by an "interface "condition on common boundary Γ. For instance, if $u_i^{(1)}$, $\sigma^{(1)}$ and $u_i^{(2)}$, $\sigma^{(2)}$ are the fields characterizing respectively the regions B_1 and B_2 it may be asked that

$$\begin{aligned} u_i^{(1)}(y,t) &= u_i^{(2)}(y,t) \quad y \in \Gamma,\ t>0 \\ \sigma^{(1)}(y,t) &= \sigma^{(2)}(y,t) \quad y \in \Gamma,\ t>0 \end{aligned} \tag{32}$$

Other interface conditions can be assumed too.

By using the integral representations , a global integral boundary condition can be obtained for $u_i^{(1)}$ and $\sigma^{(1)}$ to be used for instance in FEM as a natural boundary condition for the variational formulation of the problem to solve in B_1 (see Predeleanu 1979).

REFERENCES

Biot M.A. (1961) General Theory of the Three Dimensional Consolidation.
J. Appl. Phys. 12 : 155-164.

Biot M.A. (1972) Theory of Finite Deformations of Porous Solids.
Indiana University Mathematics Journal, 21, 7 : 597-620.

Derski W. (1965) Equations of the Consolidation Theory for The Case of a Source of Fluid.
Bull. Acad. Polon. Sci. Sér. Sci. Tech 13, 1 : 37-43.

Gurtin M.E. and Sternberg E. (1962) On the linear theory of Viscoelasticity.
Arch. Rational Mech. Anal. 11, 4 : 291-356.

Ionescu-Cazimir V. (1964) Problem of Linear Coupled Thermoelasticity. Some Applications of the Theorems of Reciprocity for the Dynamic Problem of Coupled Thermoelasticity.
Bull. Acad. Polon. Sci, Sér. Sci. Tech, 9, 12

Lewis R.W. and Schrefler B. (1978) A Fully Coupled Consolidation Model of the Subsidence of Venice .
Water Resources Research 14, 2 : 223-230.

Nowacki W.(1966) Dynamic Problems of Thermoelasticity.
Noordhoff International Publishing Leyden, P.W.N. Polish Scientific Publishers, Warszawa.

Predeleanu M. (1968) Reciprocal Theorem in the Consolidation Theory of the Porous Media (Roum).
Anal. Univ. Bucuresti, Seria Stiintele Naturü Matematica-Mecanica 17, 2 : 69-74.

Predeleanu M. and Nan L.(1968) Formulas for the Variation of Volume in the Porous Media.(Roum).
Anal. Univ. Bucuresti Seria Stüntele Naturii Matematica-Mecanica, 17, 2 : 75-79.

Predeleanu M. (1979) On Boundary Solution Approach for Dynamic Problem of Thermo-Viscoelasticity Theory .
International Conference on Numerical Methods in Thermal Problems. Swansea 2-6 July 1979. (in press).

Terzaghi K. (1923) Ann. Oster. Akad. Wissensch Wien Kl. Abt. IIa : 132.

Zienkiewicz O.C. - Hinton E. - Leung K.H. - Taylor R.L. (1980) Staggered Time Marching Schemes in Dynamic Soil Analysis and a Selective Explicit Extrapolation Algorithm.
Innovative Numerical Analysis for the Engineering Sciences Ed. R. Shaw, University Press of Virginia. Charlottesville.

=-=-=-=-=-=

Section IV
Material Problems

NUMERICAL ANALYSIS OF CYCLIC PLASTICITY USING THE BOUNDARY INTEGRAL EQUATION METHOD

M. Brunet

Laboratoire de Mécanique des Solides - INSA Lyon - Bât. 304
69621 Villeurbanne Cédex - France

INTRODUCTION

Reliability criteria for the design of engineering components and structures subjected to complex service conditions often include their resistance to fatigue failure. Low cycle fatigue may be part of the design criteria and a major drawback in life predictions has been the inaccuracy in relating nominal loads to local stresses and strains, particularly if local plastic deformation occurs. Finite element methods (FEM) are mostly being used for such problems and the cyclic plastic analyses usually require large amounts of computer time. The boundary element method (BEM) is less used and it is being explored for this class of non-linear problems. In this paper a computational procedure is presented for the solution of such problems in planar bodies subjected to arbitrary loading histories. Linear and non-linear constitutive relations are used in the calculations.

PLASTICITY RELATIONS

The plastic behaviour of a material can be described by a yield function such as :

$$f(\sigma_{ij} - \alpha_{ij}) - \sigma_0^2 = 0 \qquad (1)$$

σ_{ij} is the stress tensor, σ_0 the yield stress and α_{ij} is a tensor defining the total translations of the yield surface. The flow law is defined by the normality condition of the plastic strain increment $\dot{\varepsilon}^p_{ij}$, to the associate yield surface at the stress point.

Therefore :

$$\dot{\varepsilon}^p_{ij} = \dot{\lambda} \frac{\partial f}{\partial \sigma_{ij}} \qquad (2)$$

in which $\dot{\lambda}$ is a positive scalar determined by the condition that the stress point remains on the yield surface during plastic flow, such as :

$$\dot{f} = \frac{\partial f}{\partial \sigma_{ij}} \dot{\sigma}_{ij} + \frac{\partial f}{\partial \alpha_{ij}} \dot{\alpha}_{ij} = \frac{\partial f}{\partial \sigma_{ij}} (\dot{\sigma}_{ij} - \dot{\alpha}_{ij}) = 0 \qquad (3)$$

It is to be noted from the incompressibility of plastic strains that :

$$\delta_{ij} \dot{\varepsilon}^p_{ij} = 0 \qquad (4)$$

In Prager's rule, the increment of translation of the yield surface is in the direction of the plastic strain increment. Therefore,

$$\dot{\alpha}_{ij} = C \dot{\varepsilon}^p_{ij} \qquad (5)$$

Ziegler's modification of Prager's model suggests instead of (5) the relation :

$$\dot{\alpha}_{ij} = \dot{\mu}(\sigma_{ij} - \alpha_{ij}) \qquad (6)$$

With constant coefficients, Equations (5) - (6) describe linear kinematic hardening and when these models are applied to cases of cyclic stressing, they predict closed cyclic loops after one cycle of stress. This prediction is not corroborated by numerous experimental data and a more realistic description can be obtained when accumulation effects accompany plastic cycles. Mroz [1] and others postulate instead of (5) and (6) :

$$\dot{x}_{ij} = C(\bar{\varepsilon}^p) \dot{\varepsilon}^p_{ij} - d(\bar{\varepsilon}^p) \dot{\bar{\varepsilon}}^p \varepsilon^p_{ij} \qquad (7)$$

with

$$\dot{x}_{ij} = \dot{\alpha}_{ij} - \frac{1}{3} \delta_{ij} \dot{\alpha}_{mm} \quad \text{and} \quad \dot{\bar{\varepsilon}}^p = \left[\frac{2}{3} \dot{\varepsilon}^p_{ij} \dot{\varepsilon}^p_{ij}\right]^{1/2} \qquad (8)$$

where $C(\dot{\bar{\varepsilon}}^p)$ and $d(\dot{\bar{\varepsilon}}^p)$ are two material functions which could be determined from uniaxial cyclic tests.

Marquis [2] modified (7) and proposed the hardening rule :

$$\dot{x}_{ij} = \frac{2}{3} \beta_0 \left[\dot{\varepsilon}^p_{ij} - \frac{2}{3} (\phi_0 + \psi e^{-\omega \bar{\varepsilon}^p}) \dot{\bar{\varepsilon}}^p x_{ij} \right] \qquad (9)$$

σ_0, β_0, ϕ_0, ψ and ω are five parameters characterizing each material obtained by trial and error from uniaxial experimental data . This law can account for the cyclic transient behaviour of most structural materials. When hardening develops, the term

$\psi e^{-\omega\bar{\varepsilon}^p}$ vanishes in (9) and we get the cyclic steady behaviour when $\psi = \omega = 0$.

In order to complete the constitutive relationships, assume that the total strain increment can be decomposed into an elastic part and a plastic part :

$$\dot{\sigma}_{ij} = D^e_{ijkl}\,\dot{\varepsilon}^e_{kl} = D^e_{ijkl}(\dot{\varepsilon}_{kl} - \dot{\varepsilon}^p_{kl}) \qquad (10)$$

in which D^e_{ijkl} is the elastic material tensor :

$$D^e_{ijkl} = \frac{E}{2(1+\nu)}\left[\delta_{ik}\delta_{jl} + \delta_{il}\delta_{jk} + \left(\frac{2\nu}{1-2\nu}\right)\delta_{ij}\delta_{kl}\right] \qquad (11)$$

where E = Young's modulus and ν = Poisson's ratio.

Using equation (5), the following relationships are obtained

$$\dot{\lambda} = \frac{1}{\alpha+\beta}\left[\frac{\partial f}{\partial\sigma_{ij}}\,D^e_{ijkl}\,\dot{\varepsilon}_{kl}\right] \qquad (12)$$

with

$$\alpha + \beta = \left[C\,\frac{\partial f}{\partial\sigma_{kl}}\,\frac{\partial f}{\partial\sigma_{kl}} + \frac{\partial f}{\partial\sigma_{ij}}\,D^e_{ijkl}\,\frac{\partial f}{\partial\sigma_{kl}}\right] \qquad (13)$$

and

$$\dot{\varepsilon}^p_{ij} = G_{ijkl}\,\dot{\varepsilon}_{kl} \qquad (14)$$

$$\dot{\sigma}_{ij} = D^{ep}_{ijkl}\,\dot{\varepsilon}_{kl} \qquad (15)$$

in which

$$G_{ijkl} = \frac{1}{\alpha+\beta}\left[\frac{\partial f}{\partial\sigma_{ij}}\,D^e_{ijkl}\,\frac{\partial f}{\partial\sigma_{kl}}\right] \qquad (16)$$

$$D^{ep}_{ijkl} = D^e_{ijkl} - D^e_{ijmn}\,G_{mnkl} \qquad (17)$$

We explicitly formulate equation (15) for the special case of Von Mises yield condition and plane-stress state.

$f = \frac{1}{2}\,S_{ij}S_{ij} - \frac{1}{3}\,\sigma_0^2 = 0$ in which S_{ij} is the translated deviatoric stress, defined by :

$$S_{ij} = (\sigma_{ij} - \alpha_{ij}) - \frac{1}{3}\,\delta_{ij}(\sigma_{mm} - \alpha_{mm}) \qquad (18)$$

$$\begin{bmatrix}\dot{\sigma}_{11}\\ \dot{\sigma}_{22}\\ \dot{\sigma}_{12}\end{bmatrix} = \frac{E}{Q}\begin{bmatrix} S_{22}^2 + 2P & -S_{11}S_{22} + 2\nu P & -(S_{11}+\nu S_{22})\frac{S_{12}}{1+\nu} \\ & S_{11}^2 + 2P & -(S_{22}+\nu S_{11})\frac{S_{12}}{1+\nu} \\ \text{SYM.} & & \frac{R}{2(1+\nu)} + A(1-\nu)\end{bmatrix}\begin{bmatrix}\dot{\varepsilon}_{11}\\ \dot{\varepsilon}_{22}\\ 2\dot{\varepsilon}_{12}\end{bmatrix} \tag{19}$$

where

$$R = S_{11}^2 + 2\nu S_{11}S_{22} + S_{22}^2 \quad \text{and} \quad P = A + S_{12}^2/(1+\nu)$$

$$Q = R + 2(1-\nu^2)\,P$$

with

* $A = \frac{2}{9}\sigma_0^2\, H'/E$ for Prager's rule (or Ziegler's rule) where H' is the plastic slope of the uniaxial stress-strain curve.
* $A = \frac{2}{3}\beta_0\left[2\sigma_0^2 - (\phi_0 + \psi e^{-\omega\bar{\varepsilon}^p})\sigma_0 T\right]$

$$T = 2(S_{11}x_{11} + S_{12}x_{12} + S_{22}x_{22}) + S_{22}x_{11} + S_{11}x_{22}$$

for the complex non-linear hardening rule (9). These cases are incorporated in the solution procedure analyzed with the boundary element method.

BOUNDARY ELEMENT FORMULATION

The boundary integral formulation for the two dimensional elasto-plastic problem without body forces are presented in this section following the results of Swedlow and Cruse [3] and Mukherjee [4]. The range of subscript indices in all the following equations is 1, 2.

$$C_{ij}\dot{u}_j(P) = \int_\Gamma U_{ij}(P,Q)\dot{t}_j(Q)d\Gamma - \int_\Gamma T_{ij}(P,Q)\dot{u}_j(Q)d\Gamma + \int_\Omega \Sigma_{ijk}(P,q)\dot{\varepsilon}^p_{jk}(q)\, d\Omega \tag{20}$$

where u_j is the displacement vector, t_j the traction vector, Ω is the surface of the planar body, Γ its boundary, P and Q are boundary points, p and q are interior points respectively. This expression allows us to determine the boundary unknowns and once this is done, the displacement increments are given by (20) with $C_{ij} = \delta_{ij}$ and $P = p$

The kernels U_{ij} and T_{ij} are known singular solutions due to a point load in an infinite elastic solid and are available in many references such as C.A. Brebbia [5] . The kernel Σ_{ijk} for the case of plane stress is given by (21) with ν replaced by $\bar{\nu} = \nu/(1+\nu)$

$$\Sigma_{ijk} = - \frac{1}{4\pi(1-\nu)r} \left[(1-2\nu)(\delta_{ij}r_{,k} + \delta_{ki}r_{,j} - \delta_{jk}r_{,i}) + 2\, r_{,i}r_{,j}r_{,k} \right] \quad (21)$$

The kernel Σ is strongly singular and in order to perform the derivative, H.D. Bui [6] has arrived at the correct solution by employing Mikhlin's concept of derivatives of singular integrals.

Internal stresses can then be calculated and for the plane stress state we obtain [7] :

$$\dot{\sigma}_{ij}(p) = \int_\Gamma D_{ijk}(p,Q)\dot{t}_k(Q)d\Gamma - \int_\Gamma S_{ijk}(p,Q)\dot{u}_k(Q)d\Gamma + \int_\Omega \Sigma_{ijkl}(p,q)\dot{\varepsilon}^p_{kl}(q)d\Omega - \frac{G}{4(1-\bar{\nu})}(2\dot{\varepsilon}^p_{ij} + \dot{\varepsilon}^p_{ll}\,\delta_{ij}) \quad (22)$$

where $D_{ijk} = - \Sigma_{ijk}$ (23)

and

$$S_{ijk} = \frac{G}{2\pi(1-\bar{\nu})r^2} \left\{ 2\frac{\partial r}{\partial n} \left[(1-2\bar{\nu})\,\delta_{ij}r_{,k} + \bar{\nu}(\delta_{ik}r_{,j} + \delta_{jk}r_{,i}) - 4r_{,i}r_{,j}r_{,k} \right] + 2\bar{\nu}(n_i r_{,j}r_{,k} + n_j r_{,i}r_{,k}) + (1-2\bar{\nu})(2n_k r_{,i}r_{,j} + n_j\,\delta_{ik} + n_i\delta_{jk}) - (1-4\bar{\nu})\,n_k\delta_{ij} \right\} \quad (24)$$

The last integral in (22) is to be taken in the Cauchy principal value sense in which :

$$\Sigma_{ijkl} = \frac{G}{2\pi(1-\bar{\nu})r^2} \{ 2(1-\bar{\nu})(\delta_{ij}\, r_{,k}\, r_{,l} + \delta_{kl}\, r_{,i}\, r_{,j})$$

$$+ 2\,\bar{\nu}(\delta_{li}\, r_{,j}\, r_{,k} + \delta_{jk}\, r_{,l}\, r_{,i} + \delta_{ik}\, r_{,l}\, r_{,j}$$

$$+ \delta_{jl}\, r_{,i}\, r_{,k}) - 8\, r_{,i}\, r_{,j}\, r_{,k}\, r_{,l} + (1-2\bar{\nu})(\delta_{ik}\,\delta_{lj}$$

$$+ \delta_{jk}\,\delta_{li}) - (1-4\,\bar{\nu})\,\delta_{ij}\,\delta_{kl}\} \qquad (25)$$

NUMERICAL IMPLEMENTATION

The boundary of the region is discretized into a series of elements which can be constant or linear in the program developed. With the linear elements, the tractions must be allowed to have discontinuities across a node at corner. Additional equations may be obtained using the symmetry of the stress tensor and the compatibility between strains and displacements. In both cases, the integrals in equation (20) have been evaluated in closed form for straight boundary elements. We assume the incremental values of internal plastic strains to remain constant over rectangular or triangular internal cells. The integrals in equation (22) are calculated analytically for straight boundary elements and rectangular internal cells and numerically for triangular internal cells using three or six Gauss-Cowper points.

Equation (20) can then be written in matrix form :

$$[H]\{\dot{u}\} = [G]\{\dot{t}\} + [\Sigma]\{\dot{\varepsilon}^p\} \qquad (26)$$

After applying the boundary conditions the system can be reordered and we obtain :

$$[A]\{\dot{x}\} = \{\dot{f}\} + [\Sigma]\{\dot{\varepsilon}^p\} \qquad (27)$$

where $\{\dot{x}\}$ contains the unknown displacement and traction increments and $\{\dot{f}\}$ is obtained by multiplying the known displacement and traction increments with the appropriate columns of matrices $[H]$ and $[G]$.

Similarly, equation (22) gives :

$$\{\dot{\sigma}\} = [G']\{\dot{t}\} - [H']\{\dot{u}\} + [D' + C']\{\dot{\varepsilon}^p\} \qquad (28)$$

in which $[C']$ represents the independent terms and $[D']$ stands for the plastic strain integral. As it was shown in [7], the principal value can be computed by applying equation (28) to represent a state of constant plastic strains.

ITERATIVE SOLUTION PROCEDURE

An incremental solution algorithm based on equations (27) and (28) can be developed as follows :

The first increment is scaled up to magnitude at which first yield is encountered at the centroïd of an internal cell. Starting with $\{\dot{\varepsilon}^p\} = 0$ at each internal cells, the following steps are common for each load increment.

1. We apply a loading increment $\{\dot{f}\}$ and we solve

$$\{\dot{x}\} = [A]^{-1}\{\dot{f}\} \tag{29}$$

We obtain the incremental values of unknown boundary displacements and tractions and update the boundary vectors.

2. By using equation (28) and noticing that :

$$\{\dot{\sigma}^e\} = \{\dot{\sigma}\} + \{\dot{\sigma}^p\} \tag{30}$$

We calculate the stress increment $\{\dot{\sigma}^e\}$ and with the current stress state $\{\sigma\} + \{\dot{\sigma}^e\}$ we compile a list of yielded cells. For the points in the plastic range, we recalculate the correct stresses $\{\dot{\sigma}\}$ with the aid of the elasto-plastic stress-strain relation (19) where the total strain increment $\{\dot{\varepsilon}\}$ is imposed by the elastic calculation using $\{\dot{\sigma}^e\}$.

Therefore : $$\{\dot{\varepsilon}\} = [D^e]^{-1}\{\dot{\sigma}^e\} \tag{31}$$

and $$\{\dot{\sigma}\} = [D^{ep}]\{\dot{\varepsilon}\} \tag{32}$$

The current stress state necessary to evaluate $[D^{ep}]$ may be taken as the average during the increment such that no important departure from the true yield surface is observed.

3. We update stresses and strains and compute the plastic strain increment as :

$$\{\dot{\varepsilon}^p\} = \{\dot{\varepsilon}\} - [D^e]\{\dot{\sigma}\} \tag{33}$$

and $$\{\dot{\sigma}^p\} = \{\dot{\sigma}^e\} - \{\dot{\sigma}\} \tag{34}$$

The hardening parameters such as the equivalent plastic strain $\{\dot{\bar{\varepsilon}}^p\}$ which is a state variable defined by (8) and the translation of the yield surface $\{\alpha\}$ are calculated. If necessary some corrections of the stress level can be made to maintain the stresses on the true yield surface.

4. The non-linear "nodal boundary contribution" of the plastic-strain increment is computed as

$$\{\dot{b}^p\} = [\Sigma]\{\dot{\varepsilon}^p\} \qquad (35)$$

Then we solve :

$$\{\dot{x}\} = [A]^{-1}\{\dot{b}^p\} \qquad (36)$$

and update the boundary displacement vector.

5. The process is continued until some tests of convergence ensure sufficient accuracy for termination. An efficient criterion formulated on the basis of the equivalent plastic strain increment has been found. The criterion takes the form for the (k+1) iteration :

$$\text{If} \quad \frac{||\dot{\bar{\varepsilon}}^p||_\infty^{k+1} - ||\dot{\bar{\varepsilon}}^p||_\infty^{k}}{||\dot{\bar{\varepsilon}}^p||_\infty^{k+1}} \leqslant 10^{-4} \text{ or } 10^{-5}$$

go to step 1, otherwise go to step 2.

As a result, it has been found from numerical experience that this method is convergent and stable for even large size increments.

NUMERICAL EXAMPLES

The first example is a rectangular thin strip modelled by 20 linear boundary elements (or 40 constant B.E.) with 16 rectangular internal cells as shown in figure 1. Axial loads were applied at the free end of the strip and the other end was restrained against any displacement, such that a two-dimensional stress distribution exists. The iterative procedure is seen to converge and for example, 6 iterations in each load increment are required using the linear B.E. and 8 iterations with the constant B.E. As shown, good correlation between the B.E. method and the exact solution [8] is achieved.

Figure 2 shows numerical results of the same two dimensional strip. The material is stainless steel A 316 L (Nuclear engineering) idealized by the non-linear hardening rule (9). The figure displays the capability of the present non-linear hardening model with the B.E. method in describing the cyclic transient behaviour and the cyclic steady behaviour.

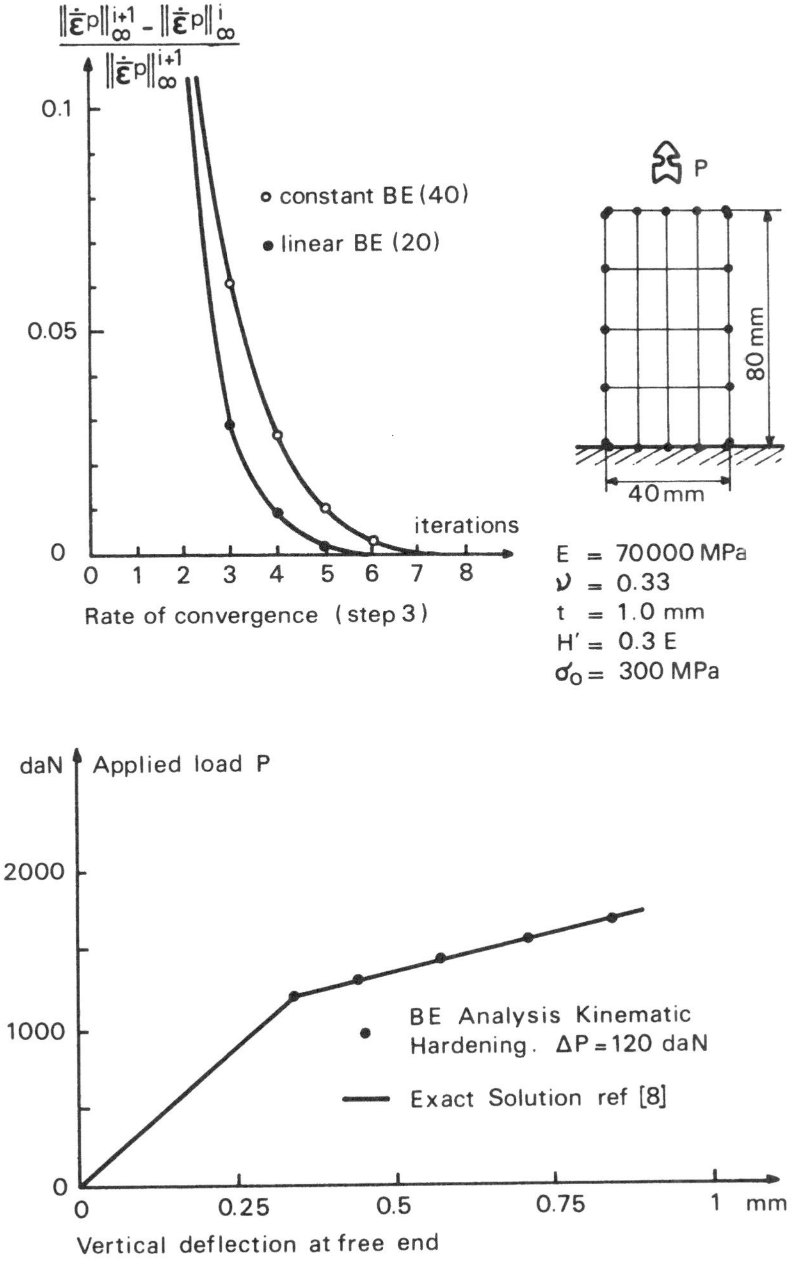

$$\frac{\|\dot{\bar{\varepsilon}}^p\|^{i+1}_\infty - \|\dot{\bar{\varepsilon}}^p\|^i_\infty}{\|\dot{\bar{\varepsilon}}^p\|^{i+1}_\infty}$$

0.1

o constant BE (40)

• linear BE (20)

0.05

iterations

0

0 1 2 3 4 5 6 7 8

Rate of convergence (step 3)

E = 70000 MPa
ν = 0.33
t = 1.0 mm
H' = 0.3 E
σ_0 = 300 MPa

80 mm

40 mm

P

daN Applied load P

2000

1000

BE Analysis Kinematic
Hardening. ∆P = 120 daN

—— Exact Solution ref [8]

0

0 0.25 0.5 0.75 1 mm

Vertical deflection at free end

Figure 1

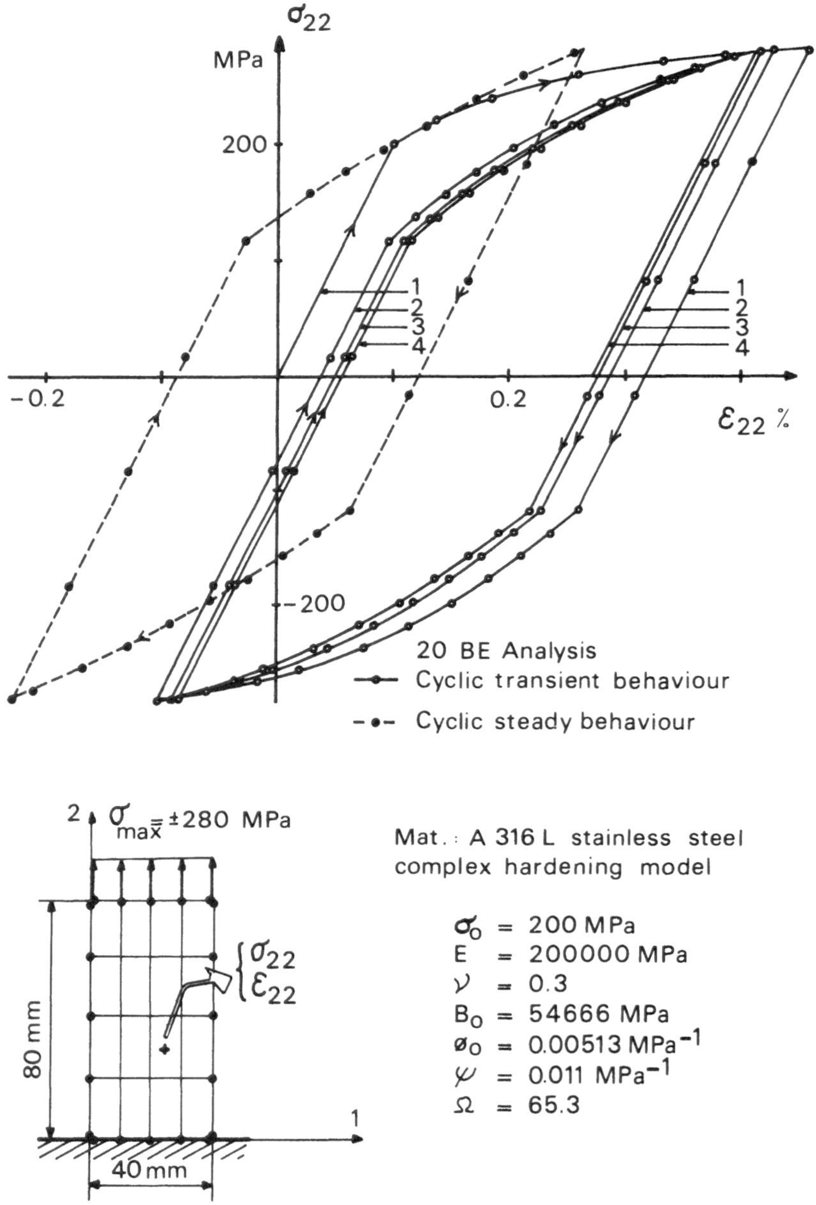

σ22
MPa
200
-200
-0.2
0.2
ε22 %
1
2
3
4
20 BE Analysis
Cyclic transient behaviour
Cyclic steady behaviour
2
σmax = ±280 MPa
σ22
ε22
80 mm
40 mm
1
Mat.: A 316 L stainless steel
complex hardening model
σo = 200 MPa
E = 200000 MPa
ν = 0.3
Bo = 54666 MPa
øo = 0.00513 MPa-1
ψ = 0.011 MPa-1
Ω = 65.3

Figure 2

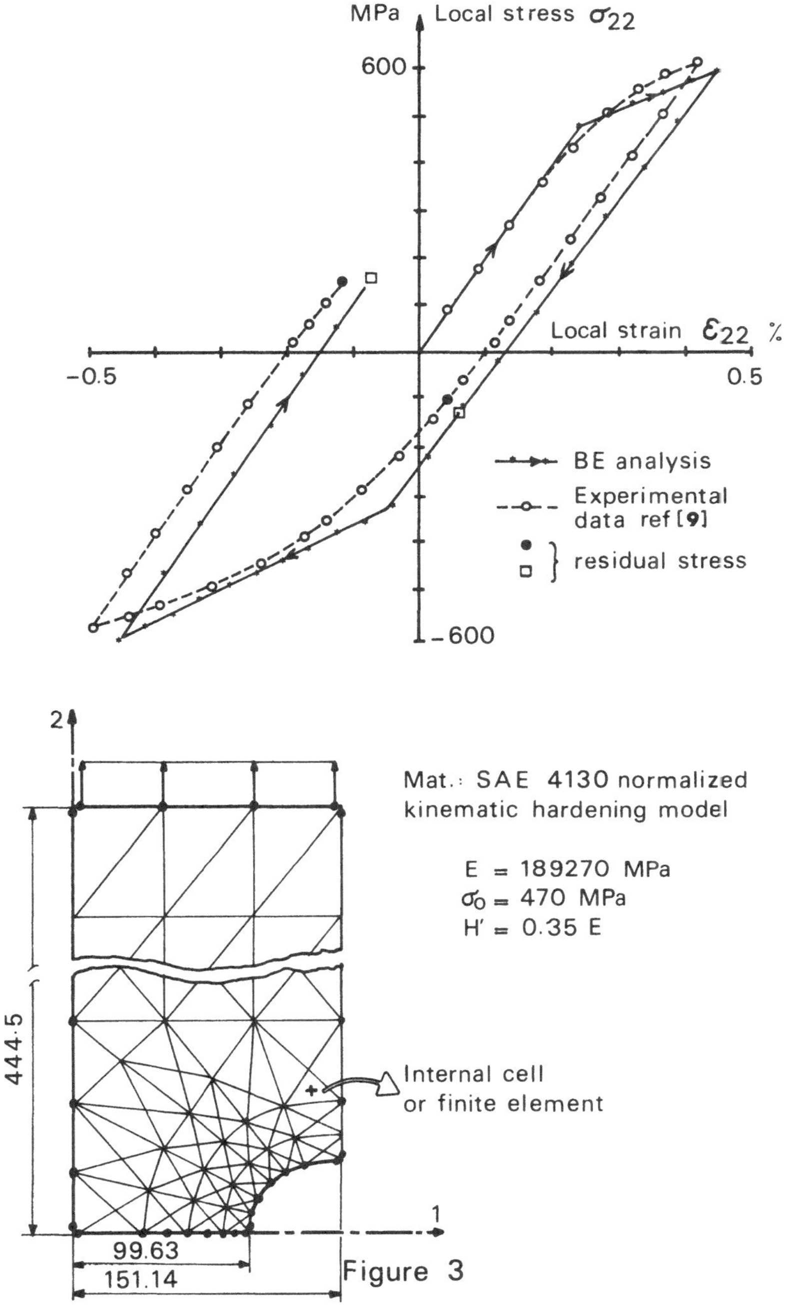
MPa
Local stress σ_{22}
600
-600
Local strain ε_{22} %
-0.5
0.5
BE analysis
Experimental data ref [9]
residual stress
2
1
444.5
99.63
151.14
Mat.: SAE 4130 normalized
kinematic hardening model
E = 189270 MPa
σ_0 = 470 MPa
H' = 0.35 E
Internal cell
or finite element

Figure 3

The third example is a notched bar shown in figure 3 used for the cyclic elasto-plastic analysis of an experimental study carried out by Crews [9]. The boundary element model includes 31 linear B.E. and the finite element model includes 158 degrees of freedom and 127 constant strain triangular elements which serve as internal cells in the B.E. computation. The material is steel alloy SAE 4130 idealized by the Prager-Ziegler hardening model with a hardening slope of 0,35 E. The calculated results with the B.E.M. and the F.E.M. using the initial-stress procedure reasonably correlate with the test data.

As an example of the required computer time, the CPU time using an IRIS 80 computer for the notched bar is 15,2 mn for the F.E. analysis (32 load increments with 3-5 iterations for each load increment and 29 yielded elements) and 18.4 mn for the B.E. analysis (32 load increments with 3-4 iterations for each load increment and 30 yielded elements). However, with the F.E.M., during the process of iteration the stiffness matrix can be kept constant as above (modified Newton-Raphson) or can be updated as often as desired and in this case we have obtained a CPU time of 10.3 mn. But, using a domain discretization only where plasticity occurs, the B.E.M. requires a CPU time of 12.6 mn with 62 triangular internal cells instead of 127.

If for the last example, the computational costs are about 20 % higher than the corresponding solutions obtained by using the F.E.M., it is important to note that for the B.E. and the F.E. subdivision in figure 3, the stress concentration factor in the elastic range was obtained to K_T = 2.10 with the B.E.M. and 1.97 with the F.E.M. compared with the theoretical value of K_T = 2.11, so the B.E.M. seems to be better.

CONCLUSION

An elasto-plastic boundary element approach for cyclic plasticity has been presented. Two kinds of constitutive equations, linear and non-linear hardening with strain memory effect, have been used in the calculation and the computational costs can be reduced by further careful programming effort. From the results presented in this paper, it appears that the B.E.M. is a promising computational procedure for the solution of such complicated non-linear problems. One of the advantages of the B.E.M. applied to plasticity problems is that the domain discretization is necessary only in regions where plasticity occurs.

REFERENCES

1 Mroz Z., Shrivastava H.P., Dubey R.N. (1976) A non-linear hardening model and its applications to cyclic loading Acta Mechanica 25, 51-61

2 Marquis D. (1979) Modèlisation et identification de l'écrouissage anisotrope des métaux. Thèse Université Paris

3 Swedlow, T.A. Cruse (1971) Formulation of Boundary integral Equations for three-dimensional Elasto-plastic flow. Int. J. Solids Structures - Vol. 7, 1673-1683

4 Mukherjee S. (1977) Corrected Boundary Integral Equations in Planar thermoelastoplasticity. Int. J. Solids Structures - Vol. 13. 331-335

5 Brebbia C.A. (1978) Boundary element Method for Engineers Pentech Press. Plymouth and London.

6 Bui H.D. (1978) Some remarks about the formulation of three dimensional thermoelastoplastic problems by Integral Equations. Int. J. Solids Structures. Vol. 14, 935-939

7 Telles J.C.F., Brebbia C.A. (1979) On the application of the boundary element method to plasticity. Appl. Math. Modelling - Vol. 3, December

8 Kalev I., Gluck J. (1977) Elasto-plastic finite element analysis. Int. J. Numerical Meth. Engineering. Vol. 11, 875-881

9 Crews J.H. (1969) Elasto-plastic stress-strain behavior at notch roots in sheet specimens under constant amplitude loading. NASA T.N. D. 5253

NEW DEVELOPMENTS IN ELASTOPLASTIC ANALYSIS

J.C.F. Telles and C.A. Brebbia

Department of Civil Engineering, University of Southampton, Southampton, SO9 5NH, U.K.

INTRODUCTION

This paper presents further developments on the application of boundary elements to plasticity on the lines previously presented by the authors (Telles and Brebbia, 1979, 1980a, 1980b).

The application of boundary elements (or boundary integral equations) to nonlinear material problems started with Riccardella (1973) who presented a valid plasticity formulation but used a not entirely successful numerical procedure. Later on Mukherjee and Kumar (1978) produced a formulation based on original constitutive relations which involved plasticity as well. All the implementations were then restricted to piecewise constant interpolation for the plastic strains.

In 1979 Telles and Brebbia produced the complete and correct formulation for two and three dimensional bodies, including the implementation of higher order interpolation functions for the plastic strains. In 1980 (a and b) they applied this formulation to study two-dimensional plasticity problems showing how different numerical procedures could be properly used; namely initial strain, initial stress and fictitious tractions and body forces. Emphasis was given to the accuracy and efficiency of the numerical formulation, firmly establishing the potentialities of boundary elements in plasticity.

One of the advantages of the implementation which was not pointed out in the past is that it overcomes the well-known disadvantage of finite elements first reported by Nagtegaal et al. (1974); i.e. the impossibility of representing properly the limit load due to the severe constraints created by the incompressibility of the plastic strains.

This difficulty is always present when internal strains are computed from the derivatives of the interpolated displacements within the finite elements (or internal cells), and is likely to occur in the approximate boundary element formulation presented by Banerjee and Cathie (1980).

In the present paper the plasticity formulation is extended to solve half-plane problems using the elastic fundamental solution of Melan's problem. This author (Melan, 1932) presented the stresses due to point loads within the semi-infinite plane. Telles and Brebbia (1981) extended this work to compute displacements, strains and surface tractions, presenting the complete fundamental solution needed in generalized boundary elements. This solution is the one used in the last two examples shown here.

In addition to the known advantages of the Kelvin elastoplastic formulation, the main benefit of the half-plane implementation is that boundary elements are not needed over the traction-free surface of the semi-plane. This property is of great significance in problems connected with soil-structure interaction, mining engineering, tunnels, etc. and has not been presented before for plasticity.

BASIC EQUATIONS

The incremental equilibrium equations for three-dimensional bodies in terms of displacement rates ($\dot{u}_i$) and plastic strain rates ($\dot{\varepsilon}^p_{ij}$), are

$$\dot{u}_{j,\ell\ell} + \frac{1}{1-2\nu}\,\dot{u}_{\ell,\ell j} = 2\left(\dot{\varepsilon}^p_{ij,i} + \frac{\nu}{1-2\nu}\,\dot{\varepsilon}^p_{kk,j}\right) - \frac{\dot{b}_j}{G} \tag{1}$$

where $\dot{b}_j$ represents the body force rate components, ν is Poisson's ratio and G is the shear modulus. Time derivatives (or simply increments) are indicated by a dot and space derivatives by a comma.

The traction boundary conditions for equation (1) can be written as

$$\dot{p}_i + 2G\left(\dot{\varepsilon}^p_{ij}n_j + \frac{\nu}{1-2\nu}\,\dot{\varepsilon}^p_{kk}n_i\right) = \frac{2G\nu}{1-2\nu}\,\dot{u}_{\ell,\ell}n_i + G(\dot{u}_{i,j} + \dot{u}_{j,i})n_j \tag{2}$$

where $\dot{p}_i$ represents the traction rate components and n_j stands for the direction cosines of the outward normal to the boundary of the body.

If plastic strains are considered as initial strains, the application of Hooke's law to the elastic part of the total strain rate tensor leads to the following expression for the stress rate components,

$$\dot{\sigma}_{ij} = 2G(\dot{\varepsilon}_{ij} - \dot{\varepsilon}^{p}_{ij}) + \frac{2G\nu}{1-2\nu}(\dot{\varepsilon}_{kk} - \dot{\varepsilon}^{p}_{kk})\delta_{ij} \tag{3}$$

in which δ_{ij} is the Kronecker delta.

Alternatively, equation (3) can be written in terms of initial stresses,

$$\dot{\sigma}_{ij} = 2G\dot{\varepsilon}_{ij} + \frac{2G\nu}{1-2\nu}\dot{\varepsilon}_{kk}\delta_{ij} - \dot{\sigma}^{p}_{ij} \tag{4}$$

where σ^{p}_{ij} stands for the components of the "initial stresses" given by

$$\dot{\sigma}^{p}_{ij} = 2G\dot{\varepsilon}^{p}_{ij} + \frac{2G\nu}{1-2\nu}\dot{\varepsilon}^{p}_{kk}\delta_{ij} \tag{5}$$

Expressions (1) to (5) are valid for plane strain ($i,j,\ell = 1,2$; $k = 1,2,3$) and also for plane stress ($i,j,k,\ell = 1,2$) if ν is replaced by $\bar{\nu} = \nu/(1+\nu)$.

HALF-PLANE FUNDAMENTAL SOLUTION

The complete fundamental solution for half-plane problems can be obtained by adding to the well known two-dimensional Kelvin fundamental solution a complementary part. This provides satisfaction of the traction-free condition over the surface of the half-plane. In order to save space, only the complementary part of the solution will be presented, but the complete fundamental solution is obtained by the following relation,

$$(\ )^{*} = (\ )^{k} + (\ )^{c} \tag{6}$$

where $(\ )^{k}$ and $(\ )^{c}$ stand for Kelvin part and complementary part respectively.

In the present work the traction-free surface of the half-plane is assumed to be represented by the axis $x_1 = 0$.

With reference to figure 1, the complementary part of the components of the displacements u^{*}_{ij}, corresponding to the displacement in j direction due to a unit force acting in i direction will be shown. For plane strains they are given by

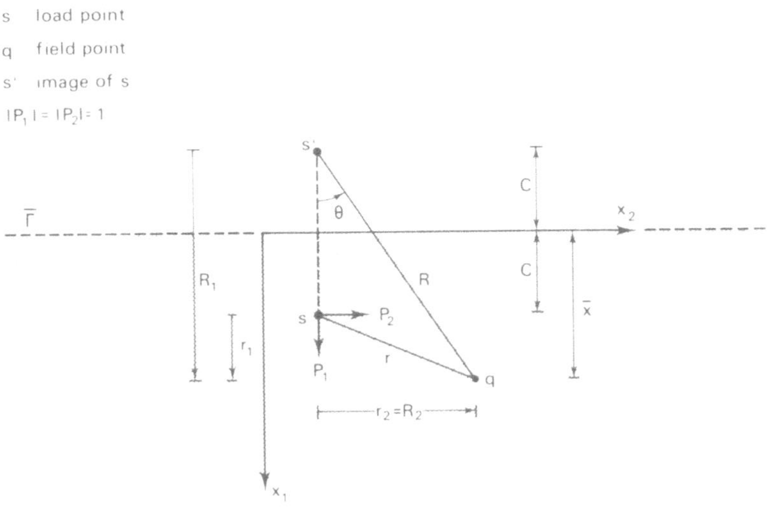

Figure 1. Unit point loads applied within the half-plane

$$u_{11}^{c} = K_d\left\{ -\left[8(1-\nu)^2 - (3-4\nu)\right] \ln R + \frac{\left[(3-4\nu)R_1^2 - 2c\,\bar{x}\right]}{R^2} + \frac{4c\,\bar{x}\,R_1^2}{R^4}\right\}$$

$$u_{12}^{c} = K_d\left\{\frac{(3-4\nu)r_1 r_2}{R^2} + \frac{4c\,\bar{x}\,R_1 r_2}{R^4} - 4(1-\nu)(1-2\nu)\theta\right\} \qquad (7)$$

$$u_{21}^{c} = K_d\left\{\frac{(3-4\nu)r_1 r_2}{R^2} - \frac{4c\,\bar{x}\,R_1 r_2}{R^4} + 4(1-\nu)(1-2\nu)\theta\right\}$$

$$u_{22}^{c} = K_d\left\{ -\left[8(1-\nu)^2 - (3-4\nu)\right] \ln R + \frac{\left[(3-4\nu)r_2^2 + 2c\,\bar{x}\right]}{R^2} - \frac{4c\,\bar{x}\,r_2^2}{R^4}\right\}$$

where the following notation was used,

$$\begin{aligned}
\theta &= \arctan(R_2/R_1) \\
r &= (r_i r_i)^{\frac{1}{2}} \\
R &= (R_i R_i)^{\frac{1}{2}} \\
r_i &= x_i(q) - x_i(s) \\
R_i &= x_i(q) - x_i(s') \\
c &= x_1(s) \geq 0 \\
\bar{x} &= x_1(q) \geq 0 \\
K_d &= 1/\left[8\pi G(1-\nu)\right]
\end{aligned} \tag{8}$$

The complementary part of the stresses due to unit forces in i direction (represented by σ^*_{jki}) are written as,

$$\sigma^c_{111} = -K_s \left\{ \frac{(3\bar{x} + c)(1-2\nu)}{R^2} + \frac{2R_1(R_1^2 + 2c\bar{x}) - 4\bar{x}r_2^2(1-2\nu)}{R^4} - \frac{16c\bar{x}\,R_1 r_2^2}{R^6} \right\}$$

$$\sigma^c_{121} = -K_s r_2 \left\{ -\frac{(1-2\nu)}{R^2} + \frac{2\left[x^2 - 2c\bar{x} - c^2 + 2\bar{x}R_1(1-2\nu)\right]}{R^4} + \frac{16c\bar{x}\,R_1^2}{R^6} \right\}$$

$$\sigma^c_{221} = -K_s \left\{ \frac{(\bar{x} + 3c)(1-2\nu)}{R^2} + \frac{2\left[R_1(r_2^2 + 2c^2) - 2cr_2^2 + 2\bar{x}r_2^2(1-2\nu)\right]}{R^4} + \frac{16c\bar{x}\,R_1 r_2^2}{R^6} \right\} \tag{9}$$

$$\sigma^c_{112} = -K_s r_2 \left\{ \frac{(1-2\nu)}{R^2} - \frac{2\left[c^2 - \bar{x}^2 + 6c\bar{x} - 2\bar{x}R_1(1-2\nu)\right]}{R^4} + \frac{16c\bar{x}\, r_2^2}{R^6} \right\}$$

$$\sigma^c_{122} = -K_s \left\{ \frac{(3\bar{x} + c)(1-2\nu)}{R^2} + \frac{2\left[(2c\bar{x} + r_2^2)R_1 - 2\bar{x}R_1^2(1-2\nu)\right]}{R^4} - \frac{16c\bar{x}R_1 r_2^2}{R^6} \right\}$$

$$\sigma^c_{222} = -K_s r_2 \left\{ \frac{3(1-2\nu)}{R^2} + \frac{2\left[r_2^2 - 4c\bar{x} - 2c^2 - 2\bar{x}R_1(1-2\nu)\right]}{R^4} + \frac{16c\bar{x}\, R_1^2}{R^6} \right\}$$

where

$$K_s = 1/\left[4\pi(1-\nu)\right] \tag{10}$$

The corresponding tractions and strains can be computed from (9) by using the relations,

$$p^c_{ij} = \sigma^c_{jki}\, n_k \tag{11}$$

and

$$\varepsilon^c_{jki} = \frac{1}{2G}\left[\sigma^c_{jki} - \nu\, \sigma^c_{\ell\ell i}\, \delta_{jk}\right] \tag{12}$$

For plane stress ν is replaced by $\bar{\nu}$ in all the above formulae.

It is interesting to note that the complementary part of the fundamental solution does not present any singularities within the domain $x_1 \geq 0$ when $c > 0$. Therefore, many expressions already obtained for the Kelvin elasto-plastic implementation, can still be used in the present formulation.

BOUNDARY ELEMENT FORMULATION

The extension of the elastic half-plane boundary element formulation (Telles and Brebbia, 1981) to elastoplastic problems, follows the same procedure as for the Kelvin implementation.

If we consider the plastic strains to be incompressible, the starting equation for the initial strain formulation is given by,

$$c_{ij}\,\dot{u}_j = \int_{\Gamma} u^*_{ij}\,\dot{p}_j\,d\Gamma - \int_{\Gamma'} p^*_{ij}\,\dot{u}_j\,d\Gamma + \int_{\Omega} u^*_{ij}\,\dot{b}_j\,d\Omega + \int_{\Omega} \hat{\sigma}^*_{jki}\,\dot{\varepsilon}^p_{jk}\,d\Omega \tag{13}$$

where Γ represents the total boundary of the body, Γ' the part of Γ in which $x_1 > 0$ and Ω stands for the domain of the body that has to be always located within the half-plane $x_1 \geq 0$. Note that the second boundary integral is performed over Γ' because the fundamental solution is zero on the surface of the half-plane and hence the integral vanishes there.

For plane strain problems the complementary part of the tensor that multiplies the plastic strain rates is (Mukherjee, 1977),

$$\hat{\sigma}^c_{jki} = \sigma^c_{jki} - \nu\,\sigma^c_{\ell\ell i}\,\delta_{jk} \tag{14}$$

whereas for plane stress

$$\hat{\sigma}^c_{jki} = \sigma^c_{jki} \tag{15}$$

Equation (13) is valid for any location of the load point ($s \in \Gamma$ or $s \in \Omega$), provided c_{ij} and the boundary integral over Γ' are properly interpreted as known from the previous elastic application of the half-plane fundamental solution.

By suitably modifying the plastic strain rate integral in equation (13), an initial stress formulation without the condition of incompressibility of the plastic strains (Telles and Brebbia, 1980b) can be equally obtained,

$$c_{ij}\,\dot{u}_j = \int_{\Gamma} u^*_{ij}\,\dot{p}_j\,d\Gamma - \int_{\Gamma'} p^*_{ij}\,\dot{u}_j\,d\Gamma + \int_{\Omega} u^*_{ij}\,\dot{b}_j\,d\Omega + \int_{\Omega} \varepsilon^*_{jki}\,\dot{\sigma}^p_{jk}\,d\Omega \tag{16}$$

where $\dot{\sigma}^p_{jk}$ was given in expression (5).

It is interesting to note that in contrast with the initial strain formulation, equation (16) can now be used for plane strain and plane stress problems with the replacement of ν by $\bar{\nu}$ being the only modification.

STRESSES AT INTERNAL POINTS

It is well-known that one of the advantages of boundary elements over "domain type" techniques in elastic analysis, is that stresses at internal points can be accurately computed by using the appropriate integral equation. The elastoplastic implementation of the technique is by all means no exception, and from the beginning the present authors have used the correct integral expression for computing the internal values of the stress rate tensor (Telles and Brebbia, 1979). In spite of this, some boundary element researchers have found the internal stresses by computing displacements at internal points, and differentiating them numerically as it is done in finite differences or finite elements (Banerjee and Cathie, 1980 and Cathie, 1980). This procedure possesses some drawbacks which are going to be commented on in other sections of this paper.

In order to compute accurately the stress rates at internal points, the derivatives of (13) are combined to produce the expressions for the total strain rate and then substituted into equation (3). Here one notices that, due to the non singular nature of the complementary tensors, the derivatives of the plastic strain rate integral create exactly the same singularities obtained for the single Kelvin implementation. Hence, for plane strains one has,

$$\dot{\sigma}_{ij} = \int_{\Gamma} u^*_{ijk}\,\dot{p}_k\,d\Gamma - \int_{\Gamma'} p^*_{ijk}\,\dot{u}_k\,d\Gamma + \int_{\Omega} u^*_{ijk}\dot{b}_k\,d\Omega$$

$$+ \int_{\Omega} \hat{\sigma}^*_{ijk\ell}\,\dot{\varepsilon}^p_{k\ell}\,d\Omega + f_{ij}(\dot{\varepsilon}^p) \qquad (17)$$

in which the plastic strain rate integral is to be computed in the principal value sense and f_{ij} is the same independent term obtained for the Kelvin formulation. i.e.,

$$f_{ij} = -\frac{G}{4(1-\nu)}\left[2\,\varepsilon^p_{ij} + (1-4\nu)\dot{\varepsilon}^p_{\ell\ell}\delta_{ij}\right] \qquad (18)$$

In addition,

$$\hat{\sigma}^{c}_{ijk\ell} = \sigma^{c}_{ijk\ell} - \nu\, \sigma^{c}_{ijmm}\, \delta_{k\ell} \tag{19}$$

and

$$\sigma^{c}_{ijk\ell} = G \left(\frac{\partial \sigma^{c}_{k\ell i}}{\partial x_j} + \frac{\partial \sigma^{c}_{k\ell j}}{\partial x_i} \right) + \frac{2G\nu}{1-2\nu} \frac{\partial \sigma^{c}_{k\ell m}}{\partial x_m} \delta_{ij} \tag{20}$$

where the derivatives are taken with reference to the coordinates of the load point. These derivatives together with u^{c}_{ijk} and p^{c}_{ijk} are given in (Telles and Brebbia, 1981).

An interesting feature of the half-plane implementation is that if the problem to be analysed satisfies the traction-free condition ($\dot{p}_k = 0$) over some part of the boundary $\Gamma - \Gamma'$, stresses at points located along this part of the boundary can be computed as if they were internal points. In order to validate equation (17) for such cases, only the expression of f_{ij} needs to be modified to take into consideration the limiting case c = 0. This expression can be easily obtained as follows; let us assume a semi-circular free body, of radius ρ, whose straight boundary is contained by the surface of the half-plane. If body forces are not considered, the application of a uniform plastic strain field ($\bar{\varepsilon}^{p}_{ij}$) to this body, will produce only displacements, internal stresses and tractions remain zero throughout the process.

The application of equation (17) to represent the stresses at the centre of the semi-circle leads directly to,

$$\int_{\Gamma'} p^{*}_{ijk}\, \bar{u}_k\, d\Gamma = \bar{\varepsilon}^{p}_{k\ell} \int_{\Omega} \hat{\sigma}^{*}_{ijk\ell}\, d\Omega + f_{ij}(\bar{\varepsilon}^{p}) \tag{21}$$

moreover, from the condition of existence of the principal value (Mikhlin, 1962), one can prove that,

$$\bar{\varepsilon}^{p}_{k\ell} \int_{\Omega} \hat{\sigma}^{*}_{ijk\ell}\, d\Omega = 0 \tag{22}$$

hence,

$$f_{ij}(\bar{\varepsilon}^{p}) = \int_{\Gamma'} p^{*}_{ijk}\, \bar{u}_k\, d\Gamma \tag{23}$$

where the relevant boundary displacements (neglecting rigid boundary movements) can be computed by (Telles and Brebbia, 1979),

$$\bar{u}_i = \rho(\bar{\varepsilon}^p_{ij} - \nu\,\bar{\varepsilon}^p_{kk}\,\delta_{ij})n_j \tag{24}$$

in which n_j represents the direction cosines of the outward normal to the curved boundary.

Equation (23) therefore provides the required expression for f_{ij} when $c = 0$,

$$f_{11} = f_{12} = 0$$
$$f_{22} = -\frac{G}{2(1-\nu)}(\dot{\varepsilon}^p_{22} - \dot{\varepsilon}^p_{11}) \tag{25}$$

For the initial stress formulation the procedure is entirely similar and the equation equivalent to (17) is of the following form,

$$\dot{\sigma}_{ij} = \int_\Gamma u^*_{ijk}\,\dot{p}_k\,d\Gamma - \int_{\Gamma'} p^*_{ijk}\,\dot{u}_k\,d\Gamma + \int_\Omega u^*_{ijk}\,\dot{b}_k\,d\Omega$$

$$+ \int_\Omega \varepsilon^*_{ijk\ell}\,\dot{\sigma}^p_{k\ell}\,d\Omega + g_{ij}(\dot{\sigma}^p) \tag{26}$$

where $\varepsilon^c_{ijk\ell}$ is obtained from (20) by substituting σ^c_{jki} by ε^c_{jki},

$$g_{ij} = -\frac{1}{8(1-\nu)}\left[2\dot{\sigma}^p_{ij} + (1-4\nu)\dot{\sigma}^p_{\ell\ell}\,\delta_{ij}\right] \quad \text{for } c > 0 \tag{27}$$

and

$$g_{11} = g_{12} = 0$$
$$g_{22} = -\frac{1}{4(1-\nu)}\left[\dot{\sigma}^p_{22} - \dot{\sigma}^p_{11}\right] \quad \text{for } c = 0 \tag{28}$$

Plane stress problems can be handled by equations (17) and (26) if Poisson's ratio is modified as before, $\hat{\sigma}^*_{ijk\ell} = \sigma^*_{ijk\ell}$ and expression (18) is replaced by

$$f_{ij} = -\frac{G}{4(1-\bar{\nu})}\left[2\dot{\varepsilon}^p_{ij} + \dot{\varepsilon}^p_{\ell\ell}\,\delta_{ij}\right] \tag{29}$$

Notice that expressions (25), (27) and (28) still remain valid.

Although body forces have been considered in the equations presented so far, in the following sections this term will not be included for simplicity.

NUMERICAL IMPLEMENTATION

For numerical implementation of the equations presented in the previous sections, linear boundary elements and linear triangular cells were used in much the same way as described by Telles and Brebbia (1980a,b) for the initial strain and initial stress formulations. It is worth mentioning that integration of the complementary part of the expressions presents no singularities when $c > 0$. Consequently, simple quadrature formulae can be used for the boundary and domain integrals. The limiting case $c = 0$ can be shown to produce singularities of the same order as those present in Kelvin's, therefore the two parts of the fundamental solution are added up and integrals are properly computed using the same integration scheme normally employed for the Kelvin part.

In addition, computation of stress rates at boundary nodes ($s \in \Gamma'$ and $s \in$ non traction-free part of $\Gamma - \Gamma'$) is included into the implementation through simple expressions which do not require any integration and involve the interpolated traction and displacement rates over the boundary elements.

DIFFERENT SOLUTION TECHNIQUES

At present, the initial strain formulation has been implemented with the von Mises criterion and employs Mendelson's successive elastic solutions method (Mendelson, 1968). This solution technique, also called "elastic predictor-radial corrector method" by Schreyer et al. (1979), has proved to be very efficient and stable with reference to the load increment size. Therefore it is recommended for von Mises material problems. The initial stress implementation, on the other hand, is more general and can handle four different yield criteria (Mises, Tresca, Mohr-Coulomb and Drucker-Prager), with two different iterative routines. The first is a pure incremental technique, comparable to what was used by Zienkiewicz et al. (1969) for finite elements. The second deals with accumulated values of the initial stresses, in a similar fashion to the initial strain implementation.

A common feature of the alternative implementations is that they are all incremental-iterative processes, capable of performing iterations by using a single recursive expression relating stresses to the plastic strains and the initial elastic solution. In what follows, it will be seen that the equivalent relations are not so straightforward when internal displacements are interpolated for computation of the internal stresses.

Recalling the procedure presented by Cathie (1980), one notices that in order to implement linear interpolation functions for the initial stresses, triangular cells with quadratic interpolation for the displacements must be used. Consequently, the number of internal points to which the cell is connected (also called internal points by the author), is doubled to six. By differentiating the interpolated displacements, a linear variation of the total strains is obtained and used to calculate the initial stresses at three so-called stress points. All operations are performed in incremental form.

For comparing the number of operations during one iteration cycle, one can consider, for simplicity, the case of a unique cell located within the body. Following Cathie's procedure, the computation of internal displacement increments at the six internal points would be carried out by using a 12 × 9 matrix (called $\underset{\sim}{H}_2$ in the reference), which multiplies the increments of initial stress at the three stress points. On top of this the derivatives still have to be performed to obtain the corresponding strains.

For the pure incremental procedure of the initial stress formulation, the equivalent operation will only involve the product of a 9 × 9 matrix by the initial stress increments, leading directly to the increments of elastic stress. By elastic stresses one means the result from the application of Hooke's law to the total strains, which already saves one operation when applying the elastoplastic constitutive relations (Telles and Brebbia, 1980b). Note that for quadrilateral cells the same argument still applies. Furthermore, as the number of cells increases, the difference between the two procedures becomes even more pronounced.

In conclusion, the above demonstrates that for accurate and efficient elastoplastic BE implementation, where the influence of the plastic term should be interpolated at least in linear piecewise form, computation of internal stresses by using the proper integral equation should always be preferred.

EXAMPLES

In the present section, the results of some applications of the technique described are compared with numerical and analytical solutions presented in the literature. The first example was solved by the elastoplastic implementation of the Kelvin fundamental solution. It is here included to highlight the improvement obtained when the integral equation is used for computation of internal stresses. In the second and third applications, the half-plane fundamental solution is employed throughout and the capability of the formulation is demonstrated.

Thick cylinder

In the first example the plane strain expansion of a thick cylinder subjected to internal pressure is studied. Ideal plasticity under von Mises yield criterion is assumed with the

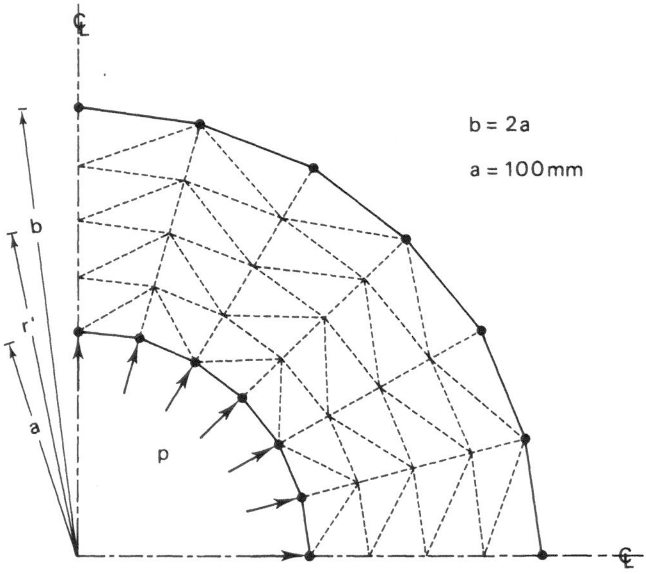

Figure 2. Thick cylinder problem. Boundary element and internal cell discretization

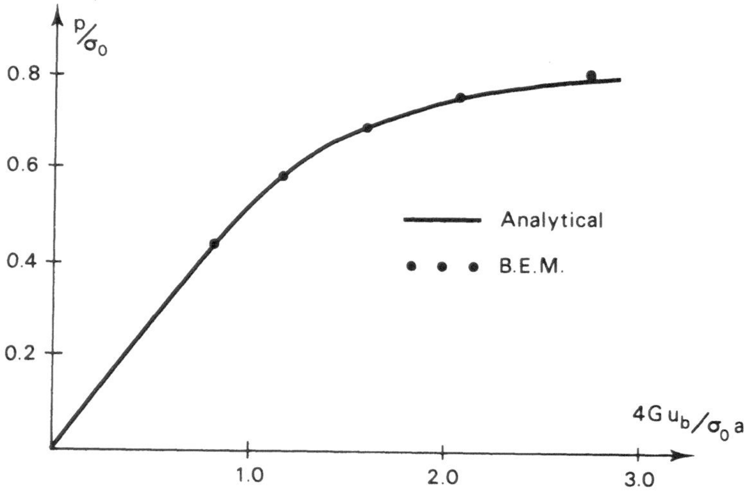

Figure 3. Outer surface displacements for thick cylinder problem

following material parameters,

$E = 12000.\ dN/mm^2$

$\sigma_o = 24.\ dN/mm^2$ (uniaxial yield stress)

$\nu = 0.3$

Boundary element results computed without boundary discretization of the symmetry axes as depicted in figure 2, are here compared with the analytical solution produced by Hodge and White (Prager and Hodge, 1968).

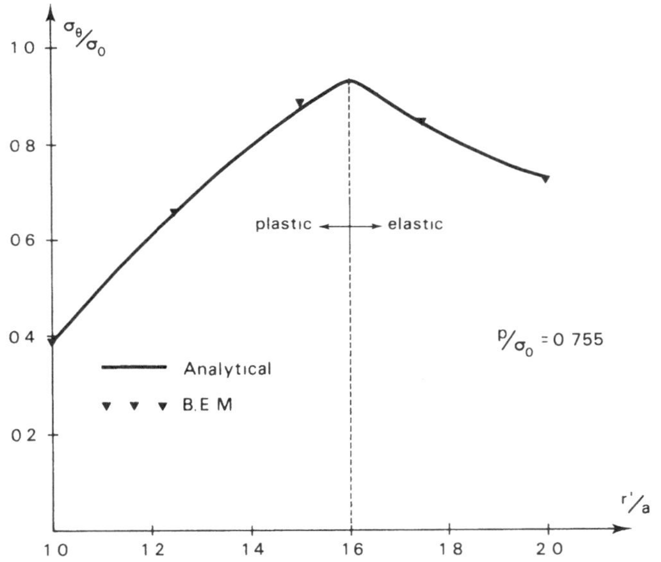

Figure 4. Circumferencial stress distribution in thick cylinder. Plastic front at r' = 1.6a

Radial displacements over the outer boundary and circumferential stress distribution (plastic front at r' = 1.6a) exhibit good agreement with the analytical solution as shown in figures 3 and 4 respectively.

It is worth mentioning that the example presented here has been solved by Banerjee and Cathie (1980) using constant boundary elements and linear piecewise displacements over internal cells. For comparing results, they have employed a coarse mesh which had the same number and pattern of cells shown in figure 2 and a fine mesh with 192 cells (twice the number of boundary elements). Although radial displacements were in good agreement with the analytical solution, the circumferential stresses presented for the refined mesh still did not match the accuracy exhibited in figure 4.

Strip footing

In this example, the plane strain analysis of a flexible strip footing under uniform loading is presented. The finite soil stratum is discretized taking full advantage of both, symmetry and free-surface condition, using the reduced number of 14 boundary elements and 42 internal points as shown in figure 5.

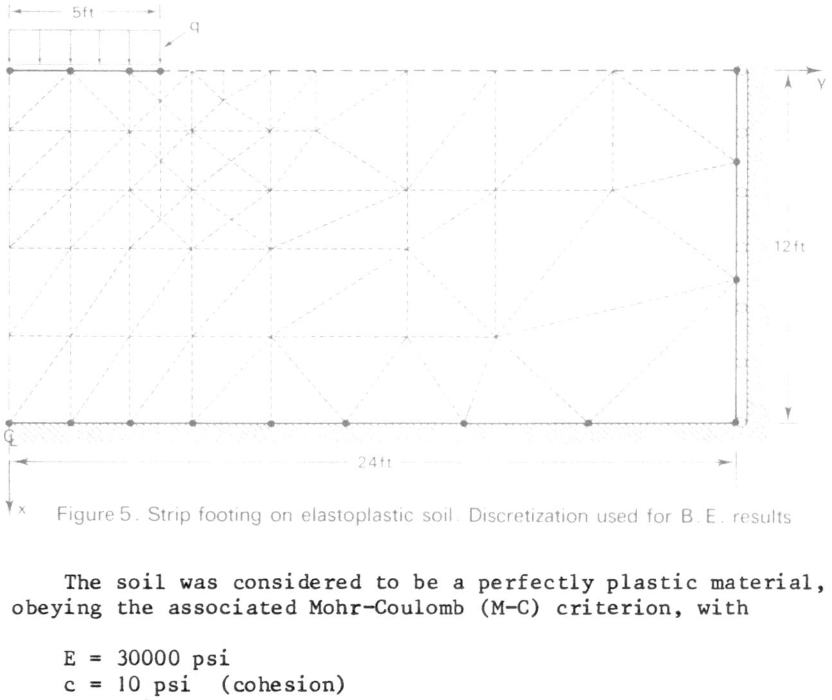

Figure 5. Strip footing on elastoplastic soil. Discretization used for B. E. results

The soil was considered to be a perfectly plastic material, obeying the associated Mohr-Coulomb (M-C) criterion, with

$E = 30000$ psi
$c = 10$ psi (cohesion)
$\phi = 20^o$ (angle of internal friction)
$\nu = 0.3$

An alternative solution was also carried out using the associated Drucker-Prager (D-P) yield criterion (extended von Mises) given by,

$$\frac{3\tan\phi}{(9+12\tan^2\phi)^{\frac{1}{2}}}\,\bar{p} + \sqrt{J_2} - \frac{3c}{(9+12\tan^2\phi)^{\frac{1}{2}}} = 0$$

where $\bar{p}$ is the hydrostatic component of the stress tensor and J_2 is the second invariant of the deviatoric stresses.

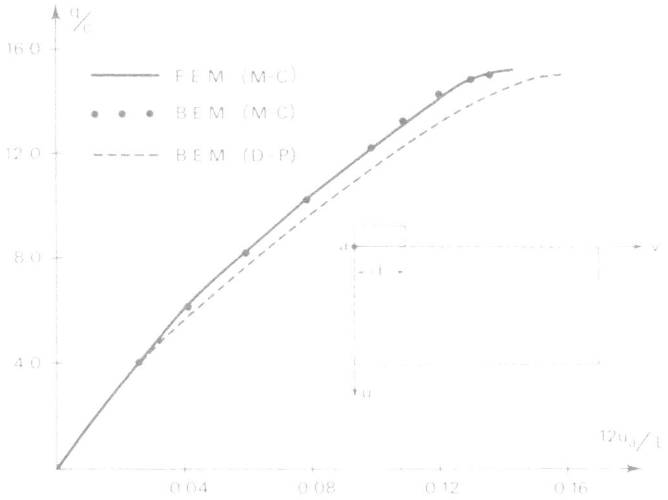

Figure 6. Load – displacement curves for strip footing problem

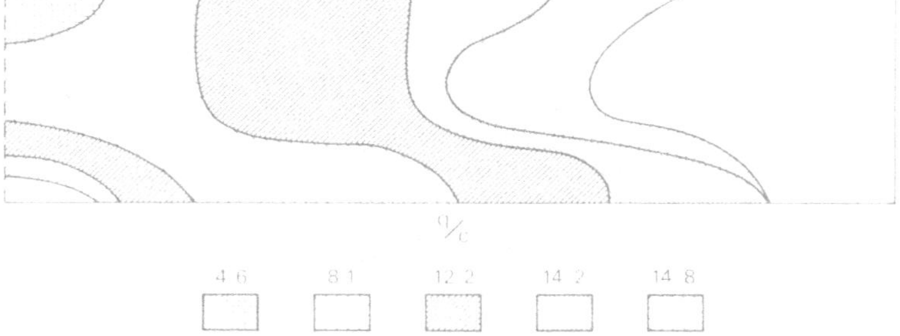

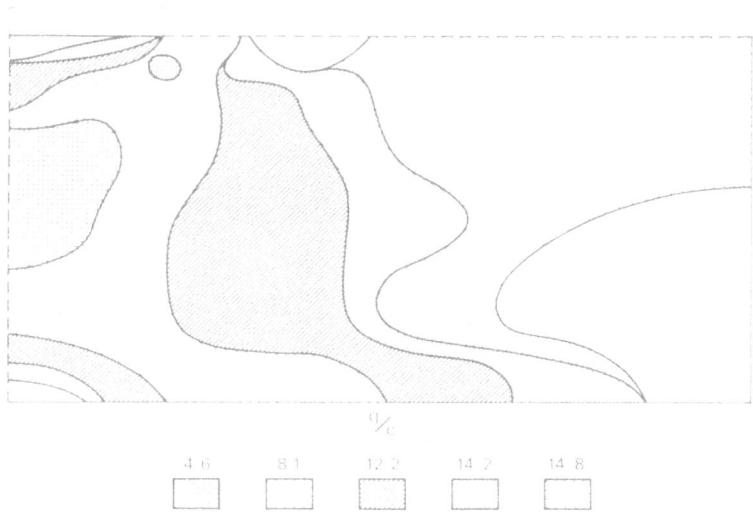

Figure 7. Spread of plastic zones at different load levels.
Mohr – Coulomb yield criterion.

Ground surface displacements are presented in figure 6. Also included is the equivalent M-C finite element solution obtained by Zienkiewicz et al. (1975) using quadratic isoparametric elements with 121 nodal points. The collapse loads achieved by the boundary element and finite element techniques (M-C) are $q/c = 14.9$ and $q/c = 15.1$ respectively, which agree well with the Prandtl solution (Chen, 1975) $q/c = 14.8$. As for the D-P results, it is seen that although the displacements were larger, the maximum load obtained was still not far from the previous ones.

Zones of yielding defined by the M-C solution are shown in figure 7. These zones compare well with the reported finite element computations.

Shallow tunnel

In a recent paper the present authors have applied the elasto-plastic boundary element technique to solve the problem of a deep circular excavation of radius r' in an infinite medium. The great advantages of the technique were then pointed out when comparing results with different finite element solutions. Herein, an analogous problem is studied by considering the tunnel to be shallow, located within the semi-infinite domain and with its centre at a depth of 5r'.

The rock-like material was assumed to follow the D-P yield criterion presented in the last example, with the following characteristics

$E = 500$ ksi
$c = 0.28$ ksi
$\phi = 30^{o}$
$\nu = 0.2$

In order to produce a more realistic analysis, the semi-infinite medium was assumed to be initially under the in-situ linearly varying stress field given by the formulae,

$$\sigma_v = \bar{\sigma}_v + \gamma h \qquad \text{(vertical stress)}$$

$$\sigma_h = 0.4\sigma_v \qquad \text{(horizontal stresses)}$$

where $\bar{\sigma}_v$ is a uniform pressure that may be due to an overburden of water or very weak material, γ is the unit weight of the rock and h is the distance from the ground surface.

To simulate the stress state adopted for the deep tunnel problem ($\sigma_v = 1$ ksi) at the depth of the excavation axis, the following values were chosen,

$\bar{\sigma}_v = 0.3$ ksi
$\gamma = 8.9 \times 10^{-2}$ lb/in^3
$r' = 131$ ft

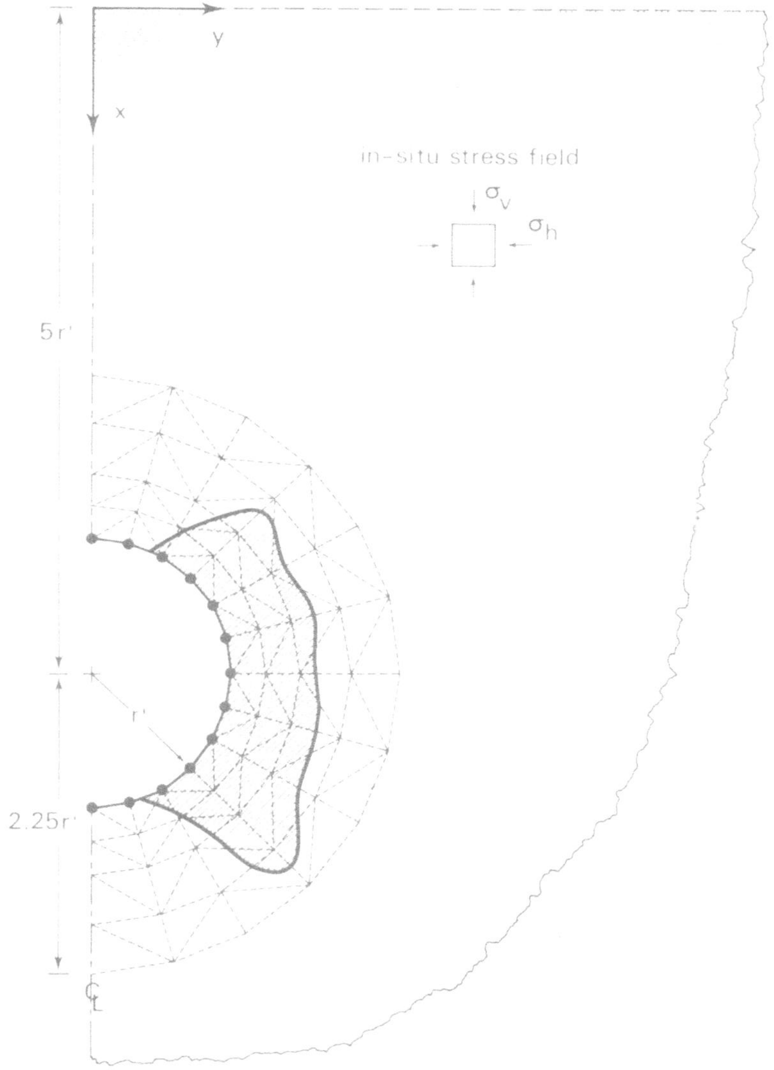

Figure 8. Shallow circular tunnel problem. Discretization used for B.E. results and total spread of plastic zone

The plane strain analysis was carried out by applying increments of external loads, corresponding to the relaxation of the in-situ stresses, over the boundary of the cavity. The discretization employed is depicted in figure 8 where the total extent of the plastic zone is also shown.

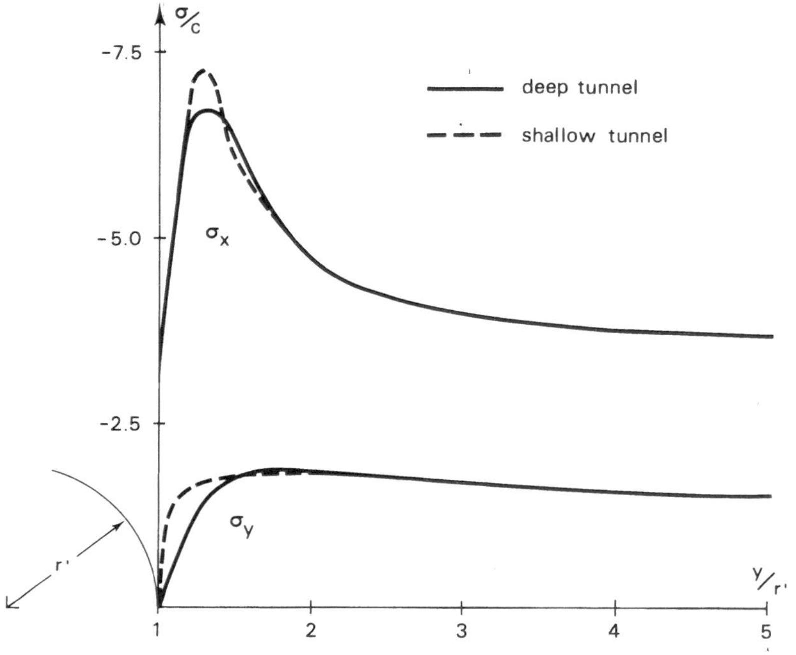

Figure 9. Final stresses along the horizontal section through the medium

Final stresses along the horizontal section are presented in figure 9 with the equivalent results from the deep tunnel case (Telles and Brebbia, 1980b) included for comparison. It should be noted that stress values outside the internal cell region were computed at simple internal points.

The above example clearly indicates the powerfulness of the half-plane implementation. Problems of this sort can only be satisfactorily solved by using this technique, which requires neither ground surface nor outer boundary discretization.

CONCLUSIONS

In this paper the complete formulation for the half-plane boundary element technique applied to plasticity has been presented. The correct integral equations for stresses at internal points and stresses at the traction-free surface were derived following the same procedures previously established by the authors.

The efficiency of computing stresses at internal points by using the proper integral equation in preference to employing interpolated displacements over internal cells is demonstrated.

The practical applicability of the formulation is illustrated by examples, which show that semi-infinite or finite sized problems can be analysed without boundary elements over the traction-free surface. This is, without doubt, a great advantage of the elastoplastic boundary element technique for solving problems concerning the semi-plane.

REFERENCES

Banerjee, P.K. and Cathie, D.N. (1980) A Direct Formulation and Numerical Implementation of the Boundary Element Method for Two-Dimensional Problems of Elastoplasticity, Int. J. Mech. Sci., 22: 233-245.

Brebbia, C.A. and Walker, S. (1979) The Boundary Element Techniques in Engineering. Butterworths, London.

Cathie, D.N. (1980) On the Implementation of Elastoplastic Boundary Element Analysis, Proc. 2nd Int. Seminar on Recent Advances in Boundary Element Methods, (edited by C.A. Brebbia), Univ. of Southampton, U.K.: 318-334.

Chen, W.F. (1975) Limit Analysis and Soil Plasticity. Elsevier Scientific Publishing Co., Amsterdam.

Melan, E. (1932) Der Spannungszustand der durch eine Einzelkraft in Innern beanspruchten Halbscheibe, Z. Angew. Math. Mech., 12: 343-346.

Mendelson, A. (1968) Plasticity: Theory and Application. MacMillan, New York and London.

Mikhlin, S.G. (1962) Singular Integral Equations. Amer. Math. Soc. Trans., Series 1, 10: 84-197.

Mukherjee, S. (1977) Corrected Boundary Integral Equation in Planar Thermoelastoplasticity, Int. J. Solids Structures, 13: 331-335.

Mukherjee, S. and Kumar, V. (1978) Numerical Analysis of Time-Dependent Inelastic Deformation in Metallic Media Using the Boundary Integral Equation Method, trans. ASME, J. Appl. Mech., 45: 785-790.

Nagtegaal, J.C.; Parks, D.M. and Rice, J.R. (1974) On Numerically Accurate Finite Element Solutions in the Fully Plastic Range, Comp. Meth. Appl. Mech. Engng., 4: 153-177.

Prager, W. and Hodge, P.G. (1968) Theory of Perfectly Plastic Solids. Dover, New York.

Riccardella, P.C. (1973) An Implementation of the Boundary Integral Technique for Planar Problems in Elasticity and Elasto-plasticity. Report No. SM-73-10, Dept. Mech. Engng., Carnegie Mellon Univ., Pittsburgh.

Schreyer, H.L.; Kulak, R.F. and Kramer, J.M. (1979) Accurate Numerical Solutions for Elastic-Plastic Problems, trans. ASME, J. Pressure Vessel Technology, 101: 226-234.

Telles, J.C.F. and Brebbia, C.A. (1979) On the Application of the Boundary Element Method to Plasticity, Appl. Math. Modelling, 3: 466-470.

Telles, J.C.F. and Brebbia, C.A. (1980a) The Boundary Element Method in Plasticity, Proc. 2nd Int. Seminar on Recent Advances in Boundary Element Methods, (edited by C.A. Brebbia), Univ. of Southampton, U.K.: 295-317.

Telles, J.C.F. and Brebbia, C.A. (1980b) Elastoplastic Boundary Element Analysis, Proc. U.S.-Europe Workshop on Nonlinear Finite Element Analysis in Structural Mechanics, Ruhr-University Bochum, W. Germany.

Telles, J.C.F. and Brebbia, C.A. (1981) Boundary Element Solution for Half-Plane Problems, Int. J. Solids Structures, (in press).

Zienkiewicz, O.C.; Valliappan, S. and King, I.P. (1969) Elasto-plastic Solutions of Engineering Problems, Initial Stress Finite Element Approach, Int. J. Num. Meth. Engng., 1: 75-100.

Zienkiewicz, O.C.: Humpheson, C. and Lewis, R.W. (1975) Associated and Non-Associated Visco-Plasticity and Plasticity in Soil Mechanics, Géotechnique, 25: 671-689.

THE BOUNDARY ELEMENT METHOD FOR THE SOLUTION OF NO - TENSION MATERIALS

W.S. Venturini & C. Brebbia

University of Southampton, England

1. INTRODUCTION

The Boundary Element Method has developed considerably since its beginning in 1967 for plane elastostatics[17]. The technique has been applied to a large number of problems in engineering such as plasticity, elastodynamics, viscoelasticity, and others.

The purpose of the present paper is to extend the Boundary Element Method to no-tension materials such as those present in underground and surface excavation and discussed in ref. (11). Unlike Finite Elements, the Boundary Element Method does not require the imposition of artificial outer boundaries at a finite distance, since the technique automatically satisfies the boundary conditions at infinity and prescribed displacements are not required.

In this paper the boundary of the underground tunnel or surface excavation is divided into a number of linear elements over which the displacements and tractions are considered to vary linearly from node to node. The applied loads are computed by considering the residual stresses produced by removing the excavated material.

The no-tension solution is achieved using an iterative process which consists of applying at each step a series of initial stresses to compensate the tensile stresses. The initial stress components are approximated over the domain by linear interpolation functions.

In the following, the boundary integral equations are presented for elastostatics and they are extended to cover no-tension materials using an initial stress formulation. In addition, the way in which prestressing forces can be applied is discussed.

In Section 3, the necessary steps to formulate the matrix equations are presented and following that, the iterative technique required to obtain the no-tension solution is also shown.

Finally, some problems are solved to illustrate the application of boundary elements for no-tension materials. The examples were selected to compare the boundary elements solution with already published finite elements results and point out the advantages of using the former.

2. BOUNDARY INTEGRAL EQUATIONS FOR ELASTOSTATICS

2.1 Governing Equations

An isotropic linear elastic problem is completely defined by the following set of governing equations.

i) Equilibrium equations

$$\sigma_{ij,j} + b_i = 0 \qquad \text{in } \Omega \tag{2.1}$$

where σ_{ij} and b_i are the stress and body forces components and Ω is the domain.

ii) Displacement-strain relationships

$$\varepsilon_{ij} = \frac{1}{2}(u_{i,j} + u_{j,i}) \qquad \text{in } \Omega \tag{2.2}$$

where $u_{i,j}$ and ε_{ij} are the displacement derivatives and strain components respectively.

iii) Constitutive equations, in this case they are the Hooke's law

$$\sigma_{ij} = \frac{2G\nu}{1-2\nu}\,\varepsilon_{kk}\,\delta_{ij} + 2\,G\,\varepsilon_{ij} - \sigma^o_{ij} \tag{2.3}$$

where G and ν are the shear modulus and Poisson's ratio respectively and σ^o_{ij} are the initial stress components[10]

iv) Boundary conditions

$$\begin{aligned} u_i(Q) &= \bar{u}_i(Q) \qquad & Q \in \Gamma_1 \\ p_i(Q) &= \bar{p}_i(Q) \qquad & Q \in \Gamma_2 \end{aligned} \tag{2.4}$$

$\Gamma = \Gamma_1 + \Gamma_2$ is the whole boundary of the structure; $\bar{u}_i$ and $\bar{p}_i$ are prescribed components of the displacements and tractions on Γ.

2.2 Boundary integral equations

Using Betti's theorem[2,3] or the weighted residual technique,[1] displacements at any point in the domain Ω can be expressed in terms of boundary values u_k and p_k, and body forces b_k, i.e.

$$u_\ell = -\int_\Gamma p^*_{\ell k}\, u_k\, d\Gamma + \int_\Gamma u^*_{\ell k}\, p_k\, d\Gamma + \int_\Omega u^*_{\ell k}\, b_k\, d\Omega \tag{2.5}$$

The above equation can be extended to the case of having an initial stress field as follows.

$$u_\ell = -\int_\Gamma p^*_{\ell k}\, u_k\, d\Gamma + \int_\Gamma u^*_{\ell k}\, p_k\, d\Gamma + \int_\Omega u^*_{\ell k}\, b_k\, d\Omega + \int_\Omega \varepsilon^*_{\ell mk}\, \sigma^o_{mk}\, d\Omega \tag{2.6}$$

$p^*_{\ell k}$, $u^*_{\ell k}$ and $\varepsilon^*_{\ell mk}$ are Kelvin's fundamental solutions[4] for displacements, tractions and strain respectively, and their values are given by:

$$u^*_{\ell k} = \left[r_{,\ell}\, r_{,k} - (3-4\nu)\ln r\, \delta_{\ell k}\right]/8\pi G(1-\nu)$$

$$p^*_{\ell k} = -\frac{\left[\frac{\partial r}{\partial n}\{(1-2\nu)\delta_{\ell k} + 2r_{,k} r_{,\ell}\} - (1-2\nu)(r_{,\ell}\eta_k - r_{,k}\eta_\ell)\right]}{4\pi(1-\nu)r} \tag{2.7}$$

$$\varepsilon^*_{\ell mk} = -\frac{\left[(1-2\nu)\left[r_{,k}\delta_{\ell m} + r_{,m}\delta_{\ell k}\right] + \delta_{mk} r_{,\ell} - 2r_{,\ell} r_{,m} r_{,k}\right]}{8\pi G(1-\nu)r}$$

where δ_{ij} is the Kronecker delta; $r_{,i}$ are the derivatives of r with respect to x_i (which is a coordinate of field point); and η_i is a direction cosine (fig. 1)

The expression (2.6) gives the displacements for an interior point. If the point is considered to be on the boundary the expression becomes,

$$\vec{\eta}_K = \frac{\partial \vec{n}}{\partial x_K}$$

$$r_{,K} = \frac{\partial r}{\partial x_K} = \frac{x_K^Q - x_K^P}{r}$$

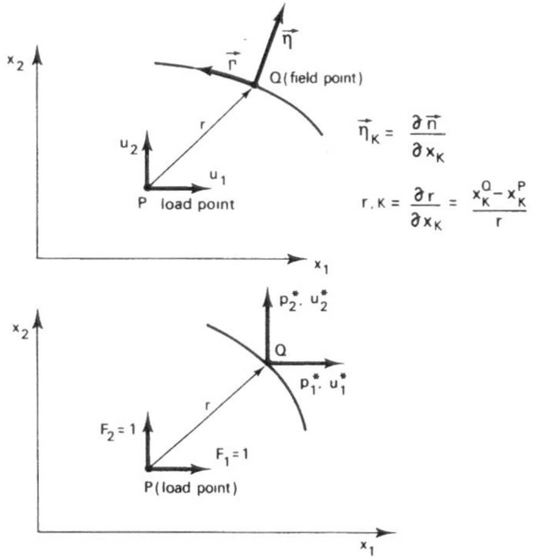

Fig. 1 Geometric parameters. Fundamental solutions

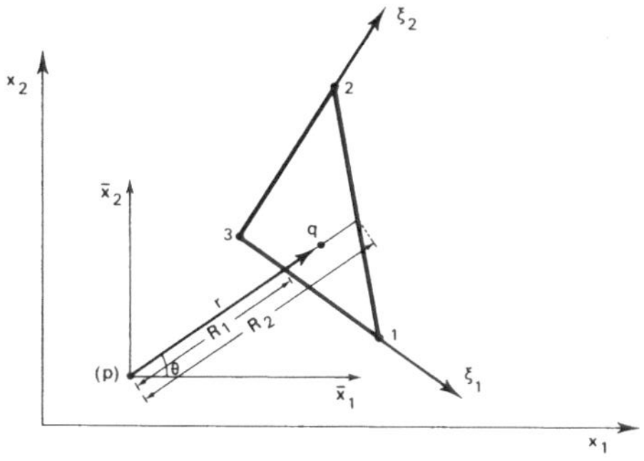

Fig. 2 Triangular coordinates

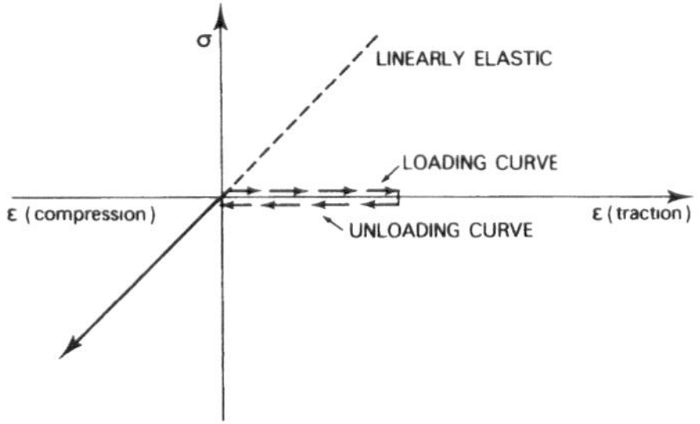

Fig. 3 No-tension as an elastic material

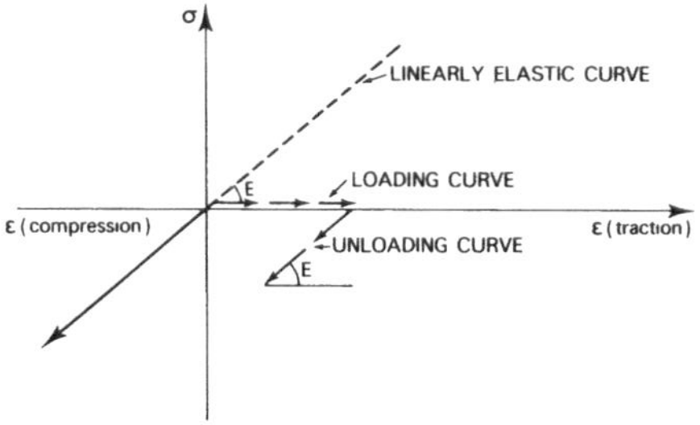

Fig. 4 No-tension as a plastic material

$$u_{\ell k}\, c_{\ell k} = -\int_\Gamma p^*_{\ell k}\, u_k \, d\Gamma + \int_\Gamma u^*_{\ell k}\, p_k \, d\Gamma + \int_\Omega u^*_{\ell k}\, b_k \, d\Omega +$$

$$+ \int_\Omega \varepsilon^*_{\ell mk}\, \sigma^o_{mk}\, d\Omega \qquad (2.8)$$

the value $c_{\ell k}$ has been determined by different authors[1,2].

The value of stress at an internal point can be obtained by differentiating the equation (2.6) and applying the Hooke's law , this gives

$$\sigma_{ij} = -\int_\Gamma S_{ijk}\, u_k\, d\Gamma \int_\Gamma D_{ijk}\, p_k\, d\Gamma + \int_\Omega D_{ijk}\, b_k\, d\Omega + \int_\Omega F_{ijmk}\, \sigma^o_{mk}\, d\Omega$$

$$- \sigma^o_{mk} \int_{\Gamma_1} \phi_{mkij}\, d\Gamma_1 - \sigma^o_{ij} \qquad (2.9)$$

where S_{ijk} and D_{ijk} are presented below; F_{ijmk} and $\int_{\Gamma_1} \phi_{mkij}\, d\Gamma_1$ are given in the Appendix A.

$$D_{ijk} = \frac{\{(1-2\nu)\left[\delta_{ki} r_{,j} + \delta_{kj} r_{,i} - \delta_{ij} r_{,k}\right] + 2 r_{,i} r_{,j} r_{,k}\}}{4\pi(1-\nu) r}$$

$$S_{ijk} = \Big\{ 2\frac{\partial r}{\partial n}\left[(1-2\nu)\delta_{ij} r_{,k} + \nu(\delta_{ik} r_{,j} + \delta_{jk} r_{,i}) - 4 r_{,i} r_{,j} r_{,k}\right]$$

$$+ 2\nu(\eta_i r_{,j} r_{,k} + \eta_j r_{,i} r_{,k}) + (1-2\nu)(2\eta_k r_{,i} r_{,j}$$

$$+ \eta_j \delta_{ik} + \eta_i \delta_{jk}) - (1-4\nu)\eta_k \delta_{ij} \Big\} \frac{2G}{4\pi(1-\nu) r^2}$$

$$(2.10)$$

The stress values at boundary points can be calculated from surface tractions and displacement derivatives as in Appendix B.

Prestressing forces can be introduced as body, line or boundary forces. For the last case we effectively increase the boundary of the domain to be analysed. For the other two cases, the force can be treated as a concentrated load. Notice that at the load point a singularity will arise.

Considering prestressing forces as concentrated loads the equation 2.5 can be written as follows,

$$u_\ell = -\int_\Gamma \overset{*}{p}_{\ell k}\, u_k \, d\Gamma + \int_\Gamma \overset{*}{u}_{\ell k}\, p_k \, d\Gamma + \int_\Omega \overset{*}{u}_{\ell k}\, b_k \, d\Omega + \overset{*}{u}_{\ell k}\, P_{v_k} \tag{2.11}$$

Notice that when this equation is applied on the boundary the term $c_{\ell k}$ must multiply the displacement u_ℓ.

For stress determination, we also have to add the prestressing term $D_{ijk}\, P_{v_k}$ in the equation (2.9).

3. BOUNDARY ELEMENT METHOD

In the last section we have presented the internal equations for displacements and and stresses in two-dimensional elasticity. Now as was presented in ref. (1), we are going to transform them into a set of algebraic equations by discretizing the boundary into elements and expressing the values of displacements and tractions in terms of interpolation functions and nodal values.

In this case, linear interpolation functions (eq. 3.1) for displacements and tractions on the boundary are considered. They can be written in matrix form[6]

$$\begin{aligned} \underset{\sim}{u} &= \underset{\sim}{\Phi}^T \underset{\sim}{U}^\alpha \\ \underset{\sim}{p} &= \underset{\sim}{\Phi}^T \underset{\sim}{P}^\alpha \end{aligned} \qquad \text{on } \Gamma$$

$$\underset{\sim}{b} = \underset{\sim}{\psi}^T \underset{\sim}{b}^\alpha \qquad \text{in } \Omega \tag{3.1}$$

and

$$\underset{\sim}{\sigma}^o = \underset{\sim}{\psi}^T \underset{\sim}{\sigma}^{o\alpha} \qquad \text{in } \Omega$$

where $\underset{\sim}{U}^\alpha$, $\underset{\sim}{P}^\alpha$, $\underset{\sim}{b}^\alpha$ and $\underset{\sim}{\sigma}^{o\alpha}$ are the nodal values for displacements, tractions, body forces and initial stresses respectively. $\underset{\sim}{\Phi}^T$ and $\underset{\sim}{\psi}^T$ are interpolation functions such as those used in finite elements[7].

Notice that we need to consider $\underset{\sim}{U}^\alpha$ and $\underset{\sim}{P}^\alpha$ only on the boundary in order to solve the integral equations, while $\underset{\sim}{B}^\alpha$ and $\underset{\sim}{\sigma}^{o\alpha}$ have to be defined over the whole domain.

The integral equation can be written in discretized form by expressing each integral as a summation of element and cell integrals, thus one has,

$$\underset{\sim}{U}\,\underset{\sim}{C} + \sum_{j=1}^{NE}\left(\int_{\Gamma_j} \underset{\sim}{P}^{*}\,\underset{\sim}{\phi}^{T}\,d\Gamma_j\right)\underset{\sim}{U} = \sum_{j=1}^{NE}\left(\int_{\Gamma_j} \underset{\sim}{U}^{*}\,\underset{\sim}{\phi}^{T}\,d\Gamma_j\right)\underset{\sim}{P}^{\alpha}$$

$$+ \sum_{j=1}^{NCELL}\left(\int_{\Omega_j} \underset{\sim}{U}^{*}\,\underset{\sim}{\psi}^{T}\,d\Omega_j\right)\underset{\sim}{b}^{\alpha} + \sum_{j=1}^{NCELL}\left(\int_{\Omega_j} \underset{\sim}{\varepsilon}^{*}\,\underset{\sim}{\psi}^{T}\,d\Omega_j\right)\underset{\sim}{\sigma}^{o\alpha}$$

$$+ \sum_{j=1}^{NP} \underset{\sim}{U}^{*}\,\underset{\sim}{P}_{v_j} \qquad (3.2)$$

The above integrations over elements and cells can be carried out and the final results written in matrix form as follows,

$$\underset{\sim}{C}\,\underset{\sim}{U} + \underset{\sim}{\hat{H}}\,\underset{\sim}{U} = \underset{\sim}{G}\,\underset{\sim}{P} + \underset{\sim}{D}\,\underset{\sim}{b} + \underset{\sim}{E}\,\underset{\sim}{\sigma}^{o} + \underset{\sim}{V}\,\underset{\sim}{P}_{u} \qquad (3.3)$$

The integrals over each element can be done analytically or numerically. Only over elements close to the singular point, the integral which gives $\hat{H}$ terms must be done analytically.

The domain integral (integral of body forces and initial stress) can be carried out numerically. In what follows the procedure to obtain the initial stress integral is discussed.

The initial stress integral over any single cell j can be written as

$$I = \int_{\Omega_j} \underset{\sim}{\varepsilon}^{*}\,\underset{\sim}{\psi}^{T} d\Omega \qquad (3.4)$$

For triangular cells and linear interpolation the following function can be used

$$\psi^{T} = \left[\,\xi^{1} \quad \xi^{2} \quad \xi^{3}\,\right] \qquad (3.5)$$

with

$$\xi^{\alpha} = \frac{1}{2A}\left[2A^{o}_{\alpha} + b_{\alpha}x + a_{\alpha}y\right] \tag{3.6}$$

where A is the area of cell and

$$\begin{aligned} a_i &= x_1^k - x_1^j \\ b_i &= x_2^j - x_2^k \\ 2A^o_\alpha &= x_1^j\,x_2^k - x_1^k\,x_2^j \quad \text{with } i = 1,2,\ \ j = 2,3 \text{ and } k = 3,1 \end{aligned} \tag{3.7}$$

This interpolation function can also be written in a polar coordinate (fig. 2) i.e.

$$\xi^{\alpha} = \xi^{\alpha}(p) + \frac{1}{2A}(b_{\alpha}\cos\theta + a_{\alpha}\sin\theta)r \tag{3.8}$$

where p is the load point and

$$\xi^{\alpha}(p) = \frac{1}{2A}\left[2A^{o}_{\alpha} + b_{\alpha}x_1(p) + a_{\alpha}X_2(p)\right] \tag{3.9}$$

The fundamental solution $\varepsilon^{*}_{\ell mk}$ can be written as,

$$\varepsilon^{*}_{\ell mk}(p,\theta,r) = e_{\ell mk}(p,\theta).\frac{1}{r} \tag{3.10}$$

and each cell integral becomes,

$$I = \int_{\Omega} e_{\ell mk}(p,\theta)\left[\xi^{\alpha}(p) + \frac{1}{2A}(b_{\alpha}\cos\theta + a_{\alpha}\sin\theta)r\right|drd\theta \tag{3.11}$$

This can be integrated over r remaining only to carry out the integral over the angle θ. The θ integral is performed numerically using a four or six Gauss integration rule. Once these integrals are computed over cells one obtains the matrix $\underset{\sim}{E}$.

The expression (3.12), after integration over r, becomes

$$I = \int_{\theta} e_{\ell mk}(p,\theta)\left[\xi^{\alpha}(p)(R_2-R_1) + \frac{1}{2A}(b_{\alpha}\cos\theta + a_{\alpha}\sin\theta) \times \left(\frac{R_2-R_1}{2}\right)(R_2+R_1)\right]d\theta \tag{3.12}$$

R_1 and R_2 are given in function of θ and cell geometry and can be easily calculated using expression (3.9) over an appropriate side of the cell.

The final system of matrix equations is obtained by interchanging columns in the matrices shown in eq.(3.4) in accordance with the applied boundary conditions[1]. So the unknowns are written in a vector $\underset{\sim}{X}$ on the left hand side and all known terms are accumulated on the right hand side.

After grouping together the right hand side matrices, one can write,

$$\underset{\sim}{A}\,\underset{\sim}{X} = \underset{\sim}{T} + \underset{\sim}{E}\,\underset{\sim}{\sigma}^{o} \tag{3.13}$$

where T contains the effects of all prescribed loads and displacements, the initial stress term has not been added to it as it will be needed during the iterative process.

Solving equation (3.14) one obtains,

$$\underset{\sim}{X} = \underset{\sim}{M} + \underset{\sim}{K}\,\underset{\sim}{\sigma}^{o} \tag{3.14}$$

where

$$\begin{aligned} \underset{\sim}{M} &= \underset{\sim}{A}^{-1}\,\underset{\sim}{T} \\ \underset{\sim}{K} &= \underset{\sim}{A}^{-1}\,\underset{\sim}{E} \end{aligned} \tag{3.15}$$

The stresses of internal points can also be expressed in matrix form after integration, i.e.

$$\underset{\sim}{\sigma} = \underset{\sim}{D}\,\underset{\sim}{P} - \underset{\sim}{S}\,\underset{\sim}{U} + \underset{\sim}{D} + \underset{\sim}{W}\,\underset{\sim}{P}_v + \underset{\sim}{F}\,\underset{\sim}{\sigma}^{o} - \underset{\sim}{\sigma}^{o} \tag{3.16}$$

We can rearrange the matrices in function of the prescribed values U and P, i.e. all the unknown values are in $\underset{\sim}{X}$ and the values prescribed on the boundary are given by $\underset{\sim}{Y}$. After rearranging the appropriate matrices in the above equation, one obtains,

$$\underset{\sim}{\sigma} = \underset{\sim}{\hat{D}}\,\underset{\sim}{Y} - \underset{\sim}{\hat{S}}\,\underset{\sim}{X} + \underset{\sim}{D}\,\underset{\sim}{b} - \underset{\sim}{W}\,\underset{\sim}{P}_v + \underset{\sim}{\hat{F}}\,\underset{\sim}{\sigma} \tag{3.17}$$

with

$$\underset{\sim}{\hat{F}} = \underset{\sim}{F} - \underset{\sim}{I} \tag{3.18}$$

In a more compact form

$$\underset{\sim}{\sigma} = \underset{\sim}{\hat{T}} - \underset{\sim}{S}\,\underset{\sim}{X} + \underset{\sim}{\hat{F}}\,\underset{\sim}{\sigma}_o \tag{3.19}$$

substituting $\underset{\sim}{X}$ value one obtains

$$\underset{\sim}{\sigma} = \underset{\sim}{N} + \underset{\sim}{B}\,\underset{\sim}{\sigma}^o \tag{3.20}$$

where

$$\begin{aligned} \underset{\sim}{N} &= \underset{\sim}{\hat{T}} - \underset{\sim}{\hat{S}}\,\underset{\sim}{M} \\ \underset{\sim}{B} &= \underset{\sim}{\hat{F}} - \underset{\sim}{\hat{S}}\,\underset{\sim}{K} \end{aligned} \tag{3.21}$$

Stresses on the boundary are calculated in the same way and the matrices $\underset{\sim}{N}$ and $\underset{\sim}{B}$ can be obtained as in Appendix B.

4. ANALYSIS OF NO-TENSION MATERIALS

In many practical applications, tunnels can be built only by excavation, and no structural lining is required to guarantee the cavity stability. Careful analysis of the rock medium is required in this case. Considering the medium as a linear elastic material is generally unsafe, as the tensile strength of rock is usually small and numerous faults and joints are to be expected.

In the present analysis, the rock will be considered as a no-tension material. The solution will be iterated until structural equilibrium is reached with only compressive stresses. This stage of stress is usually possible for underground structures, and the limit load is reached when the mass removal is completed.

Problems with no-tension material can be solved in one of the two following ways, i) by considering the material can not withstand any traction as shown in fig. 3; loading and unloading occur along a straight line, i.e. its behaviour is assumed to be elastic. ii) by assuming that the material behaves as a plastic one, when in tension this will produce a different unloading path as shown in fig. 4.

The second case is a special case of plasticity, and can be solved using load increment small enough to achieve the required accuracy.

The numerical procedure adopted here to reach the no-tension solution is similar for the two cases described above. The load can be divided into a series of increments for the plastic analysis or applied in one step for the first case.

The procedure consists of the following steps:

i) Elastic stress and shown determination. The elastic stress due to a load increment or an initial stress field can be calculated as

$$\underset{\sim}{\sigma}_e = \underset{\sim}{B}\,\underset{\sim}{\sigma}_i + (\underset{\sim}{\sigma}_{e\ell})$$

where $\underset{\sim}{\sigma}_{e\ell}$ corresponds to the elastic response for the load increment. The total strain $\underset{\sim}{\varepsilon}_e$ can be obtained using the elastic stress values.

ii) The principal directions can then be determined using the strain values; and the stresses in these directions can be computed. Letting the tensile stresses be zero in those directions, the true stresses ($\underset{\sim}{\sigma}_t$) can be determined.

iii) Determination of the initial stresses for the next step which can be given by

$$\underset{\sim}{\sigma}_i = \underset{\sim}{\sigma}_{e\ell} - (\underset{\sim}{\sigma}_t - \underset{\sim}{\sigma}_{to})$$

where $\underset{\sim}{\sigma}_{to}$ is the previous value of the $\underset{\sim}{\sigma}_t$.

iv) If the values of tensile stresses are not equal to zero within the required accuracy, the first three steps must be repeated. When convergence has been achieved another load increment can be applied.

5. APPLICATIONS

In this section, some practical examples, assuming no-tension material, i.e. that material cannot withstand any tensile stress, are presented.

Whenever possible results are compared with published finite elements solutions.

Steep Valley

The case of a steep valley was solved by Valliappan [11], using finite elements and no-tension material. The load applied to this structure is only due to the removal of residual compressive stresses which are taken to be equal to $-\gamma y$ and $-\gamma K_0 y$ in vertical and horizontal directions respectively (fig. 5).

The sloped valley boundary is discretized into a series of straight elements over which tractions corresponding to the residual stresses are applied. In figure 6 the mesh used to solve the problem is presented. The convergence of the results was tested by using a finer mesh which produced practically the same values. Notice that for this problem the domain tends to infinity and no displacement boundary conditions have to be prescribed.

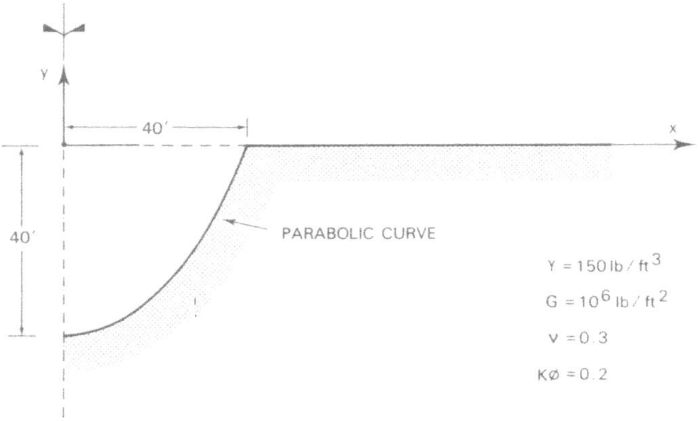

Fig. 5 Valley to be excavated

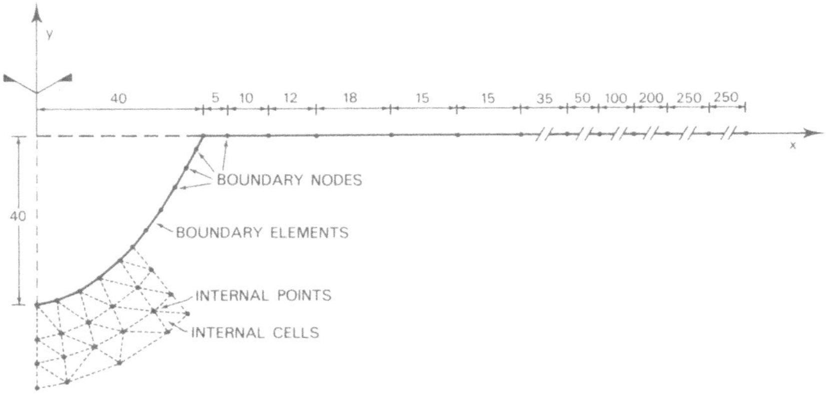

Fig. 6 Steep valley. Discretization

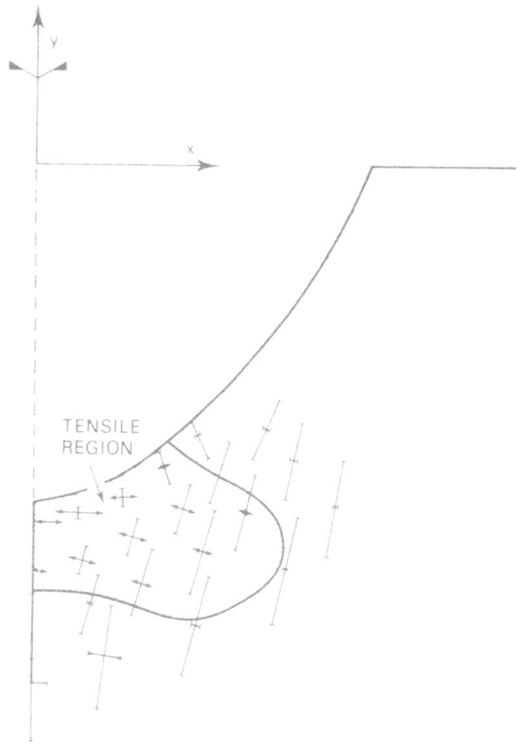

Fig. 7 Steep valley. Elastic solution

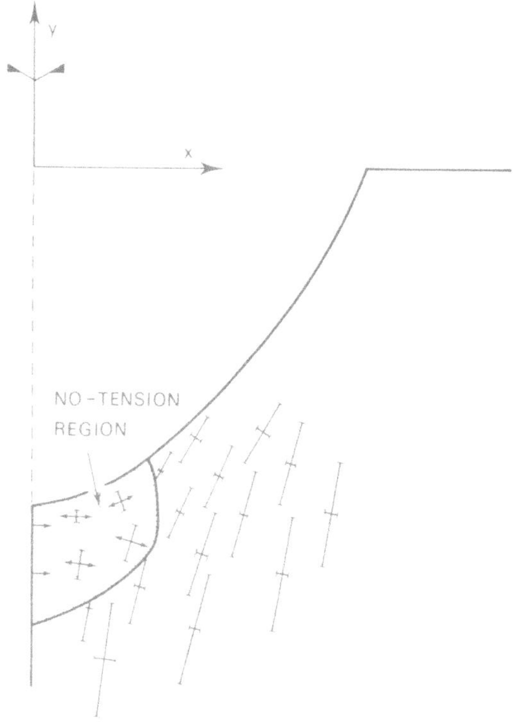

Fig. 8 Steep valley. No-tension solution

Figure 7 shows the elastic initial solution and figure 8 the final no-tension results in function of the principal stresses. The results agree well with a published finite element solution[11], for which a 247 elements and 144 nodes mesh was used. The boundary element solution only required 24 nodes and 23 boundary elements. For integrating the initial stress term 29 internal cells were used.

This example was solved using the two no-tension solution criteria previously discussed (see figs 3 and 4). It was found during the solution that the discretization of the curved boundary into straight elements produces small tractions forces which cannot be completely eliminated. This problem occurs with finite elements as well, but in both methods, it can be solved using curved elements.

Semi-circular tunnel

The semi-circular tunnel, fig. 9 is a comparison example previously solved by Bath[12,13] using an isoparametric quadratic finite elements program. The loading and the boundary conditions were chosen to make this example an optimum numerical test rather than try to reproduce an actual tunnel.

The load consists of a large vertical load P which is applied to the ground surface. Residual stresses are taken as constant during the loading process and are calculated using a specific weight equal to 120 lb/ft^3 and a maximum height equal to 116 ft.

The results for no-tension can be seen in figure 10. The no-tension zones are plotted for load corresponding to 28.8%, 40% and 100% of the total load respectively.

Figure 11 presents the displacements for points A and B showing their non-linear behaviour. The non-linearities started at 4% of loading and continued up to 60% of it. From then to the final load the structure behaves again linearly, once the final no-tension zone has already been reached.

Deep tunnel

The tunnel example shown in fig. 11 was used for a hydro-electric power station and taken from Valliappan[11]. He analysed it on a no-tension problem using 500 constant elements and 276 nodes mesh with a finite element program. The fig. 12 and fig. 13 show the discretization of this tunnel for the boundary element solution, both with and without prestressing forces. A 48 nodes and 39 elements boundary mesh was used for both cases, the internal cells for initial stress integration are plotted in the same figures. For the first case the only applied forces are those corresponding to the removal of the material. Fig. 14 shows the initial elastic and final no-tension solutions. The results are very similar to those presented in (11) using the Finite Element Method.

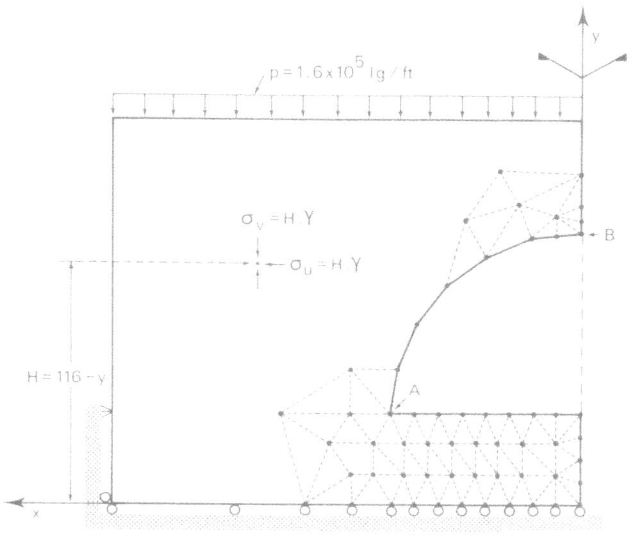

Fig. 9 Semi-circular tunnel. Residual stresses, Loading, Boundary discretization and internal cells.

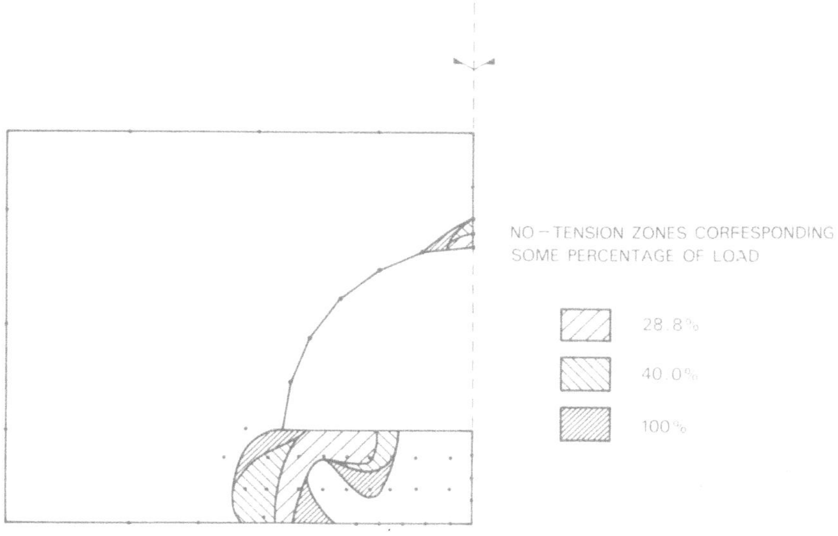

Fig. 10 No-tension zones

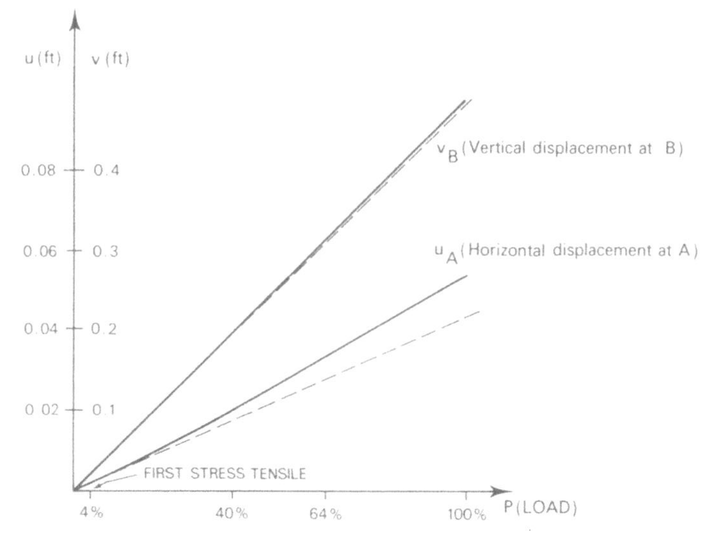

Fig. 11 Displacements at the structure

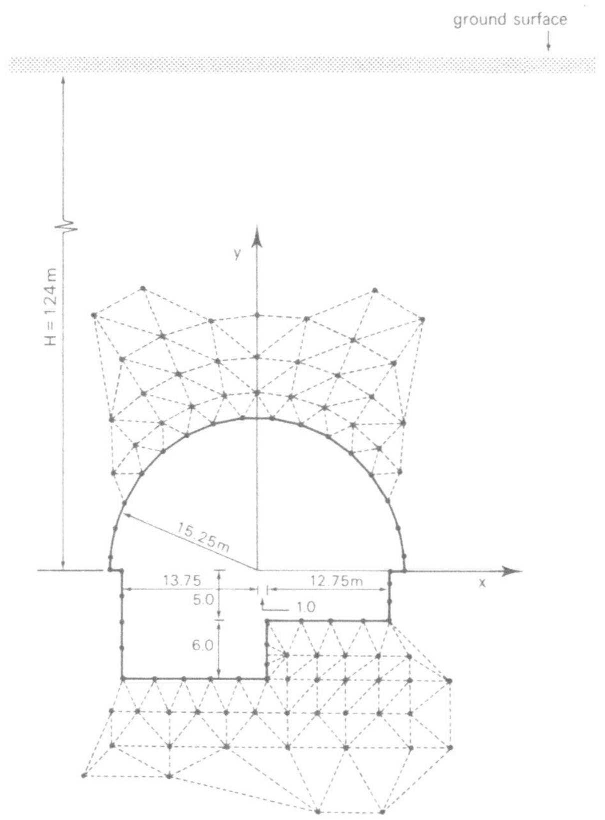

Fig. 12 Boundary discretization and internal cells

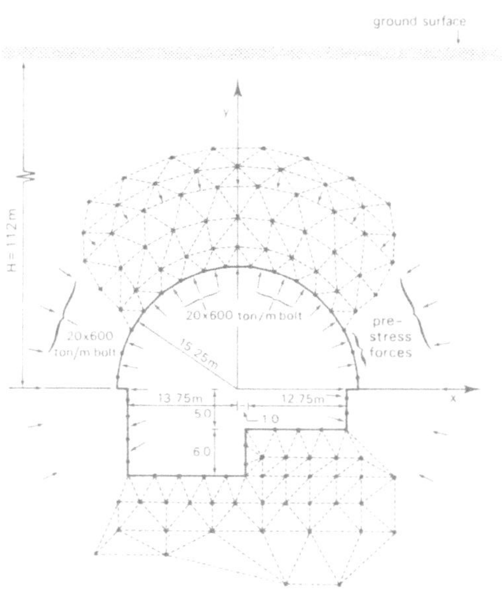

Fig. 13 **Boundary discretization, internal cells and prestressing forces at the deep tunnel**

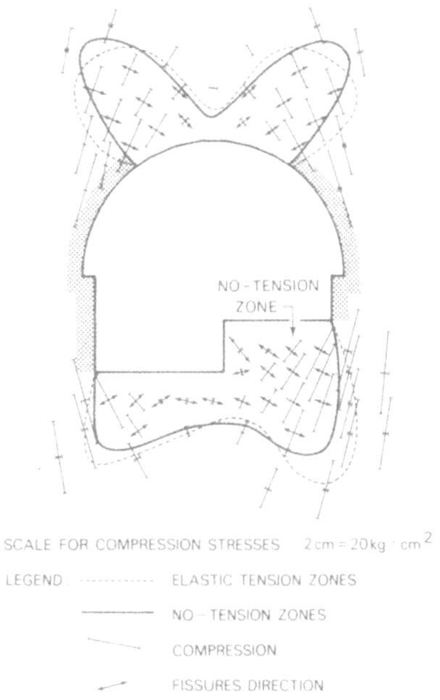

Fig. 14 **Elastic and No-tension results. Tunnel without prestressing forces**

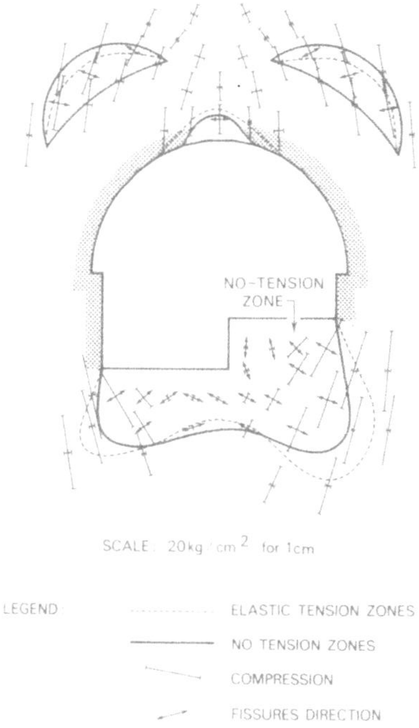

Fig. 15 **Prestressed tunnel. Elastic and No-tension results**

Notice the large no-tension zone present over the roof which requires some additional structural component or prestressing to avoid the rock ralling down. Prestressing using rock bolts has recently been extensively used in tunnels in place of lining and this technique has been adopted for the present example.

The boundary discretization for the analysis of the case with prestressing is the same as it was shown in fig. 13, but the internal cell configuration has been changed to take into account the expected new tension free zones. The prestressing forces shown in the same figure are applied in 20 boundary points and 20 internal points which are distributed along a circular path.

The results obtained with the Boundary Element Method are shown in fig. 15, where the initial elastic zone and the final no-tension zone are plotted. As can be seen, the tension free zones in the roof have been considerably reduced due to prestressing effects. These results are in close agreement with the finite element solution presented in (11).

6. CONCLUSION

In this paper the Boundary Element Method is applied, for the first time, to solve no-tension material problems in tunnels and excavations.

The technique is of considerable advantage for this type of problem as only the boundary of the domain has to be discretized, except where the boundary goes to infinity, for which no discretization is required. Finite elements, on the other hand, require the definition of an artificial outer boundary where displacements must be prescribed, in addition to the discretization of all the domain. The artificial outer boundary is by defininition an approximation, and is a source of inaccuracy which is compounded by the discretization of the domain. Besides, these approximations, the finite elements method requires considerable increase in the amount of data needed to run the problem.

The iterative process used to obtain the no-tension solution proved to be very stable and has a fast convergence. Notice that during the whole process the same system of equations is used and the initial stresses only modify the right hand side vector. Because of the way the right hand side vector is formed, the effect of cell discretization is reduced and the numerical solutions are only slightly affected by the coarseness of the mesh.

This paper also presents, for the first time, the way in which concentrated prestressing forces can be introduced in the Boundary Element Method.

1. REFERENCES

1. BREBBIA, C.A. "The Boundary Element Method for Engineers" Pentech Press, 1978.

2. CRUSE, T.A. "Mathematical Foundations of the Boundary Integral Equation Method in Solid Mechanics" Air Force Office of Scientific Research, Report No. AFOSR-7R-1002, 1977.

3. CRUSE, T.A. "Boundary Integral Equation Method in Solid Mechanics", Report SM-73-17, 1973.

4. LOVE, A.E.H. Treatise on the Mathematical Theory of Elasticity, Dover, 1944.

5. TELLES, J.C.F. and BREBBIA, C.A. "On the Application of the Boundary Element Method to Plasticity" Research Note. Applied Mathematical Modelling, Vol. 3, 1979.

6. TELLES, J.C.F. and BREBBIA, C.A. "The Boundary Element Method in Plasticity" Second International Seminar on Recent Advances in Boundary Element Methods, University of Southampton, March, 1980.

7. BREBBIA, C.A. and CONNOR, J.J. "Fundamentals of Finite Element Techniques for Structural Engineers" Butterworths, (1973).

8. LACHAT, J.O. and WATSON, J.O. "Progress in the use of Boundary Integral Equations illustrated by Examples", Computer method in applied mechanics and engineering 10 (1972) 273-289.

9. ZIENKIEWICZ, O.C., VALLIAPPAN, S. and KING, I.P. "Stress Analysis of Rock as a No-tension Material", Geotechnique, Vol. 18, 56-66, 1968

10. ZIENKIEWICZ, O.C. and VALLIAPPAN, S. "Analysis of Real Structures for Creep, Plasticity and other complex Constitutive Laws" Conference on Materials in Civil Eng., University of Southampton, 1969.

11. VALLIAPPAN, S. "Non-linear Stress Analysis of Two Dimensional Problems with Special Reference to Rock and Soil Mechanics" Thesis submitted to the University of Wales, March 1968, Swansea.

12. BATHE, K.J. "An Assessment of Current Finite Element Analysis of Non-linear Problems in Solid Mechanics" Symposium on "Numerical solution of differential equations III" University of Maryland, 1975.

13. BATHE, K.J., OZPEMIR, H. and WILSON, E.L. "Static and Dynamic Geometric and Material Non-linear Analysis" Report No. UCSESM74-4, Department of Civil Engineering, University of California, Berkeley, February 1974.

14. RABCEWICZ, L.V. "The New Austrian Tunnelling Method" Water Power, No.v, Dec. 1964 and January 1965.

15. RICCARDELLA, P.C. "An Implementation of the Boundary Integral Technique for Planar Problems of Elasticity and Elastoplasticity" Ph.D. Thesis presented to Carnegie-Mellon University 1973.

16. MIKHLIN, S.G. "The Problem of the Minimum of a Quadratic Functional" Translated by A. Feinstein, Holden-day Inc. (1965).

17. F.J. RIZZO "An Integral Equation Approach to Boundary Value Problems of Classical Elastostatics" Quart. Appl. Math. 25, 83-95 (1967).

APPENDIX A SINGULAR INTEGRAL IN THE DOMAIN

The initial stress integrals involve multiplication of initial stress components by the strain fundamental solution. Special care has to be taken when calculating these integrals due to the singular characteristics of some of their terms.

Considering equation (2.6) with only initial stress term, i.e.

$$u_i(p) = \int \varepsilon^*_{imk}(p,q)\sigma^o_{mk}(q)\ d\Omega(q) \tag{A.1}$$

stresses are related to displacement derivatives as follows

$$\sigma_{ij}(p) = \frac{2G\nu}{1-2\nu}\,\delta_{ij}\,\frac{\partial u_\ell(p)}{\partial x\ (p)} + G\left(\frac{\partial u_j(p)}{\partial x_j(p)} + \frac{\partial u_j(p)}{\partial x_j(p)}\right) \tag{A.2}$$

Notice that the initial stresses are being interpolated on the cell, one can write

$$u_i(p) = \left(\int \varepsilon^*_{imk}(p,q)\ \psi^\alpha\ d\Omega(q)\right)\sigma^{o\alpha}_{mk} \tag{A.3}$$

As shown in equation (2.5) the fundamental solution for $\varepsilon^*_{imk}(p,q)$ can be written as

$$\varepsilon^*_{imk}(p,q) = f_{imk}(p,\theta)/r \tag{A.4}$$

Hence any polynomial interpolation function $\psi(r,\theta)$ can be written as

$$\psi^\alpha = \psi^\alpha_o(\theta) + \psi^\alpha_1(\theta)r + \ldots\ \psi^\alpha_n(\theta).r^n \tag{A.5}$$

For the case of linear interpolation functions such as those shown in equations (3.9) one has

$$\psi^\alpha = \xi^\alpha = \xi^\alpha(p) + \frac{1}{2A}\,(b_\alpha\ \cos\theta + a_\alpha\ \sin\theta)r \tag{A.6}$$

Substituting these relationships into A.3 and using Hooke's law, derivatives of the singular integrals arise. The singularity is due to the first term in equation A.6 which will now be considered in more detail by analysing the following integral,

$$D = \frac{\partial}{\partial x_j(p)} \int_\Omega f_{imk}(p,\theta)\ \xi^\alpha(p)/r\ d\Omega \tag{A.7}$$

One can treat this derivatives in two different ways. The first way consists of performing the integral and then the derivatives . Assuming triangular cells as shown in fig. A.1 and considering, for instance, the point p at corner 1, one has

$$\xi^\alpha(p) = 1 \tag{A.8}$$

and the derivatives become

$$D = \frac{\partial}{\partial x_j(p)} \int f_{imk}(p,\theta)\ dr\ d\theta \tag{A.9}$$

Integrating over r one can write

$$D = \frac{\partial}{\partial x_j(p)} \int f_{imk}(p,\theta)\ R(\theta) d\theta \tag{A.10}$$

with

$$R(\theta) = \frac{-2A}{b_1\ \cos\theta + a_1\ \sin\theta} \tag{A.11}$$

Substituting the equation A.11 and the value of $f_{imk}(p,\theta)$ into (A.3) one can carry out the integral and then find the derivatives. The derivatives on the linear term imply no singularities and their values can be computed numerically performing the integration after derivation.

The second way to carry out the value of (A.7) is using the treatment due to Mikhlin[16], in which the derivatives can be introduced into the integral and another term is added, so that the equation is written as

$$D = \int_\Omega \frac{\partial}{\partial x_j(p)} \left(\frac{f_{imk}(p,\theta)}{r}\right) \xi^\alpha(p) d\Omega - \xi^\alpha(p) \int_0^{2\pi} f_{imk}(p,\theta) \cos\theta_j\, d\theta \tag{A.12}$$

where $\cos\theta_j$ is $\cos\theta$ or $\cos(\theta - \frac{\pi}{2})$ for $j = 1$ or $j = 2$ respectively.

In this way the derivatives of all terms can be introduced into the integral and the whole expression for initial stress can be written as follows,

$$\sigma_{ij}(p) = \frac{2G\nu}{1-2\nu}\,\delta_{ij}\left[\int_{\Omega} \frac{\partial}{\partial x_{\ell}(p)}\left\{\varepsilon^{*}_{\ell mk}(p,q)\right\}\psi^{\alpha} d\Omega(p)\right]\sigma^{o\alpha}_{mk} +$$

$$G\left[\int_{\Omega} \frac{\partial}{\partial x_{j}(p)}\left(\varepsilon^{*}_{imk}(p,q)\right)\psi^{\alpha}\, d\Omega(q) + \int \frac{\partial}{\partial x_{i}(p)}\left(\varepsilon^{*}_{jmk}(p,q)\right)\right.$$

$$\left.\psi^{\alpha} d\Omega\,(q)\right]\sigma^{o\alpha}_{mk} - \frac{2G\nu}{1-2\nu}\,\delta_{ij}\left[\xi^{\alpha}(p)\int_{o}^{2\pi} f_{\ell mk}(p,\theta)\cos\theta_{\ell}\right]\sigma^{o\alpha}_{mk}$$

$$- G\,\xi^{\alpha}(p)\left\{\int_{o}^{2\Pi}\left[f_{imk}(p,\theta)\cos\theta_{j} + f_{jmk}(p,\theta)\cos\theta_{i}\right]d\theta\right\}\sigma^{o\alpha}_{mk} \tag{A.13}$$

After carrying out the necessary steps and performing the last three integrals the whole expression for the stresses (2.9) can be written as,

$$\sigma_{ij}(p) = \int_{\Gamma} D_{kij}\, P_{k}\, d\Gamma - \int_{\Gamma} S_{kij}\, u_{k}\, d\Gamma + \int_{\Omega} D_{kij}\, b_{k}\, d\Omega +$$

$$\int_{\Omega} F_{ijmk}\, \sigma^{o}_{mk}\, d\Omega + \frac{1}{8(1-\nu)}\left[(6-8\nu)\sigma^{o}_{ij} - (1-\nu)\sigma^{o}_{kk}\,\delta_{ij}\right]$$

$$- \sigma^{o}_{ij} \tag{A.14}$$

where the term F_{ijmk} was achieved after performing all derivatives of $\varepsilon^{*}(p,q)$ indicated in (A.13). Hence, one can write

$$F_{ijmk} = \frac{1}{8\pi(1-\nu)r^{2}}\left\{2(1-2\nu)\left[\delta_{kj}\delta_{im} + \delta_{ki}\delta_{mj} - \delta_{mk}\delta_{ij} + 2\delta_{ij}r_{,m}r_{,m}\right]\right.$$

$$+ 4\left[\delta_{mk}r_{,k}r_{,j} + \nu(\delta_{nk}r_{,k}r_{,i} + \delta_{jk}r_{,i}r_{,m} + \delta_{im}r_{,k}r_{,j}\right.$$

$$\left.\left. + \delta_{ik}r_{,m}r_{,j})\right] - 16r_{,i}r_{,j}r_{,m}r_{,k}\right\} \tag{A.15}$$

Notice that the integral of F_{ijmk} has to be performed in the sense of Cauchy principle and the correct value can be reached only after all the cells adjacent to the point p have been integrated (see ref. 6).

The value obtained in equation (A.14) after carrying out the last three integrals at (A.13) can also be found by integrating the expression (A.9) before any derivation.

If one removes the singularities and includes the initial stress value, the integral can be written as,

$$I = \frac{\partial}{\partial x_j(p)} \left[\int_{(\Omega-\Omega_\varepsilon)} \left(\frac{f_{imk}(p,\theta)}{r} \right) \sigma^o_{mk}(q)\, d\Omega(p) + \int_{\Omega_\varepsilon} \left(\frac{f_{imk}(p,\theta)}{r} \right) \sigma^o_{mk}(q)\, d\Omega(q) \right] \tag{A.15}$$

where Ω_ε is a circular area whose centre is the singular point.

The derivatives can be passed inside the first integral and one can write

$$I = \int_{\Omega-\Omega_\varepsilon} \frac{\partial}{\partial x_j(p)} (\varepsilon^*_{imk}(p,q))\sigma^o_{mk}(q)\, d\Omega(q) + \frac{\partial}{\partial x_j(p)} \int_{\Omega_\varepsilon} \varepsilon^*_{imk}(p,q)\sigma^o_{mk}(q)\, d\Omega(q) \tag{A.16}$$

or

$$I = \int_{\Omega-\Omega_\varepsilon} F_{ijmk}(p,q)\sigma^o_{mk}(q)\, d\Omega(q) + \frac{\partial}{\partial x_j(p)} \int_{\Omega_\varepsilon} \varepsilon^*_{imk}(p,q)\sigma^o_{mk}(q)\, d\Omega(q) \tag{A.17}$$

The limit when Ω_ε tends to zero can be carried out at the first integral taking into account that it has to be performed in the sense of Cauchy principle value.

Considering the density $\sigma^o_{mk}(q)$ satisfies a Lipschitz condition with positive exponent, in the second term of (A.17) when $\Omega_\varepsilon \rightarrow 0$, one can write,

$$I = \int_{\Omega} F_{ijmk}(p,q)\sigma^o_{mk}(q)\, d\Omega(q) + \sigma^o_{mk}(p) \frac{\partial}{\partial x_j(p)} \int_{\Omega_\varepsilon} \varepsilon^*_{imk}(p,q)\, d\Omega(q)$$

Now the integrals for all components of ε^*_{imk} must be carried out, then the derivatives and the limit when $\Omega_\varepsilon \rightarrow 0$

can be performed. After these steps one can find the same constant values as those given in (A.14).

APPENDIX B CALCULATION OF STRESSES ON THE BOUNDARY

The simplest way to calculate the stresses on the boundary is by differentiating the displacements.

The displacement and traction at any boundary element can be written in a local or global system of coordinates (fig. B.1).

In the local system, fig. (B.1) the tractions and displacements are,

$$\bar{\underset{\sim}{P}} = \underset{\sim}{A}\,\underset{\sim}{P} \tag{B.1}$$

and

$$\bar{\underset{\sim}{U}} = \underset{\sim}{A}\,\underset{\sim}{U} \tag{B.2}$$

where

$$\underset{\sim}{A} = \begin{bmatrix} \sin\theta & -\cos\theta \\ \cos\theta & \sin\theta \end{bmatrix} \tag{B.3}$$

On the boundary one of the relationships between stresses and tractions can be written as,

$$\begin{aligned} \bar{\sigma}_{11} &= \bar{p}_1 = p_1 \sin\theta - p_2 \cos\theta \\ \sigma_{12} &= \bar{p}_2 = p_1 \cos\theta - p_2 \sin\theta \end{aligned} \tag{B.4}$$

The third stress can be obtained by using expressions for strains deduced from the boundary displacements, i.e.

$$\bar{\varepsilon}_{22} = \frac{\partial \bar{u}_2}{\partial \bar{x}_2} = \frac{1}{\ell}\,(\bar{u}_2^2 - \bar{u}_2^1) \tag{B.5}$$

$$\bar{\varepsilon}_{22} = \frac{1}{\ell}\left\{\cos\theta\left[u_1^2 - u_1^1\right] + \sin\theta\left[u_2^2 - u_2^2\right]\right\} \tag{B.6}$$

Using Hooke's law and including initial stresses, one has

$$\sigma_{22} = 2G\,\bar{\varepsilon}_{22} + \frac{\nu}{1-\nu}\,\bar{\sigma}_{11} - \frac{\nu}{1-\nu}\,\sigma^{o}_{11} - \sigma^{o}_{22} \tag{B.7}$$

Once the value of $\underset{\sim}{\sigma}^{o}$ is known in the global system of coordinates, one has to transform these values to the local system to obtain $\bar{\sigma}^{o}$, then obtain $\underset{\sim}{\sigma}$ and finally transform back to the global system.

After all these transformations have been performed one is able to assemble the matrices, $\underset{\sim}{D}$, $\underset{\sim}{S}$ and $\underset{\sim}{F}$ using the p, u and σ^{o} nodal value coefficients respectively.

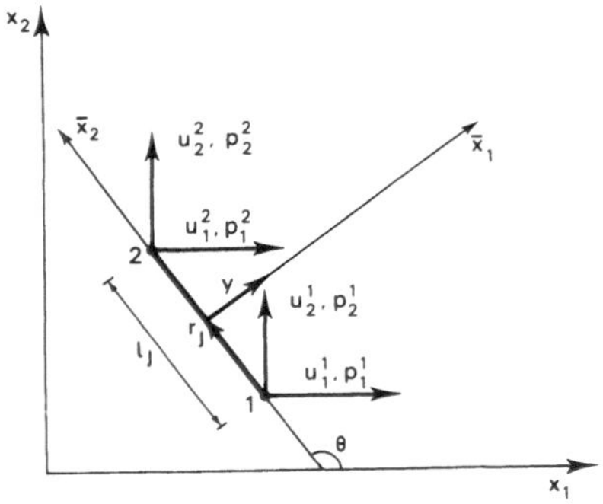

Fig. B (1a) Traction and displacement according to global system of coordinates

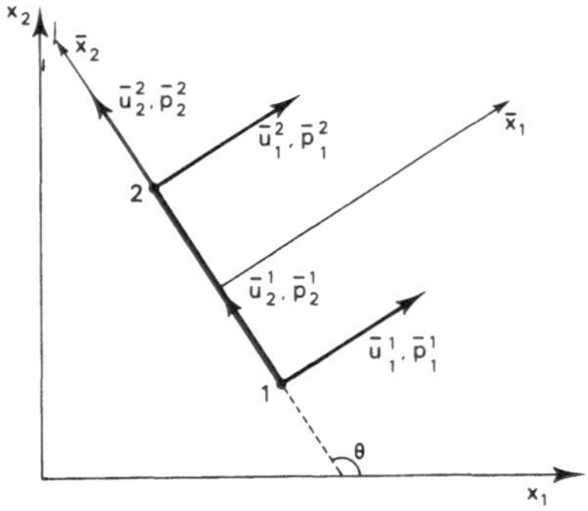

Fig. B (1b) Traction and displacement according to local system of coordinates

Section V

Numerical Techniques and Mathematical Principles

SOME THEORETICAL ASPECTS OF BOUNDARY INTEGRAL EQUATIONS

M. A. Jaswon ,Department of Mathematics,
The City University, London

1. Introduction

Harmonic functions in a domain may be represented as simple-layer or double-layer potentials, generated by hypothetical source density distributions on the boundary which satisfy Fredholm integral equations. An alternative formulation is via Green's formula, in which the boundary data themselves function as source density distributions, so yielding Fredholm integral equations satisfied by the boundary data. The two approaches are theoretically equivalent, and we show how the second provides some insight into the first.

Vector potentials analogous to scalar simple-layer and double-layer potentials were first introduced by Kupradze. These provide a representation of linear elastostatic displacement fields analogous to the representation of harmonic functions by scalar potentials. In consequence, the fundamental boundary-value problems of linear elastostatics may be formulated by vector integral equations analogous to the Fredholm integral equations. From this point of view Somigliana's formula is the exact vector analogue of Green's formula, and it yields vector integral equations analogous to those obtained on the basis of Green's formula. We show how the scalar and vector theories may be unified by means of a suitable symbolism.

2. Scalar potential theory

A continuous distribution of simple sources extending over a closed Liapunov surface ∂B, of surface density $\sigma(\underline{q})$ at $\underline{q} \in \partial B$, generates the simple-layer potential

$$V(\underline{p}) = \int_{\partial B} g(\underline{p},\underline{q})\sigma(\underline{q})dq \qquad (1)$$

at any point $\underline{p}$ in space, where $g(\underline{p},\underline{q}) = |\underline{p}-\underline{q}|^{-1}$ and dq denotes the area element at $\underline{q}$. It is sometimes convenient to refer to $\underline{q}$ as the source point and $\underline{p}$ as the field point. This potential is continuous everywhere and is differentiable everywhere except at ∂B. Also it satisfies Laplace's equation everywhere except at ∂B, and it is therefore a harmonic function everywhere except at ∂B. Its normal derivative at $\underline{p} \in \partial B$ is given by

$$\frac{\partial V(p)}{\partial n} = \int_{\partial B} g'(\underline{p},\underline{q})\sigma(\underline{q})dq - 2\pi\sigma(\underline{p}) \; ; \quad \underline{p} \in \partial B \tag{2}$$

where $g'(\underline{p},\underline{q})$ denotes the normal derivative of g at $\underline{p}$ keeping $\underline{q}$ fixed. However, we preferably write

$$\frac{\partial V}{\partial n_i} = \int_{\partial B} g_i'(\underline{p},\underline{q})\sigma(\underline{q})dq - 2\pi\sigma(\underline{p}) \; ; \; \underline{p} \in \partial B \tag{3}$$

$$\frac{\partial V}{\partial n_e} = \int_{\partial B} g_e'(\underline{p},\underline{q})\sigma(\underline{q})dq - 2\pi\sigma(\underline{p}) \; ; \; \underline{p} \in \partial B \tag{4}$$

where $g_i'(\underline{p},\underline{q})$, $g_e'(\underline{p},\underline{q})$ respectively denote the interior and exterior normal derivatives of g at $\underline{p}$. These are connected by

$$g_i'(\underline{p},\underline{q}) + g_e'(\underline{p},\underline{q}) = 0 \; , \tag{5}$$

from which follows the useful relation

$$\frac{\partial V}{\partial n_i} + \frac{\partial V}{\partial n_e} = -4\pi\sigma \; . \tag{6}$$

These properties of V may be exploited to provide a simple, but efficient, formulation of the classical boundary-value problems of harmonic function theory. First, the Dirichlet problem requires us to construct a harmonic function ϕ inside or outside ∂B given its values (assumed continuous) on ∂B. A straight-forward procedure is to write

$$\phi(\underline{p}) = \int_{\partial B} g(\underline{p},\underline{q})\sigma(\underline{q})dq \; ; \quad \underline{p} \in B_i \tag{7}$$

$$\phi(\underline{p}) = \int_{\partial B} g(\underline{p},\underline{q})\sigma(\underline{q})dq \; ; \quad \underline{p} \in B_e \tag{8}$$

where σ appears as a *hypothetical* simple source density on ∂B, B_i denotes the interior domain bounded by ∂B and B_e denotes the infinite exterior domain bounded internally by ∂B. Now both the right-hand side and left-hand side of (7), (8) are continuous at ∂B, and therefore each of them yields the boundary relation

$$\phi(\underline{p}) = \int_{\partial B} g(\underline{p},\underline{q})\sigma(\underline{q})dq \;;\quad \underline{p} \in \partial B \;, \tag{9}$$

which is a Fredholm integral equation of the first kind for σ in terms of ϕ on ∂B. This has a unique solution, which may be utilised to generate ϕ in B_i, or in B_e subject to the regular behaviour

$$\phi = |\underline{p}|^{-1} \int_{\partial B} \sigma(\underline{q})dq + 0|\underline{p}|^{-2} \quad \text{as } |\underline{p}| \to \infty \;. \tag{10}$$

The most important example of the Dirichlet problem is $\phi = 1$ on ∂B, which defines the capacitance problem of electrostatics. In this case we replace (7) by

$$1 = \int_{\partial B} g(\underline{p},\underline{q})\lambda(\underline{q})dq \quad ; \quad \underline{p} \in \partial B \;, \tag{11}$$

where $\lambda > 0$ and $\int_{\partial B} \lambda(\underline{q})dq$ measures the electrostatic capacitance of ∂B.

The Neumann problem requires us to construct a harmonic function inside or outside ∂B given its normal derivatives (assumed continuous) on ∂B. As before we write ϕ in the form (7) or (8), from which there follows

$$\phi_i'(\underline{p}) = \int_{\partial B} g_i'(\underline{p},\underline{q})\sigma(\underline{q})dq - 2\pi\sigma(\underline{p}) \quad ; \quad \underline{p} \in \partial B \;, \tag{12}$$

$$\phi_e'(\underline{p}) = \int_{\partial B} g_e'(\underline{p},\underline{q})\sigma(\underline{q})dq - 2\pi\sigma(\underline{p}) \quad ; \quad \underline{p} \in \partial B \;, \tag{13}$$

where ϕ_i' or ϕ_e' is given on ∂B. These are Fredholm integral equations of the second kind for σ in terms of ϕ_i' or ϕ_e', and equation (13) has a unique solution which generates a regular ϕ in B_e. Equation (12) only has a solution if

$$\int_{\partial B} \phi_i'(\underline{p})dp = 0 \;, \tag{14}$$

i.e. the Gauss condition on harmonic functions, in which case

$$\sigma = \sigma_0 + k\lambda \tag{15}$$

where ϕ_e is a particular solution and k is an arbitrary constant. This non-uniqueness stems essentially from the fact that $\phi = 1$ on ∂B implies $\phi = 1$ in B_i, which implies $\phi_e' = 0$ on ∂B, which implies that λ in (10) satisfies the homogeneous equation

$$0 = \int_{\partial B} g_i'(\underline{p},\underline{q})\lambda(\underline{q})dq - 2\pi\lambda(\underline{p}) \quad ; \quad \underline{p} \in \partial B . \qquad (16)$$

Accordingly the capacitance problem may also be formulated as a Neumann problem, since the non-trivial solution of (16) may be normalised to satisfy (11) and then utilised to compute

$$\int_{\partial B} \lambda(\underline{q})dq .$$

Fredholm integral equations of the first kind generally present considerable theoretical and numerical difficulties, though equation (10) has been successfully exploited to compute the capacitance of various closed surfaces. A classically preferred formulation of the Dirichlet problem is to represent ϕ as a double-layer potential, generated by a *hypothetical* double source density on ∂B, i.e. we write

$$\phi(\underline{p}) = \int_{\partial B} g(\underline{p},\underline{q})_i' \;\mu(\underline{q})dq \quad ; \quad \underline{p} \in B_i , \qquad (17)$$

$$\phi(\underline{p}) = \int_{\partial B} g(\underline{p},\underline{q})_i' \;\mu(\underline{q})dq \quad ; \quad \underline{p} \in B_e . \qquad (18)$$

where μ is the hypothetical double source density and $g(\underline{p},\underline{q})_i'$ denotes the inward normal derivative of $g(\underline{p},\underline{q})$ at $\underline{q}$ keeping $\underline{p}$ fixed. Here $g(\underline{p},\underline{q})_i'$ functions as the dipole potential at $\underline{p}$ generated by a unit dipole source at $\underline{q}$, this interpretation being implied by the fact that the normal derivative operation moves with the source point $\underline{q}$, by contrast with $g_i'(\underline{p},\underline{q})$ in (12) where the normal derivative operation is taken at the field point $\underline{p}$. It would of course be perfectly acceptable to replace $g(\underline{p},\underline{q})_i'$ by $g(\underline{p},\underline{q})_e'$ in (17) and (18). Now the right-hand sides of (17),(18) jump by amounts $-2\pi\mu(\underline{p})$, $2\pi\mu(\underline{p})$ at $\underline{p} \in \partial B$, whilst ϕ remains continuous, so yielding the boundary relations

$$\phi(\underline{p}) = \int_{\partial B} g(\underline{p},\underline{q})_i' \;\mu(\underline{q})dq + 2\pi\mu(\underline{p}) \quad ; \quad \underline{p} \in \partial B \qquad (19)$$

$$\phi(\underline{p}) = \int_{\partial B} g(\underline{p},\underline{q})_i' \;\mu(\underline{q})dq - 2\pi\mu(\underline{p}) \quad ; \quad \underline{p} \in \partial B \qquad (20)$$

respectively. These are Fredholm integral equations of the second kind for μ in terms of ϕ on ∂B, and equation (19) has a unique solution which generates ϕ in B_i by virtue of (17). Equation (20) only has a solution if

$$\int_{\partial B} \phi(\underline{p})\lambda(\underline{p})dp = 0. \qquad (21)$$

in which case

$$\mu = \mu_0 + k \qquad (22)$$

where μ_0 is a particular solution and h is an arbitrary constant. Condition (21) essentially arises from the fact that the representation (18) implies

$$\phi = 0|\underline{p}|^{-2} \quad \text{as} \quad (\underline{p}) \to \infty \quad . \qquad (23)$$

whereas the classical existence theorem envisages the more general behaviour (10). Bearing in mind that

$$\int_{\partial B} \phi(\underline{p})\lambda(\underline{p})dp = \int_{\partial B} \sigma(\underline{q})dq \quad , \qquad (24)$$

we see that condition (21) eliminates the leading term of (10) and converts it into (23). As regards (22), k satisfies the homogeneous equation

$$0 = \int_{\partial B} g(\underline{p},\underline{q})_i' \, k \, dq - 2\pi k \quad ; \quad \underline{p} \in \partial B \, , \qquad (25)$$

which is a disguised form of the Gauss flux theorem, more usually expressed as

$$\begin{aligned} \int_{\partial B} g(\underline{p},\underline{q})_i' dq &= 4\pi \quad ; \quad \underline{p} \in B_i \\ &= 2\pi \quad ; \quad \underline{p} \in \partial B \\ &= 0 \quad ; \quad \underline{p} \in B_e \quad . \end{aligned} \qquad (26)$$

3. Green's formula

All the above formulations are indirect in that they introduce hypothetical source densities on the boundary. However, there also exist direct formulations in which the boundary data themselves function as source densities. Thus, given a harmonic function ϕ in B_i which assumes continuous values $\phi(\underline{q})$ and normal derivatives $\phi_i'(\underline{q})$ at each $\underline{q} \in \partial B$, we have identically

$$\int_{\partial B} g(\underline{p},\underline{q})'_i \, \phi(\underline{q}) dq - \int_{\partial B} g(\underline{p},\underline{q}) \, \phi'_i(\underline{q}) dq = 4\pi\phi(\underline{p}) \; ; \; \underline{p} \in B_i . \quad (27)$$

This is Green's formula. It exhibits ϕ in B_i as the superposition of a double-layer potential, generated by a source density $\phi/4\pi$ on ∂B and of a simple-layer potential generated by a source density $-\phi'_i/4\pi$ on ∂B. However, according to the fundamental existence theorems, either ϕ alone or ϕ'_i alone essentially suffice to determine ϕ in B_1 . Thus, the application of Green's formula requires more boundary information than would be available in any well posed boundary-value problem. An effective way forward is to note that the second integral in (27) remains continuous at ∂B, whilst the first integral jumps by an amount $-2\pi(\phi/4\pi)$, i.e. $-\phi/2$, at ∂B so yielding the boundary formula

$$\int_{\partial B} g(\underline{p},\underline{q})'_i \phi(\underline{q}) dq - \int_{\partial B} g(\underline{p},\underline{q}) \phi'_i(\underline{q}) dq = 2\pi\phi(\underline{p}) ; \underline{p} \in \partial B . \quad (28)$$

This differs fundamentally from (27) in that it provides a functional relation between ϕ , ϕ'_i on ∂B which ensures their compatibility as boundary data. If, for instance, values of ϕ and ϕ'_i are separately prescribed on ∂B, they could not be utilised in (27) unless (28) were satisfied. Well posed boundary conditions provide partial information which may be extended by coupling it with (28). Some examples will now be outlined.

Given ϕ on ∂B (Dirichlet problem), relation (28) becomes a Fredholm integral equation of the first kind for ϕ'_i in terms of ϕ . This has a unique solution, which automatically satisfies the Gauss condition (14). Conversely, given ϕ'_i on ∂B (Neumann problem), relation (28) becomes an integral equation of the second kind for ϕ in terms of ϕ'_i . This only has a solution if ϕ'_i satisfies (14), in which case

$$\phi = \phi_0 + k \quad (29)$$

as in (22). The Dirichlet and Neumann conditions are examples of a prescribed linear relation

$$\alpha\phi + \beta\phi'_i = \gamma \quad (30)$$

at each point of ∂B, i.e. $\alpha = 1$, $\beta = 0$ and $\alpha = 0$, $\beta = 1$ respectively, where γ is a given continuous function on ∂B. An existence-uniqueness theorem is available for the Robin problem of heat conduction, i.e.

$$\alpha < 0, \; \beta = 1, \; \gamma \text{ given on } \partial B. \quad (31)$$

Finally, an existence-uniqueness theorem is available for the difficult mixed boundary conditions

$$\left.\begin{array}{l}\alpha = 1,\ \beta = 0 \ \text{ on } \partial B_1 \quad \text{and} \quad \alpha = 0,\ \beta = 1 \ \text{ on } \partial B_2 \\ \partial B = \partial B_1 + \partial B_2\ ,\ \gamma \ \text{given on } \partial B\end{array}\right\} \quad (32)$$

Essentially, in all these cases, we couple the local relation (30) with the global relation (28) to compute a compatible ϕ, ϕ_i' on ∂B. From the numerical point of view, hardly any distinction needs to be made between one type of boundary condition and another.

To each interior problem there corresponds an exterior problem subject to regular behaviour at infinity. Thus, Green's exterior formula is

$$\int_{\partial B} g(\underline{p},\underline{q})_e' \phi(\underline{q})dq - \int_{\partial B} g(\underline{p},\underline{q})\phi_e'(\underline{q})dq = 4\pi\phi(\underline{p}) \ ;\ \underline{p}\in B_e, \quad (33)$$

which yields the exterior boundary formula

$$\int_{\partial B} g(\underline{p},\underline{q})_e' \phi(\underline{q})dq - \int_{\partial B} g(\underline{p},\underline{q})\phi_e'(\underline{q})dq = 2\pi\phi(\underline{p}) \ ;\ \underline{p}\in \partial B\ . \quad (34)$$

Given ϕ on ∂B (exterior Dirichlet problem), relation (34) becomes a Fredholm integral equation of the first kind for ϕ_e' in terms of ϕ. This has a unique solution, which does not necessarily satisfy (14). Conversely, given ϕ_e' on ∂B (exterior Neumann problem), this becomes a Fredholm integral equation of the second kind for ϕ which always has a unique solution. With a compatible ϕ, ϕ_e' known on ∂B, we may generate ϕ in B_e via (33). The exterior problems corresponding to (31), (32) present no special analytical difficulties compared with the interior problems.

Nothing in potential theory is more fundamental than Green's formula, and we now prove that it covers the indirect formulations mentioned in section 2. First, relation (28) yields

$$\int_{\partial B} g(\underline{p},\underline{q})_i' \ \phi(\underline{q})dq - \int_{\partial B} g(\underline{p},\underline{q}) \ \phi_i'(\underline{q})dq = 0 \ ;\ \underline{p}\in B_e \quad (35)$$

as $\underline{p}$ crosses from ∂B into B_e. Similarly, relation (34) yields

$$\int_{\partial B} g(\underline{p},\underline{q})_e' \phi(\underline{q})dq - \int_{\partial B} g(\underline{p},\underline{q})\phi_i'(\underline{q})dq = 0 \ ;\ \underline{p}\in B_i \quad (36)$$

as ∂B crosses from ∂B into B_i. Changing ϕ to f in (36), where f is an arbitrary regular harmonic function in B_e, and superposing this on (27), we obtain the more general continuation formula

$$\int_{\partial B} g(\underline{p},\underline{q})'_i[\phi(\underline{q})-f(\underline{q})]dq - \int_{\partial B} g(\underline{p},\underline{q})[\phi'_i(\underline{q})+f'_e(\underline{q})]dq=4\pi\phi(\underline{p});\underline{p}\in B_e. \quad (37)$$

Two distinct possibilities for f may now be considered :

1. $f = \phi$ on ∂B, providing the representation

$$-\tfrac{1}{4}\pi \int_{\partial B} g(\underline{p},\underline{q})[\phi'_i(\underline{q}) + f'_e(\underline{q})]dq = \phi(\underline{p}) \; ; \; \underline{p}\in B_i, \quad (38)$$

which may be identified as a simple-layer potential generated by the source density

$$\sigma = -\tfrac{1}{4}\pi(\phi'_i + f'_e) \text{ on } \partial B. \quad (39)$$

This possibility hinges upon the existence of a unique regular f on B_e satisfying $f = \phi$ on ∂B, as is in fact ensured by the exterior Dirichlet existence theorem. Accordingly we have recovered the formulation (7) of Dirichlet problems by an adaption of Green's formula, and we have also shown that the integral equation (9) has a unique solution given by (39).

2. $f'_e = -\phi'_i$ on ∂B, providing the representation

$$\tfrac{1}{4}\pi \int_{\partial B} g(\underline{p},\underline{q})'[\phi(\underline{q}) - f(\underline{q})]dq = \phi(\underline{p}) \; ; \; p\in B_i \quad (40)$$

which may be identified as a double-layer potential generated by the source density

$$\mu = \tfrac{1}{4}\pi\,(\phi-f) \text{ on } \partial B. \quad (41)$$

This possibility hinges upon the existence of a unique regular f in B_e satisfying $f'_e = -\phi_i$ on ∂B, as is in fact ensured by the exterior Neumann existence theorem. Accordingly we have recovered the formulation (17) of Dirichlet problems by an adaptation of Green's formula, and we have also shown that the integral equation (19) has a unique solution given by (41).

4. Vector potential theory

Classical linear elastostatics may be formulated by a vector potential theory which closely parallels scalar potential theory (Kupradze, 1965). Thus the elastostatic displacement vector parallels the scalar harmonic function. The traction vector parallels the normal derivative. Well known global properties of the traction parallel the Gauss flow theorem. Vector potentials may be constructed which closely parallel the scalar simple - and double - layer potentials. Vector boundary integral equations parallel the scalar boundary integral equations.

Corresponding to $g(\underline{p},\underline{q})$ we introduce the fundamental displacement dyadic of the medium

$$\underline{\underline{g}}(\underline{p},\underline{q}) = \begin{bmatrix} g(\underline{p}_1\underline{q}_1) & g(\underline{p}_1\underline{q}_2) & g(\underline{p}_1\underline{q}_3) \\ g(\underline{p}_2\underline{q}_1) & g(\underline{p}_2\underline{q}_2) & g(\underline{p}_2\underline{q}_3) \\ g(\underline{p}_3\underline{q}_1) & g(\underline{p}_3\underline{q}_2) & g(\underline{p}_3\underline{q}_3) \end{bmatrix} . \quad (42)$$

Here $g(\underline{p}_\alpha\underline{q}_\eta)$ signifies the displacementin the α-direction at $\underline{p}$ generated by a unit point force active in the y-direction at $\underline{q}$, i.e. column 1 defines the displacement vector at $\underline{p}$ generated by a unit point force in the 1-direction at $\underline{q}$, etc. By virtue of the equality

$$g(\underline{p}_\alpha\underline{q}_\eta) = g(\underline{q}_\eta\ \underline{p}_\alpha)$$

$g(\underline{p}_\alpha\underline{q}_\eta)$ may also be interpreted as the displacement in the η-direction at $\underline{q}$ generated by a unit point force in the α-direction at $\underline{p}$, e.g. row 1 gives the displacement vector at $\underline{q}$ generated by a unit point force in the 1-direction at $\underline{p}$. Corresponding to $g'(\underline{p},\underline{q})$ we have the fundamental traction dyadic of the medium

$$\underline{\underline{g}}^*(\underline{p},\underline{q}) = \begin{bmatrix} g^*(\underline{p}_1\underline{q}_1) & g^*(\underline{p}_1\underline{q}_2) & g^*(\underline{p}_1\underline{q}_3) \\ g^*(\underline{p}_2\underline{q}_1) & g^*(\underline{p}_2\underline{q}_2) & g^*(\underline{p}_2\underline{q}_3) \\ g^*(\underline{p}_3\underline{q}_1) & g^*(\underline{p}_3\underline{q}_2) & g^*(\underline{p}_3\underline{q}_3) \end{bmatrix} , \quad (43)$$

where column 1 defines the traction vector at $\underline{p}$ generated by a unit point force acting in the 1-direction at $\underline{q}$, etc. Similarly the traction dyadic

$$\underline{\underline{g}}(\underline{p},\underline{q})^* = \begin{bmatrix} g(\underline{p}_1\underline{q}_1)^* & g(\underline{p}_1\underline{q}_2)^* & g(\underline{p}_1\underline{q}_3)^* \\ g(\underline{p}_2\underline{q}_1)^* & g(\underline{p}_2\underline{q}_2)^* & g(\underline{p}_2\underline{q}_3)^* \\ g(\underline{p}_3\underline{q}_1)^* & g(\underline{p}_3\underline{q}_2)^* & g(\underline{p}_3\underline{q}_3)^* \end{bmatrix}, \quad (44)$$

parallels $g(\underline{p},\underline{q})'$. Each column of (44) defines a singular elastostatic displacement vector generated by a unit traction source at $\underline{q}$ e.g. column 1 refers to a unit traction source acting in the 1-direction at $\underline{q}$, in line with the fact that $g(\underline{p},\underline{q})'$ may function as a unit dipole potential. Clearly each row of (43) defines a singular elastostatic displacement vector generated by a unit traction source at $\underline{p}$, and by the same token each row of (44) defines a traction vector generated by a unit point force at $\underline{p}$. Any individual component of the dyadics (43),(44) carries two possible interpretations, i.e. it is the traction component generated by a unit point force or it is the displacement component generated by a unit traction source. However, the interpretation will always be clear from the context. Finally we note that $g^*(\underline{p},\underline{q})$ stands for $g_i^*(\underline{p},\underline{q})$ or $g_e^*(\underline{p},\underline{q})$ as the case may be, and similarly for $g(\underline{p},\underline{q})^*$.

Corresponding to σ we introduce the vector simple-source density $\underline{\sigma}$, which enables us to write down the vector integral equation

$$\underline{\phi}(\underline{p}) = \int_{\partial B} \underline{\underline{g}}(\underline{p},\underline{q}) \cdot \underline{\sigma}(\underline{q})dq \quad ; \quad \underline{p} \in \partial B \quad (45)$$

analogous to (q), with components

$$\phi_\alpha(\underline{p}) = \int_{\partial B} \sum_{\eta=1}^{3} g(\underline{p}_\alpha \underline{q}_\eta)\, \sigma_\eta(\underline{q})dq \quad ; \quad \begin{matrix}\underline{p} \in \partial B \\ \alpha = 1,2,3\end{matrix} \quad (46)$$

where $\underline{\phi}$ is a prescribed displacement on ∂B. By analogy with (39), this has the unique solution

$$\underline{\sigma} = -\tfrac{1}{4\pi}\,(\underline{\phi}^*_i + \underset{\sim}{f}{}^*_e) \;, \quad (47)$$

where $\underline{\phi}^*$ signifies the traction vector associated with $\underline{\phi}$ in B_i and $\underset{\sim}{f}^*$ signifies the traction vector associated with $\underset{\sim}{f}$ in B_e ; here $\underline{f}$ is the unique regular displacement vector in B_e defined by $\underline{f} = \underline{\phi}$ on ∂B. Also, corresponding with the electrostatic equation (11), we have the vector equation

$$\underset{\sim}{a} + \underline{b} \wedge \underline{p} = \int_{\partial B} \underline{\underline{g}}(\underline{p},\underline{q}) \cdot \underline{\lambda}(\underline{q})dq \quad ; \quad \underline{p} \in \partial B \quad (48)$$

where $\underline{a},\underline{b}$ are arbitrary constant vectors defining an arbitrary rigid-body displacement field. It is convenient to break this down into the six independent vectors

$$\left.\begin{array}{llll} \underline{\mu}_1 = \langle 1,0,0\rangle , & \underline{\mu}_2 = \langle 0,1,0\rangle , & \underline{\mu}_3 = \langle 0,0,1\rangle \\ \underline{\mu}_4 = \langle 1,0,0\rangle \wedge \underline{p}, & \underline{\mu}_5 = \langle 0,1,0\rangle \wedge \underline{p}, & \underline{\mu}_6 = \langle 0,0,1\rangle \wedge \underline{p} \end{array}\right\} \quad (49)$$

which break down (48) into the six independent equations

$$\underline{\mu}_s(\underline{p}) = \int_{\partial B} \underline{\underline{g}}\,(\underline{p},\underline{q}) \cdot \underline{\lambda}_s(\underline{q})dq \quad ; \quad \begin{array}{l}\underline{p} \in \partial B \\ s = 1,2,\ldots,6.\end{array} \quad (50)$$

The solutions $\underline{\lambda}_s$ play an analogous role in the vector theory to that of the electrostatic source density λ .

Corresponding with (12) we have the vector integral equation

$$\underline{\phi}_i^*(\underline{p}) = \int_{\partial B} \underline{\underline{g}}_i^*(\underline{p},\underline{q}) \cdot \underline{\sigma}(\underline{q})dq - 2\pi\, \underline{\sigma}(\underline{p}) ; \; \underline{p} \in \partial B, \quad (51)$$

where the traction vector $\underline{\phi}_i^*$ is given on ∂B. This equation only has a solution if

$$\int_{\partial B} \underline{\phi}_i^*(\underline{p}) \cdot \underline{\mu}_s(\underline{p})dp = 0 \quad ; \quad s = 1,2,\ldots,6 \quad (52)$$

i.e. the tractions must be self-equilibrated with respect to both force and moment resultants on ∂B. Clearly (52) is analogous to (14). If (52) holds, then

$$\underline{\sigma} = \underline{\sigma}_o + k_s\underline{\lambda}_s \quad ; \quad s = 1,2,\ldots,6 \quad (53)$$

where $\underline{\sigma}_o$ is a particular solution and the k_s are arbitrary coefficients. This non-uniqueness stems essentially from the fact that $\underline{\phi} = \underline{\mu}_s$ on ∂B implies $\underline{\phi} = \underline{\mu}_s$ in B_i, which implies $\underline{\phi}^* = 0$ on ∂B, which implies that $\underline{\lambda}_s$ in (50) satisfy the homogeneous equations

$$0 = \int_{\partial B} \underline{\underline{g}}_i^*\,(\underline{p},\underline{q}) \cdot \underline{\lambda}_s(\underline{q})dq - 2\pi\underline{\lambda}_s(\underline{p}) \; ; \quad \begin{array}{l}\underline{p} \in \partial B \\ s = 1,2,\ldots,6\end{array} \quad (54)$$

analogous to (16). Similarly, corresponding with (20), we have the vector integral equation

$$\underline{\phi}(\underline{p}) = \int_{\partial B} \underline{\underline{g}}(\underline{p},\underline{q})_i^* \cdot \underline{\mu}(\underline{q})dq - 2\pi\underline{\mu}(\underline{p}) \; ; \; \underline{p} \in \partial B \quad (55)$$

which only has a solution if

$$\int_{\partial B} \underline{\phi}(\underline{p}) \cdot \underline{\lambda}_s(\underline{p}) dp = \int_{\partial B} \underline{\sigma}(\underline{q}) \cdot \underline{\mu}_s(q) dq = 0 , \qquad (56)$$

in which case

$$\underline{\mu} = \underline{\mu}_0 + k_s \underline{\mu}_s \quad ; \quad s = 1,2,\ldots,6 \qquad (57)$$

where $\underline{\mu}_0$ is a particular solution and the k_s are arbitrary coefficients. The homogeneous equations

$$0 = \int_{\partial B} \underline{\underline{g}}(\underline{p},\underline{q})^*_i \cdot \mu_s(\underline{q}) dq - 2\pi \underline{\mu}_s(\underline{p}) \; ; \quad \begin{matrix} \underline{p} \in \partial B \\ s = 1,2,\ldots,6 \end{matrix} \qquad (58)$$

are essentially vector generalisations of the Gauss flux theorem, alternatively expressed by

$$\begin{aligned} \int_{\partial B} \underline{\underline{g}}(\underline{p},\underline{q})^*_i \cdot \underline{\mu}_s(\underline{q}) dq &= 4\pi \underline{\mu}_s(\underline{p}) \quad ; \quad \underline{p} \in B_i \\ &= 2\pi \underline{\mu}_s(\underline{p}) \quad ; \quad \underline{p} \in \partial B \qquad (59) \\ &= 0 \quad ; \quad \underline{p} \in B_e \end{aligned}$$

for $s = 1,2,\ldots,6$. The vector analogues of equations (13), (19) hardly need to be written down explicitly, and the analogue of (19) has a unique solution given by

$$\underline{\mu} = \tfrac{1}{4}\pi(\underline{\phi} - \underline{f}) , \qquad (60)$$

where $\underline{f}$ signifies the unique regular exterior displacement field on ∂B defined by $\underset{\sim}{f}{}^*_e = -\underline{\phi}^*_i$ on ∂B.

5. Somigliana's formula

Corresponding to Green's formula (27), we have Somigliana's formula

$$\int_{\partial B} [\underline{\underline{g}}(\underline{p},\underline{q})^*_i \cdot \underline{\phi}(\underline{q}) - \underline{\underline{g}}(\underline{p},\underline{q}) \cdot \underline{\phi}^*_i(\underline{q})] \, dq = 4\pi \phi(\underline{p}) \; ; \; \underline{p} \in B_i , \qquad (61)$$

which yields the boundary formula

$$\int_{\partial B} [\underline{\underline{g}}(\underline{p},\underline{q})^*_i \cdot \underline{\phi}(\underline{q}) - \underline{\underline{g}}(\underline{p},\underline{q}) \cdot \underline{\phi}^*_i(\underline{q})] dq = 2\pi \underline{\phi}(\underline{p}) \; ; \; \underline{p} \in \partial B. \qquad (62)$$

This provides a functional relation between tractions and displacements on ∂B which ensures their compatibility

as boundary data. Given $\underline{\phi}$ on ∂B, this becomes a vector integral equation for $\underline{\phi}_i^*$ on ∂B, with a unique solution which automatically satisfies (52). Conversely, given $\underline{\phi}_i^*$ on ∂B, this becomes a vector integral for $\underline{\phi}$ on ∂B, which only has a solution if (52) holds.
In this case

$$\underline{\phi} = \underline{\phi}_o + k_s \underline{\mu}_s \quad ; \quad s = 1,2,\ldots,6 \qquad (63)$$

as in (57). There are, of course, corresponding exterior equations, and reciprocal relations analogous to (35), (36). Integral equations generated from Somigliana's formula were formulated and solved numerically by Rizzo, 1967.

List of References

Jaswon, M.A. & Symm, G.T. (1977). Integral Equation Methods in Potential Theory and Elastostatics. Academic Press : London & New York.

Kupradze, V.D. (1965). Potential Methods in the Theory of Elasticity. Israel Program for Scientific Translations : Jerusalem.

Rizzo, F.J. (1967). An integral equation approach to boundary value problems of classical elastostatics. Quart. App. Math. 25 (1), 83-95.

ON THE ASYMPTOTIC CONVERGENCE OF BOUNDARY INTEGRAL METHODS

W.L. WENDLAND

TECHNISCHE HOCHSCHULE DARMSTADT, GERMANY
and UNIVERSITY OF DELAWARE, NEWARK, DEL. 19711 U.S.A.

INTRODUCTION

The numerical treatment and corresponding error analysis of boundary integral equations hinges on the type of discretization due to the shape and type of trial functions used for the approximation of the unknown functions, due to the type of test functionals replacing the integral equations - which hold everywhere on the boundary - by a finite number of equations and due to the numerical integration. In reality further errors accumulate from round off effects. Here we are concerned with an error analysis which only takes into account the effects of the first three kinds. Since it seems to be a too pretentious task to find computable error bounds we consider so called asymptotic estimates.

For the trial functions we use finite element functions on the boundary manifold, i.e. boundary elements. We consider the boundary manifold to be given by local representations and regular partitions of the parameter domain into subintervals or triangles, such that they are mapped onto corresponding partitions of the boundary curve or surface Γ , respectively. To one partition we assign a meshwidth h , i.e. the longest subinterval or side length of the partitioning triangles. On the partition in the parameter domains we use as trial functions regular finite elements, e.g. piecewise polynomials. Then the local representation of Γ transplants these finite element functions onto Γ . For the error analysis we consider not only one partition but a whole family with corresponding h tending to zero. For this whole family we choose a regular $(m+1,m)$ system in the sense of [6]. Then asymptotic error estimates are formulated in terms of powers of h if $h \to 0$.

Here we present such estimates for the two most popular methods of discretization, the Galerkin procedure and the standard collocation together with corresponding discrete versions due to numerical integrations.

For Galerkin's procedure, the error analysis can be develcped along the lines of finite element methods for variational problems. Theoretically Galerkin's method requires a time consuming double integration over Γ for the computation of every element of the stiffness matrix. Thus we choose the corresponding numerical quadrature formulas with respect to a fast computation, i.e. by evaluating the kernels of the integral operators as seldom as possible. To this end we assume that the principal part of the equation is given by a convolutional integral. After converting the principal part into a standard form, the corresponding weights in the stiffness matrix can be evaluated from two vectors of weights forming a Toeplitz matrix provided the finite elements are defined by shifting and stretching subject to a regular uniform grid. The vector of weights can be evaluated independently of the boundary and independently of the meshwidth. For all remaining smooth terms we use numerical integrations with node points connected with the finite element grid. For $n=2$, i.e. boundary elements on a curve we perform this method in §2. It results in a very simple and effective modified collocation scheme which we call Galerkin-collocation. Numerical experiments with examples from conformal mapping, plate bending and Stokes flows show high accuracy and efficiency of this method.

On the other hand almost all computer programs inplementing the boundary integral method are executing the standard collocation method numerically. Under the same assumption on the principal part as before it can be shown that the convergence of the standard collocation follows from that of the Galerkin procedure. Then the convergence can be transmitted to the discretized versions if the numerical integrations are accurate enough.

In all these cases the asymptotic error estimates show an optimal order t of h^t, i.e. the same order as the best approximation to the actual solution in the sense of $(m+1,m)$ finite element systems. The class of linear boundary integral equations providing these results is characterized by a generalization of the positive definiteness of energy forms, i.e. coerciveness in the sense of the Gårding inequality. In addition we assume that the principal parts are convolutional operators. This class is characterized in §1. It turns out that most problems of the applications belong to this class. The reason is that those boundary integral equations that correspond to boundary value problems are defined via the fundamental solution which depends on the difference between the integration point and the observation point only. In §2 we formulate the Galerkin procedure and the corresponding asymptotic error estimates as well as the Galerkin-collocation. For the latter we present a numerical example from [30]. In §3 we formulate corresponding estimates for the standard collocation.

The whole lecture is essentially extracted from [32] where also the proofs can be found which here are omitted completely. For brevity we consider here only one single equation although in the applications one mostly finds systems of integral equations. The error analysis applied to systems also [32]. The whole analysis can also be extended to the integral equations of mixed boundary value problems which correspond to crack and punch problems [33].

1. STRONGLY ELLIPTIC INTEGRAL EQUATIONS

All boundary element methods can be considered as to be approximations of corresponding boundary integral equations. Conversely, all boundary value problems of mathematical physics with elliptic differential equations can be transformed into equivalent boundary integral equations provided the fundamental solution of the differential operator is available. The boundary integral equations are by no means uniquely determined by the boundary value problem. In general there are several different integral equations being equivalent to the original problem. E.g. the Dirichlet problem for the Laplacian can be solvedwith the classical Fredholm integral equation of the second kind via Gaussians double layer potential [27] or with the Fredholm integral equation of the first kind ("Symm's equation' in two dimensions) via Green's identity and the "direct method" [16]. A third method using complex function theory leads to a singular integral equation with Cauchy's kernel [20]. (For singular integral equations see also [19].)

The Neumann problem for the Laplacian can also be solved with everyone of the above types of boundary integral equations. In addition, one can even formulate a corresponding integro-differential equation with nonintegrable kernel [12].

All these types of equations or corresponding systems appear in applications as in classical potential theory, scattering theory and inviscid incompressible ideal flows in conformal mappings, viscous flows, electrostatics and elastomechanics, in elasticity, thermoelasticity and electromagnetic field theory.

Although these types of equations have very different properties in classical theory of integral equations it turns out that if they are considered as so called <u>pseudodifferential operators</u> [28] they have a very strong, common property. Namely the equations of practical interest are "strongly elliptic". In order to formulate this property one needs the Sobolev spaces $H^s(\Gamma)$ of generalized functions on Γ, their interpolation spaces and their dual spaces. For the definitions we refer to [2] (in particular p. 214). Then each of the above mentioned operators A defines a continuous linear mapping $A : H^s \to H^{s-2\alpha}$ for a whole scale of real s (depending on the smoothness of Γ). 2α is called the <u>order</u> of the pseudodifferential operator A [28]. (G. Richter calls -2α in [26] "smoothing index".) For our examples we have $2\alpha = 0; -1; 1$, respectively. The boundary integral equation we write in short

$$Au = f \quad \text{on} \quad \Gamma . \tag{1.1}$$

The announced common property is the <u>coerciveness</u> in form of the Gårding inequality:

$$\operatorname{Re}(Av,v) = \operatorname{Re}\int_\Gamma vAv\,ds \geq \gamma\|v\|^2_{H^\alpha} - |k[v,v]| \quad \text{for all} \quad v \in H^\alpha(\Gamma) \tag{1.2}$$

where $\gamma > 0$ is a constant independent of v_α and where $k[u,v]$ denotes a compact bilinear form on $H^\alpha \times H^\alpha$. In some cases k equals zero, then Equation (1.2) corresponds to energy estimates.

We further assume that the principal part of A is a convolutional operator and - for brevity - that Γ is a simple closed curve. Let Γ be given by a regular parameter representation

$$\Gamma : x = x(t), \quad t \in [0,1] \tag{1.3}$$

with $x(t)$ a 1-periodic sufficiently smooth vector valued function satisfying

$$\left|\frac{dx}{dt}\right| = \rho(t) \geq \rho_0 > 0 \quad \text{for all } t , \tag{1.4}$$

where ρ denotes the Jacobian. Then the operator A with a convolution operator as principal part has the form

$$Au|_t = \text{p.v.} \int_{|t-\tau|<\frac{1}{2}} [p_1(t-\tau) + \log|t-\tau|p_2(t-\tau)]u(t)\rho(t))dt + \int_{|t-\tau|<\frac{1}{2}} L(\tau,t)(u(t)\rho(t))dt = f(t) . \tag{1.5}$$

Here $p_1(\zeta)$ and $p_2(\zeta)$ for $\zeta \neq 0$ are homogeneous functions of degree $\beta = 1-2\alpha-n$. In this case the coerciveness (Equation (1.2)) is a consequence of the so called <u>strong ellipticity</u> of the operator A which is defined by means of the Fourier transformed kernel function $a_0(\xi)$ of the principal part of A, i.e.

$$a_0(\xi) := F(p_1(\cdot) + \log|\cdot|p_2(\cdot))|_\xi \quad . \tag{1.6}$$

A is called <u>strongly elliptic</u> if there exists a complex valued $\Theta \neq 0$ and a constant γ' such that

$$\operatorname{Re} \Theta a_0(\xi) \geq \gamma' > 0 \quad \text{for all} \quad |\xi| = 1, \quad \xi \in \mathbb{R}, \tag{1.7}$$

without loss of generality we may multiply the Equation (1.1) by Θ then in Equation (1.7) and furtheron it suffices to consider the case $\Theta \equiv 1$.

For singular integral equations with the Cauchy kernel, the above definition and in particular the special form (Equation (1.5)) of A are too restrictive. We leave this detail to [32].

From now on we consider strongly elliptic integral equations of the form of Equation (1.5) and we further assume that the remaining terms collected in $L(\tau,t)$ define a sufficiently smooth function of τ and t. Otherwise we again split into two terms, where the first contains the singularity and has to be treated similarly to the principal part. Then strong ellipticity implies coerciveness (Equation (1.2)) [17].

To our knowledge, all examples of boundary integral equations in applications belong to the class of strongly elliptic integral equations or systems, for references see [11] and [30].

Since in Equation (1.5) only ρ depends on Γ we consider Equation (1.5) as an integral equation over $[0,1]$ for the 1-periodic new unknown function

$$v(t) := \rho(t)u(t) . \tag{1.8}$$

Note that the principal part in Equation (1.5) then becomes independent of the special choice of the curve Γ.

2. THE GALERKIN METHOD AND THE GALERKIN-COLLOCATION

Now let us assume that the linear boundary integral equation

$$Au = f \quad \text{on} \quad \Gamma \tag{2.1}$$

is strongly elliptic and has no eigensolution, i.e. the solution of Equation (2.1) is unique and therefore it is uniquely solvable due to Fredholm's alternative which is valid because of the estimate (1.2). In case of eigensolutions to Equation (2.1) the problem can easily be modified [32].

By $\tilde{H}_h$ let us denote the finite dimensional space of finite element functions on Γ belonging to the meshwidth $h > 0$ and belonging to the regular $(m+1,m)$ system [6]. Let μ_j, $j=0,1,\dots,N$ denote a basis of $\tilde{H}_h$. Then the well known Galerkin procedure for Equation (2.1) is to find the coefficients γ_j of the approximate solution

$$\hat{v}(x) = \sum_{j=0}^{N} \gamma_j \mu_j(x) , \quad x \in \Gamma \tag{2.2}$$

by solving the finite system of linear equations,

$$\sum_{j=0}^{N} (A\mu_j,\mu_k)\gamma_j = (f,\mu_k) , \quad k=0,\dots,N . \tag{2.3}$$

For the convergence of this procedure we have well known results going back to S. Michlin [18], S. Hildebrandt and E. Wienholtz [10]. Here we use the version known as Céa's lemma, [8] p. 104.

Theorem 2.1: Let Equation (2.1) with A be a strongly elliptic equation with unique solution $u \in H^{\alpha}$ to any $f \in H^{-\alpha}$. Then there exists $h_o>0$ such that Equations (2.3) are uniquely solvable for every $0<h\leq h_o$. Moreover there exists a constant c independent of h and f such that

$$\|\hat{v} - u\|_{\alpha} \leq c \inf_{\chi\in \tilde{H}_h} \|u - \chi\|_{\alpha} \tag{2.4}$$

For convenience, in the following asymptotic error analysis we are always using $c, c', \ldots$ as generic constants which might change their size and meaning at different places.

It should be mentioned that Theorem 2.1 is not restricted to our finite element approximations but applies to a rather wide class of Galerkin methods as e.g. for the projection methods using trigonometric polynomials as in [25].

For the proof of Theorem 2.1 we refer to [32]. The proof is based on the coerciveness(estimate (1.2)) which also holds for $v \in \tilde{H}_h$. That means that A can be decomposed as

$$A = D + C \tag{2.5}$$

where D is a positive definite operator and $C : H^{\alpha} \dot{\rightarrow} H^{-\alpha}$ is compact.

Our boundary elements have been defined by the transplantation of a regular (m+1,m) system in the parameter domains onto Γ with local parameter representations of Γ. For calculations, the integrals can be evaluated by using the local coordinates. In those the finite elements appear as simple functions over the parameter domains. This construction of finite elements on Γ requires that the parameter representations are fully available. For the two-dimensional case this is a sensible requirement. In the space, however, the boundary surface has also to be approximated [21], [22]. The regular (m+1,m) systems in the parameter domains in connection with regular representations of the boundary curve provide the approximation property:

Let $-m-1 \leq t \leq s \leq m+1$, $-m \leq s, t \leq m$. Then to any given $u \in H^s(\Gamma)$ and any $\tilde{H}_h$ of our family there exists a finite element function $u_h \in \tilde{H}_h$ such that [7]

$$\|u - u_h\|_{H^t} \leq ch^{s-t}\|u\|_{H^s} \tag{2.6}$$

The constant c is independent of u, h and u_h. Moreover, it provides the inverse assumption:

For $-m \leq t \leq s \leq m$ and any $\mu \in \tilde{H}_h$ there holds an estimate

$$\|\mu\|_{H^s} \leq ch^{t-s}\|\mu\|_{H^t} \tag{2.7}$$

where the constant c is independent of μ and h [23]. Using the regular finite elements for Galerkin's method (Equations (2.2), (2.3)) and inserting Formula (2.6) into Formula (2.4) we immediately find optimal order of convergence. The result can easily be improved with the inverse assumption (Formula (2.7)). But with all our assumptions, also the Aubin-Nitsche lemma can be applied and yields super approximation [15]. Collecting all these results we find the following convergence theorem [15],[26].

Theorem 2.2: Let A in Equation (2.1) be strongly elliptic and Equation (2.1) be uniquely solvable. Let $\hat{v}$ denote the Galerkin solution of Equations (2.2), (2.3) subject to regular finite element spaces providing the ineaualities (2.6) and (2.7). Define $\alpha' := \min\{\alpha, 0\}$ and suppose $2\alpha - m - 1 \le t \le s \le m + 1$, $-m \le \alpha \le m$, $\alpha \le s, t \le m$. Then we have the asymptotic error estimate

$$\| u - \hat{v} \|_{H^t} \le c h^{s-t} \| u \|_{H^s} . \tag{2.8}$$

In addition, if we consider the discrete equations (2.3) in $L_2(\Gamma)$ then we find for the conditioning the asymptotic estimate

$$\| \hat{v} \|_{L_2(\Gamma)} \le c h^{2\alpha'} \| f \|_{L_2(\Gamma)} . \tag{2.9}$$

For $t = -m-1+2\alpha$ and $s = m+1$ in Equation (2.8) one finds super approximation of order $h^{2m+2-2\alpha}$ which yields corresponding super approximation of the desired potentials away from Γ.

For the numerical implementation of Galerkin's procedure (Equations (2.3)), the weights

$$a_{jk} := (A\mu_j, \mu_k) , \quad j,k = 0,\dots,N \tag{2.10}$$

have to be evaluated. Since A is given by an integral operator (in the usual or the generalized) sense, the computation of a_{jk} requires a double integration over $\Gamma \times \Gamma$. If this is done numerically, the kernels of the integral operators must be computed at all combinations of grid points on Γ. In addition, special care must be taken of the singular integrals. Therefore we shall adapt numerical integration to the special integrals in Equation (1.5).

The principal part in the standard form (Equation (1.5)) will be handled independently of the special boundary Γ yielding a Toeplitz matrix whose elements are given by a vector. This vector can be computed exactly up to the desired accuracy once for all independent of Γ as well as of h for any fixed type of elements. It should be pointed out that the accuracy of the numerical results depends significantly on how to compute the approximate principal part.

The Galerkin weights due to the smooth remaining parts will be treated numerically by appropriate quadrature formulas depending on the particular finite elements to be used. In them we use only grid points in a regular grid connected with the finite elements such that the kernel functions are to be evaluated as seldom as necessary. This leads to simple modified collocation formulas and the computation of the corresponding stiffness matrix is extremely fast.

In order to utilize the convolution in the principal part we use regular finite elements on a uniform grid of [0,1] defined with shifts and stretched variables from one shape function $\mu(\eta)$. The latter we define as in [5] Chap. 4 by suitable piecewise polynomials of order m with $\mu \in C^{m-1}$. For $m = 0,1,2$ e.g. we have

$\mu(\eta) =$ (2.11)

$m = 0$	$m = 1$	$m = 2$	for
1	η	$\frac{1}{2}\eta^2$	$0 \le \eta < 1$
0	$2 - \eta$	$-\eta^2 + 3\eta - 3/2$	$1 \le \eta < 2$
0	0	$\frac{1}{2}\eta^2 - 3\eta + 9/2$	$2 \le \eta < 3$
0	0	0	elsewhere

With μ we define a basis of $\tilde{H}_h$ by

$$\mu_j(t) := \mu\left(\frac{t}{h} - j\right) \text{ for } hj \le t \le 1+hj,\ j=0,\dots,N,\ h=1/(N+1) \tag{2.12}$$

and their 1-periodic extensions

$$\mu_j(t+\ell) := \mu_j(t) \quad \text{for integer } \ell \,.$$

For v in Equation (1.8) we use the approximation

$$v_h(t) := \sum_{j=0}^{N} \gamma_j \mu_j(t) \,. \tag{2.13}$$

Inserting Equations (2.12) into Equations (2.10) we find for the terms due to the first expression in Equation (1.5)

$$d_{jk} = \int_0^1 \text{p.v.} \int_{|t-\tau| \le \frac{1}{2}} [p_1(t-\tau) + \log|t-\tau|\, p_2(t-\tau)]\mu_j(t)dt\mu_k(\tau)d\tau$$

$$= h^{2+\beta}\{\int_{\tau'=0}^{m+1} \text{p.v.} \int_{t'=0}^{m+1} [p_1(t'-\tau'+(j-k)) + p_2 \cdot \log|t'-\tau'+(j-k)|]$$

$$\cdot\, \mu(t')\mu(\tau')dt'd\tau'$$

$$+ \log h \int_{\tau'=0}^{m+1} \text{p.v.} \int_{t'=0}^{m+1} p_2(t'-\tau'+(j-k))\mu(t')\mu(\tau')dt'd\tau'\} \,,$$

$$d_{jk} = h^{2+\beta}\{W_{1\rho} + W_{2\rho} \log h\} \text{ with } \rho = j-k \in \mathbb{Z} . \tag{2.14}$$

Here the two vectors of weights

$$W_{1\rho} = \int_{\tau'=0}^{m+1} \text{p.v.} \int_{t'=0}^{m+1} [p_1(t'-\tau'+\rho)+p_2 \log|t'-\tau'+\rho|]\mu(t')\mu(\tau')dt'd\tau' \tag{2.15}$$

$$W_{2\rho} = \int_{\tau'=0}^{m+1} \text{p.v.} \int_{t'=0}^{m+1} p_2(t'-\tau'+\rho)\mu(t')\mu(\tau')dt'd\tau', \rho \in \mathbf{Z} \tag{2.16}$$

can be computed once for all independent of Γ and h. For more details see [32] and [11]. For all the remaining smooth terms in the Galerkin equations to Equation (1.5) we use numerical integration.

Since in the corresponding integrals

$$\int_{\text{supp } \mu_j} f(t)\mu_j(t)dt = h \int_{\sigma=0}^{m+1} f(h(j+\sigma))\mu(\sigma)d\sigma \tag{2.17}$$

the finite element functions appear as factors, the numerical integrations are chosen accordingly to the respective reference function μ such that polynomials f up to the order $2M+1$ are <u>integrated exactly</u>. This leads to formulas like

$$\int_{\text{supp } \mu_j} f(t)\mu_j(t)dt = h \sum_{\ell=-M}^{M} b_\ell f(z_{j\ell}) + R \tag{2.18}$$

where

$$z_k := z(h(k+\frac{m+1}{2})) \text{ and } z_{j\ell} := z(h(j+\frac{m+1}{2} + \ell\cdot\gamma_\ell)) \tag{2.19}$$

are the gridpoints subject to the boundary elements and, correspondingly subject to the integration formula. R denotes the error term which is of order $2M+2$. The simplest choice $\gamma_\ell = 1$ yields $z_{j\ell} = z_{j+\ell}$ and weights $b_\ell = b_{-\ell}$ as follows:

	m = 0		m = 1		m = 2	
	b_o	b_1	b_o	b_1	b_o	b_1
M = 0 :	1	0	1	0	$\frac{1}{3}$	0
M = 1 :	$\frac{11}{12}$	$\frac{1}{24}$	$\frac{5}{6}$	$\frac{1}{12}$	$\frac{3}{4}$	$\frac{1}{8}$

.

For $\gamma_\ell = \frac{1}{2}$ and $M = 2$ one has

m = 1			m = 2		
b_o	b_1	b_2	b_o	b_1	b_2
$\frac{13}{30}$	$\frac{4}{15}$	$\frac{1}{60}$	$\frac{2}{5}$	$\frac{7}{30}$	$\frac{1}{15}$

.

Using Equation (2.18) for the smooth terms of the weights in Equation (1.5) we obtain

$$\int_{\tau=0}^{1} \int_{|\tau-t|\le\frac{1}{2}} L(\tau,t)\mu_j(t)dt\ \mu_k(\tau)d\tau = h^2 \sum_{\ell,i=-M}^{M} b_i b_\ell L(z_{ki}, z_{j\ell}) + R \tag{2.20}$$

with the error term

$$|R| \le h^{s+2} c\{\max|\frac{\partial^s L}{\partial\tau^s}| + \max|\frac{\partial^s L}{\partial t^s}|\},\ 0\le s\le 2M+2\ . \tag{2.21}$$

Now we are ready to formulate the Galerkin-collocation equations by using Equations (1.8), (2.2), (2.14) and (2.20). They read as

$$\begin{aligned}\sum_{j=0}^{N} a_{hjk}\gamma_j := \sum_{j=0}^{N} \{h^{2+\beta}(W_{1,\rho(j,k)} + \log h\ W_{2,\rho(j,k)}) \\ + h^2 \sum_{\ell,i=-M}^{M} b_i b_\ell L(z_{ki}, z_{j\ell})\}\gamma_j \\ = h \sum_{i=-M}^{M} b_i f(z_{ki}) =: F_k \quad k=0,\dots,N\ .\end{aligned} \tag{2.22}$$

For saving computing time, the values of L and f at the grid points should be evaluated only once at the beginning and then be stored for further use as to build up the stiffness matrix in Equations (2.22).

This suggests a choice $\gamma_\ell = 1$ or $\frac{1}{2}$ or $\frac{1}{3}$ etc. in the numerical integration formulas.

For the examples treated in [11],[12],[13],[30] with $\alpha = -\frac{1}{2} = \alpha'$ the choice $m=2$, $M=1$ and $\gamma_\ell = 1$ provided excellent numerical results in combination with short computing times. For the asymptotic error due to the Galerkin-collocation we shall use the already established error estimates (Formula (2.8)) for Galerkin's method. To this end we abbreviate the Equations (2.22) by

$$\sum_{j=0}^{N} a_{hjk}\gamma_j = F_k\ ,\ k=0,\dots,N \tag{2.23}$$

as mappings in $\tilde{H}_h$. If

$$W_h = \sum_{j=0}^{N} \alpha_j \mu_j \tag{2.24}$$

then the mapping $\tilde{A}$ associated with Equation (2.23),

$$\sum_{\ell=0}^{N} \beta_\ell \mu_\ell = \tilde{A} w_h \tag{2.25}$$

will be defined by the linear equations for the coefficients β_ℓ,

$$\sum_{\ell=0}^{N} \beta_\ell (\mu_\ell, \mu_k) = \sum_{j=0}^{N} a_{hjk} \alpha_j \ , \quad k=0,\dots,N. \tag{2.26}$$

Since the Gram matrix (μ_ℓ, μ_k) is regular, $\tilde{A}$ in Equation (2.25) is well defined. Correspondingly we define $\tilde{F} \in \tilde{H}_h$ by

$$(\tilde{F}, \mu_k) = F_k \quad \text{for} \quad k = 0,\dots,N \ . \tag{2.27}$$

Then the Galerkin Equations (2.3) and the Galerkin-collocation Equations (2.22) take the form

$$P_h A P_h \hat{v} = P_h f \quad \text{and} \quad \tilde{A} v_h = \tilde{F} \ , \quad \hat{v}, v_h \in \tilde{H}_h \subset L_2 \cap H^\alpha \ , \tag{2.28}$$

respectively. One easily obtains the estimate

$$\| \hat{v} - v_h \|_{L_2} \le \| \tilde{A}^{-1} \|_{L_2 L_2} \{ \| (\tilde{A} - P_h A P_h) \hat{v} \|_{L_2} + \| P_h f - \tilde{F} \|_{L_2} \} \ . \tag{2.29}$$

This estimate shows clearly that we need estimates for stability, i.e. $\| \tilde{A}^{-1} \|_{L_2 L_2}$, consistency, i.e. $\| (\tilde{A} - P_h A P_h) \hat{v} \|_{L_2}$ and the truncation error $\| P_h f - \tilde{F} \|_{L_2}$. Let us begin with the consistency. With Formula (2.21) one can prove the following:

<u>Theorem 2.3: Let the weights</u> $W_{1\rho}, W_{2\rho}$ <u>be accurate to an order</u> h^a <u>and let</u> $(\frac{\partial}{\partial \tau})^{2M+2} L$ <u>and</u> $(\frac{\partial}{\partial t})^{2M+2} L$ <u>be continuous. Then we have the consistency</u>

$$| (\tilde{A}\mu, \nu) - (A\mu, \nu) | \le \lambda(h) \| \mu \|_{L_2} \| \nu \|_{L_2} \quad \underline{\text{for all}} \ \mu, \nu \in \tilde{H}_h \tag{2.30}$$

<u>where</u>

$$\lambda(h) \le c_1 |\log h| h^{a-2\alpha-1} + c_2 h^{2M+2} \ . \tag{2.31}$$

From the estimates (2.30) and (2.9) one easily obtains stability.

<u>Theorem 2.4: Let the assumptions of Theorem 2.3 be fulfilled and in addition let</u> $a > 1+2(\alpha-\alpha')$, $M > -\alpha'-1$. <u>Then we have stability</u>, i.e. <u>there exists</u> $h_o > 0$ <u>such that</u>

$$\| \tilde{A}^{-1} \|_{L_2 L_2} \le c h^{2\alpha'} \tag{2.32}$$

<u>where</u> c <u>is independent of</u> h <u>for all</u> $0 < h \le h_o$.

Finally, the estimation of the error term R in Equation (2.18) in connection with Equations (2.27) yields for the truncation error:

Theorem 2.5: For $F_k = h \sum_{\ell=-M}^{M} b_\ell f(z_{k\ell})$ in Equations (2.27) there holds

$$\|P_h f - \tilde{F}\|_{L_2} \le ch^\sigma \|f\|_{H^\sigma} \quad \text{with} \quad 1 \le \sigma \le 2M+2 \,. \tag{2.33}$$

Collecting the foregoing estimates and using Formulae (2.29) and (2.8) we find the following estimates for our Galerkin-collocation.

Theorem 2.6: For $a > m+2+2(\alpha'-\alpha)$ and $M \ge \frac{m-1}{2} - \alpha'$ we find an error estimate

$$\|v - v_h\|_{L_2} \le ch^s \{\|u\|_{H^s} + \|f\|_{H^{s-2\alpha'}}\} \tag{2.34}$$

with $1 + 2\alpha' \le s \le m+1$ and $0 \le s$.

For $a > 2m+3-2\alpha'$ and $M \ge m-\alpha-\alpha'$ we have even the super approximation

$$\|v - v_h\|_{H^t} \le ch^{s-t}\{\|u\|_{H^s} + \|u\|_{L_2} + \|f\|_{H^{s-t-2\alpha'}}\} \tag{2.35}$$

provided $2\alpha-m-1 \le t \le s \le m+1$, $s-t \ge 1-2\alpha'$.

As we can see from the foregoing error estimates, it seems that the Galerkin-collocation (Equations (2.22)) combines the theoretical advantages of Galerkin's method with the practical advantages of the collocation methods. For illustration we present some numerical results from [13] which have been obtained for the computation of slow plane viscous flows around elliptical obstacles. Let us consider the exterior flow around the ellipse

$$\Gamma : \vec{z}(t) = \{\cos 2\pi t \,,\, \frac{1}{2} \sin 2\pi t\} = \{z_1(t),\, z_2(t)\}$$

governed by the dimensionless Navier Stokes equations

$$\Delta\vec{q} - \nabla p = R(\vec{q}\cdot\nabla)\vec{q} \quad \text{and} \quad \nabla\cdot\vec{q} = 0 \tag{2.36}$$

in the exterior of Γ with the boundary conditions

$$\vec{q} = 0 \ \text{ on } \Gamma \ \text{ and } \ \vec{q}(\vec{x}) - \{1,0\} \to 0 \ \text{ for } |x| \to \infty \,.$$

$\vec{q}(\vec{x})$ denotes the dimensionless velocity, p is a dimensionless pressure and R the Reynolds number.

Following Hsiao and Mac Camy [14] we computed the first two terms of the Stokes expansion

$$\vec{q}(\vec{x}) \sim \vec{q}_1(\vec{x})(\log R)^{-1} + \vec{q}_2(\vec{x})(\log R)^{-2} \tag{2.37}$$

by solving the following system of integral equations:

$$\int_\Gamma \log|\vec{x}-\vec{z}(t)|\, g_{j\beta}(t)\rho(t)dt + \sum_{\alpha=1}^{2} \{ \frac{(x_\alpha - z_\alpha(t))(x_\beta - z_\beta)}{|\vec{z}(t) - \vec{x}|^2} + (\frac{1}{2} - 2\xi)\delta_{\alpha\beta}\} g_{j\alpha}\rho\, dt - \omega_{j\beta} = 0, \quad \vec{x}\in\Gamma, \quad \beta=1,2, \tag{2.38}$$

$$\int_\Gamma \vec{g}_j \rho dt = \vec{A}_j \; ; \quad j = 1,2 .$$

Here $\vec{x} = \{x_1, x_2\}$ and $\xi = \frac{1}{2}(\frac{3}{2} + \log 4 - \gamma)$ with γ Euler's constant,

$$\vec{A}_1 = -\{1,0\} \quad \text{and} \quad \vec{A}_2 = \{\omega_{12}, -\omega_{11}\} .$$

Then the velocity terms in the expansion (2.37) are given by

$$\vec{q}_j(\vec{x}) = \int_\Gamma \nabla_x [(\vec{x}-\vec{z}(t)) \cdot \vec{g}_j(t)\rho(t)(\log|\vec{x}-\vec{z}(t)| + (\frac{1}{2} - 2\xi))]dt - \vec{\omega}_j \quad \text{for} \quad x \notin \Gamma . \tag{2.39}$$

The computations for Equation (2.38) have been made by our Galerkin-collocation with $m = 2$, i.e. piecewise quadratic polynomials and with $N = 39$, i.e. 40 grid points. A test computation (like in [12]) showed accuracy of $\vec{\omega}_j$ (and $\vec{g}_j(t)\rho$) up to 6 decimal digits. In particular at the points $\{1.5,0\}$, $\{1.5, 0.5\}$ and $\{3,0\}$ we found the following velocities:

R	q_1	q_2	q_1	q_2	q_1	q_2
0.0025	0.0399	0	0.0641	-0.0330	0.1232	0
0.005	0.0426	0	0.0655	-0.0353	0.1317	0
0.0075	0.0442	0	0.0711	-0.0366	0.1368	0
0.01	0.0454	0	0.0729	-0.0376	0.1402	0
	{1.5; 0}		{1.5; 0,5}		{3;0}	

The results show clearly the acceleration of the flow for increasing R . Further results including plots of the velocity field will be presented in [13]. Other numerical results for modifications of Equation (2.38) belonging to plate bending and to conformal mapping can be found in [11],[12] and [30].

§3 STANDARD COLLOCATION AND ITS DISCRETISATION

Although most numerical implementations of boundary integral methods are done with the standard collocation there are yet known only a few results on the asymptotic errors except in the case of Fredholm integral equations of the second kind. (See references in [32].) For the more general case there are results available only for the special case of Symm's equation, i.e. $p_1 \equiv 0$ and $p_2 \equiv 1$ in Equation (1.5) [1],[3] p.119 and [29].

For singular integral equations with the Cauchy kernel S. Prössdorf and G. Schmidt have proved that the collocation method with piecewise linear functions (m=1) only converges if A is strongly elliptic [24].

We announce here that, conversely, for strongly elliptic equations with convolutional principal part the convergence of Galerkin's procedure implies convergence of the standard collocation.

The standard collocation method can be formulated as to find

$$\tilde{v} = \sum_{j=0}^{N} \tilde{\gamma}_j \mu_j \in \tilde{H}_h \tag{3.1}$$

such that the collocation equations at the grid points,

$$\sum_{j=0}^{N} \tilde{\gamma}_j A\mu_j(z_k) = f(z_k) \ , \quad k=0,\dots,N \tag{3.2}$$

are satisfied. In order to formulate the convergence result let us introduce the interpolation operator I_h mapping continuous functions onto $\tilde{H}_h$:

$$I_h f = \sum_{j=0}^{N} \alpha_j \mu_j \quad \text{with} \quad \sum_{j=0}^{N} \alpha_j \mu_j(z_k) = f(z_k), \ k=0,\dots,N \ . \tag{3.3}$$

For I_h we have the approximation property [6]

$$\| I_h f - f \|_{H^t} \le c h^{s-t} \| f \|_{H^s} \quad \text{for} \quad 0 \le t \le s \le m+1, \ \frac{1}{2} \le s, \ t \le m \ . \tag{3.4}$$

Then Equations (3.2) can also be written as

$$I_h \, A \, P_h \, \tilde{v} = I_h f = \tilde{f} \in \tilde{H}_h \ . \tag{3.5}$$

Based on the special form (Equation (1.5)) and on estimates for R in Formulae (2.18) with $M = 0$ and by making use of the collectively compact operator theory by P. Anselone [4] we can prove asymptotic error estimates.

The key result is the stability.

Theorem 3.1: [31] : The Equations (3.5) in $\tilde{H}_h$ are stable with

$$\| (I_h A P_h)^{-1} \|_{L_2,L_2} \le c h^{2\alpha'} \tag{3.6}$$

where c is independent of $h \le h_o$ with a suitable $h_o > 0$.

Estimate (3.6) is in accordance with [1] and [29]. The stability in connection with convergence (inequalities (2.6) and (3.4)) and with the inverse assumption (Formula (2.7)) provides us the following convergence result:

Theorem 3.2 : The collocation equations (3.2) are uniquely solvable for $0 < h \le h_o$ and the approximate solutions converge as

$$\| \tilde{v} - v \|_{L_2} \le c h^{s+2(\alpha'-\alpha)} \| v \|_{H^s} \quad \text{if} \quad 0 \le s \le m+1,\ \frac{1}{2} + 2\alpha < s,\ \frac{1}{2} + 2\alpha < m. \tag{3.7}$$

For the discretisation of Equations (3.2) we again insert Equations (1.8), (3.1) into Equation (1.5) and use Equation (2.18) for the integral of the second term in Equation (1.5). Neglecting the error term we find the discrete version of the collocation equations:

$$\sum_{j=0}^{N} \Big(h^{1+\beta} \{ W_{1\rho}^{x} + W_{2\rho}^{x} \log h \} + h \sum_{\ell=-M}^{M} b_\ell L(z_k, z_{j\ell}) \Big) \gamma_j^{x} = f(z_k) , \tag{3.8}$$

$$k = 0, \dots, N .$$

Here the weights $W_{1\rho}^{x}$, $W_{2\rho}^{x}$ with $\rho = j-k$ also can be computed once for all independently of h and Γ by the vectors of weights

$$W_{1\rho}^{x} = \text{p.v.} \int_{t'=0}^{m+1} \mu(t') [p_1 (t' - \frac{m+1}{2} + \rho) + p_2 \log | t' - \frac{m+1}{2} + \rho |] dt', \tag{3.9}$$

$$W_{2\rho}^{x} = \text{p.v.} \int_{t'=0}^{m+1} \mu(t') [p_2 (t' - \frac{m+1}{2} + \rho)] dt' . \tag{3.10}$$

Since the neglected error terms of estimate (2.8) are of the same order as those in Formula (2.21) for the Galerkin-collocation, one easily finds that under the same assumptions as in Theorem 2.3 there holds between Equations (3.2) and Equations (3.8) a consistency estimate like inequality (2.30). Hence one eventually has:

Theorem 3.3: Let $W_{1\rho}^{x}$, $W_{2\rho}^{x}$ be accurate to an order h^a with $a > m+2 + 2(\alpha'-\alpha)$ and let $2M > m-1 - 2\alpha'$. Let $(\frac{\partial}{\partial\tau})^{2M+2}L$ and $(\frac{\partial}{\partial t})^{2M+2}L$ be continuous. Then we find an error estimate

$$\| v^x - v \|_{L_2} \leq c \cdot h^s \{ \| f \|_{H^{s-2\alpha}} + \| f \|_{H^{s-2\alpha'}} \} \qquad (3.11)$$

with $1 + 2\alpha' \leq s \leq m+1$ and $0 \leq s$.

In practice the numerical integrations in Equation (2.18) are often executed with Gaussian quadrature. Then the b_ℓ are defined by the Gaussian weights multiplied by $\mu_j(z_{j\ell})$ and the γ_ℓ correspond to q Gauss-Lobatto points in each of the intervals $[\kappa,\kappa+1]$, respectively, with $0 \leq \kappa \leq m$, $\kappa \in \mathbb{N}_0$. In this case the estimate (3.11) remains valid if one replaces $2M$ by q in the assumption.

ACKNOWLEDGEMENTS :

The author wants to thank Dr. P. Kopp who made the numerical computations.

This research was supported by the "Deutsche Forschungsgemeinschaft" under the project number We 659 and also partially supported by the "Applied Mathematics Institute" and the Department of Mathematical Sciences at the University of Delaware.

REFERENCES :

[1] Abou El-Seoud, M.S. (1979) Numerische Behandlung von schwach singulären Integralgleichungen erster Art. Dissertation, Technische Hochschule Darmstadt, Germany.

[2] Adams, R.A. (1975) Sobolev Spaces. Academic Press, New York.

[3] Aleksidze, M.A. (1978) The Solution of Boundary Value Problems with the Method of the Expansion with Respect to Nonorthonormal Functions. Nauka, Moscow (Russian).

[4] Anselone, P.M. (1971) Collectively Compact Operator Approximation Theory. London, Prentice Hall.

[5] Aubin, J.P. (1972) Approximation of Elliptic Boundary-Value Problems. Wiley-Interscience, New York.

[6] Babuška, I. and Aziz, A.K. (1972) Survey lectures on the mathematical foundations of the finite element method. In: "The Mathematical Foundation of the Finite Element Method with Applications to Partial Differential Equations" (Aziz, A.K. Ed.) 3 - 359, New York, Academic Press.

[7] Bramble, J. and Schatz, A. (1970) Rayleigh-Ritz-Galerkin methods for Dirichlet's problem using subspaces without boundary conditions. Comm. Pure Appl. Math. 23:653-675.

[8] Ciarlet, P.G. (1978) The Finite Element Method for Elliptic Problems. North Holland, Amsterdam.

[9] Giroire, J. and Nedelec, J.C. (1978) Numerical solution of an exterior Neumann problem using a double layer potential. Math. of Comp. 32:973-990.

[10] Hildebrandt, St. and Wienholtz, E. (1964) Constructive proofs of representation theorems in separable Hilbert space. Comm. Pure Appl. Math. 17:369-373.

[11] Hsiao, G.C., Kopp, P. and Wendland, W.L. (1980) A Galerkin collocation method for some integral equations of the first kind. Computing 25:89-130.

[12] Hsiao, G.C., Kopp, P. and Wendland, W.L. (1980) The synthesis of the collocation and the Galerkin method applied to some integral equations of the first kind. In: C.A. Brebbia (ed.): New Developments in Boundary Element Methods. CML Publ. Southampton: 122-136.

[13] Hsiao, G.C., Kopp, P. and Wendland, W.L.: Some applications of a Galerkin-collocation method for integral equations of the first kind. In preparation.

[14] Hsiao, G.C. and Mac Camy, R.C. (1973) Solution of boundary value problems by integral equations of the first kind. SIAM Review 15:687-705.

[15] Hsiao, G.C. and Wendland, W.L.: The Aubin-Nitsche lemma for integral equations. To appear.

[16] Jaswon, M.A. and Symm, G.T. (1977) Integral Equation Methods in Potential Theory and Elastostatics. Academic Press London.

[17] Kohn, J.J. and Nirenberg, L. (1965) On the algebra of pseudodifferential operators. Comm. Pure Appl. Math. 18:269-305.

[18] Michlin, S.G. (1962) Variationsmethoden der Mathematischen Physik, Akademie-Verlag Berlin.

[19] Michlin, S.G. and Prößdorf, S. (1980) Singuläre Integraloperatoren. Akademie-Verlag Berlin.

[20] Muskhelishvili, N.I. (1953) Singular Integral Equations. Noordhoff Groningen.

[21] Nédelc, J.C. (1976) Curved finite element methods for the solution of singular integral equations on surfaces in $\mathbb{R}^3$. Comp. Math. Appl. Mech. Engin. 8:61-80.

[22] Nédelec, J.C. (1980) Formulations variationelles de quelques équations intégrales faisant intervenir des parties finies. In: R. Shaw et al. (ed.): Innovative Numerical Analysis for the Engineering Sciences. Univ. Press of Virginia, Charlottesville: 517-524.

[23] Nitsche, J.A. (1970) Zur Konvergenz von Näherungsverfahren bezüglich verschiedener Normen. Num. Math. 15:224-228.

[24] Prössdorf, S. and Schmidt, G.: A finite element collocation method for singular integral equations. To appear.

[25] Prössdorf, S. and Silbermann, B. (1977) Projektionsverfahren und die näherungsweise Lösung singulärer Gleichungen. B.G. Teubner Leipzig.

[26] Richter, G.R. (1978) Numerical solution of integral equations of the first kind with nonsmooth kernels. SIAM J. Numer. Anal. 17:511-522.

[27] Riesz, F. and Nagy, B.Sz. (1956) Vorlesungen über Funktionalanalysis. Dt. Verl. Wiss. Berlin.

[28] Treves, F. (1980) Introduction to Pseudodifferential and Fourier Integral Operators I. Plenum Press, New York and London.

[29] Voronin, V.V. and Cecoho, V.A. (1974) An interpolation method for solving an integral equation of the first kind with a logarithmic singularity. Dokl. Akad. Nauk SSR 216; Soviet Math. Dokl. 15:949-952.

[30] Wendland, W.L. (1980) On Galerkin collocation methods for integral equations of elliptic boundary value problems. In: J. Albrecht and L. Collatz (ed.): Numerical Treatment of Integral Equations. Intern. Ser. Num. Math., Birkhäuser Basel, 53:244-275.

[31] Wendland, W.L.: On the asymptotic convergence of the collocation method for a class of strongly elliptic integral equations. In preparation.

[32] Wendland, W.L.: Asymptotic accuracy and convergence. In: C. Brebbia (ed.): Progress in Boundary Elements, Vol. 1, Pentech Press, London, 1981.

[33] Wendland, W.L., Stephan, E. and Hsiao, G.C. (1979) On the integral equation method for the plane mixed boundary value problem of the Laplacian. Math. Meth. in the Appl. Sci. 1:265-321.

BOUNDARY METHODS. C-COMPLETE SYSTEMS FOR THE BIHARMONIC EQUATIONS

Hervé Gourgeon and Ismael Herrera

Instituto de Investigaciones en Matemáticas Aplicadas y en Sistemas (IIMAS). Universidad Nacional Autónoma de México (UNAM), Apdo. Postal 20-726, México 20, D. F. MEXICO

ABSTRACT

A boundary method for solving the biharmonic equation is presented. It is based on the use of systems of solutions of the homogeneous equations, which are complete. A convenient criterium for the completeness of such systems, is the notion of c-completeness. Using a convenient representation of solutions for the biharmonic equation a procedure for constructing c-complete systems for this equation is developed. Examples of such systems are constructed.

1. INTRODUCTION

In recent years, by a boundary method, it is usually understood, a numerical procedure in which a subregion or the entire region, is left out of the numerical treatment, by use of available analytical solutions (or more generally, previously computed solutions). Boundary methods reduce the dimensions involved in the problem leading to considerable economy in the numerical work and constitute a very convenient manner of treating adequately unbounded regions by numerical means. Generally, the dimensionality of the problem is reduced by one, but even when part of the region is treated by finite elements, the size of the discretized domain is reduced [Zienkiewicz, 1977, Zienkiewicz, et al., 1977].

There are two main approaches for the formulation of boundary methods; one is based on the use of boundary integral equations and the other one, on the use of complete systems of solutions. In numerical applications, the first one of these methods has received most of the attention [Brebbia, 1978]. This is in spite of the fact that the use of complete systems of solutions presents important numerical advantages; e.g., it avoids the introduction of singular integral equations and it does not require the construction of a fundamental solution. The latter is especially relevant in connection with complicated problems, for which, it may be extremely laborious to build up a fundamental solution. This is illustrated by the fact that there are methods for synthetizing fundamental solutions starting from plane waves, which can be shown to be a complete system [Sánchez-Sesma, Herrera and Aviles, 1981].

One may advance some possible explanations for this situation. Although, the principle of superposition, is a standard procedure for building up solutions of linear equations, many of its applications have been based on the method of separation of variables; this has lead to the frequent, but false, belief that complete systems of solutions have to be constructed specifically for a given region. Of course, this is not the case; indeed,most frequently systems of solutions are complete independently of the detailed shape of the region considered [Herrera and Sabina, 1978], and the systems developed here for the biharmonic equation possess this property.

Also, in some fields of application, procedures which constitute particular cases of the approximation by complete systems of solutions, have presented severe restrictions and inconveniences. For the case of acoustics and electromagnetic field computations, a survey of such difficulties, was carried out by Bates [1975]. For this kind of studies, the so called "Rayleigh hypothesis", restricts drastically the applicability of the method. However, work by Millar [1973], implies that these difficulties are due, mainly, to lack of clarity, since he avoided Rayleigh hypothesis, altogether, by adopting a different point of view.

Motivated by this situation, one of the authors, started a systematic research of the subject [Herrera and Sabina, 1978; Herrera, 1977a, 1979b, 1980e], oriented to clarify the theoretical foundations of the method, allowing its systematic and reliable use. The aims of the research have been satisfactorily achieved to a large extent and have just been reported [Herrera, 1981b,c]. This has been possible due to the progress that has been made in the understanding of partial differential equations [Lions and Magenes, 1972; Temam, 1977]. The methodology also owes much to work of Amerio, Fichera, Picone, Kupradze and Trefftz [Miranda, 1955; Kupradze, 1967; Trefftz, 1926]. The systematic development of the procedure, in a manner which is applicable to any linear problem, was made possible, however, by an abstract theory that has been developed by Herrera [1979a,b, 1980b,c,d,e, 1981a].

The numerical solution of Stokes and Navier-Stokes equations, is a problem of great practical interest at present, and it is not our purpose to review it, since recent surveys are available [Glowinski and Pironneau,1978; Temam, 1977]. Taking this interest for granted, we explain briefly the method mentioned before, in connection with the biharmonic equation and supply an efficient procedure for developing c-complete systems for this equation, starting from c-complete systems for Laplace equation.

2. THE BOUNDARY METHOD USED

Consider the biharmonic equation

$$\nabla^4 u = 0 \quad \text{in } \Omega \tag{2.1}$$

This equation must be satisfied in the sense of distributions by elements of some spaces of functions. In general, we ask u to be in a linear subspace $D \subset H^2(\Omega)$, so that the equation (2.1) is between elements of $H^{-2}(\Omega)$.†

On this assumption the biharmonic problem (2.1) is equivalent to the formulation

$$u \in \text{Ker } P \ , \ P: H^2(\Omega) \to (H^2)^*$$

defined by

$$\langle Pu,v\rangle = \int_\Omega \nabla^4 u \ v \ dx \quad \forall\ u \in H^2(\Omega), \ \forall\ v \in H^2(\Omega) \tag{2.2}$$

Integration by parts gives

$$\langle Pu,v\rangle = \int_\Omega \nabla^2 u \ \nabla^2 v \ dx + \int_{\partial\Omega} \{\frac{\partial \Delta u}{\partial n} v - \Delta u \frac{\partial v}{\partial n}\} dx \tag{2.3}$$

In (2.3) four different boundary values occur. We note that [Lions and Magenes, 1972]

$$u \in H^2(\Omega) \Rightarrow \quad u \in H^{3/2}(\partial\Omega)$$
$$\frac{\partial u}{\partial n} \in H^{1/2}(\partial\Omega)$$
$$\Delta u \in H^{-1/2}(\partial\Omega)$$
$$\frac{\partial \Delta u}{\partial n} \in H^{-3/2}(\partial\Omega)$$

Let us associate with the operator P, an antisymmetric operator A by

$$A = P - P^*$$

$$\langle Au,v\rangle = \langle Pu,v\rangle - \langle Pv,u\rangle \tag{2.4}$$

$$\forall\ u,v \in D \quad \langle Au,v\rangle = \int_{\partial\Omega} \{v \frac{\partial}{\partial n}\Delta u - \Delta u \frac{\partial v}{\partial n} + \Delta v \frac{\partial u}{\partial n} - u \frac{\partial \Delta v}{\partial n}\} dx \tag{2.5}$$

in which, only boundary values appear. Let us introduce the Boundary operators B and B':

---

† We use the usual notation for Sobolev spaces.

$$\forall\ u,v \in D\ \langle Bu,v\rangle = \int_{\partial\Omega} v \frac{\partial}{\partial n} \Delta u\ dx \tag{2.6}$$

$$\langle B'u,v\rangle = + \int_{\partial\Omega} \Delta v \frac{\partial u}{\partial n} dx \tag{2.7}$$

Then

$$A = B + B' - B'^* - B^* \tag{2.8}$$

In fact we can directly define A by (2.5) in a different space D:

$$D = \{u \in H^{1/2}(\Omega) \mid u \in H^o(\partial\Omega);\ \frac{\partial u}{\partial n} \in H^o(\partial\Omega)\ ;$$
$$\Delta u \in H^o(\partial\Omega)\ ;\ \frac{\partial \Delta u}{\partial n} \in H^o(\partial\Omega)\}$$

D is not a Sobolev space, but we note the inclusions:

$$H^{7/2}(\Omega) \subset D \subset H^{1/2}(\Omega) \tag{2.9}$$

and that $C^\infty(\Omega)$ is dense in D.

Define

$$I_1 = \mathrm{Ker}(B + B') \text{ and } I_2 = \mathrm{Ker}(B^* + B'^*) \tag{2.10}$$

Then it can easily be shown that $\{I_1, I_2\}$ is a canonical decomposition of D, in the sense defined in [Herrera, 1980b]; i.e., I_1, I_2 are completely regular:

$$\langle Au,v\rangle = 0 \quad \forall\ v \in I_1 \Leftrightarrow u \in I_1 \tag{2.11}$$

$$\langle Au,v\rangle = 0 \quad \forall\ v \in I_2 \Leftrightarrow u \in I_2 \tag{2.12}$$

and

$$I_1 + I_2 = D \qquad I_1 \cap I_2 = \mathrm{Ker}\ A$$

This implies that every $u \in D$ can be written as $u=u_1+u_2$, with $u_1 \in I_1$ and $u_2 \in I_2$, and this representation is unique, except for elements of Ker A.

Another canonical decomposition would be

$$I_1' = \mathrm{Ker}\ (B - B'^*)\ ;\ I_2' = \mathrm{Ker}(B^* - B')\quad . \tag{2.13}$$

Notice that the boundary values of elements of D, can be characterized as follows:

$$u \in D \rightarrow [u_a, u_b, u_c, u_d] \in D/\mathrm{Ker}\ A \subset [H_o(\partial\Omega)]^4$$

with

$$u_a = u \quad u_b = \frac{\partial u}{\partial n} \quad u_c = \Delta u \quad u_d = \frac{\partial}{\partial n} \Delta u \tag{2.14}$$

on $\partial\Omega$. Now, associated with the canonical decomposition $\{I_1, I_2\}$, we have

$$D/\mathrm{Ker}\ A = I_1/\mathrm{Ker}\ A \oplus I_2/\mathrm{Ker}\ A$$

so if $u \in D/\text{Ker } A \ \exists \ \{u_1, u_2\} \ni u = u_1 + u_2$

$$u_1 \in \text{Ker}(B + B') \quad u_1 = [u_a , u_c] \tag{2.15}$$

$$u_2 \in \text{Ker}(B^* + B'^*) \quad u_2 = [u_d, u_b] \tag{2.16}$$

with these notations then it is easy to exhibit the identity:

$$\langle Au,v\rangle = (u_2 , v_1) - (u_1 , v_2) \tag{2.17}$$

where it is understood that if $[a,b]$, $[c,d] \in [H_o(\partial\Omega)]^2$

$$([a,b] , [cd]) = \int_{\partial\Omega} (ac + bd)dx \tag{2.18}$$

Using the other decomposition we would have similarly

$$D/\text{Ker } A = I_1'/\text{Ker } A \oplus I_2'/\text{Ker } A$$

$u \in D/\text{Ker } A \qquad u = u_1' + u_2'$

$$u_1' = [u_a, u_b] \quad u_2' = [u_d, -u_c] \tag{2.19}$$

and the identity:

$$\langle Au,v\rangle = (u_2' , v_1') - (u_1' , v_2') \tag{2.20}$$

Let N_P be the subspace of solutions of the biharmonic equation (2.1) and define

$$I_P = N_P + N_A \tag{2.21}$$

Then I_P is completely regular; i.e.

$$\langle Au,v\rangle = 0 \quad \forall\, v \in I_P \Leftrightarrow u \in I_P \tag{2.22}$$

This comes straight forwardly from some results of existence of solution of the biharmonic equation with compatible boundary conditions, the density of $C^\infty(\Omega)$ in D and results reported previously [Herrera, 1980b]. Define $\mathcal{I}_P = I_P/\text{Ker } A$. The results of uniqueness imply that $N_P \cap \text{Ker } A = \{0\}$ so that $\mathcal{I}_P$ is naturally imbedded in D/Ker A, with the notations introduced. The following definition is relevant, for our discussion.

<u>Definition</u>: *A denumerable set* $B = \{w_1, w_2, \ldots\}$ *of* N_P *is c-complete (complete in connectivity), with respect to* A *if* $\langle Au, w_\alpha\rangle = 0 \quad \forall\, \alpha \in N \Leftrightarrow u \in \mathcal{I}_P$.

For any canonical decomposition $\{I_1, I_2\}$, $\mathcal{I}_P$ is decomposed as $\mathcal{I}_P = \mathcal{I}_{1P} \oplus \mathcal{I}_{2P}$, $\mathcal{I}_{1P} \subset I_1, \mathcal{I}_{2P} \subset I_2$

The following can be proved [Herrera, 1980e, 1981b,c].

<u>Proposition 1</u>. *The 3 statements are equivalent*

i) $\mathcal{B}$ is c-complete in $\mathcal{I}_P$, with respect to A.

ii) $\mathcal{B}_1 = \{w_{1\alpha}\}_{\alpha\in N}$ spans $\mathcal{I}_{1P}$.

iii) $\mathcal{B}_2 = \{w_{2\alpha}\}_{\alpha\in N}$ spans $\mathcal{I}_{2P}$.

Let us suppose that we have a c-complete system

$$\mathcal{B} = \{w_\alpha\}_{\alpha=1,\ldots} \subset N_P$$

and a canonical decomposition $\{I_1, I_2\}$. Consider the problem; find $u \in D$ such that

$$\nabla^4 u = 0 \quad \text{in} \quad \Omega$$

$$u_1 = u_\beta \text{ given on the boundary}$$

Assume $u_1 \in \mathcal{I}_{1P}$, in order to have existence of a solution to the problem. In view of the Proposition, we know that $\{w_{1\alpha}\}_{\alpha\in N}$ spans $\mathcal{I}_{1P}$ so that any element of $\mathcal{I}_{1P}$ can be approximated by linear combinations of $\{w_{1\alpha}\}$; more precisely, one can choose coefficients $\{a_\alpha^N\}_{\alpha=1,\ldots N}$ such that

$$u_1^N = \sum_1^N a_\alpha^N w_{1\alpha} \tag{2.24}$$

has the property that $u_1^N \to u_1$ in $[H^0(\partial\Omega)]^2$.

Then

$$u^N = \sum_1^N a_\alpha^N w_\alpha \tag{2.25}$$

is the biharmonic function in Ω, such that the boundary value u_1^N approximates the data. Therefore, u^N is an approximation to a solution of the problem and as $N \to \infty$ one has $u^N \to u$ in the sense (at least) of $H^{1/2}(\Omega)$ [Lions and Magenes, 1972].

If the missing boundary value u_2 is required, and if it is known to be in $[H^0(\partial\Omega)]^2$ in $\mathcal{I}_{2P}$, we can indeed approximate it by

$$u_2^N = \sum_{\alpha=1}^N a_\alpha^N w_{2\alpha} \tag{2.26}$$

but we can also compute it using (2.17). Indeed

$$(u_2, w_{1\alpha}) = (u_1, w_{2\alpha}) \quad \forall\, \alpha \in N$$

If $\{w_{1\alpha}\}$ is orthonormal (and if it is not, we can orthonormalize it by the well known Gram-Schmidt orthonormalization process), then, u_2 is given by

$$u_2 = \sum_{\alpha=1}^{\infty} (u_2, w_{1\alpha}) w_{1\alpha} = \sum_{\alpha=1}^{\infty} (u_1, w_{2\alpha}) w_{1\alpha} \tag{2.27}$$

3. C-COMPLETE SYSTEMS FOR THE BIHARMONIC EQUATION

In this section we give a general procedure for constructing c-complete systems for the biharmonic equation, whenever a c-complete system for Laplace's equation is known.

Proposition 2. *Let $\{\psi_1,\psi_2,...\}$ be harmonic functions such that they are a c-complete system for Laplace equation in the region Ω. Assume $\{\phi_1,\phi_2,...\}$ are also harmonic and such that*

$$\frac{\partial\phi_\alpha}{\partial x} = \psi_\alpha \quad ; \quad \alpha = 1,2,... \tag{3.1}$$

Then the system $\{\psi_1,\psi_2,...\} \cup \{x\phi_1,x\phi_2,...\}$ are biharmonic and c-complete for equation (2.1).

Proof. Consider the canonical decomposition (2.15), (2.16), then

$$I_{1P} = \{u_1 = [u,\Delta u] \mid u \in I_P\} = [H^o(\partial\Omega)]^2 \tag{3.2}$$

Here, u, Δu refer to the boundary values on $\partial\Omega$. The biharmonic problem with u_1 given, is the biharmonic equation (2.1), subjected to the boundary conditions

$$u = f_1 \quad , \quad \text{on } \partial\Omega \tag{3.3a}$$

$$\Delta u = f_2 \quad , \quad \text{on } \partial\Omega \tag{3.3b}$$

where f_1 and f_2 are given functions of $H^o(\partial\Omega)$. Equivalently, one can solve

$$\Delta p = 0 \quad , \quad \text{in } \Omega \tag{3.4a}$$

$$p = f_2 \quad , \quad \text{on } \partial\Omega \tag{3.4b}$$

and

$$\Delta u = p \quad , \quad \text{in } \Omega \tag{3.5a}$$

$$u = f_1 \quad , \quad \text{on } \partial\Omega \tag{3.5b}$$

In view of Proposition 1, it is enough to prove that the system of boundary values $\{[\psi_1,\Delta\psi_1], [\psi_2,\Delta\psi_2],...\} \cup \{[x\phi_1,\Delta x\phi_1], [x\phi_2,\Delta x\phi_2],...\}$ spans $I_{1P} = [H_o(\partial\Omega)]^2$. To this end, notice that

$$\Delta\psi_\alpha = 0 \quad ; \quad \Delta(x\phi_\alpha) = 2\psi_\alpha \ , \quad \alpha=1,2,... \tag{3.6}$$

Therefore, given $[f_1,f_2] \in [H^o(\partial\Omega)]^2$, consider the following approximating sequence

$$u^N = \sum_{\alpha=1}^{N} a_\alpha^N \psi_\alpha + \sum_{\alpha=1}^{N} b_\alpha^N x \phi_\alpha \tag{3.7}$$

Define

$$p^N = \sum_{\alpha=1}^{N} b_\alpha^N \, \Delta x\phi_\alpha^N = 2 \sum_{\alpha=1}^{N} b_\alpha^N \, \psi_\alpha \tag{3.8}$$

and choose b_α^N, so that (as $N\to\infty$)

$$p^N \to f_2 \quad , \quad \text{on } H^0(\partial\Omega) \tag{3.9}$$

This is possible, because $\{\psi_1,\psi_2,\ldots\}$ is c-complete. Relation (3.9), implies that there exists $v\in H^{5/2}(\partial\Omega)$ such that as $N\to\infty$,

$$p^N \to v \quad ; \quad \text{in } H^{5/2}(\Omega) \tag{3.10}$$

Therefore

$$p^N \to v \quad , \quad \text{in } H^2(\partial\Omega)\subset H^0(\partial\Omega) \tag{3.11}$$

Choose now a_α^N so that

$$\sum_{\alpha=1}^{N} a_\alpha^N \, \psi_\alpha \to f_1 - v \quad , \quad \text{on } H^0(\partial\Omega) \tag{3.12}$$

This is again possible because $\{\psi_1,\psi_2,\ldots\}$ is c-complete. Hence, clearly

$$[u^N, \Delta u^N] \to [f_1,f_2] \quad , \quad [H^0(\partial\Omega)]^2 \tag{3.13}$$

and the proof of Proposition 2, is complete.

As an example of the application of Proposition 2, we exhibit a polynomial system which is c-complete for biharmonic equation in any bounded region Ω.

Proposition 3. *Let* $(\alpha=1,2,\ldots)$

$$\psi_\alpha = \operatorname{Re} z^{(\alpha-1)/2} \quad \textit{when } \alpha \text{ is odd} \tag{3.14a}$$

$$\psi_\alpha = \operatorname{Im} z^{\alpha/2} \quad \textit{when } \alpha \textit{ is even} \tag{3.14b}$$

Define

$$\phi_\alpha = \psi_{\alpha+2} \tag{3.15}$$

Then $\{\psi_1,\psi_2,\ldots\} \cup \{x\phi_1,x\phi_2,\ldots\}$ *is c-complete for the biharmonic equation, in any bounded region* Ω.

Proof. It has been shown [Herrera and Sabina, 1978], that $\{\psi_1,\psi_2,\ldots\}$ is c-complete for Laplace's equation in any bounded region. In addition, it is easy to see that equation (3.1) is satisfied.

We recall finally, that a c-complete can be used to approximate any other boundary value problem prescribed by means of regular subspace; this, by virtue of Proposition 1.

4. THE EXTERIOR DOMAIN

Let Ω be the exterior of a bounded domain. A c-complete system for Laplace's equation, which satisfies a radiation condition, in Ω, is given [Herrera and Sabina, 1978], by $\{\psi_1, \psi_2, \ldots\}$

$$\psi_1 = \text{Re Log } z \quad ; \qquad (4.1a)$$

$$\psi_\alpha = \text{Re } z^{-(\alpha-1)/2} \quad ; \ \alpha \text{ odd} \geq 3 \qquad (4.1b)$$

$$\psi_\alpha = \text{Im } z^{-\alpha/2} \quad ; \ \alpha \text{ even} \qquad (4.1c)$$

Applying Proposition 2, it can be seen that system $\{\psi_1, \psi_2, \ldots\} \cup \{x\phi_1, x\phi_2, \ldots\}$, where

$$\phi_1 = \text{Re}(z \log z - z) \ ; \ \phi_2 = \text{Im} \log z \qquad (4.2a)$$

$$\phi_\alpha = \psi_{\alpha-2} \quad ; \ \alpha \geq 3 \qquad (4.2b)$$

is a c-complete system for the exterior problem.

REFERENCES

Bates, R.H.T. (1975) Analytic Constraints on Electro-magnetic Field Computations. IEEE Trans. on Microwave Theory of Techniques. 23, 605-623.

Brebbia, C.A. (1978) The Boundary Element Method for Engineers. Pentech, Press. London.

Glowinski, R., and Pironneau, O. (1978) On Numerical Methods for the Stokes Problem. Chapter 13 of Energy Methods in Finite Element Analysis. R. Glowinski, E.Y. Rodin and O.C. Zienkiewicz, ed., Wiley and Sons.

Herrera, I. (1977a) General Variational Principles Applicable to the Hybrid Element Method. Proc. Nat. Acad. Sci. USA, 74, 7:2595-2597.

Herrera, I. (1979a) On the Variational Principles of Mechanics. Trends in Applications of Pure Mathematics to Mechanics. II, 115-128, H. Zorsky, ed., Pitman Publishing Limited (Invited general lecture)

Herrera, I. (1979b) Theory of Connectivity: A Systematic Formulation of Boundary Element Methods. Applied Mathematical Modelling. 3, 2:151-156.

Herrera, I. (1980b) Variational Principles for Problems with Linear Constraints. Prescribed Jumps and Continuation Type Restrictions. Jour. Inst. Maths. Applics. 25, 67-96.

Herrera, I. (1980c) Boundary Methods in Flow Problems. Proc. Third International Conference on Finite Elements in Flow Problems, Banff, Canada, 10-13 June, 30-42. (Invited general lecture).

Herrera, I. (1980d) Boundary Methods in Water Resources. Finite Elements in Water Resources, S.Y. Wang, et al., ed., The University of Mississippi, 58-71. (Invited general lecture).

Herrera, I. (1980e) Boundary Methods. A Criterion for Completeness. Proc. Nat. Acad. Sci. USA, 77, 8:4395-4398.

Herrera, I. (1981a) An Algebraic Theory of Boundary Value Problems. Comunicaciones Técnicas, IIMAS-UNAM. (In press).

Herrera, I. (1981b) Boundary Methods in Fluids. Finite Elements in Fluids, Volume IV, R.H. Gallagher, ed., John Wiley and Sons. (In press).

Herrera, I. (1981c) Boundary Methods. Theoretical Foundations for Numerical Applications of Complete Systems of Solutions. Comunicaciones Técnicas, IIMAS-UNAM.

Herrera, I., and Sabina, F.J. (1978) Connectivity as an Alternative to Boundary Integral Equations. Construction of bases. Proc. Nat. Acad. Sci. USA, 75, 5:2059-2063.

Kupradze, V.D. (1967) On the Approximate Solution of Problems in Mathematical Physics. Russian Math. Surveys. 22, 2:58-108. (Uspehi Mat. Nauk. 22, 2:59-107).

Lions, J.L., and Magenes, E. (1972) Non-homogeneous Boundary Value Problems and Applications. Springer-Verlag, New York.

Millar, R.F. (1973) The Rayleigh Hypothesis and a Related Least-Squares Solutions to Scattering Problems for Periodic Surfaces and other Scatterers. Radio Science. 8, 785-796.

Miranda, C. (1970) Partial Differential Equations of Elliptic Type. 2nd Ed., Springer-Verlag, New York. (Translation of Equazioni alle Derivate Parziali di Tipo Ellitico, 1955).

Sánchez-Sesma, F.J., Herrera, I., and Avilés, J. (1981) Boundary Methods for Elastic Wave Diffraction-Application to Scattering of SH Waves by Surface Irregularities. Comunicaciones Técnicas, IIMAS-UNAM. (In press).

Temam, R. (1977) Navier Stokes Equations: Theory and Numerical Analysis. North-Holland, Amsterdam.

Trefftz, E. (1926) Ein Gegenstruck zum Ritzschen Vergaren. Proc. 2nd. Int. Congress Appl. Mech. Zurich.

Zienkiewicz, O.C. (1977) The Finite Element Method in Engineering Science. Mc-Graw Hill, New York.

Zienkiewicz, O.C., Kelly, D.W., and Bettess P., (1977) The Coupling of the Finite Element Methods and Boundary Solution Procedures. Int. J. Num. Math. Eng. 11, 355-377.

THE EFFECT OF MESH REFINEMENT IN THE BOUNDARY ELEMENT SOLUTION OF LAPLACE'S EQUATION WITH SINGULARITIES

H.L.G. Pina, J.L.M. Fernandes & C.A. Brebbia

ABSTRACT

It is known that the non-smoothness of boundary data causes the order of convergence of the numerical solutions of partial differential equations [11] to be less than optimal. In this paper we assess the efect of mesh grading to overcome this difficulty in the context of the Boundary Element Method (BEM). As test cases we employed two potential problems proposed by Schultz [10]. We conclude that the BEM yields for a given mesh smaller errors than those obtained by the Finite Difference Methods (FDM) of [10], but at the expense of a greater computational effort. Also a judicious choice of mesh grading can improve significantly the actual error and recover the optimal order of convergence.

INTRODUCTION

It has been well known for some time that potential problems in a domain $\Omega \subset R^n$ can be formulated via integral equations on the boundary Γ of Ω. Since this reduces the dimensionality of the problem from R^n to R^{n-1}, it is expected that the numerical solution of this class of problems will become easier. The application of finite element concepts to the discretization of the boundary integral equations has brought renewed interest in this field leading to the Boundary Element Method (BEM).

A description of the BEM can be found in [8] where numerous references are given. Brebbia [4] shows how to implement the BEM for several classes of problems found in elasticity and fluid flow. The paper by Zienkiewicz et al. [15] discusses the relative merits of the Finite Element Method (FEM) and BEM and develops ways of taking advantage of both techniques. In [2,7] we can find theoretical analysis of the BEM in the context of singular Fredholm integral equations.

In this paper we examine the performance of the BEM as

compared with the Finite Diference Method (FDM) and FEM using as test cases two problems proposed by Schultz [10]. Several meshes are tried, to assess the order of convergence and we investigate the effect of mesh grading on the magnitude and order of convergence of the error.

THE BOUNDARY INTEGRAL EQUATIONS

Let Ω be a domain (i.e., an open connected bounded set) of R^2 with boundary Γ. We wish to solve the following problem for the unknown function u:

$$\text{lap } u = 0 \quad \text{in } \Omega, \tag{1a}$$
$$u = g \quad \text{on } \Gamma_1, \tag{1b}$$
$$\partial_\nu u = h \quad \text{on } \Gamma_2, \tag{1c}$$

where Γ_1 and Γ_2 are complementary parts of Γ. The symbol $\partial_\nu(.)$ denotes the derivative along the exterior unit normal ν to Γ. We assume that the given functions g and h and the boundary Γ possess the necessary smoothness properties to ensure the existence and uniqueness of a solution u to problem (1).

The method to transform (1) into a boundary integral equations is well known. If we represent by $G(\underset{\sim}{x};\underset{\sim}{y})$ the fundamental solution of (1a); that is,

$$\text{lap } G(\underset{\sim}{x};\underset{\sim}{y}) + \delta(\underset{\sim}{x};\underset{\sim}{y}) = 0 \tag{2}$$

where $\delta(\underset{\sim}{x};\underset{\sim}{y})$ is the Dirac distribution centered at $\underset{\sim}{y}$, then we find that [4,8] u must satisfy the integral equation

$$c(\underset{\sim}{x})\, u(\underset{\sim}{x}) - \int_{\Gamma_1} G(\underset{\sim}{y};\underset{\sim}{x})\, \partial_\nu u(\underset{\sim}{y})\, d\Gamma_1(\underset{\sim}{y}) + \int_{\Gamma_2} u(\underset{\sim}{y})\, \partial_\nu \,.$$
$$. \, G(\underset{\sim}{y};\underset{\sim}{x})\, d\Gamma_2(\underset{\sim}{y}) = - \int_{\Gamma_1} g(\underset{\sim}{y})\, \partial_\nu G(\underset{\sim}{y};\underset{\sim}{x})\, d\Gamma_1(\underset{\sim}{y}) +$$
$$+ \int_{\Gamma_2} G(\underset{\sim}{y};\underset{\sim}{x})\, h(\underset{\sim}{y})\, d\Gamma_2(\underset{\sim}{y}) \,. \tag{3}$$

In this equation $c(\underset{\sim}{x})$ is defined by

$$c(\underset{\sim}{x})\, u(\underset{\sim}{x}) = \int_\Omega u(\underset{\sim}{y})\, \delta(\underset{\sim}{x};\underset{\sim}{y})\, d\Omega(\underset{\sim}{y}) \,, \tag{4}$$

and by the known properties of the Dirac distribution $c(\underset{\sim}{x}) = 1$ if $\underset{\sim}{x}$ is an interior point of Ω, $c(\underset{\sim}{x}) = 1/2$ if $\underset{\sim}{x}$ is a smooth point of Γ and $c(\underset{\sim}{x}) = \omega/2\Pi$ if $\underset{\sim}{x}$ is a corner point of Γ with an interior solid angle of value $\tilde{\omega}$ at $\underset{\sim}{x}$. The fundamental solution G for the Laplace equation in R^2 is

$$G(\underset{\sim}{x},\underset{\sim}{y}) = \frac{1}{2\Pi} \ln \frac{1}{|\underset{\sim}{x}-\underset{\sim}{y}|} \,, \tag{5}$$

where $|\underset{\sim}{x}-\underset{\sim}{y}|$ denotes the Euclidean distance between points $\underset{\sim}{x}$ and $\underset{\sim}{y}$.

Now ∂_ν u is unknown on Γ_1 and u on Γ_2. We introduce functions ϕ and F such that

$$\phi(\underset{\sim}{x}) = u(\underset{\sim}{x}) \quad \text{and } F(\underset{\sim}{x};\underset{\sim}{y}) = -\partial_\nu G(\underset{\sim}{x};\underset{\sim}{y}) \quad \text{for } \underset{\sim}{x}\in\Gamma_2, \tag{6a}$$
$$\phi(\underset{\sim}{x}) = \partial_\nu u(\underset{\sim}{x}) \quad \text{and } F(\underset{\sim}{x};\underset{\sim}{y}) = G(\underset{\sim}{x},\underset{\sim}{y}) \quad \text{for } \underset{\sim}{x}\in\Gamma_1. \tag{6b}$$

Denoting by $p(x)$ the right hand side of equation (3) we can write this expression as

$$c(x)u(x) - \int_\Gamma F(y;x)\ \phi(y)\ dP(y) = p(x). \tag{7}$$

We notice that $F(x;y)$, like $G(x;y)$, and $\partial_\nu G(x;y)$ have a singularity as $x \to y$. However the integrals in (3) or (7) do exist [7,9]. Since $p(x)$ is a known function depending on the boundary values this expression could be used to find u at any point x of the domain Ω if $\phi(x)$ were known. Therefore the next step is to find methods to determine this function.

If we set the point x on the boundary Γ equation (7) yields the following integral equations

$$c(x)\ \phi(x) - \int_\Gamma F(y,x)\ \phi(y)\ d\Gamma(y) = p(x) \text{ for } x \in \Gamma_2, \tag{8a}$$

$$- \int_\Gamma F(y,x)\ \phi(y)\ d\Gamma(y) = p(x) - c(x)\ g(x) \text{ for } y \in \Gamma_1. \tag{8b}$$

These are coupled linear integral equations of Fredholm type of the second kind on Γ_2 and of the first kind on Γ_1, respectively.

Now we show how to solve this problem numerically by the BEM. Let us construct via finite elements a function space over Γ which we denote by $V_h(\Gamma)$ with $\dim V_h(\Gamma) = N_h < \infty$ and having the set of function ψ_{hi}, $i = 1,\dots, N_h$ as a basis. Then we seek an approximate solution ϕ_h to (8) belonging to $V_h(\Gamma)$. As

$$\phi_h(x) = \sum_{i=1}^{N_h} \phi_{hi}\ \psi_{hi}(x)\ , \tag{9}$$

and introducing this relation in (8) we obtain

$$\sum_{i=1}^{N_h} \left(c(x)\ \psi_{hi}(x) - \int_\Gamma F(y;x)\ \psi_{hi}(y)\ d\Gamma(y)\right)\phi_{hi} = $$
$$= p(x), \text{ for } x \in \Gamma_2 \tag{10a}$$

$$\sum_{i=1}^{N_h} \left(- \int_\Gamma F(y;x)\ \psi_{hi}(y)\ d\Gamma(y)\right)\phi_{hi} = p(x)-c(x)\ g(x)\ ,$$
$$\text{for } x \in \Gamma_1\ . \tag{10b}$$

To determine the coefficients ϕ_{hi} we solve these equations by collocation at the N_h nodes associated with the finite element mesh. Thus for any such node x_j we have

$$\sum_{i=1}^{N_h} \left(c(x_j)\ \delta_{ij} - \int_\Gamma F(y;x_j)\ \psi_{hi}(y)\ d\Gamma(y)\right)\phi_{hi} = $$
$$= p(x_j) \text{ for } x_j \in \Gamma_2\ , \tag{11a}$$

$$\sum_{i=1}^{N_h} \left(- \int_\Gamma \breve{F}(\underset{\sim}{y},\underset{\sim}{x}_j)\ \psi_{hi}(\underset{\sim}{y})\ d\Gamma(\underset{\sim}{y}) \right) \phi_{hi} = p(\underset{\sim}{x}_j) - c(\underset{\sim}{x}_j)\ g(\underset{\sim}{x}_j)$$

for $\underset{\sim}{x}_j \in \Gamma_1$. (11b)

In this way we have formed a linear system of N_h equations with N_h unknowns which we can write in matrix notation as

$$\underset{\sim}{A}_h\ \underset{\sim}{\phi}_h = \underset{\sim}{b}_h \qquad (12)$$

In spite of the fact that the basis functions ψ_{hi} have local support, the formulation of the BEM leads to fully populated matrices $\underset{\sim}{A}_h$, contrary to what happens with the standard finite element method.

THE TWO TEST CASES

To test the error and order of convergence of the BEM we chose two problems proposed by Schultz [10]. These consist of solving the Laplace equation in the square with mixed boundary condition, which we state as follows.

Problem 1

$$\text{lap } u = 0 \text{ in } \Omega = \{ (x,y) : x \in (0,\pi),\ y \in (0,\pi) \}, \qquad (13a)$$

$$u = 0 \text{ on } \Gamma_1' = \{ (x,y) : x \in [\tfrac{\pi}{2},\pi],\ y = 0\}, \qquad (13b)$$

$$u = \bar{u}_1 \text{ on } \Gamma_1'' = \{ (x,y) : x \in [0,\pi],\ y = \pi\}, \qquad (13c)$$

$$\partial_\nu u = 0 \text{ on } \Gamma_2' = \{ (x,y) : x = 0 \text{ or } x = \pi \text{ and } y \in [0,\pi] \}, \qquad (13d)$$

$$\partial_\nu u = 0 \text{ on } \Gamma_2'' = \{ (x,y) : x \in [0,\tfrac{\pi}{2}],\ y = 0 \}. \qquad (13e)$$

The function $\bar{u}_1$ was calculated by fitting the numerical data of [10].

Problem 2

$$\text{lap } u = 0 \text{ in } \Omega = \{ (x,y) : x \in (0,\pi),\ y \in (0,\pi) \}, \qquad (14a)$$

$$u = 0 \text{ on } \Gamma_1' = \{ (x,y) : x=0 \text{ or } x = \pi \text{ and } y \in [0,\pi]\}, \qquad (14b)$$

$$u = 0 \text{ on } \Gamma_1'' = \{(x,y) : x \in [\tfrac{\pi}{2},\pi],\ y = 0 \}, \qquad (14c)$$

$$\partial_\nu u = 0 \text{ on } \Gamma_2 = \{ (x,y) : x \in (0,\tfrac{\pi}{2}),\ y = 0 \}. \qquad (14d)$$

The function $\bar{u}_2$ was also computed by fitting the numerical data given in [10].

We note that the derivatives of u show a singularity at point $(\frac{\pi}{2},0)$. As Schultz [10] pointed out with regard to the FDM and we shall confirm for the BEM, this has a detrimental effect on the accuracy of the aproximate solutions.

RESULTS AND COMMENTS

Using the linear and parabolic elements as described in the appendices we computed the BEM approximated solutions to Problems 1 and 2 for several meshes. A typical sample of the BEM results is compiled in Tables 1 and 2. Table 3 shows the results obtained by the FEM for Problem 2.

The computer program developed employed everywhere single precision arithmetic (32-bit words). The final system of equations (12) was solved by Gauss condensation with full pivoting using a standard subroutine (GELG in [13]).

Figures 1a and 1b present typical meshes. We note the presence of double nodes that allows one to have u and $\partial_\nu u$ simultaneously unknowns at some points. This is possible since either these nodes are placed at corners of Γ_1 where the normal ν suffers a discontinuity and so does $\partial_\nu u$, or they are placed at the singularity $(\frac{\pi}{2},0)$ where one can form by collocation one equation of type (11a) and another of type (11b). In both cases this procedure gives rise to two distinct rows of matrix $\tilde{A}_h$ in (12).

The type of singularity is known for the two test problems and the development of special elements incorporating functions modelling the respective behaviour is therefore possible. Nevertheless, we used a standard program, so we had to resort to mesh grading to investigate if this way we could improve the order of convergence of the error and namely if the optimal order of convergence could be recovered.

Mesh grading was accomplished by decreasing the distance between consecutive nodes on the face y = 0 (see Figure 1b) in a geometric series of ratio r. This is in contrast with the arithmetic spacing employed by Symm [12], the change being motivated by the fact that this latter type of grading tends to generate uniform meshes when the number of nodes becomes great. Since we wanted to determine how mesh grading would improve the order of convergence of the BEM, the arithmetic spacing would therefore confuse the issue. On the other hand however, when the number of nodes increases, the geometric grading leads to a strongly non-uniform mesh. As an example, for $\Pi/32$ and $r = 0.6$ the ratio of greatest distance to smallest distance between consecutive nodes becomes equal to $0.6^{16} \simeq 3500$, which is very large indeed and caused ill-conditioning of matrix $\tilde{A}_h$ of system (12). Those runs that were unsuccessful due to this phenomenon are marked with an asterisk (*) in Tables 1 and 2.

Let us analyse briefly the results presented in Tables 1 and 2.

a) <u>Problem 1, linear element</u>

The results for the uniform mesh show a clear o(h) convergence similar to the FDM. However the magnitude of the error of the

BEM is about half of that obtained with the FDM of Schultz [10].

When grading with r = 0.8 is used the order of convergence improves significantly to $o(h^2)$ which is the optimal for linear elements. The magnitude of the error decreases also at all points shown except for the point $(\frac{\Pi}{4},0)$ and meshes $h = \frac{\Pi}{4}$ and $h = \frac{\Pi}{8}$. This we attribute to the fact that this point is no longer a node when mesh grading is employed and the respective values of the unknown have to be obtained by interpolation. For meshes with a large mesh size h this may offset the benefit from the grading.

The results for grading with r = 0.6 show a similar behaviour convergence of $o(h^2)$ everywhere except at a point $(\frac{\Pi}{4},0)$ where the values obtained do not seem conclusive. We remark that mesh $h = \frac{\Pi}{32}$ is strongly non-uniform and ill-conditioning prevented us from obtaining the respective results.

That pattern is even more evident in the case of r = 0.4. Now even the convergence order seems to be less than $O(h^2)$. As ill-conditioning must be even greater than before we suspect that round-off errors now play an important part.

b) Problem 1, parabolic element

The uniform mesh yields an o(h) convergence as before, with the magnitude of the error marginally smaller than for the linear element.

When grading r = 0.8 is employed results show a very marked improvement, with a sizeable decrease in the error and convergence order approaching the optimal $o(h^3)$.

For stronger grading r = 0.6 and r = 0.4 the results shown are too scarce to allow conclusions. Nevertheless case r = 0.6 still exhibits an $o(h^3)$ convergence except perhaps at point $(\frac{\Pi}{4},0)$ and the error is still less than for the other previous meshes. For grading r = 0.4 we still have gains at some points but at the expense of greater errors at others and an order of convergence probably less than optimal.

c) Problem 2, linear element

For the uniform mesh we have also a o(h) order of convergence. For points away from the singularity however this seems to increase slightly. The error is smaller than for methods 1 and 2 of Schultz, method 3 performing better near the singularity.

For grading r = 0.8 we get an order of convergence roughly close to $o(h^2)$ with a smaller error magnitude, which is more marked the finer the mesh.

Results for grading r = 0.6 confirm this behaviour. As before ill-conditioning prevented us from obtaining results for mesh $h = \frac{\Pi}{32}$.

Again the values presented for grading r = 0.4 do show some improvement but the order of convergence seems to be less than optimal.

d) Problem 2, parabolic element
The uniform mesh exhibits a sharp o(h) convergence with an error magnitude somewhat smaller than that for the corresponding mesh with the linear element.

Grading with r = 0.8 shows an improvement of the order of convergence but less than optimal, except for points away from the singularity.

The case r = 0.6 seems to yield an optimal order of convergence of $O(h^3)$ with a marked reduction in the magnitude of the error.

The results for grading r = 0.4 do not show a systematic behaviour. In fact, at points near the singularity the errors for mesh $h = \frac{\Pi}{16}$ are worse than for the mesh $h = \frac{\Pi}{8}$; we guess that round-off due to ill-conditioning is responsible for this situation.

From the evidence presented in this paper we draw the following conclusions. The BEM tends to yield smaller errors than the corresponding FDM schemes, requiring however a greater computational effort for the same accuracy. Mesh grading, judiciously employed, can allow us to recover the optimal order of convergence in cases where singularities are present. The development of special elements may not be necessary.

As compared with the FEM (Table 3) our results point to a roughly similar accuracy for the first order elements. For the second order elements the BEM seems to provide greater precision. In both cases the FEM required more computing time than the corresponding BEM runs.

APPENDIX A. The linear Isoparametric Element

In this appendix we summarize briefly the main aspects of the linear isoparametric element [4,14] implementation used to obtain the results presented above.

Base functions. The reference element is the interval [-1,1] equipped with the basis functions

$$N_1(\xi) = \frac{1}{2}(1 - \xi), \quad N_2(\xi) = \frac{1}{2}(1 + \xi). \qquad \text{(A.1a,b)}$$

The coordinate transformation is $\underset{\sim}{x} = \underset{\sim}{N}(\underset{\sim}{\xi})$ with

$$N : [-1,1] \to \Gamma_\alpha \in R^2 \qquad \text{(A.2a)}$$

and is given explicity by the relations

$$x = \sum_{i=1}^{2} N_i(\ )\, x_i, \quad y = \sum_{i=1}^{2} N_i(\xi)\, y_i, \qquad \text{(A.2b,c)}$$

where (x_i, y_i) are the global coordinates of node i of element α. The Jacobian J of transformation (A.1) is

$$J(\xi) = \left| \left(\frac{dx}{d\xi}\right)^2 + \left(\frac{dy}{d\xi}\right)^2 \right|^{1/2} . \qquad \text{(A.3)}$$

The linearity of equations (A.1) imply that this Jacobian is constant in each element and Γ_α is a segment of a straight line.

Computation of regular integrals. Let $I(\phi)$ be the integral over Γ_α of a bounded function ϕ. Then

$$I(\phi) = \int_{\Gamma_\alpha} \phi(\underset{\sim}{x})\, d\Gamma = \int_{-1}^{1} \hat{\phi}(\xi)\, J(\xi)\, d\xi = J \int_{-1}^{1} \hat{\phi}(\xi)\, d\xi, \quad \text{(A.4a)}$$

with

$$\hat{\phi}(\xi) = \phi(\underset{\sim}{x}) = \phi\,(N(\xi)) \quad \text{(A.4b)}$$

The integral on the right hand side of expression (A.4a) can be evaluated numerically. We employed a two-point Gauss-Legendre quadrature rule [1] to form the linear system of equations (12). Later to compute values at interior points and in order to overcome the difficulties already mentioned we resorted to an adaptive integration scheme based on the following heuristic approach. Let h_α be length of element α and define $\delta_\alpha(\underset{\sim}{x})$ by

$$\delta_\alpha(\underset{\sim}{x}) = \frac{1}{2}\,(\,|\,\underset{\sim}{x} - \underset{\sim}{x}_1| + |\,\underset{\sim}{x} - \underset{\sim}{x}_2|\,)\,, \quad \text{(A.5)}$$

where $\underset{\sim}{x}$ are the coordinates of the interior point under consideration and $\underset{\sim}{x}_1$ and $\underset{\sim}{x}_2$ are the coordinates of nodes 1 and 2 of element α. In a sense $\delta_\alpha(\underset{\sim}{x})$ measures the mean distance from point $\underset{\sim}{x}$ to element α. We compare this distance $\delta_\alpha(\underset{\sim}{x})$ with h_α and determine the number $n_\alpha(\underset{\sim}{x})$ of Gauss-Legendre points to use to integrate on Γ_α by the formula

$$n_\alpha(\underset{\sim}{x}) = \min\,(2 + 5k_\alpha(\underset{\sim}{x}),\, 5) \text{ with } k_\alpha(\underset{\sim}{x}) = \frac{\delta_\alpha(\underset{\sim}{x})}{h_\alpha}\,. \quad \text{(A.6)}$$

This criteria resembles that employed by Hess [6]. Remembering the well known properties of the ellipse it can be easily shown that the points satisfying

$$\delta_\alpha(\underset{\sim}{x}) < k\, h_\alpha \quad \text{(A.7)}$$

lie within an ellipse with focii on the nodes 1 and 2 of element α and semi-axes $k\, h_\alpha$ and $(k^2 - 1/4)^{1/2}\, h_\alpha$.

Computation of integrals with a logarithmic singularity. As shown above we need to evaluate integrals of the form

$$I_\alpha(\phi) = \int_{\Gamma_\alpha} \ln r(\underset{\sim}{x})\, \phi(\underset{\sim}{x})\, d\Gamma, \quad \text{(A.8)}$$

where $r(x)$ is the distance from a node of element to a generic point $\underset{\sim}{x}$ on the same element. This integral can be transformed to

$$I_\alpha(\phi) = \int_{-1}^{1} \ln r(N(\xi))\, \hat{\phi}(\xi)\, J(\xi)\, d\xi\,, \quad \text{(A.9)}$$

Now, for the sake of definiteness, suppose that the singularity sits on node 1 of element α. Then (A.9) can be transformed to

$$I_\alpha(\phi) = \int_{-1}^{1} \ln 2 \frac{\hat{r}(\xi)}{1+\xi} \hat{\phi}(\xi) J(\xi) d\xi - \int_{-1}^{1} \ln(\frac{1+\xi}{2}) \cdot \hat{\phi}(\xi) J(\xi) d\xi . \quad (A.10)$$

The integrand in the first integral of this expression can be easily shown to be bounded and therefore we are in condition to use the Gauss-Legendre quadrature points as before. The second integral in (A.10) has an integral with a logarithmic singularity at $\xi = -1$. Using the transformation

$$\bar{\xi} = \frac{1+\xi}{2} , \quad (A.11)$$

it can be put in the form

$$\int_{-1}^{1} \ln(\frac{1+\xi}{2}) \hat{\phi}(\xi) J(\xi) d\xi = 2\int_{0}^{1} \ln \bar{\xi} \bar{\phi}(\xi) \bar{J}(\bar{\xi}) d\bar{\xi} . \quad (A.12)$$

We evaluate this integral by applying to the right hand side of (A.12) the quadrature rule of Berthod-Zaborowiski [9] with two integration points.

If the singularity sits on node 2 of element α we use a similar procedure with obvious adaptations.

APPENDIX B. The Parabolic Isoparametric Element

Here we present briefly the main aspects of the parabolic isoparametric element implementation [4,14] employed to obtain the results presented in this paper.

Base functions. The reference element in this interval [-1,1] equipped with the basis functions

$$N_1(\xi) = \frac{1}{2} (1 - \xi) \xi , \quad (B.1a)$$

$$N_2(\xi) = 1 - \xi^2 , \quad (B,1b)$$

$$N_3(\xi) = \frac{1}{2} (1 + \xi) \xi . \quad (B.1c)$$

The coordinate transformation is $\underset{\sim}{x} = \underset{\sim}{N}(\underset{\sim}{\xi})$ with

$$N: [-1,1] \rightarrow \Gamma_\alpha \in R^2 \quad (B.2a)$$

and is given explicity by the relations

$$x = \sum_{i=1}^{3} N_i(\xi) x_i , \quad (B.2b)$$

$$y = \sum_{i=1}^{3} N_i(\xi) y_i , \quad (B.2c)$$

where, as before, (x_i, y_i) are the global coordinates of node i of element α. The Jacobian J of transformation (B.1) is also given by (A.3) but, contrary to the linear element case, is no longer a constant, and Γ_α can be in general a segment of a curve whose parametric equation are (B.2). However the value of the Jacobian J at a point ξ depends on the set of nodal coordinates (x_i, y_i), i=1 to 3. For some choices of these coordinates the Jacobian vanishes and the transformation (B.2) ceases to be one to one [3,5]. This fact has been used in the FEM to model the behaviour of the singularity.

Computation of regular integrals. The computation of regular integrals proceeds as in the case of the linear element, but now we employ three Gauss-Legendre integration points [1] instead of two. To compute values at interior points the adaptive integration scheme described in the previous appendix was changed slightly to adapt to this type of element as follows.

$$\delta_\alpha(\underset{\sim}{x}) = \frac{1}{2} \left(|\underset{\sim}{x}-\underset{\sim}{x}_1| + |\underset{\sim}{x}-\underset{\sim}{x}_3| \right) , \qquad \text{(B.3)}$$

where x_1 and x_3 are the coordinates of the extreme nodes of element α. The number of integration points is now chosen according to

$$n_\alpha(x) = \min(3 + 5k_\alpha(\underset{\sim}{x}), 5) . \qquad \text{(B.4)}$$

with $k_\alpha(x)$ given by

$$k_\alpha(\underset{\sim}{x}) = \frac{\delta_\alpha(x)}{\bar{h}_\alpha} \qquad \text{(B.5a)}$$

where $\bar{h}_\alpha$ is the length of the straight line segment connecting nodes 1 and 3.

Computation of integrals with a logarithmic singularity. The evaluation of integrals with logarithmic singularity proceeds along the same lines as for the linear element with the necessary adaptations. We employed now a quadrature rule of Berthod-Zaborowiski [9] with three integration points.

AKNOWLEDGEMENT

This work was supported by CTAMFUL, Universidade Técnica de Lisboa and by NATO Research Grant No. 1964.

REFERENCES

1. M. ABRAMOVITZ and I.A. STEGUN, Handbook of Mathematical functions, Dover (1968).

2. K.E. ATKINSON, A Survey of Numerical Methods for the Solution of Fredholm Integral Equations of the Second Kind, SIAM (1976).

3. R.S. BARSOUM, On the Use of Isoparametric Finite Elements in Linear Fracture Mechanics, Int. J. Num. Engng, 10, 25-37 (1976).

4. C.A. BREBBIA, The Boundary Element Method for Engineers, Pentech Press, (1978).

5. R.D. HENSCHELL and K.G. SHAW, Crack Tip Elements Are Unnecessary, Int. J. Num. Engng., 9, 495-507 (1975).

6. J.L. HESS and A.M.O. SMITH, Potential Flow About Arbitrary Bodies, in Progress in Aeronautical Sciences, vol. 8, Pergamon (1967).

7. Y. IKEBE, The Galerkin Method for the Numerical Solution of Fredholm Integral Equations of the Second Kind, SIAM Review, 14, 3, 465-491 (1972).

8. M.A. JASWON and G.T. SYMM, Integral Equations Methods in Potential Theory and Elastostatics, Academic Press (1977).

9. H. MINEUR, Techniques de Calcul Numérique, Dunod (1966).

10. D.H. SCHULTZ, Two Test Cases for the Numerical Solution of Harmonic Mixed Boundary Value Problems, J. Inst. Maths. Applics, 15, 1-8 (1975).

11. G. STRANG and G.F. FIX, An Analysis of the Finite Element Method, Prentice-Hall (1973).

12. G.T. SYMM, The Robin Problem for Laplace's Equation, in New Developments in Boundary Element Method, Proceedings of the Second International Seminar on Recent Advances in Boundary Element Methods, ed. by C.A. Brebbia, CML Publications (1980).

13. System/360 Scientific Package, version III, Programmer's Manual, IBM (1970).

14. O.C. ZIENKIEWICZ, The Finite Element Method in Engineering and Science, McGraw-Hill (1971).

15. O.C. ZIENKIEWICZ, D.W. KELLY and P. BETTESS, The Coupling of the Finite Element Method and Boundary Solution Procedures, Int. J. Num. Meth. Engng., 11, 355-375 (1977).

TABLE 1 - RESULTS FOR PROBLEM 1

(Point) Exact Value	Mesh Size h	BEM - Linear Element				BEM - Parabolic Element			
		r = 1.	r = 0.8	r = 0.6	r = 0.4	r = 1.	r = 0.8	r = 0.6	r = 0.4
(π/4,0) 1.5286	π/4	0.133	0.152	0.175	0.208	-	-	-	-
	π/8	0.077	0.078	0.045	0.074	0.071	0.044	-0.014	0.040
	π/16	0.042	0.025	0.023	0.054	0.037	0.015	-0.009	0.023
	π/32	0.022	0.003	0.019 §	*	0.019	0.002	-0.012 §	*
(π,3π/4) 2.9570	π/4	0.026	0.021	0.014	0.006	-	-	-	-
	π/8	0.012	0.008	0.004	$4x10^{-4}$	0.010	0.006	0.002	-0.001
	π/16	0.006	0.002	$-2x10^{-4}$	-0.001	0.005	0.002	$-2x10^{-4}$	-0.007
	π/32	0.003	$1x10^{-4}$	$-4x10^{-4}$ §	*	0.002	$1x10^{-4}$	$-4x10^{-4}$ §	*
(π/2,π/2) 2.2530	π/4	0.074	0.064	0.052	0.037	-	-	-	-
	π/8	0.036	0.024	0.014	0.009	0.028	0.018	0.008	0.002
	π/16	0.018	0.007	0.003	0.004	0.014	0.005	0.001	$5x10^{-4}$
	π/32	0.009	0.001	0.001 §	*	0.007	0.001	$1x10^{-4}$ §	*
(π/2,π/4) 1.4220	π/4	0.128	0.113	0.093	0.068	-	-	-	-
	π/8	0.063	0.043	0.025	0.015	0.050	0.031	0.014	0.004
	π/16	0.031	0.012	0.004	0.006	0.024	0.009	0.002	$-1x10^{-4}$
	π/32	0.015	0.002	0.002 §	*	0.012	0.001	0.0 §	*
(π/2,π/8) 0.9452	π/4	0.184	0.158	0.135	0.103	-	-	-	-
	π/8	0.103	0.065	0.037	0.019	0.076	0.048	0.022	0.005
	π/16	0.055	0.018	0.006	0.006	0.036	0.013	0.002	-0.001
	π/32	0.023	0.003	0.002 §	*	0.018	0.002	$-1x10^{-4}$ §	*

§ Solution of system (12) in double-precision.
* Elements too small. Underflow during formation of matrices H and G of small elements.

TABLE 2 - RESULTS FOR PROBLEM 2

(Point) Exact Value	Mesh Size h	BEM - Linear Element				BEM - Parabolic Element			
		r = 1.	r = 0.8	r = 0.6	r = 0.4	r = 1.	r = 0.8	r = 0.6	r = 0.4
(π/2,π/8) 1.0924	π/4	0.165	0.150	0.129	0.102	-	-	-	-
	π/8	0.084	0.058	0.033	0.018	0.066	0.040	0.018	0.002
	π/16	0.040	0.016	0.005	0.006	0.030	0.011	0.002	-0.003
	π/32	0.019	0.003	0.002 §	*	0.015	0.002	$1x10^{-4}$ §	*
(π/2,π/4) 1.9520	π/4	0.109	0.097	0.083	0.065	-	-	-	-
	π/8	0.048	0.033	0.020	0.013	0.034	0.021	0.009	0.001
	π/16	0.022	0.099	0.004	0.005	0.016	0.006	0.001	-0.002
	π/32	0.010	0.001	0.002 §	*	0.008	0.001	0.0	*
(π/2,π/2) 4.7054	π/4	0.058	0.052	0.046	0.039	-	-	-	-
	π/8	0.020	0.015	0.010	0.007	0.013	0.008	0.004	0.001
	π/16	0.009	0.004	0.002	0.003	0.006	0.002	$3x10^{-4}$	$-3x10^{-4}$
	π/32	0.004	0.001	0.001 §	*	0.003	$3x10^{-4}$	$-1x10^{-4}$ §	*
(π/2,3π/4) 10.5032	π/4	0.037	0.035	0.032	0.030	-	-	-	-
	π/8	0.010	0.007	0.006	0.005	0.004	0.002	$1x10^{-4}$	-0.001
	π/16	0.003	0.001	$4x10^{-4}$	0.001	0.002	$4x10^{-4}$	$-3x10^{-4}$	$-4x10^{-4}$
	π/32	0.001	$-2x10^{-4}$	$-1x10^{-4}$ §	*	0.001	$-3x10^{-4}$	$-4x10^{-4}$ §	*

TABLE 3 - RESULTS FOR PROBLEM 2 WITH FINITE ELEMENTS

(Point coord.) EXACT VALUE	MESH SIZE				
$(\pi/2, \pi/8)$ 1.0924	$\pi/4$	0.0926	0.2269	0.42315	0.3336
	$\pi/8$	0.1041	0.0695	0.26922	0.1072
	$\pi/16$	0.0594	0.0390	0.17758	0.0143
	$\pi/32$	0.0302	0.0189	0.12428	0.0211
$(\pi/2, \pi/4)$ 1.9520	$\pi/4$	-0.0472	0.2211	0.393	0.0726
	$\pi/8$	0.0195	0.0789	0.2639	0.001
	$\pi/16$	0.0212	0.0299	0.1209	0.0239
	$\pi/32$	0.0139	0.0126	0.0912	0.008
$(\pi/2, \pi/2)$ 4.7054	$\pi/4$	-0.2861	0.3739	0.7054	-0.501
	$\pi/8$	-0.0639	0.1002	0.1101	0.0133
	$\pi/16$	0.0097	0.0291	0.055	0.0059
	$\pi/32$	0.0009	0.0094	0.042	0.0027
$(\pi/2, 3\pi/4)$ 10.5032	$\pi/4$	-0.3708	0.4412	-0.0318	0.5893
	$\pi/8$	-0.0928	0.1082	0.0602	-0.0498
	$\pi/16$	-0.0218	0.0282	0.0212	$2x10^{-4}$
	$\pi/32$	-0.0048	0.0072	0.0152	$2x10^{-4}$

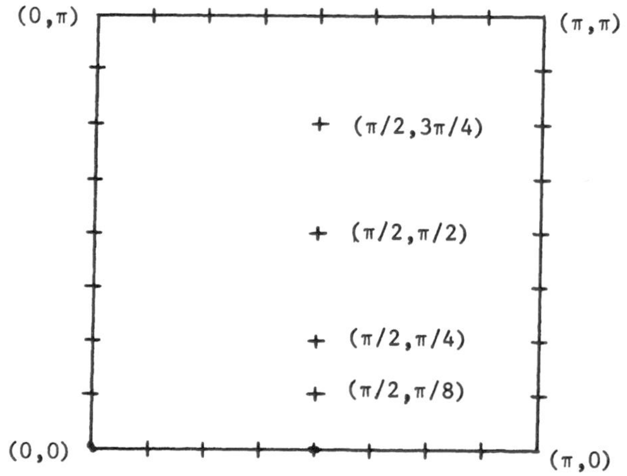

a) Uniform mesh (h=π/8)

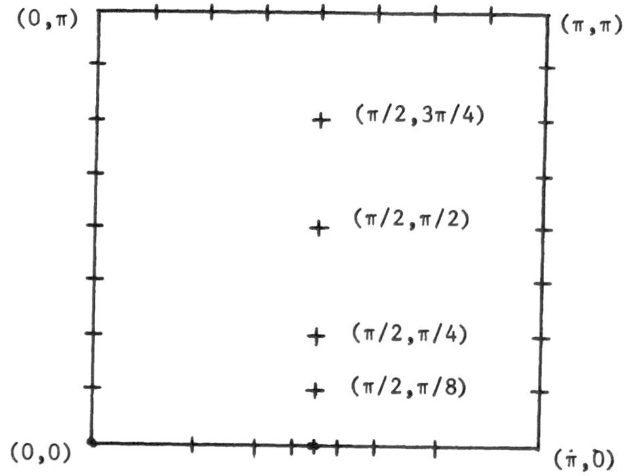

b) Mesh (h=π/8) with grading r=0.6

Figure 1. Typical meshes for Problem 2

BOUNDARY ELEMENT & LINEAR PROGRAMMING METHOD IN OPTIMIZATION OF PARTIAL DIFFERENTIAL EQUATION SYSTEMS

T. Futagami

Hiroshima Institute of Technology, Itsukaichi, Hiroshima, Japan

ABSTRACT

By combining boundary element method with linear programming, a new optimization technique (boundary element & linear programming method, or, the BE&LP method) is developed and systematized in order to control partial differential equation systems with both equality or inequality constraints and an objective function. The BE&LP method is applied to optimal control problems in heat conduction phenomena. Minimization of total of the controllable loads to meet with temperature requirements is performed by using simplex method of linear programming. The tractability in both the boundary conditions and the equality or inequality constraints makes sure that the method becomes a powerful technique for several new types of boundary value problems.

INTRODUCTION

By combining finite element method with linear programming, finite element & linear programming method (the FE&LP method) has been developed in order to control partial differential systems (Futagami (7, 8, 9)). The combined use of finite difference method with linear programming has been studied in the field of ground-water management by Aguado and Remson (1).

In this research, by combining boundary element method with linear programming, boundary element & linear programming method (the BE&LP method) is developed and systematized in order to control partial differential equation systems with both equality or inequality constraints and an objective function. Such systems are frequently encountered in various engineering and scientific problems of control and optimal design.

Boundary element method, increasingly used in Europe, is a powerful numerical method for the solution of partial differential equations because of its generality with respect to geometry, its simplicity of the input data required, and its numerical accuracy (2, 3, 4). Linear programming is one of the most frequently used mathematical methods of operations research (5,10). In the development of the BE&LP method the concepts of the decision variable and the state variable are adopted as in Bellman's dynamic programming and Pontryagin's maximum principle. The BE&LP method utilizes the advantages of established numerical techniques of both boundary element method and linear programming.

In order to show the applicability of the BE&LP method, the method is applied to optimal control problems in heat conduction phenomena. Numerical examples are performed in a simple model. The method may become a useful technique for problems of control and optimal design in various fields of engineering and science.

BOUNDARY ELEMENT & LINEAR PROGRAMMING METHOD

Systems of Basic Differential Equations

Boundary element method & linear programming method (the BE&LP method) is developed to solve the following systems of differential equations with both equality or inequality constraints and an objective function (see Fig. 1).

Objective Function (throughout the whole domain Ω)

$$Z = \underset{\{\theta\}}{\text{Opt.}}\ f(\{\phi\}, \{q\}, \{\theta\}) = \begin{cases} \underset{\{\theta\}}{\text{Max.}}\ f(\{\phi\}, \{q\}, \{\theta\}) \\ \underset{\{\theta\}}{\text{Min.}}\ f(\{\phi\}, \{q\}, \{\theta\}) \end{cases} \qquad (1)$$

subject to:

Equilibrium Equations

Governing Differential Equation (in the whole domain Ω)

$$\text{D.E.}\ (x_k, \phi, \frac{\partial \phi}{\partial x_k}, \cdots, \frac{\partial^n \phi}{\partial x_k^n}, \theta) = 0 \qquad (2)$$

Boundary Conditions (on the boundaries Γ)

$$e\ (X_k, \phi, \frac{\partial \phi}{\partial n}) = 0 \qquad (3)$$

Constraints (in the sub-domains Ω^r)

$$r\ (x_k, \phi, q, \theta) \lesseqgtr 0 \qquad (4)$$

in which ϕ = the state variable; θ = the decision variable; x_k = Cartesian coordinate (x, y, z); X_k = Cartesian coordinate of the boundary (X, Y, Z); and q = the flux.

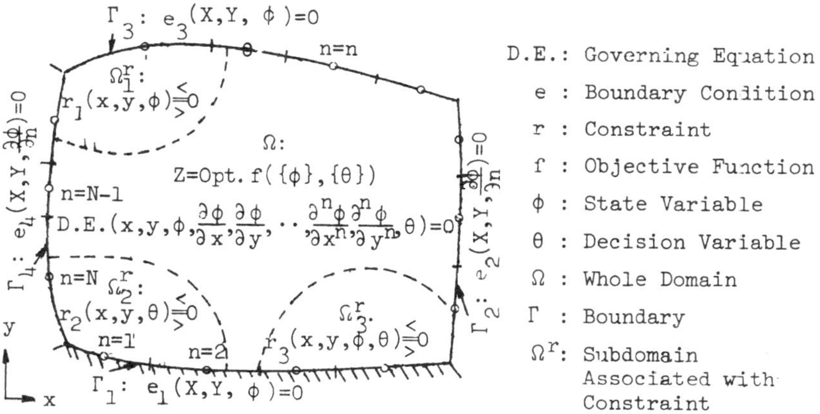

Fig. 1. General Concepts of the BE&LP method

In systems governed by two-dimensional second-order differential equations for example, Eq. 2 is expressed as follows:

$$c_1\frac{\partial^2\phi}{\partial x^2} + c_2\frac{\partial^2\phi}{\partial x\partial y} + c_3\frac{\partial^2\phi}{\partial y^2} + c_4\frac{\partial\phi}{\partial x} + c_5\frac{\partial\phi}{\partial y} + c_6\phi + \theta + b = 0 \qquad (2)'$$

The examples of the boundary conditions are as follows:

$$\phi(x=X,\ y=Y) = \hat{\Phi}(X,Y), \qquad \frac{\partial\phi}{\partial n}\Big|(x=X,\ y=Y) = -\hat{q} \qquad (3)'$$

As for the constraints, the following simple inequalities are frequently encountered.

$$\underline{\Phi} \leq \phi \leq \overline{\Phi}, \text{ or, } \begin{cases}\phi \geq \underline{\Phi}\\ \phi \leq \overline{\Phi}\end{cases}$$
$$\underline{\Theta} \leq \theta \leq \overline{\Theta}, \text{ or, } \begin{cases}\theta \geq \underline{\Theta}\\ \theta \leq \overline{\Theta}\end{cases} \qquad (4)'$$

in which $\underline{\Phi}$ = the lower limit of the state variable; $\overline{\Phi}$ = the upper limit of of the state variable; $\underline{\Theta}$ = the lower limit of the decision variable; and $\overline{\Theta}$ = the upper limit of the decision variable.

<u>Formulation of Boundary Element & Linear Programming Method</u>

Boundary element method is used in order to discretize the above-mentioned systems as systems of linear algebraic equations. (As for the details, see the next chapter). Then, the following matrix-vector forms of the BE&LP method are obtained and the application of linear programming is possible.

Objective Function (throughout the whole domain Ω)

$$Z = \underset{\{\theta_i\}}{\text{Opt.}}\ f(\{x_n\}, \{\theta_i\}) = Z = \underset{\{\theta_i\}}{\text{Opt.}}\ f(\{\phi_n\}, \{q_n\}, \{\theta_i\}) \quad (5)$$

subject to:

Equilibrium Equations (N-Eqs.)

$$\underset{N\times N}{[H]}\{\phi_n\} = \underset{N\times N}{[G]}\{q_n\} + \underset{N\times I}{[D]}\{\theta_i\} + \{p_n\}, \text{ or, } \underset{N\times N}{[A]}\{x_n\} - \underset{N\times I}{[D]}\{\theta_i\} = \{b_n\} \quad (6)$$

Constraints (L-Eqs.)

$$\underset{L\times(N+I)}{[R]} \begin{Bmatrix} \{x_n\} \\ \{\theta_i\} \end{Bmatrix} \begin{matrix} \le \\ = \\ > \end{matrix} \{B_l\} \quad (7)$$

in which x_n = nth boundary variable; ϕ_n = nth state on boundary elements; θ_i = ith decision variable; q_n = nth flux on boundary elements; [A] = matrix derived from boundary element method; [H] = the state-matrix derived from boundary element method; [G] = the flux-matrix derived from boundary element method; b_n = constant in nth equilibrium equation; p_n = constant in nth equilibrium equation; [R] = the constraint matrix, sparse matrix; B_l = constant in lth constraint; n = 1 ∿ N (N: total number of boundary elements); i = 1 ∿ I (I: total number of the decision variables); l = 1 ∿ L (L: total number of the constraints).

Therefore, the BE&LP method is one that optimizes the objective function under the conditions of the equilibrium equations and the constraints. In the sense of general linear programming, the equilibrium equations of the BE&LP method are also the constraints. The solutions of the BE&LP method are obtained by the simplex method for linear programming.

HEAT CONDUCTION CONTROL BY THE BE&LP METHOD

Systems of Basic Equations in Heat Conduction Control

Heat conduction control by the BE&LP method is studied in order to show the applicability of the method (see Fig. 2). The basic equation systems of two-dimensional heat conduction phenomena with constraints are as follows:

Objective Function (throughout the whole domain Ω)

$$Z = \underset{\{\theta\}}{\text{Opt.}}\ f(\{\phi\}, \{q\}, \{\theta\}) \simeq \underset{\{\theta\}}{\text{Min.}}\ \Sigma\,\theta \quad (8)$$

subject to:

Equilibrium Equations

Governing Differential Equation

$$K\nabla^2\phi + \theta + Q = 0 \qquad \text{in } \Omega \quad (9)$$

Boundary Conditions

$$\phi = \hat{\Phi} \qquad \text{on } \Gamma_1 \tag{10}$$

$$q = -K\frac{\partial\phi}{\partial n} = \hat{q} \qquad \text{on } \Gamma_2 \tag{11}$$

$$q = \gamma(\phi - \phi_s) \qquad \text{on } \Gamma_3 \tag{12}$$

Constraints

$$\phi \geqq \underline{\Phi} \qquad \text{in } \Omega_1^r \text{ and on } \Gamma_2 + \Gamma_3 \tag{13}$$

$$\theta \leqq \overline{\Theta} \qquad \text{in } \Omega_2^r \tag{14}$$

Non-negative Conditions

$$\phi \geqq 0 \qquad \text{in } \Omega_1^r \text{ and on } \Gamma_2 + \Gamma_3 \tag{15}$$

$$\theta \geqq 0 \qquad \text{in } \Omega_2^r \tag{16}$$

in which ϕ = the state (temperature); q = heat flow (the cut-flux (+) or the in-flux (-)); θ = the decision variable (controllable load of heat generator); $\hat{\Phi}$ = the prescribed temperature; $\hat{q}$ = the prescribed out flux; ϕ_s = the surrounding temperature; K = the conductivity coefficient; Q = the heat generated in the body (uncontrollable load); γ = the convection coefficient; $\underline{\Phi}$ = the lower limit of the state (temperature requirement); $\overline{\Theta}$ = the upper limit of the decision variable (the upper limit of the controllable load); Ω = the whole domain; $\Gamma = \Gamma_1 + \Gamma_2 + \Gamma_3$ = the boundaries; Ω_1^r = subdomain associated with the state-constraints; Ω_2^r = subdomain associated with the decision-constraints.

The boundary condition expressed in Eq. 12 is encountered when heat is lost from the surrounding media by convection. The heat flux is assumed to depend linearly on the temperature difference $(\phi - \phi_s)$.

Although, the objective function may be composed of the temperature distribution $\{\phi\}$, the heat flows $\{q\}$, and the controllable loads $\{\theta\}$ in general, the minimization of the total of the controllable loads $\Sigma\,\theta$ is sought in this study.

As for the constraints, although only the lower limit of the state $\underline{\Phi}$ and the upper limit of the controllable load $\overline{\Theta}$ are imposed in the above systems, the upper limit of the state and the lower limit of the controllable load may occasionally become necessary with respect to the conditions of the problems.

Formulation of the BE&LP Method in Heat Conduction Control

In order to discretize the above mentioned differential systems, boundary element method is used. As for the details of boundary element method, one may follow Brebbia (3). Boundary element method in heat conduction problems has been studied by many researchers (2, 4, 6). By using the weighted residual method, the equilibrium equations (Eqs. 9-12) are replaced by the following integral formulation.

$$\int_{\Omega} \{K\nabla^2\phi + \theta + Q\}\phi^* \, d\Omega + \int_{\Gamma_1} \{\phi - \hat{\phi}\}K\frac{\partial\phi^*}{\partial n} \, d\Gamma$$

$$+ \int_{\Gamma_2} \{q - \hat{q}\}\phi^* \, d\Gamma + \int_{\Gamma_3} [q - \gamma\{\phi - \phi_s\}]\phi^* \, d\Gamma = 0 \qquad (17)$$

The weighting function ϕ^* has continuous first derivatives. Later this function will also be required to satisfy the governing differential equation (Eq. 9). Integrating the first term of Eq. 17 by parts twice, we obtain:

$$\int_{\Omega} K\nabla^2\phi^*\phi \, d\Omega + \int_{\Omega} (\theta + Q)\phi^* \, d\Omega - \int_{\Gamma_1} q\phi^* \, d\Gamma - \int_{\Gamma_2} \hat{q}\phi^* \, d\Gamma$$

$$- \int_{\Gamma_3} \gamma(\phi - \phi_s)\phi^* \, d\Gamma + \int_{\Gamma_1} \hat{\phi}q^* \, d\Gamma + \int_{\Gamma_2+\Gamma_3} \phi q^* \, d\Gamma = 0 \qquad (18)$$

where

$$q = -K\frac{\partial\phi}{\partial n}, \qquad q^* = -K\frac{\partial\phi^*}{\partial n}$$

The weighting function is now taken as the fundamental solution to Laplace's equation, which is the solution corresponding to a unit point source, i.e.

$$\nabla^2\phi^* + \Lambda^n = 0 \qquad (19)$$

where Δ^n is the Dirac delta function and n is the point where the source acts. The fundamental solution ϕ^* has the following property

$$\int_{\Omega} \phi[\nabla^2\phi^* + \Delta^n] \, d\Omega = \int_{\Omega} \phi\nabla^2\phi^* d\Omega + \phi^n = 0 \qquad (20)$$

where ϕ^n represents the value of the unknown function ϕ at the point where the source is applied. The fundamental solution to two-dimensional Laplace's equation is given by

$$\phi^* = \frac{1}{2\pi} \ln \frac{1}{r} \qquad (21)$$

where r is the distance from the point of application of the unit source to the point under consideration. Substitution of Eqs. 20 and 21 into Eq. 18 yields the following equation.

$$- \pi K\phi^n + \int_{\Gamma_1} \hat{\phi} \, K\frac{\hat{\mathbf{r}}\cdot\mathbf{n}}{r^2} \, d\Gamma + \int_{\Gamma_2+\Gamma_3} \phi K\frac{\mathbf{r}\cdot\mathbf{n}}{r^2} \, d\Gamma = \int_{\Gamma_1} q \ln\frac{1}{r} \, d\Gamma$$

$$+ \int_{\Gamma_2} \hat{q} \ln\frac{1}{r} \, d\Gamma + \int_{\Gamma_3} \gamma\phi\ln\frac{1}{r} \, d\Gamma - \int_{\Gamma_3} \gamma\phi_s\ln\frac{1}{r} \, d\Gamma$$

$$- \int_{\Omega} (\theta + Q) \, d\Omega \qquad (22)$$

where **r** is the vector from the point where the source is applied to the point on the boundary and **n** is outward unit normal from the point on the boundary.

If θ (the decision variable, i.e. controllable load) or Q (constant, i.e., uncontrollable load) is not equal to zero, then the whole domain Ω has to be divided into a series of cells, so a numerical integration can be performed. But in the case all of the loads are point sources (concentrated loads), we just have to subtract $\Sigma\theta_i \ln(1/r)$ and $\Sigma Q_s \ln(1/r)$ from the right hand side of Eq. 22. θ_i means the controllable point load at internal or boundary point 'i', and Q_s means the uncontrollable point load at internal or boundary point 's'. From now on in this study we consider only such point loads.

Eq. 22 is now applied on the boundary of the domain under consideration. Although several types of boundary elements have been presented (3), constant boundary element method is used in this study (see Fig. 2). In the constant boundary element method the values of ϕ and q are assumed to be constant on each element and equal to the values at the mid-node of the element. Therefore, Eq. 22 is rewritten by the following Equation.

$$- \pi K\phi^n + \hat{\phi}\int_{\Gamma_1} K\frac{\mathbf{r}\cdot\mathbf{n}}{r^2}\, d\Gamma + \phi\int_{\Gamma_2} K\frac{\mathbf{r}\cdot\mathbf{n}}{r^2}\, d\Gamma + \phi\int_{\Gamma_3}\left(K\frac{\mathbf{r}\cdot\mathbf{n}}{r^2} - \gamma\ln\frac{1}{r}\right) d\Gamma$$

$$= q\int_{\Gamma_1}\ln\frac{1}{r}\, d\Gamma + \hat{q}\int_{\Gamma_2}\ln\frac{1}{r}\, d\Gamma - \gamma\phi_s\int_{\Gamma_3}\ln\frac{1}{r}\, d\Gamma$$

$$- \sum_{i=1}^{I}\theta_i \ln\frac{1}{r_n^i} - \sum_{s=1}^{S} Q_s \ln\frac{1}{r_n^s} \qquad (23)$$

Then, Eq. 23 for a given node 'n' becomes in the following discretized form:

$$- \pi K\phi^n + \sum_{j=1}^{N}\hat{h}_{nj}\phi_j = \sum_{j=1}^{N} g_{nj}q_j + \sum_{i=1}^{I} d_{ni}\theta_i + p_n \qquad (24)$$

where

$$\hat{h}_{nj} = \int_{\Gamma_e} K\frac{\mathbf{r}\cdot\mathbf{n}}{r^2}\, d\Gamma \qquad (\text{for 'j' on } \Gamma_1 + \Gamma_2)$$

$$\hat{h}_{nj} = \int_{\Gamma_e}\left(K\frac{\mathbf{r}\cdot\mathbf{n}}{r^2} - \gamma\ln\frac{1}{r}\right) d\Gamma \qquad (\text{for 'j' on } \Gamma_3)$$

$$g_{nj} = \int_{\Gamma_e}\ln\frac{1}{r}\, d\Gamma \qquad (\text{for 'j' on } \Gamma_1 + \Gamma_2)$$

$$g_{nj} = 0 \qquad (\text{for 'j' on } \Gamma_3)$$

$$d_{ni} = -\ln\frac{1}{r_n^i}$$

$$p_n = -\sum_{s=1}^{S} Q_s \ln\frac{1}{r_n^s} - \gamma\phi_s\int_{\Gamma_3}\ln\frac{1}{r}\, d\Gamma$$

Therefore, the whole set of the equilibrium equations for the N nodes on the whole boundary is expressed in the following matrix-vector form:

$$\underset{N\times N}{[H]}\{\phi_n\} = \underset{N\times N}{[G]}\{q_n\} + \underset{N\times I}{[D]}\{\theta_i\} + \{p_n\} \tag{25}$$

where

$$h_{nj} = \hat{h}_{nj} - \pi K \qquad (n = j)$$

$$h_{nj} = \hat{h}_{nj} \qquad (n \neq j)$$

By noting the fact that N_1 values of ϕ and N_2 values of q are known on the whole boundary, Eq. 25 is rewritten in such a way that all the unknowns are on the left hand side as follows:

$$\underset{N\times N}{[A]}\{x_n\} - \underset{N\times N}{[D]}\{\theta_i\} = \{b_n\} \tag{26}$$

where $\{x_n\}$ is the vector of unknown ϕ and q.

Then, the discretized form of the equilibrium equations are obtained in Eq. 25 or 26.

As for the state-constraints (Eq. 13), the following discretized inequalities are obtained.

$$\phi^o_l = \frac{1}{\pi K}\left(\sum_{j=1}^{N} \hat{h}^o_{lj}\phi_j - \sum_{j=1}^{N} g^o_{lj}q_j - \sum_{i=1}^{I} d^o_{li}\theta_i - p^o_l\right) \geqq \underline{\Phi}^o_l$$

$$(\text{for } 'l' \text{ point in } \Omega^r_1,\ l = 1 \sim L^\phi) \tag{27}$$

$$\phi_l \geqq \underline{\Phi}_l \qquad (\text{for } 'l' \text{ mid-node on } \Gamma_2 + \Gamma_3) \tag{28}$$

As for the decision-constraints (Eq. 14), the following inequalities are obtained for the points fitted for the controllable loads (the decision variables)

$$\theta_i \leqq \overline{\Theta}_i \qquad (\text{for } 'i' \text{ point in } \Omega^r_2,\ i = 1 \sim I) \tag{29}$$

Then, the matrix-vector forms of the BE&LP method in the heat conduction control problem are obtained as shown below:

Objective Function (throughout the whole domain Ω)

$$Z = \underset{\{\theta_i\}}{\text{Opt.}}\ f(\{x_n\}, \{\theta_i\}) = \underset{\{\theta_i\}}{\text{Opt.}}\ f(\{\phi_n\}, \{q_n\}, \{\theta_i\})$$

$$\simeq \underset{\{\theta_i\}}{\text{Min.}} \sum_{i=1}^{I} \theta_i \tag{30}$$

subject to:

Equilibrium Equations (N-Eqs.)

$$\underset{N\times N}{[H]}\{\phi_n\} = \underset{N\times N}{[G]}\{q_n\} + \underset{N\times I}{[D]}\{\theta_i\} + \{p_n\} \quad \text{, or,} \tag{31}$$

$$\underset{N\times N}{[A]}\{x_n\} - \underset{N\times I}{[D]}\{\theta_i\} = \{b_n\} \tag{31'}$$

Constraints $((L=L^{\phi}+I)$-Eqs.)

$$\{\phi^o_l\} = ([\hat{h}^o_{lj}]\{\phi_j\} - [g^o_{lj}]\{q_j\} - [d^o_{li}]\{\theta_i\} - \{p^o_l\})\frac{1}{\pi K} \geqq \{\underline{\Phi}^o_l\}$$

$$\text{(for interior points in } \Omega^r_1\text{)} \tag{32}$$

$$\{\phi_l\} \geqq \{\underline{\Phi}_l\} \qquad \text{(for mid-nodes on } \Gamma_2 + \Gamma_3\text{)} \tag{33}$$

$$\{\theta_i\} \leqq \{\overline{\Theta}_i\} \tag{34}$$

Non-negative Conditions

$$\phi_n \geqq 0 \ (n = 1 \sim N), \qquad \theta_i \geqq 0 \ (i = 1 \sim I) \tag{35}$$

NUMERICAL EXAMPLES IN A SIMPLE MODEL

In order to clarify the features of the BE&LP methods, the systems of the governing equations are written down for a simple square model (see Figs. 3, 4(a) and 4(b)). The obtained matrix-vector forms of the BE&LP method in Run 1 are as follows:

Objective Function

$$Z = \underset{\{\theta_i\}}{\text{Min.}} \sum_{i=1}^{2} \theta_i = \text{Min.}(\theta_1 + \theta_2) \tag{36}$$

subject to:

Equilibrium Equations (4-Eqs.)

$$\begin{bmatrix} -3.14 & 1.11 & 11.91 & 1.11 \\ 1.11 & -3.14 & 9.85 & 0.93 \\ 0.93 & 1.11 & -2.55 & 1.11 \\ 1.11 & 0.93 & 9.85 & -3.14 \end{bmatrix}_H \begin{Bmatrix} \phi_1 \\ \phi_2 \\ \phi_3 \\ \phi_4 \end{Bmatrix} = \begin{bmatrix} -0.59 & -8.74 & 0 & -8.74 \\ -8.74 & -0.59 & 0 & -10.98 \\ -10.98 & -8.74 & 0 & -8.74 \\ -8.74 & -10.98 & 0 & -0.59 \end{bmatrix}_G \begin{Bmatrix} 0 \\ 0 \\ q_3 \\ 0 \end{Bmatrix}$$

$$+ \begin{bmatrix} 1.10 & 1.50 \\ 1.10 & 1.21 \\ 1.10 & 0.41 \\ 1.10 & 1.21 \end{bmatrix}_D \begin{Bmatrix} \theta_1 \\ \theta_2 \end{Bmatrix} + \begin{Bmatrix} 2.03 \\ 5.49 \\ 7.52 \\ 5.49 \end{Bmatrix} \tag{37}$$

, or,

$$\begin{bmatrix} -3.14 & 1.11 & 11.91 & 1.11 \\ 1.11 & -3.14 & 9.85 & 0.93 \\ 0.93 & 1.11 & -2.55 & 1.11 \\ 1.11 & 0.93 & 9.84 & -3.14 \end{bmatrix}_A \begin{Bmatrix} \phi_1 \\ \phi_2 \\ \phi_3 \\ \phi_4 \end{Bmatrix} - \begin{bmatrix} 1.10 & 1.50 \\ 1.10 & 1.21 \\ 1.10 & 0.41 \\ 1.10 & 1.21 \end{bmatrix}_D \begin{Bmatrix} \theta_1 \\ \theta_2 \end{Bmatrix} = \begin{Bmatrix} 2.03 \\ 5.49 \\ 7.52 \\ 5.49 \end{Bmatrix} \quad (37)$$

Constraints (4-Eqs.)

$$\begin{bmatrix} 1 & 0 & 0 & 0 \\ 0 & 0 & 1 & 0 \end{bmatrix}_{R\phi} \begin{Bmatrix} \phi_1 \\ \phi_2 \\ \phi_3 \\ \phi_4 \end{Bmatrix} \geqq \begin{Bmatrix} \underline{\Phi}_1 = 10.00 \\ \underline{\Phi}_3 = 5.00 \end{Bmatrix} \quad (38)$$

$$\begin{bmatrix} 1 & 0 \\ 0 & 1 \end{bmatrix}_{R\theta} \begin{Bmatrix} \theta_1 \\ \theta_2 \end{Bmatrix} \leqq \begin{Bmatrix} \overline{\Theta}_1 = 20.00 \\ \overline{\Theta}_2 = 20.00 \end{Bmatrix} \quad (39)$$

Non-negative Conditions

$$\phi_n \geqq 0 \ (n = 1 \sim 4), \quad \theta_i \geqq 0 \ (i = 1 \sim 2) \quad (40)$$

Computed results for the boundary temperatures (ϕ_1, ϕ_2, ϕ_3, ϕ_4), the heat flux (q_3), the controllable loads (θ_1, θ_2) and the optimal objective (Z) are shown in Figs. 5(a) and 5(b). It should be noted that the distribution patterns of boundary temperatures $\{\phi_n\}$ in Figs. 5(a) and 5(b) arise from the resultant loads composed of the obtained controllable loads $\{\theta_i\}$ and the given uncontrollable load Q_1.

CONCLUSIONS

Boundary element & linear programming method (the BE&LP method) was described through the application to the optimal control of heat conduction phenomena. The tractability in both the boundary conditions and the equality or inequality constraints makes sure that the method becomes a powerful technique for various problems of optimal control and design.

Finally, the problems to be attacked from now on in the applications of the BE&LP method should be mentioned. Extension of the method to the time domain is necessary to solve transient problems. The related methods such as boundary element & non-linear programming method, and stochastic boundary element & linear programming method could be developed. The developments certainly make it possible to solve more complicated problems.

REFERENCES

1. Aguado, E. and Remson, I. (1974), "Ground-Water Hydraulics in Aquifer Management," Journal of the Hydarualics Division, ASCE, Vol. 100, No. HY1, pp. 103-118.
2. Bolteus, L. and Tullberg, O. (1980), "Boundary Element Method Applied to Two-Dimensional Heat Conduction in Non-homogeneous Media," Advances in Engineering Software, Vol.2,

No. 3, pp. 131-137.
3. Brebbia, C.A.(1978), "The Boundary Element Method for Engineers," Pentech Press, London.
4. Brebbia, C.A.(1980), "Fundamental of Boundary Elements," New Developments in Boundary Element Methods, Proceedings of the Second International Seminar on Recent Advances in Boundary Element Methods, CML Publications, pp. 3-33.
5. Dantzig, B.G. (1963), "Linear Programming and Extensions," Princeton University Press.
6. Dubois, M. and Buysse, M. (1980), "Transient Heat Transfer Analysis by the Boundary Integral," New Developments in Boundary Element Methods, Proceedings of the Second International Seminar on Recent Advances in Boundary Element Methods, CML Publications, 137-154.
7. Futagami, T. (1975), " Finite Element & Linear Programming Method and Water Pollution Control," Proceedings of 16th Congress of the International Association for Hydraulics Research, Vol. 3, c7, pp. 54-61.
8. Futagami, T., Tamai, N. and Yatsuzuka, M. (1976), "FEM Coupled with LP for Water Pollution Control," Journal of Hydraulic Division, ASCE, Vol. 102, Hy7, pp. 881-897.
9. Futagami, T. (1976), "Several Mathematical Methods in Water Pollution Control — THE FINITE ELEMENT & LINEAR PROGRAMMING METHOD and the Related Methods," HIT-C-EH-1, Department of Civil Engineering, Hiroshima Institute of Technology.
10. Gass, I.S. (1969), "Linear Programming," 3rd ed, McGraw-Hill Kogakusha.

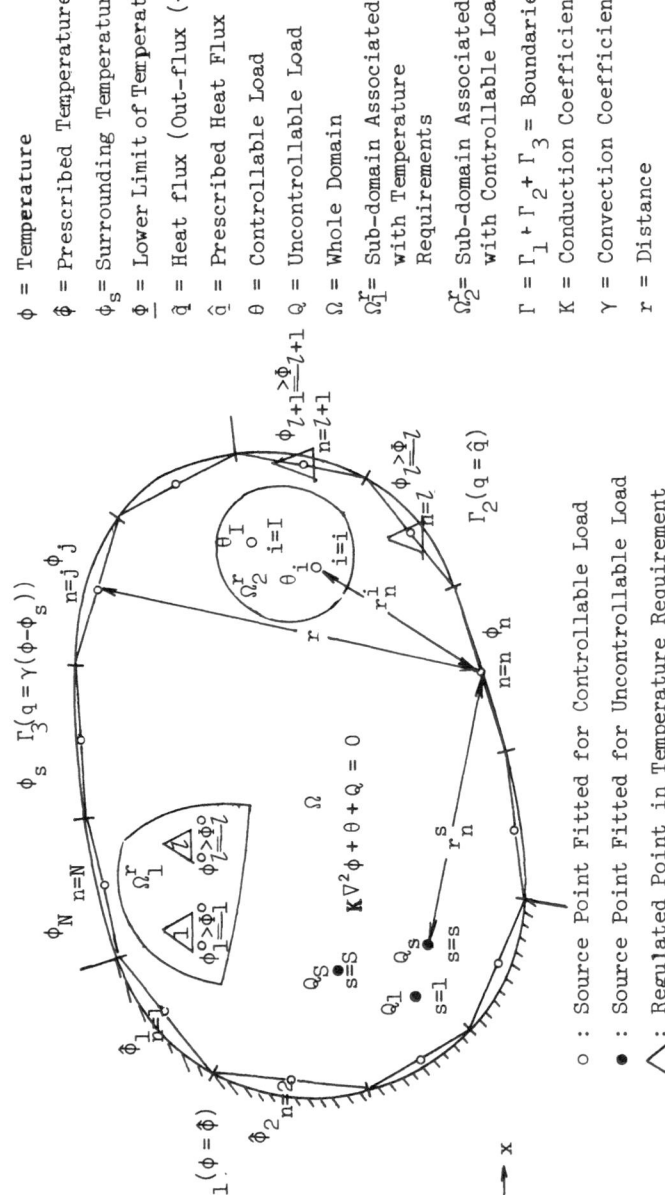

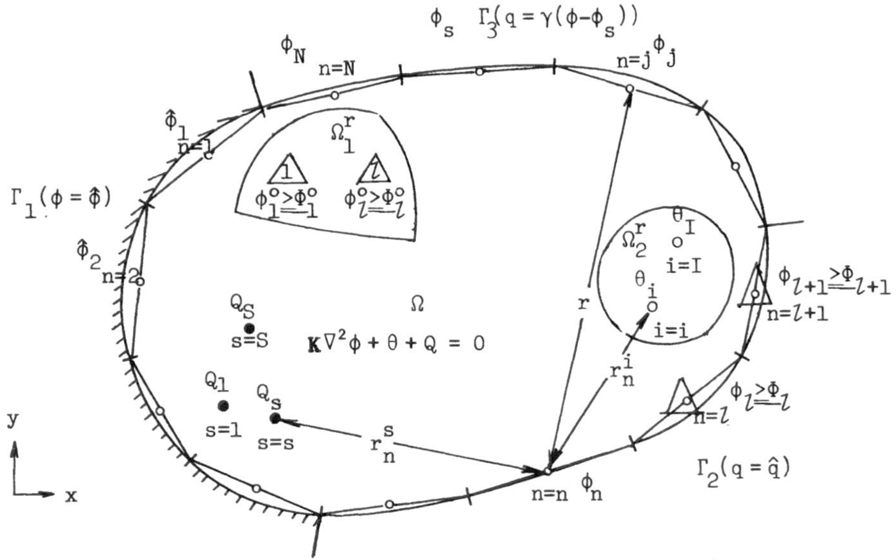

o : Source Point Fitted for Controllable Load

● : Source Point Fitted for Uncontrollable Load

△ : Regulated Point in Temperature Requirement

ϕ = Temperature

$\hat{\phi}$ = Prescribed Temperature

ϕ_s = Surrounding Temperature

$\underline{\Phi}$ = Lower Limit of Temperature

$\underset{\sim}{q}$ = Heat flux (Out-flux (+))

$\hat{\underset{\sim}{q}}$ = Prescribed Heat Flux

θ = Controllable Load

Q = Uncontrollable Load

Ω = Whole Domain

Ω_1^r = Sub-domain Associated with Temperature Requirements

Ω_2^r = Sub-domain Associated with Controllable Loads

$\Gamma = \Gamma_1 + \Gamma_2 + \Gamma_3$ = Boundaries

K = Conduction Coefficient

γ = Convection Coefficient

r = Distance

Fig. 2. The Boundary Element & Linear Programming Method in Heat Conduction Control

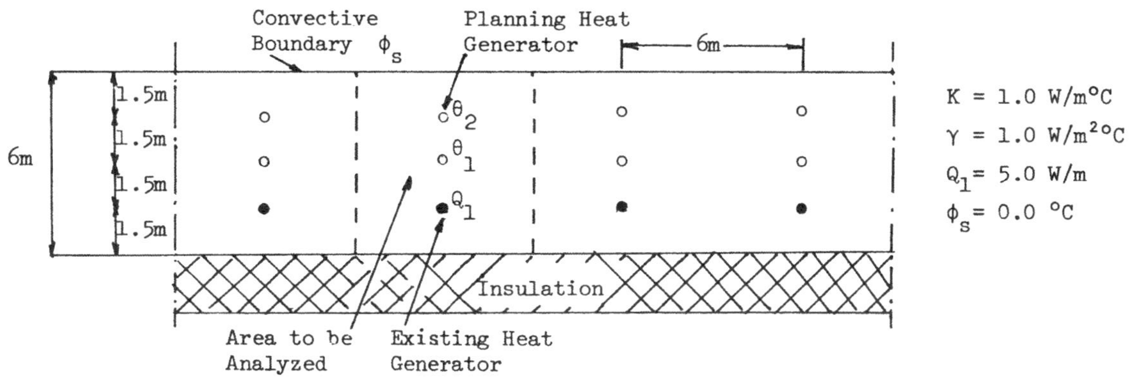

Fig. 3. Simple Model of Heat Conduction Control by Boundary Element & Linear Programming Method

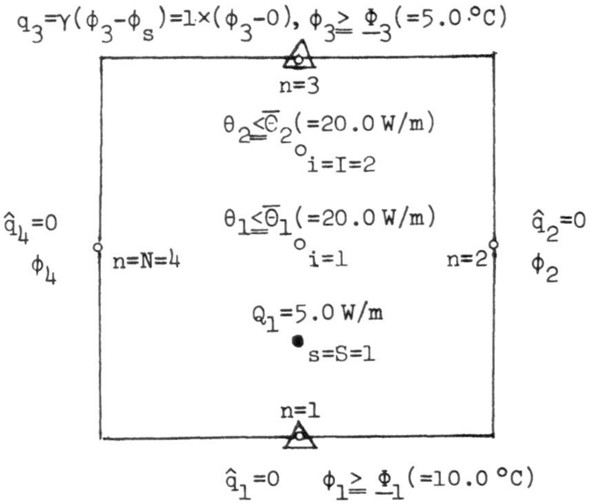

Unknowns: $\phi_1, \phi_2, \phi_3, \phi_4, q_3, \theta_1, \theta_2$ and Z

Fig. 4(a). Input Data and Unknowns in Run 1

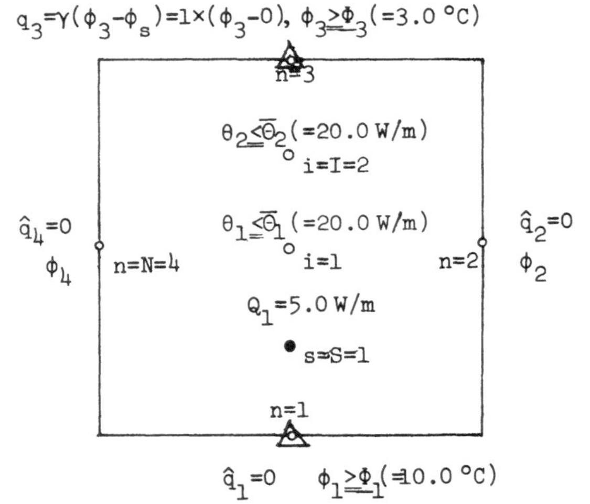

Unknowns: $\phi_1, \phi_2, \phi_3, \phi_4, q_3, \theta_1, \theta_2$ and Z

Fig. 4(b). Input Data and Unknowns in Run 2

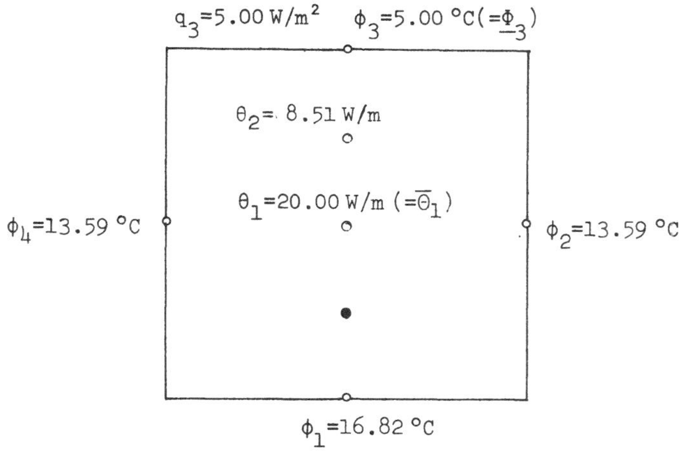

$$Z = \theta_1 + \theta_2 = 20.00 + 8.51 = 28.51 \text{ (W/m)}$$

Fig. 5(a). Results in Run 1

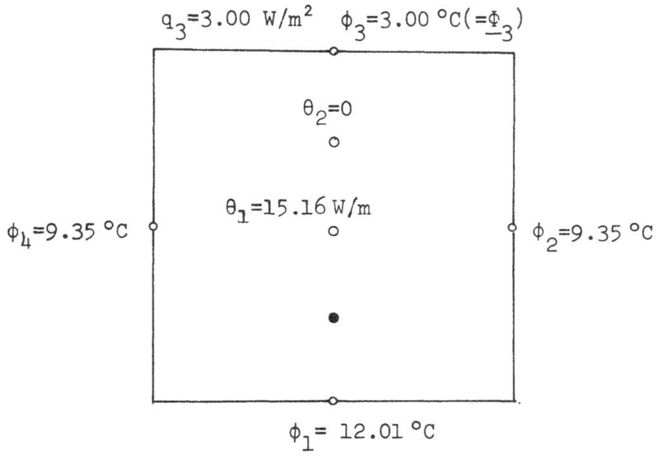

$$Z = \theta_1 + \theta_2 = 0.00 + 15.16 = 15.16 \text{ (W/m)}$$

Fig. 5(b). Results in Run 2

APPROXIMATE FUNDAMENTAL SOLUTIONS AND ALTERNATIVE FORMULATIONS

S Walker

Structural Dynamics Limited
Southampton UK

ABSTRACT

The two main problem areas associated with the application of the Boundary Element Method to the solution of physical problems are:-

1. The determination of the form of the appropriate fundamental solution.

2. The evaluation of the integral of this singular function over the elements needed for the determination of the diagonal terms in the influence matrices.

In this paper the method of 'reference points' is described which avoids the second difficulty and a method for the construction of the fundamental solution for any linear partial differential equation is given. These methods are discussed with examples from two-dimensional potential and diffraction theory in order to demonstrate simply the principles involved. All the examples were solved using an Apple II microcomputer, demonstrating the efficiency and usefulness of the Boundary Element Method.

1 INTRODUCTION

It is not proposed in this paper to present an exhaustive description of the theory of fundamental solutions, the interested reader is referred to the theoretical expositions to be found in [1-6]. In this paper we shall be concerned with applications, particularly related to potential and diffraction problems. These examples were chosen because of their familiarity and in order to bring out the essential points in the simplest possible way. It is however necessary to clarify a number of points relating to the use of fundamental solutions in the Boundary Element Method at this stage.

The Fundamental Solution

Consider a medium in which we have a physical property represented by the function u which may be a function of space and time. If we know the governing equation for u and this equation is linear we can write this equation as

$$\mathcal{L}(u) = 0 \tag{1}$$

throughout the medium which we shall for the moment consider to be infinite in extent.

Here $\mathcal{L}$ is a linear differential operator. The eigenfunctions ϕ_k corresponding to $\mathcal{L}$ are defined by

$$\mathcal{L}(\phi_k) = \lambda_k \phi_k \tag{2}$$

where λ_k is a constant called the eigenvalue corresponding to the eigenfunction ϕ_k.

The form of the eigenfunctions will depend on the coordinate system used. For rectangular Cartesian coordinates we can choose the eigenfunctions to be the harmonic functions cos (Kx), sin (Kx), cos (Ky), etc., or using complex notation $e^{\pm iKx}$. The eigenvalues λ_k are then just polynomials involving powers of 'K' and 'iK'.

For infinite regions K is a continuous parameter so we shall write ϕ_k more explicitly as $\phi(\underline{K}, \underline{x})$. These eigenfunctions are usually orthonormal in the sense that

$$\int_{\text{all space}} \phi(\underline{K}, \underline{x})\ \hat{\phi}\ (\underline{K}', \underline{x})\ d\underline{x} = \delta(\underline{K} - \underline{K}') \tag{3}$$

where the hat '^' denotes complex conjugation. δ is the Dirac delta function which is zero when $\underline{K} \neq \underline{K}'$ but infinite at $\underline{K} = \underline{K}'$. To define δ uniquely we require that

$$\int_{\Omega\infty} \phi(\underline{K}, \underline{x})\ \delta(\underline{K} - \underline{K}')\ d\underline{K} = \phi(\underline{K}', \underline{x}) \tag{4}$$

which is the replacement property of the delta function with respect to our eigenfunctions. $\Omega\infty$ represents integration over all space.

From (4):

$$\int_{\Omega\infty} \delta(\underline{K} - \underline{K}')\ d\underline{K} = 1 \tag{5}$$

and

$$\int_{\Omega\infty} \phi(\underline{K}, \underline{x})\ \hat{\phi}\ (\underline{K}, \underline{\xi}) = \delta(\underline{x} - \underline{\xi}) \tag{6}$$

This expansion enables the fundamental solution to be determined.

The fundamental solution u^* for the linear differential operator $\mathcal{L}$ may be written

$$\mathcal{L}(u^*(\underline{x}, \underline{\xi})) = \delta(\underline{x} - \underline{\xi}) \tag{7}$$

and is a function of two points, the source point $\underline{\xi}$ and the observation point $\underline{x}$. (Note $\mathcal{L}$ may be a polynomial of differentials with respect to the components of $\underline{\xi}$ or $\underline{x}$.) This fundamental solution represents the value of the field at point $\underline{x}$ due to a point source at $\underline{\xi}$ (figure 1).

Using the properties of the delta function and the normalisation properties of the eigenfunctions the fundamental solution may be written

$$u^*(\underline{x}, \underline{\xi}) = \int_{\Omega\infty} \phi(\underline{K}, \underline{x}) \frac{\hat{\phi}(\underline{K}, \underline{\xi})}{\lambda(\underline{K})} d\underline{K} \tag{8}$$

The formula given in (8) is a generalised transform. In fact in section 3 we shall use the Cartesian eigenfunctions $e^{-i\underline{K}\cdot\underline{x}}$ and (8) will then be an equation involving Fourier transforms.

2 REFERENCE POINTS

One of the main numerical problems associated with the fundamental solution is the evaluation of the diagonal terms of the influence matrices. These diagonal terms often consist of the evaluation of the integral of the fundamental solution over the singularity. Typically integrals like

$$Gii = \int_{-\ell}^{\ell} u(\underline{x}, \underline{\xi})\, ds \tag{9}$$

and

$$Hii \quad \int_{-\ell}^{\ell} \frac{\partial u(\underline{X}, \underline{\xi})}{\partial n}\, ds \tag{10}$$

are sought, where ds is an elemental length along the element considered.

These problems may be simply avoided. Consider the general second order linear partial differential equation:-

$$a'u_{xx} + 2b'u_{xxy} + c'u_{yy} + d'u_x + e'u_y + f'u = g' \tag{11}$$

where the coefficients a', b', c', d', e', f' and g' are functions of x and y only, and the subscripts x and y denote differentiation w.r.t. these variables.

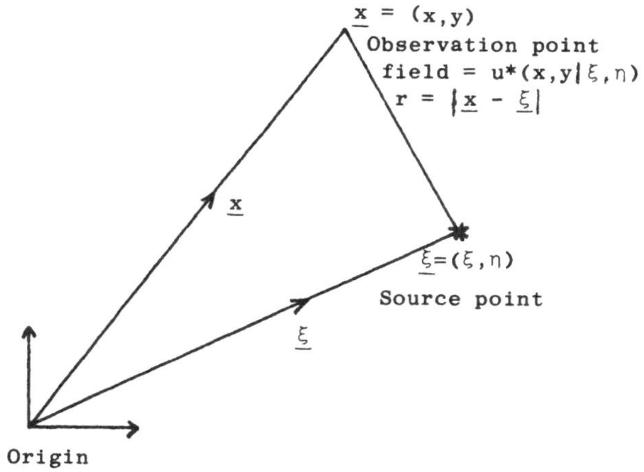

FIGURE 1 NOTATION FOR DEFINITION OF THE FUNDAMENTAL SOLUTION

If b^2-ac > 0 then the equation is elliptic, we shall consider this case. Equation (11) may always be reduced to the canonical form (see reference 6, Chapter 3)

$$\mathcal{L}(u) = u_{xx} + u_{yy} + au_x + bu_y + cu \qquad (12)$$

define the adjoint operator to $\mathcal{L}$, $\mathcal{L}^+$ such that

$$\mathcal{L}^+(v) = v_{xx} + v_{yy} - av_x - bv_y + (c - a_x - b_y)v \qquad (13)$$

Then with n as the outward normal, for u and v any sufficiently differentiable functions:-

$$\int_\Omega v\mathcal{L}(u) - u\mathcal{L}^+(v)\, dV =$$

$$= \int_\Gamma \left[v\frac{\partial u}{\partial n} - u\frac{\partial v}{\partial n} + (a\frac{\partial x}{\partial n} + b\frac{\partial y}{\partial n})\, uv \right] dS \qquad (14)$$

Ω is the problem region and Γ is the boundary.

Then if we choose v to be the fundamental solution $\underline{u}^*(\underline{x}, \underline{\xi})$ defined by:-

$$\mathcal{L}^+(u^*(\underline{x}, \underline{\xi})) = \delta(\underline{x}-\underline{\xi}) \qquad (15)$$

then we may use the selective property of the delta function to reduce (14) to:-

$$cu(\underline{x}) = \int_\Gamma \left[u^* \frac{\partial u}{\partial n} - u\frac{\partial u^*}{\partial n} \right] + \left[a\frac{\partial x}{\partial n} + b\frac{\partial y}{\partial n} \right] uu^*\, dS(\underline{\xi}) \qquad (16)$$

we have taken the integrations to be over the variable $\underline{\xi}$

where for smooth boundaries

$c = 0 \quad \underline{x} \notin \Omega$

$c = \frac{1}{2} \quad \underline{x} \in \Gamma$

$c = 1 \quad \underline{x} \in \Omega$

Notice that u* is a function of two variables and that (16) is an identity for all values of the $\underline{x}$ reference point.
Consider the above with a = b = 0 then

$$c\, u\, (\underline{x}) = \int u^*(\underline{x},\underline{\xi}) \frac{\partial u(\xi)}{\partial n} - u(\underline{\xi}) \frac{\partial}{\partial n} u^*(\underline{x},\underline{\xi}) dS(\underline{\xi}) \qquad (17)$$

Then if we choose $\underline{x}$ to be a reference point $\underline{x}_i$ outside Ω say

$\underline{x} = \underline{x}_i$

then c = 0 and (17) may be written:-

$$\int_\Gamma u^* \, (\underline{x}_i, \underline{\xi}) \, \frac{\partial u(\underline{\xi})}{\partial n} \, dS(\underline{\xi}) = \int_\Gamma u(\underline{\xi}) \, \frac{\partial}{\partial n} \, u^*(\underline{x}_i, \underline{\xi}) \, dS(\underline{\xi}) \quad (18)$$

or approximately for n elements Γ_j.

$$\sum_j \frac{\partial u(\xi)}{\partial n} \int_{\Gamma_j} u^*(\underline{x}_i, \underline{\xi}) \, dS(\underline{\xi})$$

$$= \sum_j u(\underline{\xi}) \int_{\Gamma_j} \frac{\partial}{\partial n} \, u^* \, (\underline{x}_i, \underline{\xi}) \, dS(\underline{\xi}) \quad (19)$$

If we discretise Γ into boundary elements then we can now choose n such reference points. We have now constructed n equations for the n unknown u values on the boundary.

WE NOW HAVE NO SINGULAR INTEGRALS TO EVALUATE. One disadvantage of this method is that the influence matrices become less diagonally dominant, with the resulting numerical problems, as the diagonal terms now correspond to an evaluation of u* between the ith element and i^{th} reference point. This difficulty may be overcome by choosing the reference point close to the corresponding element.

Example 2-D Laplace equation

For the two dimensional Laplace equation

$$\frac{\partial^2 u}{\partial x^2} + \frac{\partial^2 u}{\partial y^2} = \nabla^2 u = 0 \quad (20)$$

$$\nabla^2 \, u^* \, (\underline{\xi}, \, \underline{x}_i) = \delta(\underline{\xi} - \underline{x}_i) \quad (21)$$

and

$$u^* \, (\underline{\xi}, xi) = \frac{1}{2\pi} \, \ell n \, r_{ij} \quad (22)$$

where

$$r_{ij} = |\underline{\xi}_j - \underline{x}_i| \quad (23)$$

then (14) becomes

$$\sum_j \frac{\partial u(\xi)}{\partial n} \int_{\Gamma_j} \ell n \, r \, dS = \sum_j u \, (\underline{\xi}) \int_\Gamma \frac{\partial}{\partial n} \, \ell n \, r_{ij} \, dS \quad (24)$$

note that at no point on the boundary element Γ_j does r_{ij} become zero not even when i = j. In the example the reference points $\underline{x}_i$ were chosen to be on the normal to element i outside the problem region (Figure 2) distance $\alpha\ell$ from the element where ℓ is the length of the element.

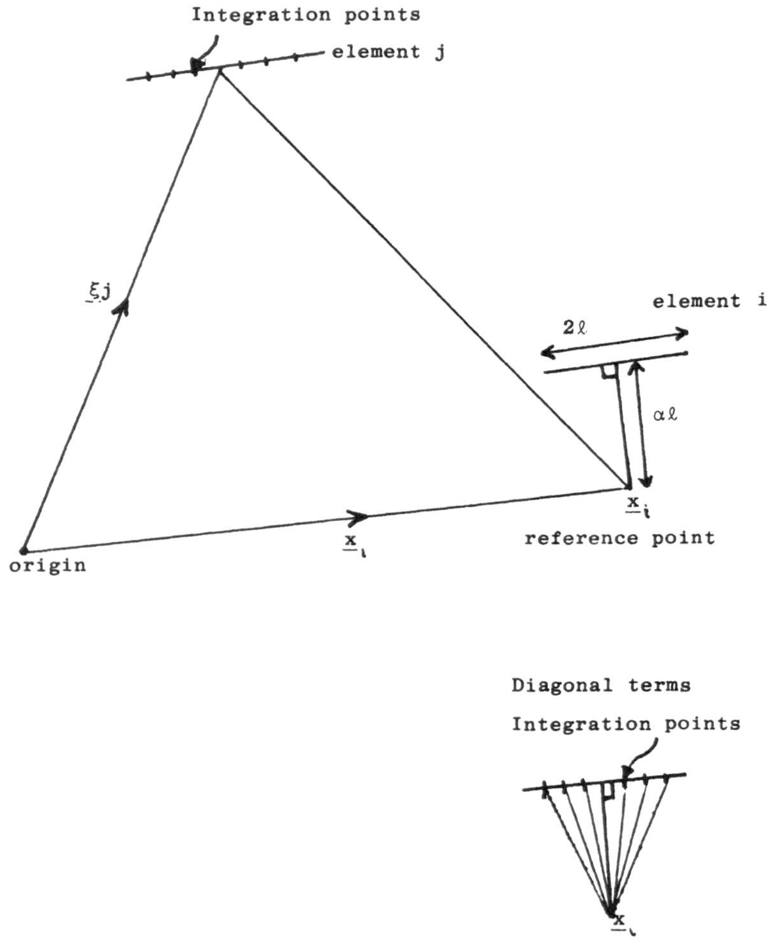

FIGURE 2 THE USE OF REFERENCE POINTS IN BOUNDARY ELEMENTS

n equations like (24) may be written down, one corresponding to each reference point position $\underline{x}_i$.

The example considered is shown in Figure 3 and follows the geometry and discretisations given in [7]. As in [7] the example was solved on a desktop microcomputer (the Apple II).

The problem was solved for a number of values of the scale parameter α which determines the distance of the reference points from their corresponding elements. $\alpha = 0$ corresponds to the conventional Boundary Element Formulation. The results are presented element by element in Figure 4. The value of α giving the best results was about 1.

For more complicated equations the evaluation of the diagonal terms will often involve the use of complicated special functions. For the two dimensional diffraction case the fundamental solution is given by

$$\nabla^2 u^*(\underline{x},\underline{\xi}) + K^2 u^*(\underline{x},\underline{\xi}) = \delta(\underline{x}-\underline{\xi}) \tag{25}$$

i.e.

$$u^*(\underline{x},\underline{\xi}) = -\frac{1}{4i} H_o^{(2)}(Kr) \tag{26}$$

where $r = |\underline{x} - \underline{\xi}|$

and $H_o^{(2)}$ is a Hankel function of order zero of the second kind.

The diagonal term is of the form

$$G_{ii} = -\frac{1}{4i} \int_{-\ell}^{\ell} H_o^{(2)}(Kr)\, dS \tag{27}$$

An integral which when evaluated involves Struve and Bessel functions. The use of reference points avoids the need for the use of Struve functions altogether.

Example

Consider the diffraction problem in Figure 5. A wave of the form $e^{i(kx-wt)}$ is incident on a circular cylinder from the right.

The zero normal flux on the cylinder determines the boundary condition on the cylinder to be:-

$$\frac{\partial u}{\partial n} = -\frac{\partial u_I}{\partial n} \tag{28}$$

Heat flow through metal plug

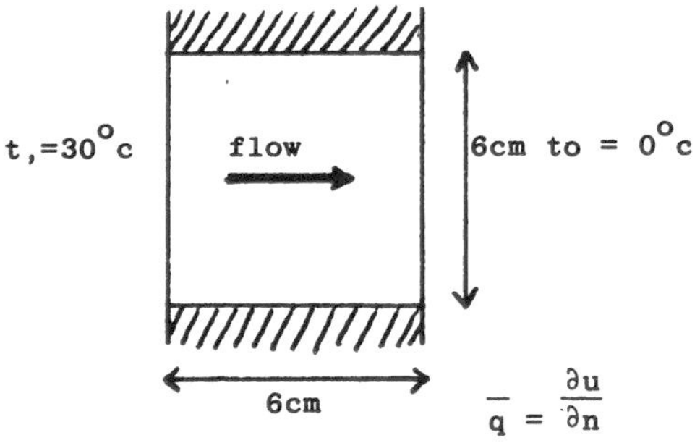

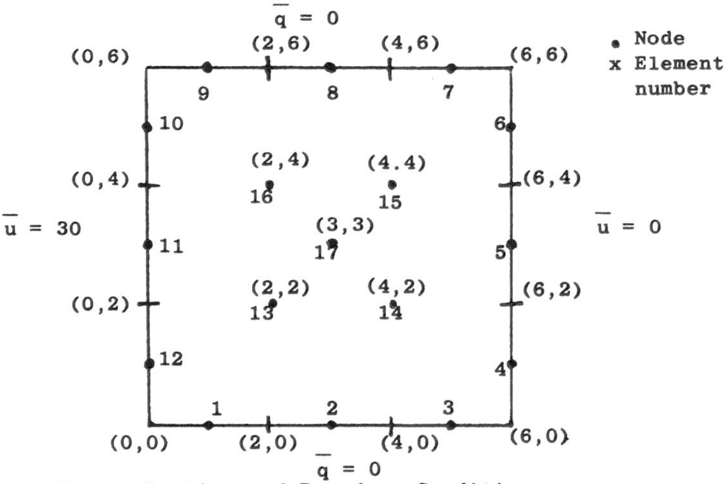

Discretisation and Boundary Conditions

FIGURE 3 METAL PLUG EXAMPLE

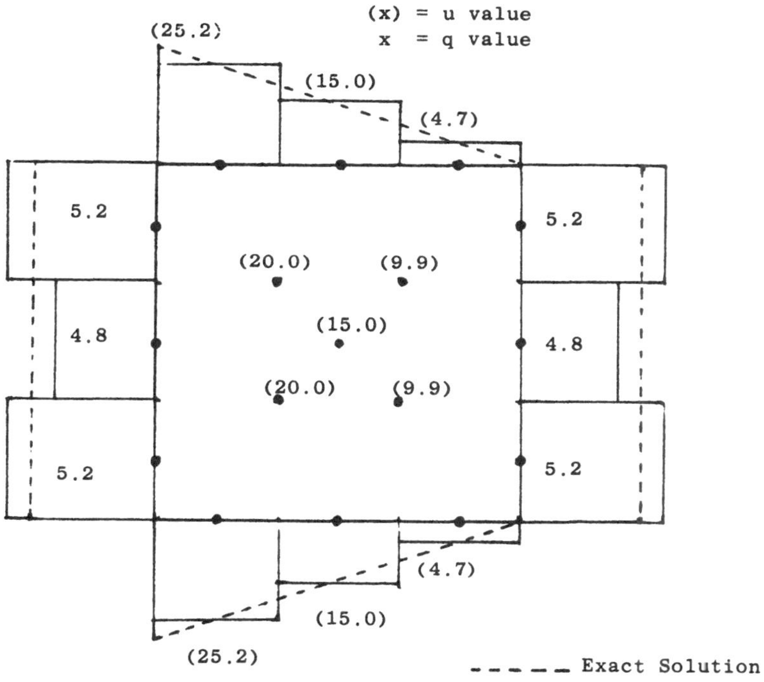

FIGURE 4 METAL PLUG EXAMPLE RESULTS

REFERENCE POINTS - COMPARISON OF RESULTS
SCALE FACTOR α

ELEMENT	0.†	0.5	1.5	3
1	(25.2)	(25.5)	(25.8)	(25.9)
2	(15.0)	(14.9)	(14.99)	(15.0)
3	(4.77)	(4.35)	(4.22)	(4.11)
4	-5.30	-4.97	-5.09	-4.92
5	-4.88	-4.75	-4.97	-5.21
6	-5.30	-4.97	-5.09	-4.92
7	(4.77)	(4.35)	(4.22)	(4.11)
8	(15.0)	(14.9)	(14.99)	(15.0)
9	(25.2)	(25.5)	(25.8)	(25.9)
10	5.30	5.16	5.09	4.92
11	4.87	4.80	4.97	5.21
12	5.30	5.16	5.09	4.92
13	(20.0)	(19.8)	(20.0)	(20.0)
14	(9.97)	(9.82)	(9.96)	(9.96)
15	(9.97)	(9.82)	(9.96)	(9.96)
16	(20.0)	(19.8)	(20.0)	(20.0)
17	(15.0)	(14.8)	(14.99)	(14.99)

† $\alpha = 0$ corresponds to the classical boundary element method

(x) u value

x q value

TABLE 1

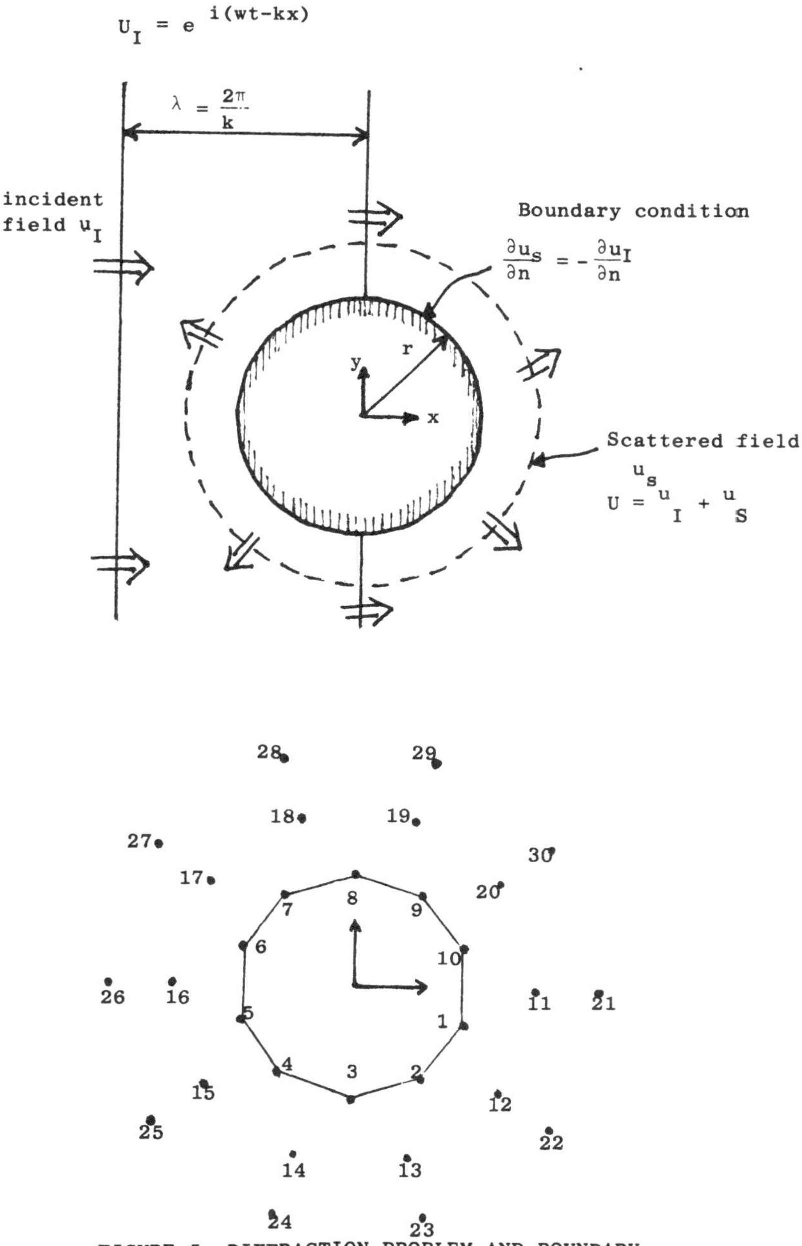

FIGURE 5 DIFFRACTION PROBLEM AND BOUNDARY ELEMENT DISCRETISATION

where $u = U_s + U_I$

and U_I is the incident field

U_S is the scattered field due to the presence of the cylinder

The results for this problem for radius = 1 and wavelength = 2 are given in Table 2 . The values given are the complex moduli of the scattered field representing the amplitude of the oscillations of this field. This field is calculated using the reference point method and the classical Boundary Element Method for comparison.

In the examples considered above it was not strictly necessary to use reference points. In some cases, however, the integrals necessary for evaluation of the diagonal terms are very cumbersome or not known in closed form. This eventuality usually necessitates the use of an elaborate and slow numerical integration scheme.

3 NUMERICAL AND APPROXIMATE FUNDAMENTAL SOLUTIONS

As an example of the methods involved in the use of numerical and approximate fundamental solutions consider the two-dimensional reduced Helmholtz equation.

$$\frac{\partial^2 u}{\partial x2} + \frac{\partial^2 u}{\partial y2} + K^2 U = 0 \tag{29}$$

using (8) and the normalised eigenfunctions given by

$$\phi\ (K_1, K_2; x, y) = \frac{1}{\sqrt{2\pi}} \exp\ (ik_1\ x + ik_2 y) \tag{30}$$

then the fundamental solution may be written

$$u^*\ (x\ y\,|\,\xi,\eta) = \frac{1}{(2\pi)}{}^2 \int_\infty^\infty \int_\infty^\infty \frac{e^{ik_1\ (\xi-x)}\ e^{ik_2(\eta-y)}}{(k^2 - K_1^{\ 2} - K_2^{\ 2})} dk_1 dk_2 \tag{31}$$

In this case the integral may be evaluated using contour integration or coordinate transformations to give

$$u^*\ (x\ y\,|\,\xi,\eta) = -\frac{1}{4i} H_o^{(2)}\ (K|\underline{x} - \underline{\xi}|) \tag{32}$$

For more complicated linear differential operators (31) becomes

DIFFRACTION PROBLEM MODULI OF SCATTERED POTENTIAL, COMPARISON OF RESULTS

ELEMENT OR NODE	SCALE FACTOR α			APPROXIMATE FUNDAMENTAL SOLUTION
	0	0.5	1.0	
1	1.23	1.29	1.24	1.28
2	0.516	0.520	0.507	0.546
3	0.482	0.506	0.455	0.480
4	0.815	0.803	0.776	0.808
5	0.891	0.894	0.924	0.917
11	1.15	1.18	1.15	1.17
12	0.326	0.358	0.340	0.346
13	0.134	0.152	0.135	0.145
14	0.446	0.437	0.429	0.432
15	0.630	0.616	0.637	0.613
21	0.975	1.00	0.976	0.996
22	0.308	0.335	0.318	0.317
23	0.137	0.153	0.136	0.146
24	0.431	0.421	0.416	0.418
25	0.516	0.504	0.524	0.504

Wavelength = 2

Radius of cylinder = 1

The omitted nodes have symmetric values $5 \rightarrow 10 = 1 \rightarrow 5$ etc

N.B. Only ten boundary elements were used for a wavelength of only twice the radius

TABLE 2

$$u^* (x,y\ \xi,\eta) = \frac{1}{(2\pi)^2} \int_{\infty}^{\infty} \int_{\infty}^{\infty} \frac{e^{ik_1 (\xi-x)}\, e^{ik_2 (\eta-y)}}{P(ik_1, ik_2)} dK_1 dK_2 \qquad (33)$$

where P is a polynomial of ik_1, and ik_2 and the following correspondences may be made between the components of the operator and the polynomial

$$\frac{\partial}{\partial x} \longleftrightarrow ik_1, \text{ and } \frac{\partial}{\partial y} \longleftrightarrow ik_2 \qquad (34)$$

The integral (33) may be evaluated numerically for all relevant points (x,y) and (ξ,η) used in the problem geometry, although the singularities corresponding to the real roots of P (k, k_1, k_2) need some considerable care. Integrals of u* over elements of the form (9) may be evaluated analytically with respect to dS by change of the order of integrations.

It now only remains to consider the evaluation of integrals of type (10) of the normal derivative of u*. Remembering that

$$\frac{\partial u^*}{\partial n} = \frac{\partial u^*}{\partial \xi}\frac{\partial \xi}{\partial n} + \frac{\partial u^*}{\partial \eta}\frac{\partial \eta}{\partial n} \qquad (35)$$

then $\frac{\partial u^*}{\partial \xi}$ and $\frac{\partial u^*}{\partial \eta}$ may be evaluated by differentiation of (33) with respect to ξ and η underneath the integral signs. $\frac{\partial \xi}{\partial \eta}$ and $\frac{\partial \eta}{\partial \eta}$ are for straight elements just dependent on the element orientation. The integrals over dS along the element may be eliminated as before. In fact

$$\int_{-\ell}^{\ell} \frac{\partial u^*}{\partial n}\, dS = \int_{-\ell \sin\Theta}^{\ell \sin\Theta} \frac{\partial u^*}{\partial \xi} - \int_{-\ell\cos\Theta}^{\ell\cos\Theta} \frac{\partial u^*}{\partial \eta}\, d\xi \qquad (36)$$

where Θ is the angle of the element makes with the horizontal. (For $\Theta = 0$ or $\frac{n\pi}{2}$ (36) is even simpler.)

The wholly numerical approach described above may not be appropriate in all cases. In fact in the example considered above an approximate fundamental solution may be used. The justification of this is that the most singular terms of the fundamental solution arise from the second derivative terms in the corresponding linear operator. In fact these are the terms which correspond to the Laplace equation which has a simple and well known fundamental solution.

Consider the limits for large and small arguments of the diffraction fundamental solution given in (26). We have:-

Small Kr   Large Kr

$$-\frac{1}{2\pi}\log Kr \sim -\frac{1}{4i} H_o^{(2)}(Kr) \sim \frac{1}{4i}\sqrt{\frac{2}{\pi kr}}\, e^{-i(kr-\frac{\pi}{4})} \qquad (37)$$

For fixed K and small Kr the diffraction fundamental solution behaves like the potential fundamental solution. For correct choice of element size the integral

$$G_{ii} = \int_{-\ell}^{\ell} u^* \, dS$$

for diffraction problems may be evaluated using the potential fundamental solution in place of u*. For widely spaced elements (with respect to the wave length considered) the approximation corresponding to Large Kr for u* may be used for the G_{ij} and H_{ij} terms.

Consider the diffraction problem in Figure 5. This time the small Kr approximation has been used in the evaluation of the diagonal terms, this is a valid approximation for elements of suitable size. The last column in Table 2 shows the small effect of this approximation on the results.

4 CONCLUSIONS

It has been shown that it is usually possible to avoid the need to integrate the fundamental solution across the singularity which occurs when the observation point and source point coincide, by the use of reference points.

In section 3 a method is outlined for solving second order partial differential equations when the explicit form of the fundamental solution is not known or very cumbersome, by numerical evaluation of specific integral representations of the solution.

Where numerical integration is not appropriate it is often possible to use an approximate fundamental solution, consequently increasing speed and decreasing complexity. This is a consequence of the fact that the singular part of most fundamental solutions is a variant of the potential fundamental solution with the same number of dimensions as the problem considered.

5 REFERENCES

1. Brebbia, C.A., Walker S Boundary Element Techniques in Engineering (1980) Newnes-Butterworths

2. Walker, S. Fundamental Solutions in 'Progress in Boundary Element Methods' (1981) Ed. C A Brebbia Pentech Press

3. Roach G F. Green's Functions-Introductory Theory with Applications (1970) Van Nostrand

4. Thome R.C. Multiple Expansions in the Theory of Surface Waves, (1972) Camb. Phil. Soc. 17

5. Titchmarsh E.C. Eigenfunction Expansions Associated with Second-Order Differential Equations (1958) (2 Vols) Clarendon Press

6. Garabedian P.R. Partial Differential Equations John Wiley & Sons

7. Waters J. J. and Nelson J.M. A Program for Solving Potential Problems on a Desktop Computer (1980) Adv. Eng. Software Vol. 2 No. 2

AN EFFICIENT ALGORITHM FOR THE NUMERICAL EVALUATION OF BOUNDARY INTEGRAL EQUATIONS

Madhukar Vable and David L. Sikarskie

Lecturer, Chairman
Department of Metallurgy, Mechanics and Materials Science
Michigan State University

ABSTRACT

The usual approach to the solution of boundary integral equations is to represent the unknown vector by a piecewise constant, linear, or quadratic function over the given mesh subdivision. These representations have the advantages of consistency, ability to integrate the equations for the given functional approximations, and, in general, improved accuracy as the degree of approximation is increased. While adequate for many problems, special requirements arise for certain nonlinear problems, e.g., plasticity, where the integral equations must be solved for each load increment. In the present paper a special numerical algorithm is outlined in which the unknown vector is represented as a combination of a Fourier series and piecewise linear function. The piecewise linear function is used only in high gradient regions of the unknown vector thus permitting an excellent representation with relatively few Fourier terms. The algorithm is compared with a linear representation alone for two problems which show the effects of multiple connectivity, sharp corners and discontinuous loading. For comparable accuracy both problems show a significant improvement in computer time required.

INTRODUCTION

The boundary integral method [3,5,7] is fast becoming a generally accepted alternative to finite element, finite difference methods at least for linear problems. The principal reason is that the problem dimension is reduced by one, e.g., a two-dimensional problem is reduced to a line integral equation on the boundary. After discretizing, the integral equation formulation results in a lower order set of algebraic equations which implies less computer time. There are also certain inherent accuracy improvements, particularly for interior information.

The issue is less clear for non-linear problems, e.g., plasticity. Since plasticity can be viewed as a linear, incremental theory, a boundary integral approach can be formulated. The practical difficulty is that a set of integral equations must be evaluated numerically for each load increment. To be competitive, the numerical algorithm for solving the integral equations must be efficient. The present paper addresses that question.

Before outlining the details of the numerical algorithm (section II), it is necessary to give a brief outline of the pertinent equations. It should be noted that there are many boundary integral formulations [1,4], some having advantages over others for particular problems. The ideas outlined herein could conceivably apply to any formulation. An indirect formulation, developed previously [2,3], will be used. One practical advantage of such an indirect formulation is that the unknown is a single traction vector around the entire boundary even for mixed boundary value problems. This permits the representation of unknown vector in a single series.

Consider a two-dimensional, linearly elastic region R with boundary B as shown in Figure 1. For prescribed data on B, i.e., tractions on B_t, displacements on B_u, the stresses (and displacements) in the region R are to be determined.

An outline of the logic involved in the formulation of the integral equation is as follows. The region R is embedded in an infinite (fictitious) plane of the same material as R for which the influence functions $H_{ij;k}(Q,P_B)$, $I_{ij;k}(Q,P_B)$ are known, see Figure 1. $H_{ij;k}(Q,P_B)$ is the ith stress component at a field point Q due to a unit line in the kth direction at a source point P_B and $I_{ij;k}(Q,P_B)$ is the displacement in the ith direction at Q again due to the unit line load at P_B. Consider now a fictitious traction $\vec{f}$ (unknown) acting along the contour B, see Figure 1. The stress and displacement fields both internal and external to B due to $\vec{f}$ are:

$$\sigma_{ij}(Q) = \int_B H_{ij;k}(Q,P_B)\, f_k(P_B)\, ds \tag{1a}$$

$$u_i(Q) = \int_B I_{i;k}(Q,P_B)\, f_k(P_B)\, ds \tag{1b}$$

where P_B is on B and ds is an element of length along B, s measured as shown in Figure 1. The functions $H_{ij;k}$ and $I_{i;k}$ are given by:

$$\begin{aligned} H_{xx;k}\, f_k = -\frac{1}{4\pi r^4}\, [&r_x(a_1 r_x^2 + a_2 r_y^2) f_x \\ &+ r_y(a_3 r_x^2 - a_2 r_y^2) f_y] \end{aligned} \tag{2a}$$

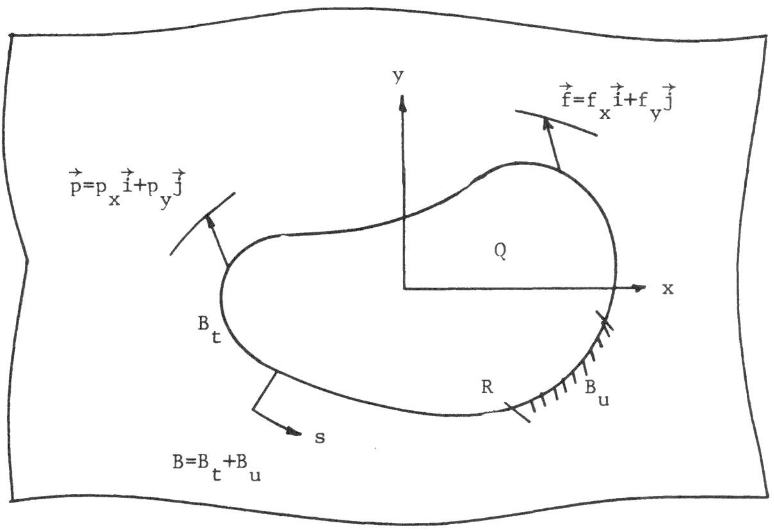

Figure 1. Region R Embedded in an Infinite Plane

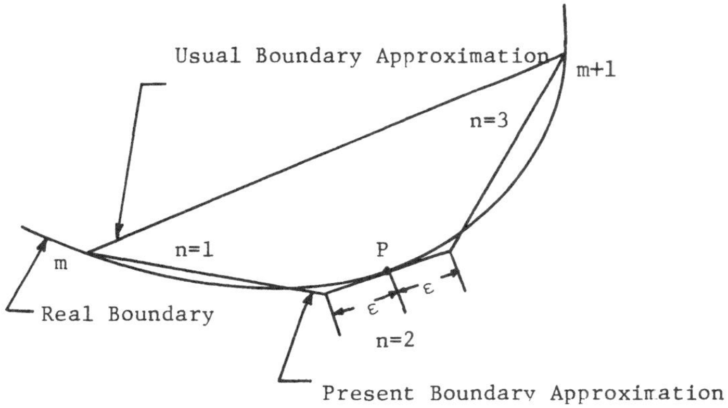

Figure 2. Boundary Division and Subdivision

$$H_{yy;k}f_k = -\frac{1}{4\pi r^4}\left[r_x(a_3 r_y^2 - a_2 r_x^2)f_x + r_y(a_1 r_y^2 + a_2 r_x^2)f_y\right] \quad (2b)$$

$$H_{xy;k}f_k = -\frac{1}{4\pi r^4}\left[r_y(a_1 r_x^2 + a_2 r_y^2)f_x + r_x(a_2 r_x^2 + a_1 r_y^2)f_y\right] \quad (2c)$$

$$I_{x;k}f_k = -\frac{1}{4\pi r^2}\left[(a_4 r^2 \log r + a_5 r_y^2)f_x - a_5 r_x r_y f_y\right] \quad (3a)$$

$$I_{y;k}f_k = -\frac{1}{4\pi r^2}\left[-a_5 r_x r_y f_x + (a_4 r^2 \log r + a_5 r_x^2)f_y\right] \quad (3b)$$

r_x, r_y are the components of the radius vector $\vec{r}$ drawn from P_B to Q. The a_i's are,

$$a_1 = \frac{3-2\nu}{1-\nu} \qquad a_4 = \frac{(3-4\nu)(1+\nu)}{E(1-\nu)}$$

$$a_2 = \frac{1-2\nu}{1-\nu} \qquad a_5 = \frac{1+\nu}{E(1-\nu)}$$

$$a_3 = \frac{1+2\nu}{1-\nu}$$

where ν is Poisson's ratio and E is Young's Modulus.

Since equations (1) represent the superposition of fundamental solutions, all equations of linear elasticity are satisfied. In order to solve the boundary value problem of interest, the boundary conditions on B are yet to be satisfied. These conditions are:

$$\sigma_{ij}(P)n_j(P) = P_i(P) \quad ; \quad P \text{ on } B_t \quad (5a)$$

$$u_i(P) = \delta_i(P) \quad ; \quad P \text{ on } B_u \quad (5b)$$

where $n_j(P)$ are the direction cosines of the normal at the point P on B and $P_i(P)$, $\delta_i(P)$ are the specified traction and displacement components, respectively. Note that B_t, B_u of Figure 1 are not mutually exclusive portions of the boundary, i.e., at a given point of B the mixed condition of specified traction in one direction and specified displacement in the other is possible.

A description of the numerical algorithm is given in the next section. Two examples have been selected to illustrate the method. The first is a multiply connected region; namely, an axisymmetric region with an axisymmetric hole loaded with a uniform external tension. This problem illustrates the advantages of the method for smooth geometries and loadings. The second problem is the problem of pure shear of a square region. This example illustrates the effects of discontinuous loading and corners.

NUMERICAL FORMULATION

Problem solution requires the determination of the fictitious traction vector f_k in equations (5).† With f_k known, stresses and displacements are available from equations (1). The usual approach to the solution of equations (5) is to first extract the singularity analytically [2]. The boundary is then subdivided into M piecewise linear segments as shown in Figure 2. f_k is then approximated by a piecewise constant [2] or linear [6] function over this linear boundary segment. After substituting the above approximation into the integral equations, the integrals can usually be evaluated analytically, resulting in a set of the usual linear algebraic equations.

A better boundary approximation used by the authors for both the piecewise linear approximation and the smooth function expansion (yet to come) is shown in Figure 2. The mth boundary segment is further subdivided into N_m segments (only three are shown in Figure 2 for ease of explanation). f_k is still linear over the mth segment, i.e., this further subdivision introduces no new unknowns. Note that the boundary conditions will be satisfied in a collocation sense at M points P which are not the nodal points m, m+1, etc.

The linear approximation for f_k is:

$$f_k(P_B) = \frac{d_k^{(m+1)} - d_k^{(m)}}{S_{m+1} - S_m} (s-S_m) + d_k^{(m)};$$

$$S_m \leq s \leq S_{m+1}$$
$$m = 1,2,\ldots,M \qquad (6)$$
$$k = x,y$$

where s = running coordinate along the fully segmented boundary,

S_m = value of s at m,

$d_k^{(m)}$ = unknown constant (nodal value for linear approximation only).

Continuity requires that

$$d_k^{(M+1)} = d_k^{(1)} \qquad (7)$$

†The actual boundary integral equations are generated by substituting equations (1) into equations (5).

Rather than substitute f_k into the integral equations resulting from substituting equations (1) into equations (5), a different procedure will be followed here. f_k, equation (6), is substituted directly into the stress and displacement equations (1). Note that these expressions are valid at a field point Q. Substituting equation (6) into equation (1a) and performing the required integrations, we obtain

$$\sigma_{ij}(Q) = \sum_{m=1}^{M} \sum_{n=1}^{N_m} \left\{ d_k^{(m)} - S_m \frac{d_k^{(m+1)} - d_k^{(m)}}{S_{m+1} - S_m} \right\} M^{(1)}_{ij;k}(Q,S_n) + \frac{d_k^{(m+1)} - d_k^{(m)}}{S_m+1 - S_m} M^{(2)}_{ij;k}(Q,S_n) \tag{8}$$

where $M^{(1)}{}_{ij;k}$ and $M^{(2)}{}_{ij;k}$ are defined in detail in [8]. Note that the $M^{(r)}{}_{ij;k}$'s are dependent on an interior point Q and subinterval geometry. A similar expression to equation (8) could be written for the displacements $u_i(Q)$. Only traction boundary value problems will be used to illustrate the numerical algorithm, hence all displacement equations will be dropped. Traction boundary conditions will be satisfied when equation (8) is substituted into (5a) with $Q \to P$ (boundary point).

Before boundary conditions are satisfied, the Fourier series expansion for f_k is introduced.

$$f_k(P_B) = a_{ko} + \sum_{q=1}^{Q_1} a_{kq} \cos \frac{2\pi qs}{S} + \sum_{q=1}^{Q_2} b_{kq} \sin \frac{2\pi qs}{S} \tag{9}$$

where Q_1, Q_2 are the number of retained cosine and sine terms respectively and S is the total subdivided boundary length. The boundary is now divided exactly as in the linear approximation case, i.e., into m intervals with further N_m subintervals. f_k, equation (9) is now substituted into equation (1a) for σ_{ij}. This results in a series of integrals which the authors could not obtain in closed form. To overcome this, the Fourier components are individually expanded in a Taylor series about the midpoint of each subinterval. Retaining the first two terms f_k can be written as

$$f_k(P_B) = A_{kn} + B_{kn}[s - \bar{s}_n]; \quad S_n \leq s \leq S_{n+1} \tag{10a}$$

where

$$A_{kn} = a_{ko} + \sum_{q=1}^{Q_1} a_{kq} \cos \frac{2\pi q\bar{s}_n}{S} + \sum_{q=1}^{Q_2} b_{kq} \sin \frac{2\pi q\bar{s}_n}{S} \tag{10b}$$

$$B_{kn} = - \sum_{q=1}^{Q_1} a_{kq} \frac{2\pi q}{S} \sin \frac{2\pi q \bar{s}_n}{S} + \sum_{q=1}^{Q_2} b_{kq} \frac{2\pi q}{S} \cos \frac{2\pi q \bar{s}_n}{S}$$

$$\bar{s}_n = (S_n + S_{n+1})/2 \qquad n = 1,2,\ldots,N_m \qquad m = 1,2,\ldots,M \tag{10c}$$

Substituting equations (10) in the expression for $\sigma_{ij}(Q)$ and performing the integrations, we obtain

$$\sigma_{ij}(Q) = \sum_{m=1}^{M} \sum_{n=1}^{N_m} \{A_{kn} M^{(1)}_{ij;k}(Q,S_n) + B_{kn}[M^{(2)}_{ij;k}(Q,S_n) - \bar{s}_n M^{(1)}_{ij;k}(Q,S_n)]\} \tag{11}$$

where $M^{(1)}_{ij;k}$ and $M^{(2)}_{ij;k}$ are the same quantities as used in equation (8).

Note as Q_1 and Q_2 increase, i.e., as more sine and cosine terms are retained, the size of the nth subdivision must become smaller if the Taylor expansion approximation is to remain valid. Usually we will restrict the number of sine and cosine terms to 3 or 4. The reason for this restriction is that the smaller subintervals required lead to larger round-off errors.

Both approximations for f_k (linear and Fourier) have problems; the linear requiring too many unknowns in regions of smooth behavior and the Fourier requiring too many unknowns in regions of large gradients. Thus, by using a combination of the two approximations, each compensates for the other's weakness.

Let

$$f_k = f_k^{(1)} + f_k^{(2)} \tag{12}$$

where $f_k^{(1)}$ is given by equations (6) and $f_k^{(2)}$ by equation (9). The stresses σ_{ij} are then the sum of equations (8) and (11).

$$\sigma_{ij}(Q) = \sum_{m=1}^{M} \sum_{n=1}^{N_m} [\{A_{kn} + d_k^{(m)} - (S_m - \bar{s}_n) \frac{d_k^{(m+1)} - d_k^{(m)}}{S_{m+1} - S_m}\} M^{(1)}_{ij;k}(Q,S_n) + \{B_{kn} + \frac{d_k^{(m+1)} - d_k^{(m)}}{S_{m+1} - S_m}\} \{M^{(2)}_{ij;k}(Q,S_n) - \bar{s}_n M^{(1)}_{ij;k}(Q,S_n)\}] \tag{13}$$

To determine the various constants in equation (13), $\sigma_{ij}(Q)$ is substituted into boundary conditions given by equation (5a). It must be pointed out that even though $H_{ij;k}(P,P_B)$ goes to infinity when $P \rightarrow P_B$, the quantities $M_{ij;k}^{(1)}$ and $M_{ij;k}^{(2)}$ are bounded. With proper care taken in constructing the boundary subdivision near the selected boundary condition point, the singularity contribution can be evaluated numerically (see Figure 2 and reference [8]). This greatly simplifies the writing of the computer code.

The summed representation of f_k has considerable flexibility. For example, non-zero values of $d_k^{(m)}$ can be permitted only in regions of anticipated high fictitious traction gradients. This occurs near sharp changes in geometry and/or loading.

The final linear algebraic equations are generated by substituting σ_{ij}, equation (13), into the boundary condition, equation (5a). The unknowns a_{ko}, a_{kq}, b_{kq}, $d_k^{(m)}$ are thus found. Stresses at any point can then be found from equation (13). The algorithm will now be demonstrated on two examples.

EXAMPLES

A. Thick-walled Cylinder (Lamé Problem)

Problem geometry and loading are shown in Figure 3. The applied boundary traction is:

$$P_x = \begin{cases} 0 & \text{for } r = 1 \\ \cos\theta & \text{for } r = 10 \end{cases} \tag{14a}$$

$$P_y = \begin{cases} 0 & \text{for } r = 1 \\ \sin\theta & \text{for } r = 10 \end{cases} \tag{14b}$$

This problem will be solved in two ways and the solution compared with respect to computer time and accuracy.

In the first solution the unknown traction vector f_k is approximated by a piecewise linear function only. Each boundary is divided into 24 intervals subtending 15° each, $M = 48$. Each interval is further subdivided into 5 subintervals, $N_m = 5$ for all m. The total number of unknowns is 96.

In the second solution, the unknown traction vector is expanded in a Fourier series, equation (9), ($Q_1 = 3$, $Q_2 = 4$) on each boundary. Each boundary is divided into 8 intervals, subtending 45°, $M = 16$. Each interval is further subdivided into 15 subintervals, $N_m = 15$, for all m. In fact, the subinterval division is identical with the first solution. The total number of unknowns is 32.

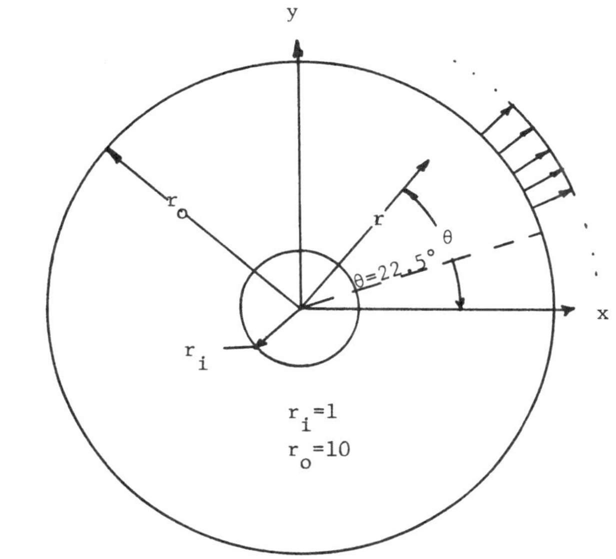

Figure 3. Lamé's Problem

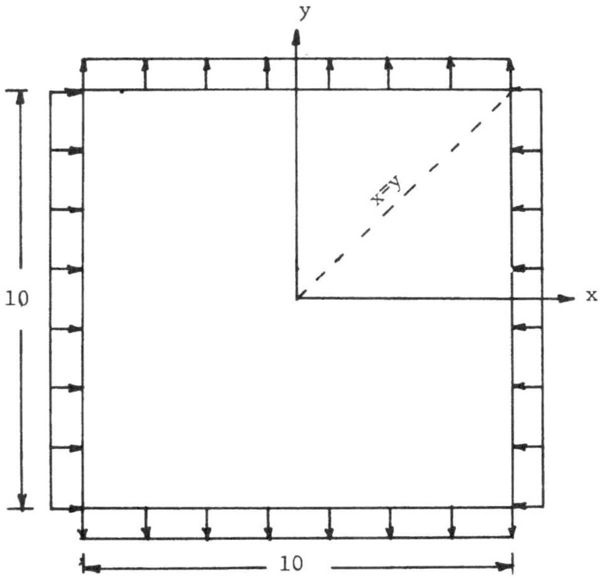

Figure 4. Pure Shear Problem

r	Percentage error in σ_r				Percentage error in σ_θ			
	$\theta = 0^\circ$		$\theta = 22.5^\circ$		$\theta = 0^\circ$		$\theta = 22.5^\circ$	
	Linear	Fourier	Linear	Fourier	Linear	Fourier	Linear	Fourier
1.05	-6.72	-7.36	4.36	-1.39	-0.61	-0.06	-0.32	0.10
1.10	-2.34	-2.97	1.59	-0.57	-0.48	0.03	-0.46	-0.02
1.20	-0.76	-1.05	-0.05	-0.73	-0.46	0.00	-0.44	-0.02
1.50	-0.48	-0.40	-0.35	-0.39	-0.41	-0.04	-0.41	-0.04
2.00	-0.41	-0.25	-0.46	-0.25	-0.39	-0.07	-0.40	-0.07
5.00	-0.38	-0.15	-0.38	-0.15	-0.38	-0.12	-0.38	-0.12
9.50	-0.44	-0.32	-0.41	-0.05	-0.47	-0.19	-0.24	-0.05
9.80	-0.19	-0.13	-0.77	-0.36	-0.72	-0.46	-0.14	-0.41
9.90	0.00	0.03	-0.85	-0.03	-0.65	-0.41	-0.41	-0.13
9.95	0.05	0.07	-0.65	-0.27	-0.41	-0.19	-0.92	-0.45

Table 1. Error in Linear and Fourier Approximations for the Lame' Problem (minus indicates approximation is less than the exact)

	Linear	Fourier
Matrix Generation (milliseconds)	8,950	10,118
Solution Procedure (milliseconds)	21,408	938
Total	30,358	11,056

Table II. Computer Time Summary for Lamé Problem

m	Linear			Fourier and Linear (combination)		
	x	y	N_m	x	y	N_m
1	5.0	-4.9	3	5.0	-4.9	3
2	5.0	-4.1	2	5.0	-4.1	10
3	5.0	-3.1	2	5.0	-0.1	11
4	5.0	-2.1	2	5.0	4.1	3
5	5.0	-1.1	5	5.0	4.9	2
6	5.0	1.1	2	4.9	5.0	3
7	5.0	2.1	2			
8	5.0	3.1	2			
9	5.0	4.1	3			
10	5.0	4.9	2			
11	4.9	5.0	3			

Table III. Boundary Subdivision for Pure Shear

The algebraic analog of the boundary integral equations for the linear boundary approximation is generated by substituting equation (8) into equation (5a) and letting Q successively take on the M discrete boundary points (collocation). The set of equations for the Fourier boundary approximation is generated in the same way except equation (11) replaces (8).

Once the fictitious tractions are found from these equations, stresses at any field point Q are found by substituting back into either equation (8) or (11). Table 1 is a comparison of stresses computed from the linear and Fourier solutions with the exact. What is shown is the percent. error of each along two representative radial lines $\theta = 0°, 22.5°$.

Note that the errors are comparable, the Fourier tending to be somewhat better near the boundaries and the linear tending to be somewhat better in the interior. The high percentage errors in σ_r near the inner boundary is expected since the exact solution approaches zero. The boundary condition is satisfied close to $\theta = 0°$ but $\theta = 22.5°$ lies between two boundary condition points for both solutions. The computer time summary is shown in Table II. Time for matrix generation is somewhat higher in the Fourier case because of some additional algebra in the approximation. Both solution procedures were identical.

In spite of the fact that this axisymmetric problem was solved in Cartesian coordinates, it can be argued that it is ideal for a Fourier series approach, i.e., both loading and geometry are smooth functions. The next example has both discontinuous loading and geometric corners.

B. Pure Shear

Problem geometry and loading are shown in Figure 4. The applied boundary traction is:

$$P_x = \begin{cases} -1 & x = 5 \quad -5 \le y \le 5 \\ 1 & x = -5 \quad -5 \le y \le 5 \\ 0 & -5 \le x \le 5 \quad y = \pm 5 \end{cases} \tag{15a}$$

$$P_y = \begin{cases} 0 & x = \pm 5 \quad -5 \le y \le 5 \\ 1 & -5 \le x \le 5 \quad y = 5 \\ -1 & -5 \le x \le 5 \quad y = -5 \end{cases} \tag{15b}$$

Again, the problem will be solved two ways. The first solution, as before, has the fictitious traction represented by a piecewise linear approximation. The second solution is

x = y	Percentage Error	
	Linear	Combination
0	1.87	2.18
1.0	1.57	2.00
4.0	1.12	2.02
4.3	.98	1.10
4.5	.11	-.08
4.6	-1.49	-1.71
4.7	-5.32	-5.47
4.8	-14.11	-14.13
4.9	-37.40	-37.35

Table IV. Percentage Error along x = y for the Pure Shear Problem

	Linear	Fourier
Matrix Generation (milliseconds)	2,918	4,358
Solution Procedure (milliseconds)	10,759	1,502
Total	13,671	5,860

Table V. Computer Time Summary for Pure Shear Problem

a combination of a Fourier expansion and a piecewise linear approximation (in the first example the Fourier expansion was used alone). The main purpose of the piecewise linear approximation in the combination is to pick up large gradients in the fictitious traction which would require many Fourier terms alone.

Table III shows the division of the boundary of one of the sides, namely $x = 5$. The rest of the sides are divided similarly. The rows under m are the node numbers where a non-zero value of $d_k^{(m)}$ is permitted. x and y are the coordinates of the node in the two cases, and N_m is the number of further subdivisions of the mth interval.

Since the boundary is already piecewise linear, no further subdivision would be required to approximate the boundary for the piecewise linear approximation of the fictitious traction. $N_m = 2$, however, is used, with one of the subdivisions very small and used for the numerical evaluation of the singularity contribution (see Figure 2 and appendix for details). The nodes near the corner are very close to one another, i.e., node 10 and 11 for the linear approximation case and nodes 5 and 6 for the combination case (see Table III). The close proximity of nodes near the corner is because of anticipated high gradients in the fictitious traction there.

In the combination approximation, further boundary subdivision is necessary because of the trigonometric function approximation (see equation (10)); note the larger values of N_m in Table III for the combination approximation.

For this example, 4 sines, 4 cosines and 20 nonzero values of $d_k^{(m)}$ were chosen to represent the boundary traction; a sum total of 56 unknowns.

For the linear case, 40 non-zero values of $d_k^{(m)}$ were chosen to represent the fictitious traction; 80 unknowns. The stresses were evaluated along $x = y$ and the percentage errors are reported in Table IV. Table V gives the time summary. The time for matrix generation was higher for the combination case as compared to the linear, because of the additional required boundary subdivision and associated algebra. Note the sizable time saving over a conventional linear approximation for comparable accuracies.

REFERENCES

1. Altiero, N. J. and Gavazza, S. D., "On a Unified Boundary-Integral Equation Method," J. Elast., 10, 1-9 (1980).

2. Altiero, N. J. and Sikarskie, D. L., "An Integral Equation Method Applied to Penetration Problems in Rock Mechanics," Boundary-Integral Equation Method: Computational Applications in Applied Mechanics, ed. T. A. Cruse and F. J. Rizzo, ASME AMD - Vol. 11, New York (1975).

3. Benjumea, R. and Sikarskie, D. L., "On the Solution of Plane, Orthotropic Elasticity Problems by an Integral Method," JAM, 39, No. 3, Series E, 801-808 (1972).

4. Heise, U., "Application of the Singularity Method for the Formulation of Plane Elastostatical Boundary Value Problems as Integral Equation," Acta Mechanica, 31, 33-69 (1978).

5. Massonnet, C. E., "Numerical Use of Integral Procedures," Stress Analysis - Recent Developments in Numerical and Experimental Methods, edited by Zienkiewicz, O. C. and Holister, G. S., Wiley, 198-235 (1965).

6. Riccardella, P. C., "An Improved Implementation of the Boundary Integral Technique for Two Dimensional Elasticity Problems," Rep SM-72-36, Carnegie-Mellon University (1972).

7. Rizzo, F. J., "An Integral Equation Approach to Boundary Value Problems of Classical Elastostatics," Q. Appl. Math., 40, 83-95 (1967).

8. Vable M. and Sikarskie D. L., "An efficient algorithum for the numerical evaluation of boundary integral equations." To appear in Computer and Structures.

SOLUTION OF THE DIRICHLET PROBLEM USING THE REDUCTION TO FREDHOLM INTEGRAL EQUATIONS

J. Caldwell

Sunderland Polytechnic, England

ABSTRACT

The numerical solution of potential problems using a method which reduces them to Fredholm integral equations of the first and second kinds is considered. The method is illustrated for potential problems of the Dirichlet type and solutions have been obtained for the case of a unit sphere. This method provides an alternative approach to the classical finite-difference methods which have been widely used in the past. It overcomes some of the difficulties associated with the classical approach.

The problem is to solve Laplace's equation

$$\nabla^2 \Phi = 0$$

in some closed volume V with the condition

$$\Phi = \Phi_p(\underline{r}_S)$$

on the boundary surface S, Φ_p being prescribed.

By using a Green's function type of solution and replacing the derivatives by simple finite differences the problem can be reduced to a Fredholm integral equation. Rather than solving a Fredholm equation of the first kind it is preferable to introduce a small parameter ε and solve a Fredholm equation of the second kind. This is done by solving for different small values of ε and then extrapolating to the limit $\varepsilon = 0$. It is possible to obtain reasonable accuracy without resorting to high order (N) Gaussian quadrature for integral evaluation.

The advantages of extrapolating in N are then considered. It is possible to use an empirical formula and fit a curve of the form

$$\Phi = A + B \cdot N^p$$

to the computed results. In this way it is possible to get an idea of the goodness of fit by comparing with the results obtained for larger N. For any r, the results can be extrapolated to $\varepsilon = 0$ which is equivalent to extrapolating to $N = \infty$ and then to $\varepsilon = 0$ by simple calculations which avoid large matrix inversion.

One of the difficulties associated with the Fredholm integral equation is the peaked nature of the kernel. This can be overcome by rearranging the equation and using an iterative approach in the solution.

INTRODUCTION

Much work has already been carried out on the solution of two and three dimensional magnetic field problems. Substantial effort has been devoted to the computerised calculation of two and three dimensional magnetic fields which involves the numerical solution of Laplace's equation. As a result relevant computerised methods have been devised by Perin and Van der Meer (1967), Halacsy (1968), Halacsy et al. (1967), Nelson et al. (1969), Erdelyi et al.(1968), Winslow (1964), Colonias (1967), Trowbridge (1972) and others.

Many problems in engineering and science involve the solution of Laplace's or Poisson's equation both for analytic and non-analytic three dimensional geometries. Among the variety of numerical techniques, the relaxation method is particularly suited to a digital computer due to the method's intrinsic simplicity. However, since the relaxation method is an iteration over a three dimensional mesh, the large number of mesh points necessary to solve typical problems has resulted in heavy demands on memory space and computer time. This applies to most of the computerised work by the authors listed above. For these reasons an alternative is considered in this paper. A summary of the initial stages of this work is provided by Caldwell (1980). The numerical solutions of potential problems in iron-free regions are found using the technique of reduction to Fredholm integral equations of the first and second kinds. The motivation of this analysis is to construct numerical solutions to potential problems of Dirichlet type in the first instance although the work could be extended to mixed boundary value type problems. This provides an alternative approach to the classical methods by the authors mentioned above.

One particular line of approach is the axi-symmetric case and, in particular, the spherical case. By assuming a Green's function type of solution for Laplace's equation and using simple difference approximations, the problems can be reduced to a Fredholm integral equation. This may be solved by a direct method using Gaussian quadrature and matrix inversion as described later.

REDUCTION OF DIRICHLET TYPE PROBLEMS TO AN INTEGRAL EQUATION

The analysis described below can be used to construct solutions to potential problems of the Dirichlet type, that is the problem of solving Laplace's equation $\nabla^2 \Phi = 0$ inside a region where Φ is specified at each point of the boundary (for example, the boundary curve of a plane region, or the surface of a three dimensional region). It can also be used for problems of the Neumann type whose values of the normal derivative $\frac{\partial \Phi}{\partial n}$ of the function are prescribed on the surface and can be further extended to solve problems of the mixed boundary value type.

We shall now show how the Dirichlet problem can be reduced to an integral equation. The problem is to solve Laplace's equation

$$\nabla^2 \Phi = 0 \tag{1}$$

in some closed volume V with the condition

$$\Phi = \Phi_p (\underline{r}_S) \tag{2}$$

on the boundary surface S, Φ_p being prescribed (see Figure 1).

A Green's function type of solution of Equation (1) is given by

$$\Phi(\underline{r}) = \int_S \frac{\sigma(\underline{r}_S')}{|\underline{r} - \underline{r}'|} dS' \tag{3}$$

where the surface charge density is

$$\sigma(\underline{r}_S) = -\frac{1}{4\pi}\left[\frac{\partial \Phi}{\partial n}\right]_S$$

$$= -\frac{1}{4\pi}\left(\frac{\partial \Phi_+}{\partial n} - \frac{\partial \Phi_-}{\partial n}\right). \tag{4}$$

Replacing the derivatives in Equation (4) by simple differences gives

$$\sigma(\underline{r}_S) = -\frac{1}{4\pi\varepsilon}\{\Phi(\underline{r}_S + \varepsilon\underline{n}) + \Phi(\underline{r}_S - \varepsilon\underline{n}) - 2\Phi_p(\underline{r}_S)\} + 0(\varepsilon). \tag{5}$$

Using Equation (3) this then reduces to

$$4\pi\varepsilon\sigma(\underline{r}_S) + \int_S \sigma(\underline{r}_S')\left\{\frac{1}{|\underline{r}_S' - \underline{r}_S - \varepsilon\underline{n}|} + \frac{1}{|\underline{r}_S' - \underline{r}_S + \varepsilon\underline{n}|}\right\} dS'$$

$$= 2\Phi_p(\underline{r}_S) + 0(\varepsilon). \tag{6}$$

This is a Fredholm integral equation of the second kind and we are interested in solutions of this equation in the limit as $\varepsilon \to 0$. Proceeding to this limit we obtain

$$\int_S \frac{\sigma(\underline{r}_S')}{|\underline{r}_S' - \underline{r}_S|} dS' = \Phi_p(\underline{r}_S) \tag{7}$$

which is a Fredholm integral equation of the first kind. Having solved Equation (7) for σ we may then compute Φ using quadrature from Equation (3).

Equation (7) and its relation to solutions of potential problems is well known (see Garabedian 1964) and is the basis for much analysis. We now show how it can be used to obtain numerical solutions of three dimensional potential problems.

The difficulties associated with the numerical solution of Equation (7) are illustrated by considering the one dimensional equation

$$\int_0^1 K(x,y)\, f(y)dy = g(x) \tag{8}$$

and the corresponding eigenvalue problem

$$\int_0^1 K(x,y)\, f(y)dy = \lambda f(x). \tag{9}$$

In Equation (8) the function g(x) and the kernel K(x,y) are prescribed and f(y) is the unknown. When we consider Equation (9) we find that it possesses a set of eigenfunctions f_i and corresponding eigenvalues λ_i which have a limit point at zero, that is

$$\lambda_i \to 0 \quad \text{as } i \to \infty.$$

This means that if we replace Equation (8), in an obvious notation, by

$$\sum_j w_j\, K_{ij} f_j = g_i \tag{10}$$

where w_j are weights corresponding to a numerical quadrature, then K_{ij} will be singular in the limit. In other words, our problem is ill-conditioned.

These difficulties do not arise in the solution of Equation (6) since it is an equation of the second kind and this suggests a possible line of approach, namely, that we solve Equation (6) for different values of ε and extrapolate to the limit. We must expect , however, that to obtain accurate results it is necessary to solve Equation (6) for small values of ε. Furthermore, as ε decreases we will find that more and more points must be included in the quadrature scheme. Note also that as $\varepsilon \to 0$ the kernel associated with Equation (6) has a sharp peak near $\underline{r}_S' = \underline{r}_S$ which will further complicate our numerical quadrature.

It should be pointed out that Bakushinskii (1965) has considered the numerical solution of Fredholm integral equations of the first kind, namely

$$\int_a^b K(x,y)\; f(y)dy = g(x)$$

by examining the limit case of the equation

$$\varepsilon\; f(x) + \int_a^b K(x,y)\; f(y)dy = g(x)$$

as $\varepsilon \to 0$.

All the difficulties outlined above are well illustrated by the problem discussed in the next section.

SOLUTION OF A SPHERICAL PROBLEM

By applying the method outlined in the previous section we can find a solution for the Dirichlet problem in the special case where the volume V is a sphere of magnetic material of unit radius. In the first instance we shall make no allowance for axi-symmetry and so the full three dimensional problem will be considered.

This means that we consider the problem

$$\left.\begin{array}{l} \nabla^2\Phi = 0 \text{ inside the unit sphere } r = 1 \\ \text{and} \\ \Phi = \Phi_p(\theta,\phi) \text{ on the surface.} \end{array}\right\} \qquad (11)$$

Now the square of the distance between $P(r,\theta,\phi)$ and $P'(r',\theta',\phi')$ is given by

$$|\underline{r} - \underline{r}'|^2 = r^2 + r'^2 - 2rr'\;\{\cos\theta\;\cos\theta' - \sin\theta\sin\theta'\;\cos(\phi - \phi')\} \qquad (12)$$

and so Equation (6) becomes

$$4\pi\varepsilon\sigma(\theta,\phi) + \int_0^{\pi}\int_0^{2\pi} \sin\theta'\;\sigma(\theta',\phi')\;\{k(\varepsilon,\theta,\theta',\phi,\phi') + k(-\varepsilon,\theta,\theta',\phi,\phi')\}\;d\theta'\;d\phi' = 2\Phi_p(\theta,\phi) \qquad (13)$$

where

$$k(\varepsilon,\theta,\theta',\phi,\phi') = \left[1 + (1-\varepsilon)^2 - 2(1+\varepsilon)\;\{\cos\theta\cos\theta' - \sin\theta\sin\theta'\;\cos(\phi - \phi')\}\right]^{-\frac{1}{2}}. \qquad (14)$$

If we now introduce a scheme based on N quadrature points and associated weights $w_i (i = 1,2,\ldots\ldots N)$ we find, in an obvious notation, that

$$4\pi\varepsilon\sigma_{ij} + \sum_{k=1}^{N}\sum_{\ell=1}^{N} A_{ijk\ell}\sigma_{k\ell} = 2(\Phi_p)_{ij}\,. \qquad (15)$$

$$(i,j = 1,2,\ldots\ldots\ldots,N)$$

In Equation (15) we have written

$$A_{ijk\ell} = w_k w_\ell \sin\theta_k \{k(\varepsilon,\theta_i, \theta_k, \phi_j, \phi_\ell) + k(-\varepsilon, \theta_i, \theta_k, \phi_j, \phi_\ell)\} . \tag{16}$$

Equation (15) may now be rewritten as

$$\sum_{k=1}^{N} \sum_{\ell=1}^{N} (A_{ijk\ell} + 4\pi\varepsilon\delta_{ik}\,\delta_{j\ell})\sigma_{k\ell} = 2(\Phi_p)_{ij} \tag{17}$$

where

$$\delta_{ij} = 0, \; i \neq j$$

$$\delta_{ii} = 1.$$

Note that the matrix equation (17) is of order $(N^2 x N^2)$ and thus to solve problems involving 8 quadrature points in each direction involves inversion of matrices of order (64 x 64).

This method will now be demonstrated by taking the axi-symmetric case as a test example. This means that we must start from Equation (13) with $\sigma = \sigma(\theta)$ and we then find that

$$4\pi\varepsilon\sigma(\theta_i) + \int_0^{\pi}\int_0^{2\pi} \sigma(\theta_j)\sin\theta_j \; f(\varepsilon,\theta_i,\theta_j,\phi)\; d\theta_j \; d\phi = 2\Phi_p(\theta_i) \tag{18}$$

where

$$f(\varepsilon,\theta_i,\theta_j,\phi) = k(\varepsilon,\theta_i,\theta_j,\phi) + k(-\varepsilon,\theta_i,\theta_j,\phi). \tag{19}$$

Note that the kernel in Equation (18) is

$$K(\theta_i,\theta_j) = \int_0^{2\pi} f(\varepsilon,\theta_i,\theta_j,\phi)\; d\phi \tag{20}$$

and is therefore independent of ϕ . We now have a one dimensional integral equation which may be solved in the usual way by a direct method. The kernel may be written as an elliptic integral but this would not appear to be useful and so it is best to use direct quadrature. Hence by introducing Gaussian quadrature Equation (20) becomes

$$K(\theta_i,\theta_j) = \pi \sum_{k=1}^{N} w_k \;\; f\{\varepsilon,\theta_i,\theta_j, \pi(1+r_k)\} \tag{21}$$

where w_k, r_k are the weights and abscissae of Gaussian quadrature.

This means that $\sigma(\theta_i)$ for i = 1,2,.......,N can be obtained by solving the integral equation

$$4\pi\varepsilon\sigma\,(\theta_i) + \int_0^{\pi} K(\theta_i,\theta_j)\,\sigma\,(\theta_j)\sin\theta_j\, d\theta_j = 2\Phi_p(\theta_i). \qquad (22)$$

Again using Gaussian quadrature to evaluate the integral in Equation (22) we then have a set of N linear equations in N unknowns, namely

$$A\,\underline{\sigma} = \underline{b}, \qquad (23)$$

where

$$A_{ij} = 4\pi\varepsilon\delta_{ij} + \frac{\pi w_i}{2} K\{\tfrac{\pi}{2}(1+r_i), \tfrac{\pi}{2}(1+r_j)\} \sin \tfrac{\pi}{2}(1+r_j), \qquad (24)$$

$$i,j = 1,2........,N$$

$$\sigma_i = \sigma\{\tfrac{\pi}{2}(1+r_i)\} \qquad i = 1,2,.........,N \qquad (25)$$

$$b_i = 2\Phi_p\{\tfrac{\pi}{2}(1+r_i)\} \qquad i = 1,2,.........,N. \qquad (26)$$

Using a computer program POTQUAD Equation (23) has been solved for various orders of Gaussian quadrature, N, in order to obtain approximate values for $\sigma(\theta_i)$. However, since increasing the number of quadrature points yields information at a different set of points, it is necessary to add to the program a numerical integration procedure to calculate the potential Φ. We could, for example, calculate Φ along the axis using Equation (3). Thus, on the axis we have

$$\Phi(\underline{r}) = \int_0^{\pi}\int_0^{2\pi} \frac{\sigma(\theta')\sin\theta'\; d\theta'\; d\phi}{(1+r^2-2r\,\cos\theta')^{\frac{1}{2}}}$$
$$= 2\pi \int_0^{\pi} \frac{\sigma(\theta')\sin\theta'\; d\theta'}{(1+r^2-2r\,\cos\theta')^{\frac{1}{2}}} \qquad (27)$$

which may be evaluated at the points r = 0(0.1)0.9 by Gaussian quadrature to give

$$\Phi(r) = \pi^2 \sum_{i=1}^{N} \frac{w_i\sigma(\theta_i)\sin\frac{\pi}{2}\cdot(1+r_i)}{\{1+r^2-2r\,\cos\frac{\pi}{2}\cdot(1+r_i)\}^{\frac{1}{2}}}\;. \qquad (28)$$

A useful check on accuracy of the above method can be obtained by considering the problem in which the potential on the surface of the sphere r = 1 is given by

$$\Phi_p(\theta) = P_n(\cos\theta) \qquad (29)$$

which has analytic solution

$$\Phi(r,\theta) = r^n\, P_n(\cos\theta). \qquad (30)$$

Results have been generated for the n=1 case, that is,

$\Phi_p(\theta) = \cos\theta$ on $r = 1$

which has exact solution $\Phi = r$ on the $\theta = 0$ axis. In this way values of $\Phi(r)$ on the axis have been obtained by using Gaussian quadrature of order $N = 2(2)16$ for $r = 0(0.1)0.9$ for the cases $\varepsilon = 0.1(0.1)0.4$. Typical results for the cases $r = 0.2$ are presented in Table 1. Clearly there is good agreement between these results and the exact values and hence these results can be used to produce more accurate results.

Table 1 Computed values of Φ for $r = 0.2$ where $\Phi_p = P_1(\cos\theta)$ on $r = 1$

N	$\Phi(0.2)$			
	$\varepsilon = 0.1$	$\varepsilon = 0.2$	$\varepsilon = 0.3$	$\varepsilon = 0.4$
2	.4209	.3629	.3219	.2920
4	.2745	.2329	.2065	.1872
6	.2341	.2079	.1904	.1769
8	.2205	.2004	.1865	.1748
10	.2128	.1965	.1845	.1739
12	.2083	.1944	.1836	.1735
14	.2052	.1931	.1831	.1733
16	.2032	.1923	.1828	.1732

Remembering that analysis has shown the error to be $O(\varepsilon)$, the results for the cases $\varepsilon = 0.1$, 0.2, 0.3 and 0.4 can be extrapolated to zero using the formula

$$\Phi_0 = \Phi_i + A\varepsilon_i + B\varepsilon^2_i + C\varepsilon^3_i$$

for $i = 1,2,3,4$. This gives

$$\Phi_0 = \Phi_1 + 3(\Phi_1 - \Phi_2) - 3(\Phi_2 - \Phi_3) + (\Phi_3 - \Phi_4). \quad (31)$$

The extrapolated results for the cases $r = 0.2$, 0.5 and 0.8 are presented in Table 2. Similar results apply for other values of r. Note that as N increases the extrapolated results tend to some fixed value which is encouraging. This suggests that we can continue the calculations with larger N and then try to obtain the results for $N = \infty$ by extrapolating in N.

Table 2 Extrapolated results to $\varepsilon = 0$

N	Φ_0		
	r = 0.2	r = 0.5	r = 0.8
2	.5018	1.2250	1.6464
4	.3394	.7382	.9772
6	.2737	.6477	.9098
8	.2508	.5986	.8732
10	.2363	.5708	.8494
12	.2277	.5536	.8375
14	.2213	.5404	.8281
16	.2170	.5320	.8221

If we continue the calculations with higher order Gaussian quadrature we find that we get very accurate results. This is clearly seen by examining the results for the case N = 20, $\varepsilon = 0.1$ which are presented in Table 3. The maximum error over the range of values of r is approximately 1%.

Table 3 Computer values of Φ for N = 20, $\varepsilon = 0.1$

r	$\Phi(r)$
0	.0000
0.1	.1004
0.2	.2007
0.3	.3007
0.4	.4004
0.5	.4995
0.6	.5981
0.7	.6960
0.8	.7934
0.9	.8903

In exactly the same way results can be obtained for the cases $\Phi_p = r^n P_n(\cos\theta)$ for higher values of n.

It should be noted that an improvement in accuracy will be obtained be replacing Equation (5) by the more accurate difference scheme

$$\sigma(\underline{r}_S) = -\frac{1}{4\pi\varepsilon}\{2\Phi(\underline{r}_S + \varepsilon\underline{n}) + 2\Phi(\underline{r}_S - \varepsilon\underline{n}) - 3\Phi_p(\underline{r}_S) - \frac{1}{2}\Phi(\underline{r}_S + 2\varepsilon\underline{n}) - \frac{1}{2}\Phi(\underline{r}_S - 2\varepsilon\underline{n})\} + 0(\varepsilon^2). \quad (32)$$

EXTRAPOLATION IN N

On examining the results presented in Table 1 in more detail we find that some interesting patterns are forming. Furthermore, we can use these to extrapolate to the limit as $\varepsilon \to 0$ and hence save in computer time and storage requirements. One of the drawbacks of using Gaussian quadrature to evaluate integrals is that the error term is not helpful in error estimation. For this reason we use an empirical formula and fit a curve of the form

$$\Phi = A + BN^p, \tag{33}$$

where A, B and p are the arbitrary constants to be determined, to the N = 2,4 and 6 results. We can then compare with the computed results for larger values of N to test the goodness of fit. This means that we can extrapolate to $N = \infty$ and then to $\varepsilon = 0$ by simple calculations which avoid large matrix inversion (see Bakushinskii 1965).

In the first instance the computed results for the N = 2,4 and 6 cases are curve fitted and compared with the computed N = 8 case. If sufficient accuracy is not achieved the process is repeated by fitting the N = 4,6 and 8 results and comparing with the N = 10 case, and so on. The equations for this procedure are as follows:

$$\Phi_{N/2} = A + BN^p$$

$$\Phi_{(N+2)/2} = A + B(N+2)^p$$

$$\Phi_{(N+4)/2} = A + B(N+4)^p$$

This gives

$$\frac{\Phi_{N/2} - \Phi_{(N+2)/2}}{\Phi_{(N+2)/2} - \Phi_{(N+4)/2}} = \frac{N^p - (N+2)^p}{(N+2)^p - (N+4)^p} = k$$

and hence

$$f(p) = k(N+4)^p - (1+k)(N+2)^p + N^p = 0. \tag{34}$$

Since Equation (34) is non-linear we use the Newton-Raphson iterative method to find p. This gives

$$p_{i+1} = p_i - \frac{f(p_i)}{f'(p_i)}$$

$$= p_i - \frac{k(N+4)^{p_i} - (1+k)(N+2)^{p_i} + N^{p_i}}{k(n+4)^{p_i}\ell n(N+4) - (1+k)(N+2)^{p_i}\ell n(N+2) + N^{p_i}\ell nN} \tag{35}$$

$$(i = 0,1,2,\ldots\ldots,)$$

Then B is found from the equation

$$B = \frac{\Phi_{N/2} - \Phi_{(N+2)/2}}{N^p - (N+2)^p} \tag{36}$$

and A from the equation

$$A = \Phi_{N/2} - BN^p \tag{37}$$

A computer subroutine POTFIT is available which reads in the computed values of Φ obtained previously for the cases $N = 2(2)16$ and calculates p to an accuracy of 6D. The process is repeated until the difference between the computed Φ from this program and the corresponding value obtained from the program POTQUAD is less than 0.0001. In this way values of Φ for higher orders of quadrature have been predicted.

By feeding in results such as those in Table 1 we find that the scheme converges at an early stage to give the accuracy specified and in the cases

$$r = 0.2, \qquad \varepsilon = 0.1(0.1)0.4$$

the values of A,B and p relating to Equation (33) are presented in Table 4. These results have then been extrapolated to $\varepsilon = 0$ to give $\Phi = 0.19793$ when $r = 0.2$.

Table 4 Values of A,B and p for the case r = 0.2

ε	$A = \Phi_0$	B	p
0.1	.19383	.61678	-1.51106
0.2	.18952	.68083	-1.98765
0.3	.18209	.95039	-2.58805
0.4	.17251	.24498	-2.24123

The problem of slow convergence may arise particularly for values of r close to 1. In such cases we may either have to relax the accuracy requirements or to feed in to the program values of Φ for orders of Gaussian quadrature beyond $N = 16$. Again this work can be generalised to consider the cases $\Phi_p = r^n P_n(\cos\theta)$ for higher values of n.

AN ITERATIVE METHOD OF SOLUTION

One of the difficulties associated with the direct method of solution of integral equations of the type (6) is the peaked nature of the kernel. This means that high order quadrature must be used and leads to inversion of large matrices. To overcome this difficulty it is possible to use an alternative method of solution of Equation (6). Firstly, we can rewrite Equation (6) in the form

$$4\pi\varepsilon\sigma(\underline{r}_S) + \sigma(\underline{r}_S)\int_S g(\underline{r}_S, \underline{r}'_S, \varepsilon\underline{n})dS' + \int_S \{\sigma(\underline{r}'_S) - \sigma(\underline{r}_S)\}\ g(\underline{r}_S, \underline{r}'_S, \varepsilon\underline{n})dS' = 2\Phi_p(\underline{r}_S) \tag{38}$$

where

$$g(\underline{r}_S, \underline{r}'_S, \varepsilon\underline{n}) = \frac{1}{|\underline{r}'_S - \underline{r}_S - \varepsilon\underline{n}|} + \frac{1}{|\underline{r}'_S - \underline{r}_S + \varepsilon\underline{n}|}. \tag{39}$$

Then we can proceed to solve Equation (38) by iteration; that is

$$\sigma(\underline{r}_S) = \sum_{M=0}^{\infty} \sigma_M(\underline{r}_S) \tag{40}$$

where

$$4\pi\varepsilon\sigma_{M+1}(\underline{r}_S) + \sigma_{M+1}(\underline{r}_S)\int_S g(\underline{r}_S, \underline{r}'_S, \varepsilon\underline{n})dS' = -\int_S \{\sigma_M(\underline{r}'_S) - \sigma_M(\underline{r}_S)\}\ g(\underline{r}_S, \underline{r}'_S, \varepsilon\underline{n})dS' \quad (M = 0,1,2,\ldots\ldots) \tag{41}$$

and

$$4\pi\varepsilon\sigma_0(\underline{r}_S) + \sigma_0(\underline{r}_S)\int_S g(\underline{r}_S, \underline{r}'_S, \varepsilon\underline{n})dS' = 2\Phi_p(\underline{r}_S). \tag{42}$$

Clearly we no longer have the difficulty of a peaked kernel since the second integral in Equation (38) has no such behaviour near $\underline{r}_S = \underline{r}'_S$ and the first integral does not involve σ. It may therefore be computed by high order integration without affecting the iteration process.

Preliminary results from a numerical investigation of Equations (40),(41) and (42) for the case of a sphere are very encouraging. It is also possible to analyse the convergence of this method together with the error accumulation.

CONCLUDING REMARKS

The previous analysis can be used to solve Dirichlet and Neumann problems for a linear elliptic partial differential equation by transforming them into integral equations. In past work the numerical solution of potential problems has largely been carried out by using the classical approach based on finite differences and then solving the resulting difference equations by relaxation methods. These methods require great computer storage capacity and a very fast access. The storage problem can only be overcome at the expense of computer time. Convergence is also always a problem in iterative methods using computers. A very large number of iterations is required and known iterative methods prove useless in accelerating convergence. Other difficulties are posed in problems having symmetry and uniformity. For example, symmetry has the effect of making the inversion matrix singular.

Many of these problems are overcome by using the technique of reducing the problem to a Fredholm integral equation of the second kind. This means that by using the direct method described we can solve for various small values of ε and then extrapolate to the limit $\varepsilon = 0$. Further improvements can be made by extrapolating in N (the order of Gaussian quadrature) and using an iterative solution to overcome the peaked nature of the kernel. The method gives fairly accurate results for reasonably small N and has the advantage over the classical method of making smaller demands on computer time and memory store. In a typical case, when using the program POTQUAD for the spherical problem values of $\Phi(r)$ for a particular ε, r = 0(0.1)0.9 and orders of Gaussian quadrature N = 2(2)16 required only 6 minutes central processor time on an ICL System 4/50 machine. The time then taken for the extrapolation in N using the program POTFIT was negligible.

ACKNOWLEDGEMENT

The author would like to thank Professor R.D. Gibson (Newcastle Upon Tyne Polytechnic, England) for his interest and assistance in this work.

REFERENCES

Bakushinskii, A.B. (1965) A Numerical Method for Solving Fredholm Integral Equations of the First Kind. Zh. vȳchisl. Mat. Fiz., 5, 4 : 744-749.

Caldwell, J. (1980) An Application of Extrapolation to the Limit. BIT, Sweden, 20, : 251-253.

Colonias, J. (1967) Calculation of Two Dimensional Magnetic Fields by Digital Display Techniques. Report No UCRL-17340, LRL Berkeley, U.S.A.

Erdelyi, E.A., Fuchs, E.F. and Ratti, U. (1968) Acceleration of Convergence Techniques for the Relaxation Solution of Nonlinear Laplace's,Poisson's and Diffusion Equations. Proc. of Discussion on Analysis of Magnetic Fields, University of Nevada, U.S.A.

Garabedian, P.R.(1964) Partial Differential Equations. Wiley.

Halacsy, A.A. (1968) Three Dimensional Calculations of Magnetic Fields. Proc. of Discussion on Analysis of Magnetic Fields, University of Nevada, U.S.A. : 109-121.

Halacsy, A.A., Clark, G and Dunks, J. (1967) Computerised Calculation of Three Dimensional Magnetic Fields. Proc. 2nd Int Conference on Magnet Technology, Oxford : 61-78.

Nelson, D., Kim,H. and Reiser,M. (1969) Computer Solutions for Three Dimensional Electromagnetic Field Geometries. I.E.E.E. Trans on Nuclear Science, NS-16, No.3.

Perin,R. and Van der Meer, S. (1967) The Program MARE for the Computation of Two Dimensional Static Magnetic Fields. Report 67-7, C.E.R.N. Geneva.

Trowbridge C.W. (1972) Progress in Magnet Design by Computer. Proc. 4th Int. Conf on Magnet Technology, Brookhaven, U.S.A.

Winslow, A.M. (1964) Numerical Solution of the Poisson Equation in a Nonuniform Triangular Mesh. Report No. UCRL 7784-T (Rev.2) TRIM, LRL Berkeley, U.S.A.

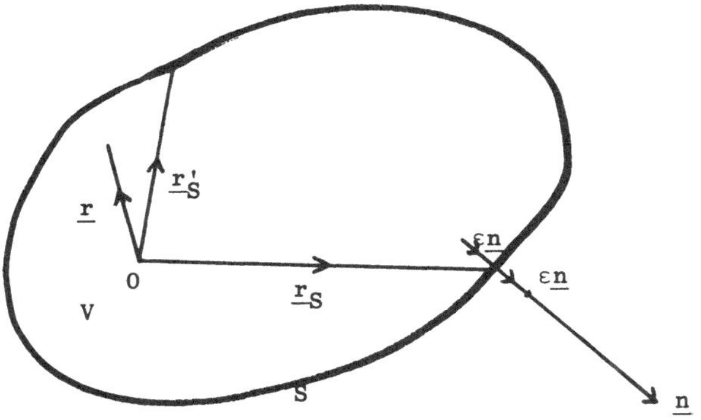

Figure 1. Volume V bounded by a surface S

BEMSTAT - A NEW TYPE OF BOUNDARY ELEMENT PROGRAM FOR TWO-DIMENSIONAL ELASTICITY PROBLEMS

Lars Bolteus and Odd Tullberg,
Dept of Structural Design and Dept of Structural Mechanics,
Chalmers University of Technology,
S-412 96 Göteborg
SWEDEN

ABSTRACT

A new type of boundary element program is presented. The equations are derived by use of integral operators. A natural way of deducing element matrices in BEM is shown, which leads to a new technique for establishing the system of equations. A simple subsidiary condition technique, which makes it possible to take discontinous tractions into account, is presented. Guidelines are given for the BE-discretization and the numerical integration. Formulas for the analytical integration of the singular terms are shown. A non-conventional method for coupling BEM and FEM is proposed. Numerical studies have been made in order to investigate the performance of different elements.

INTRODUCTION

A wide range of engineering problems in structural mechanics, fluid mechanics, and electrical engineering can be formulated with boundary integral equations (Banerjee and Butterfield, 1979). These equations are most conveniently solved by numerical methods.

The numerical techniques for integral equations are developing rapidly and many articles have been written on the subject during the last decade, for references, see Banerjee and Butterfield (1979) and Brebbia and Walker (1980). The collocation method is commonly used for the numerical solution (Jaswon and Symm, 1977), but also variational methods have been presented (Nedelec, 1977; Wendland *et al*, 1979; Jeng and Wexler, 1977).

In this paper, the two-dimensional mixed boundary value problem of elastostatics is formulated with the direct integral equations. To the authors' knowledge, no variational solution exists for these equations. Hence, the collocation method is here used for the numerical solution. The purpose of this paper is to report on BEMSTAT, a new type of boundary element program for two-dimensional eleasticity problems. The program includes a new technique for establishing the system of simultaneous equations, a simple method for handling discontinuous tractions, and a method for coupling BEM and FEM. In addition, guidelines for the numerical integration and the element discretization are given. Several numerical examples are calculated in order to investigate the behaviour of different elements.

DERIVATION OF THE BOUNDARY ELEMENT EQUATIONS

Boundary integral representations

The displacement field in a two-dimensional elasticity problem can be represented by a boundary integral formula, the Somigliana identity:

$$u(x) = \int_\Gamma U(x,\xi)t(\xi)d\xi - \int_\Gamma T^T(x,\xi)u(\xi)d\xi \qquad x\in\Omega \tag{1}$$

where

$$u = [u_x \quad u_y]^T \tag{2}$$

$$t = [t_x \quad t_y]^T \tag{3}$$

are the physical displacement and stress vector (or traction) components in a Cartesian coordinate system Oxy and the body under consideration is occupying the domain shown in Figure 1.

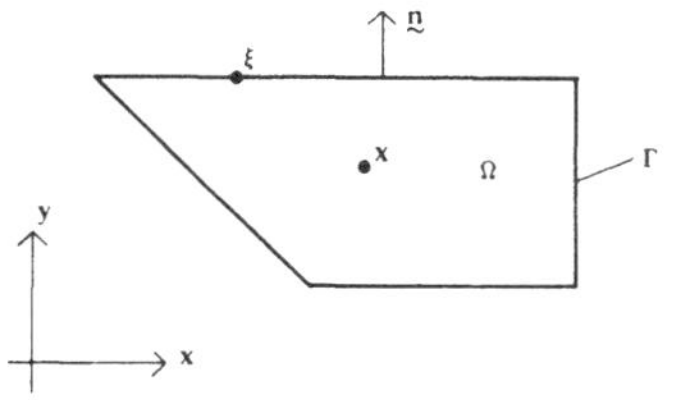

Figure 1. Considered domain

$U(x,\xi)$ is the fundamental solution matrix, which represents the displacement field due to a unit pointload in ξ with the direction vector $e(\xi)$,

$$u(x) = U(x,\xi)e(\xi) \tag{4}$$

and T is a matrix which gives the corresponding traction field (for a given normal direction),

$$t(x) = T(x,\xi)e(\xi) \tag{5}$$

T can be deduced from the fundamental matrix U. (See Appendix A.) The representation formula, Equation 1, can conveniently be written with integral operators:

$$u(x) = Kt(x) - Lu(x) \qquad x\in\Omega \tag{6}$$

These linear operators are defined as

$$Kv(x) = \int_\Gamma U(x,\xi)v(\xi)d\xi \tag{7}$$

$$Lv(x) = \int_\Gamma T^T(x,\xi)v(\xi)d\xi \tag{8}$$

When the displacements and tractions are known over the entire boundary, the interior displacement field can be generated from Equation 6. Moreover, the strains can be found through the kinematic relation (small strains and deformations)

$$\varepsilon = \tilde{\nabla} u \tag{9}$$

where the strain tensor components are represented by the matrix

$$\varepsilon = [\varepsilon_x \quad \varepsilon_y \quad \gamma_{xy}]^T \tag{10}$$

and the gradient operator is defined as

$$\tilde{\nabla} = \begin{bmatrix} \partial/\partial x & 0 & \partial/\partial y \\ 0 & \partial/\partial y & \partial/\partial x \end{bmatrix}^T \tag{11}$$

Introducing Equation 6 in Equation 9 gives the integral representation of the interior strains

$$\varepsilon(x) = \tilde{\nabla}Kt(x) - \tilde{\nabla}Lu(x) \tag{12}$$

where

$$\tilde{\nabla}Kt(x) = \int_\Gamma \tilde{\nabla}_x U(x,\xi)\,t(\xi)\,d\xi \tag{13}$$

$$\tilde{\nabla}Lu(x) = \int_\Gamma \tilde{\nabla}_x T^T(x,\xi)\,u(\xi)\,d\xi \tag{14}$$

The subscript x indicates on which argument the gradient matrix operates, and as can be seen, it only operates on the fundamental matrices. These can be calculated analytically. Hence, no approximations are involved in Equation 12. $\tilde{\nabla}U$ and $\tilde{\nabla}T$ can be found in Appendix A. When the strains have been calculated, the stresses can be found by the generalized Hooke's law

$$\sigma = S^d\varepsilon \tag{15}$$

The stress components in the Cartesian coordinate system are represented by the matrix

$$\sigma = [\sigma_x \; \sigma_y \; \tau_{xy}]^T \tag{16}$$

and S^d is the differential stiffness matrix for plane strain.

Boundary integral equations

The preceding formulas have been formulated for points in the domain. By a limiting procedure, the point can be taken to the boundary. The general boundary integral equation can then be written as

$$(E+L)u(x) - Kt(x) = 0 \qquad x\in\Gamma \tag{17}$$

where the new integral operator is defined as

$$Ev(x) = \int_\Gamma D(\xi)v(\xi)\delta(x,\xi)\,d\xi = D(x)v(x) \tag{18}$$

For a point on a smooth boundary, the matrix D is diagonal

$$D = \begin{bmatrix} 0{,}5 & 0 \\ 0 & 0{,}5 \end{bmatrix} \tag{19}$$

but on a sharp corner D is full. The general form of D for two-dimensional problems is given in Appendix A.

Equation 17 is the integral equation to be solved. In a general boundary value problem, displacements are prescribed on a part of Γ, *e.g.*, on Γ_1, and tractions on the rest, called Γ_2, thus $\Gamma = \Gamma_1 + \Gamma_2$. The system of integral equations can now formally be restated as

$$\begin{bmatrix} (E+L)_{11} & (E+L)_{12} \\ (E+L)_{21} & (E+L)_{22} \end{bmatrix} \begin{bmatrix} \bar{u}_1 \\ u_2 \end{bmatrix} - \begin{bmatrix} K_{11} & K_{12} \\ K_{21} & K_{22} \end{bmatrix} \begin{bmatrix} t_1 \\ \bar{t}_2 \end{bmatrix} = \begin{bmatrix} 0 \\ 0 \end{bmatrix} \qquad \begin{matrix} x\in\Gamma_1 \\ x\in\Gamma_2 \end{matrix} \tag{20}$$

where the first subscript indicates on which part of the boundary the pole is located, and the second over which part the integration is performed. The bar indicates prescribed function. Reordering of Equation 20 gives the final system of boundary integral equations to be solved

$$\begin{bmatrix} K_{11} & -(E+L)_{12} \\ -K_{21} & (E+L)_{22} \end{bmatrix} \begin{bmatrix} t_1 \\ u_2 \end{bmatrix} = \begin{bmatrix} f_1 \\ f_2 \end{bmatrix} \quad \begin{matrix} x \in \Gamma_1 \\ x \in \Gamma_2 \end{matrix} \tag{21}$$

where

$$\begin{aligned} f_1 &= (E+L)_{11}\bar{u}_1 - K_{12}\bar{t}_2 \\ f_2 &= -(E+L)_{21}\bar{u}_1 - K_{22}\bar{t}_2 \end{aligned} \tag{22}$$

To the authors' knowledge, the different operators in Equation 21 have not yet been mathematically investigated, which implies that a correct variational formulation cannot be performed.

Numerical method

To achieve a numerical solution, the collocation method is used. This can be looked upon as a weighted residual method with a special choice of weight functions, namely the Dirac function with origins in the collocation points. Thus the boundary integral equation will be forced to zero in the collocation points. For convenience, Equation 17 will now be used, and the boundary conditions will be introduced later. Thus,

$$(E+L)u(\tilde{x}_I) - Kt(\tilde{x}_I) = 0 \qquad I = 1,2,\ldots,N \tag{23}$$

where N is the number of collocation points and $\tilde{x}_I$ the coordinates for node I. The boundary is now discretized with boundary elements over which displacements and tractions are approximated. The approximation is of a polynomial shape. In BEMSTAT, the shape functions can be of constant, linear, or quadratic forms, and the collocation points are placed in the element function nodes. Thus, the approximation can be written as

$$u(x) = \Phi^T(x)\tilde{u} \tag{24}$$

$$t(x) = \Phi^T(x)\tilde{t} \tag{25}$$

where the shape function matrix is defined by

$$\Phi(x) = \left[\begin{array}{cc|cc|c|cc} \varphi_1(x) & 0 & \varphi_2(x) & 0 & \ldots & \varphi_N(x) & 0 \\ 0 & \varphi_1(x) & 0 & \varphi_2(x) & & 0 & N \end{array}\right]^T \tag{26}$$

and where the shape functions are defined as having only local support

$$\varphi_I(\tilde{x}_J) = \delta_{IJ} \begin{cases} = 1 \text{ if } I = J \\ = 0 \text{ if } I \neq J \end{cases} \tag{27}$$

The nodal quantities are stored in $\tilde{u}$

$$\tilde{u} = [\tilde{u}_x^1 \ \tilde{u}_y^1 \ \tilde{u}_x^2 \ \tilde{u}_y^2 \ \ldots \ \tilde{u}_x^N \ \tilde{u}_y^N]^T \tag{28}$$

The same is valid for $\tilde{t}$. As shown in Equations 24 and 25, the same shape functions are used for displacements and tractions. In a consistent formulation, tractions should be approximated by functions of one order less than the displacements. However, this can be handled in the program since different shape functions can be mixed. A more severe error in Equation 25 is the continuity requirements for the tractions. This problem is circumvented in the program by an efficient subsidiary condition technique, which makes possible discontinuous tractions. Whenever this occurs, two nodes are placed in the discontinuity point. No element is defined between the nodes. The nodal displacements are forced to be equal, *i.e.*, one of the nodes is considered to be slave of the other. The row associated with the slave is deleted in both H and G, and the associated column in H is added to the column associated with the master. Notice that the column in G associated with the slave is kept. This method is easier to use than the one proposed by Alarcon *et al.* (1979), since they define the tractions as discontinuous, *a priori*, and then generate the correct continuity requirements by a code system.

The approximation in Equations 24 and 25 is now inserted in Equation 23, which gives

$$H\tilde{u} - G\tilde{t} = 0 \tag{29}$$

where the elements in the matrices are

$$H_{IJ} = (E+L)\Phi_J(\tilde{x}_I) = \int_\Gamma [D(x)\delta(x,\tilde{x}_I) + T^T(\tilde{x}_I,x)]\Phi_J(x)dx \tag{30}$$

$$G_{IJ} = K\Phi_J(\tilde{x}_I) = \int_\Gamma U(\tilde{x}_I,x)\Phi_J(x)dx \tag{31}$$

H and G are assymmetric. In addition they are dense, although locally defined shape functions are used. When the boundary conditions are introduced, the system in Equation 29 can be reordered in the same manner as in Equation 21 to a solvable system of equations

$$A\tilde{v} = \tilde{b} \tag{32}$$

where A is a mix of H and G, which implies that it is dense and assymmetric, $\tilde{v}$ is the unknowns on the boundary, and $\tilde{b}$ includes all known boundary data.

When the boundary variables are known, requested quantities in the domain can be calculated. Interior displacements can be found by

$$u(x) = K\Phi^T(x)\tilde{t} - L\Phi^T(x)\tilde{u} \tag{33}$$

and strains by

$$\varepsilon(x) = \tilde{\nabla}K\Phi^T(x)\tilde{t} - \tilde{\nabla}L\Phi^T(x)\tilde{u} \tag{34}$$

Establishment of the system of equations

In a traditional BEM-program the system of equations is established collocation pointwise, *i.e.*, row by row. This method is different from the efficient and systematic method used in FEM, where matrices for each finite element are formed and assembled into the system matrices. Here will be shown that this procedure also can be used for boundary element methods.

The displacement in a point on the boundary is approximated as

$$u(x) = \Phi^T(x)\tilde{u} \tag{35}$$

If we define local element matrices (see also Figure 2),

$$\tilde{u}^e = [\tilde{u}_x^1 \quad \tilde{u}_y^1 \quad \tilde{u}_x^2 \quad \tilde{u}_y^2] \tag{36}$$

$$\Phi^e(x) = \left[\begin{array}{cc|cc} \varphi_1^e(x) & 0 & \varphi_2^e(x) & 0 \\ 0 & \varphi_1^e(x) & 0 & \varphi_2^e(x) \end{array}\right]^T = [\Phi_1^e \quad \Phi_2^e]^T \tag{37}$$

the approximation can be written

$$u(x) = \sum_e \Phi^{eT}(x)\tilde{u}^e \tag{38}$$

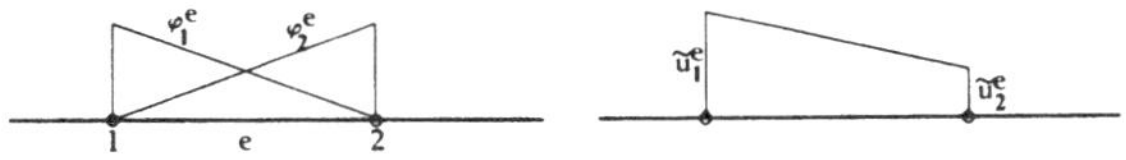

Figure 2. Locally defined element quantities

The displacements have to be continuous over the element boundaries; thus, one can define a topological incident matrix C^e for each element, which connects local and global variables

$$\tilde{u}^e = C^e\tilde{u} \tag{39}$$

or

$$\tilde{u}^e_\alpha = C^e_{\alpha I}\tilde{u}_I \qquad \alpha = 1 \text{ to } 2 \text{ (3 or more)} \qquad I = 1,N \tag{40}$$

where α is an index running from 1 to 2 for linear shape functions, 3 for quadratic, and so on. Introducing Equation 40 in Equation 38 and comparing this with Equation 35, yields

$$\Phi_I = \sum_e C^e_{I\alpha}\Phi^e_\alpha \tag{41}$$

This relation is now used in Equation 30, which leads to

$$H_{IJ} = \sum_e (E+L)\Phi^e_\alpha(\tilde{x}_I)C^e_{\alpha J} = \sum_e H^e_{I\alpha}C^e_{\alpha J} \tag{42}$$

and similar for the G matrix

$$G_{IJ} = \Sigma G^e_{I\alpha}C^e_{\alpha J} \tag{43}$$

Hence, element matrices for the collocation boundary element method have been derived

$$H^e_{I\alpha} = \int_{\Gamma^e} (D(x)\delta(\tilde{x}_I,x) + T^T(\tilde{x}_I,x))\Phi^e_\alpha(x)dx \tag{44}$$

$$G^e_{I\alpha} = \int_{\Gamma^e} U(\tilde{x}_I,x)\Phi^e_\alpha(x)dx \tag{45}$$

The matrices have the size Nx2 for linear shape functions and Nx3 for quadratic ones (block matrix formulation). By introducing the boundary conditions for the element, the final element matrices A^e and $\tilde{b}^e$, can be formed. These can then be assembled into the global system matrices, in the same manner as in FEM. Formally, this can be written

$$A = \sum_e A^e; \quad \tilde{b} = \sum_e \tilde{b}^e \tag{46}$$

where A^e is a mixture of H^e and G^e and $\tilde{b}^e$ includes the known boundary element data. By this method, element data (integration point data) are calculated only once. This is in contrast to a traditional BEM-program, where these calculations are repeated for each collocation point. Thus, the proposed method is less time consuming.

DERIVATION OF OPTIMAL INTEGRATION SCHEME

The element matrices are integrated numerically when the collocation point is located outside the element. This is here called *exterior integration.* When the collocation point is inside the element, the integration is performed analytically or with some special numerical scheme, since the integrands tends to infinity. This situation is here called *interior integration.*

Exterior integration

The exterior integrals in the element matrices are numerically integrated with a Gauss quadrature formula with unit weight function. Since the construction of the simultaneous equations may take longer than it does to solve them, an optimal integration scheme should be developed and used. In this section an integration scheme is recommended and guidelines for the discretization are given (see also Watson 1980.)

The integration is most conveniently done on a canonical element (-1,1). Hence, both geometry and functions are represented on this element, and then the geometry is mapped to the global system. Thus,

$$\underset{\sim}{x} = \Phi^{eT}(\eta)\tilde{x}^{e} \tag{47}$$

where $\underset{\sim}{x}$ is the global coordinate vector, η the intrinsic coordinate, Φ the shape function matrix, and $\tilde{x}^e$ the nodal coordinate matrix for the element. Hence, the element matrices can be evaluated as

$$\bar{H}^{e} = \int_{-1}^{1} T^{T}(x,\tilde{x})\Phi^{eT}(\eta)J^{e}(\eta)d\eta = \sum_{g=1}^{n} T^{T}(x_g,\tilde{x})\Phi^{eT}(\eta_g)J^{e}(\eta_g)w_g \tag{48}$$

$$G^{e} = \int_{-1}^{1} U(x,\tilde{x})\Phi^{eT}(\eta)J^{e}(\eta)d\eta = \sum_{g=1}^{n} U(x_g,\tilde{x})\Phi^{eT}(\eta_g)J^{e}(\eta_g)w_g \tag{49}$$

where the superbar on $\bar{H}^e$ indicates that the D-matrix is left out, »g» stands for integration point, and w is the integration weight. J^e is the Jacobian:

$$J^{e}(\eta) = \left(\left(\frac{d\Phi^{eT}}{d\eta}\tilde{x}^{e}\right)^{2} + \left(\frac{d\Phi^{eT}}{d\eta}\tilde{y}^{e}\right)^{2}\right)^{1/2} \tag{50}$$

The element matrices in Equations 48 and 49 are established in the following manner; 1) Calculate necessary data for the current integration point, 2) Calculate contributions from all collocation points, 3) Take the next integration point.

In BEMSTAT the geometry can be mapped by linear or quadratic shape functions. This choice is independent of the function approximation chosen. Hence, subparametric, isoparametric, and superparametric elements can be used.

Gaussian integration is exact for polynomials. However, the integrands consist not only of polynomials. Present are also functions like 1/r and ln(r), multiplied by complicated trigonometric functions. These functions become predominant, when the pole is close to the element considered. This implies that the Gaussian integration no longer is exact. Therefore, an empirical study has been made here, in order to investigate the optimal number of integration points necessary. In Table 1 the performance of the Gauss quadrature with 2 and 4 integration points has been studied for a characteristic integral in $\bar{H}^e$ (see also Figure 3).

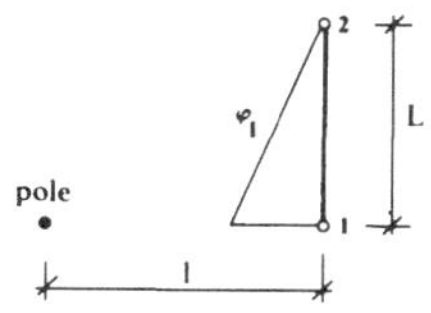

Figure 3. Position of pole and element under consideration

Table 1. Values of a characteristic integral in $\overline{H}^e$ for distance ratios between 2 and 5

l/L	»exact»	4 g.p.	2 g.p.
5	0.051311	0.051311	0.051311
4	0.063331	0.063331	0.063328
3	0.082243	0.082243	0.082235
2	0.115172	0.115172	0.115154

The »exact» values in Tables 1 and 2 have been calculated using the trapezoidal rule with 20,000 subintervals. Table 1 shows that the 2-point formula gives less than 5 correct decimals when $l/L < 4$. The 4-point formula is accurate for all $l/L > 2$, but for distance ratios less than 2, more integration points are needed. This is shown in Table 2 (see also Figure 3).

Table 2. Values of a characteristic integral in $\overline{H}^e$ for distance ratios between 0.1 and 1.0

l/L	»exact»	10 g.p.	4 g.p.
1.00	0.029552	0.029552	0.029559
0.50	0.066141	0.066141	0.065957
0.25	0.110832	0.110830	0.108785
0.10	0.169263	0.169436	0.202618

Table 2 shows that the 4-point formula gives less than 5 correct decimals when $l/L < 1$.The 10-point formula gives accurate results for distance ratios greater than 0.25. Two important questions can now be answered:

- What is the optimal number of integration points for a specific case? (The program should be able to make this choice automatically.)
- How should the BE-discretization be performed? (Since the program puts the collocation points at the boundaries or at the middle of the element, one can give guidelines concerning the element lengths.)

The answer to the first question is:

- Use 10 g.p. when $0.25 < l/L < 1$
- Use 4 g.p. when $1 < l/L < 4$
- Use 2 g.p. when $4 < l/L$

If the above integration rules are applied, the following guidelines for the BE-discretization can be given:

- for linear elements, $0.25 < L_1/L_2 < 4$
- for quadratic elements, $0.5 < L_1/L_2 < 2$

where L_1 and L_2 are lengths of two adjacent elements. Thereby, the second question is answered. The geometrical configuration in Figure 3 implies that the integrand is strongly influenced by the trigonometric functions. The formula proposed by Watson (1980) neglects this influence, since the only term considered is $1/r$.

BEMSTAT chooses dynamically the number of integration points, 2, 4, or 10, depending on the distance between the pole and element, according to the above rules. BEMSTAT has also an algorithm for automatic refinement or enlargement of the elements, which makes it easy for the user to follow the guidelines given concerning element ratios.

Interior integration

When the collocation point is located in the element under consideration, both $\bar{H}^e$ and G^e become singular. The G^e matrix is, however, only weakly singular and can be integrated numerically by a Gauss quadrature with logarithmic weight function. $\bar{H}^e$ is strongly singular, and the integral exists only in the sense of a Cauchy principle value and cannot be integrated numerically. In a finite domain with a closed boundary, this is overcome by the well-known rigid body motion trick. However, when studying problems with infinite or semi-infinite domains with an open element model, this trick does not work and the terms must be integrated analytically. This is possible when the geometry is approximated with straight line elements, and has been done by the authors' for linear and quadratic shape functions in BEMSTAT. Hence, infinite and semi-infinite domains can be treated in the program by using open element models. The contribution from the singular integrals to the element matrices for linear shape functions is shown in Appendix A.

COUPLING OF BEM AND FEM DISCRETIZED STRUCTURES

The structure in Figure 4 is divided into two regions, A and B. A is dicretized with finite elements and contains N + I number of nodes; B is discretized with boundary elements and contains M + I nodes; I is the number of interface nodes

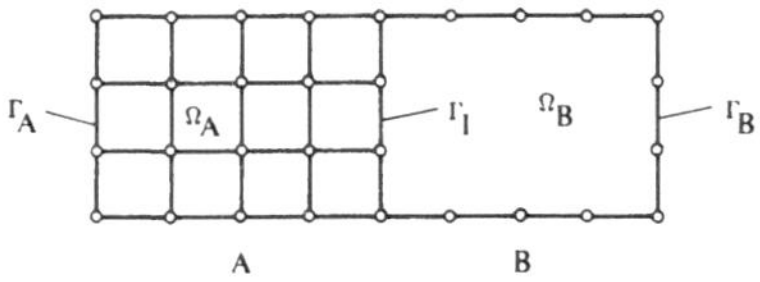

Figure 4. Structure discretized with finite elements, region A, and boundary elements, region B

The discretized finite element equations for Ω_A can formally be written

$$\underset{(N+I)\times(N+I)}{\begin{bmatrix} S_{AA} & S_{AI} \\ S_{IA} & S_{II} \end{bmatrix}} \begin{bmatrix} \tilde{u}_A \\ \tilde{u}_I \end{bmatrix} = \begin{bmatrix} \tilde{f}_A \\ \tilde{f}_I \end{bmatrix} \tag{51}$$

where S is the stiffness matrix, $\tilde{u}$ the nodal displacement matrix, and $\tilde{f}_A$ the equivalent nodal forces acting on A. $\tilde{f}_I$ is defined as

$$\tilde{f}_I = \int_{\Gamma_I} \Phi t_I^A ds \tag{52}$$

where Φ is the finite element shape function matrix for displacements on the interface and t_I^A the traction acting when A and B are uncoupled. The system in Equation 51 can be condensed to the interface variables by static condensation, which gives

$$\underset{(I\times I)}{S_{II}^{*}}\tilde{u}_I = \tilde{f}^{*} + \tilde{f}_I \tag{53}$$

where formally

$$S_{II}^{*} = S_{II} - S_{IA}S_{AA}^{-1}S_{AI}$$
$$\tilde{f}^{*} = -S_{IA}S_{AA}^{-1}\tilde{f}_A \tag{54}$$

For region B the boundary element equation can be written as

$$\underset{(M+I)\times(M+I)}{\begin{bmatrix} H_B & H_I \end{bmatrix}}\begin{bmatrix} \tilde{u}_B \\ \tilde{u}_I \end{bmatrix} = \begin{bmatrix} G_B & G_I \end{bmatrix}\begin{bmatrix} \tilde{t}_B \\ \tilde{t}_I \end{bmatrix} \tag{55}$$

When the boundary conditions on region B are introduced in Equation 55, a non-square system is obtained

$$\underset{(M+I)\times(M+2I)}{\begin{bmatrix} B & -G_I & H_I \end{bmatrix}}\begin{bmatrix} \tilde{v}_B \\ \tilde{t}_I \\ \tilde{u}_I \end{bmatrix} = \underset{(M+I)}{\begin{bmatrix} \tilde{b} \end{bmatrix}} \tag{56}$$

where B is a mix of H_B and G_B and $\tilde{v}_B$ is the unknowns on Γ_B. There are too many unknowns in this system, and the missing equations have to come from region A. Up to now, A and B have been looked upon as two uncoupled structures. They are now coupled together by forcing the displacements at the interface to be equal and the tractions to be equal in magnitude but with opposite signs, *i.e.*,

$$u_I^A = u_I^B \qquad\qquad t_I^B = -t_I^A \tag{57}$$

If the same shape functions are used in A and B, this implies that

$$\tilde{u}_I^A = \tilde{u}_I^B = \tilde{u}_I \tag{58}$$

The tractions on Γ_I are approximated in the usual BE-manner

$$t_I^B = \phi^T\tilde{t}_I \tag{59}$$

and by use of Equation 57

$$t_I^A = -\phi^T\tilde{t}_I \tag{60}$$

Introducing this in Equation 52 and rewriting Equation 53 gives

$$S_{II}^{*}\tilde{u}_I + M_{II}\tilde{t}_I = \tilde{f}^{*} \tag{61}$$

where the mass-type matrix is defined as

$$M_{II} = \int_{\Gamma_I} \phi\phi^T dS \tag{62}$$

Equations 56 and 61 can now be added together to form a square, almost dense, and non-symmetric system of equations which is solvable:

$$\begin{bmatrix} B & -G_I & H_I \\ \hline 0 & M_{II} & S^*_{II} \end{bmatrix} \begin{bmatrix} \tilde{v}_B \\ \tilde{t}_I \\ \tilde{u}_I \end{bmatrix} = \begin{bmatrix} \tilde{b} \\ \hline \tilde{f}^* \end{bmatrix} \tag{63}$$

$(M+2I) \times (M+2I)$

This method of coupling is easy to implement and no explicit matrix inversion is needed. Other methods of coupling FEM and BEM are presented in Kelly *et al* (1980).

NUMERICAL STUDIES

BEMSTAT is divided into four separate modules, and the user can choose which module should be executed, by including the proper subgroup(s) of data, as illustrated in Figure 5. A short description of the modules is given below:

Module 1: Reads and generates the geometry and topology.

Module 2: Reads and generates the boundary conditions, solves the system of equations, and prints the boundary results. This module can be executed again with new boundary conditions, without reexecuting module 1.

Module 3: Calculates requested internal quantities such as displacements, strains, stresses, and principal stresses. This module can be executed interactively, which enables the user to »zoom in» on regions of interest.

Module 4: Plots geometry with node and element numbers, boundary displacements, boundary tractions in x- and y-directions, and principal stresses.

Shear wall

The plane stress problem shown in Figure 5 is studied.

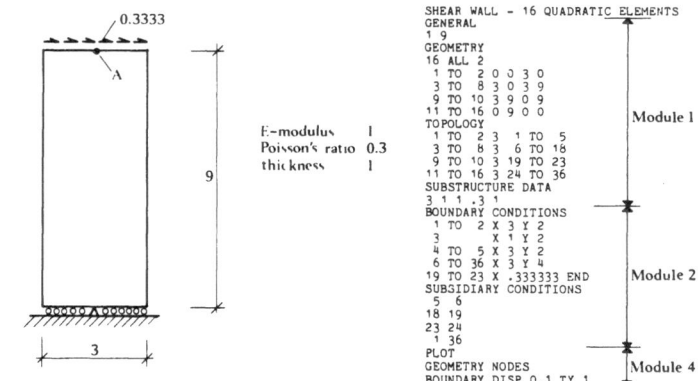

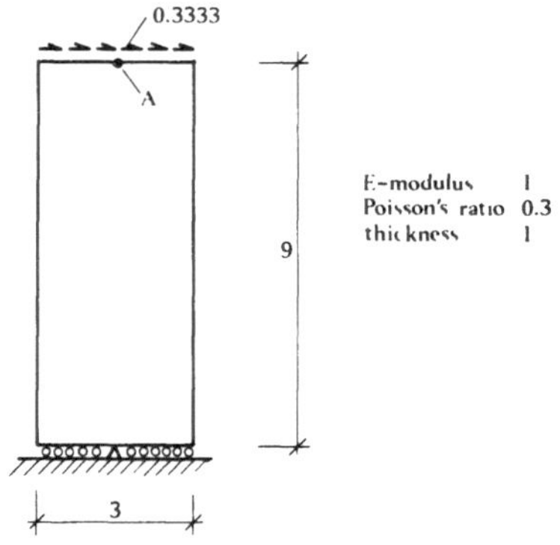


```
SHEAR WALL - 16 QUADRATIC ELEMENTS
GENERAL
1 9
GEOMETRY
16 ALL 2
 1 TO  2 0 0 3 0
 3 TO  8 3 0 3 9
 9 TO 10 3 9 0 9
11 TO 16 0 9 0 0
TOPOLOGY
 1 TO  2 3  1 TO  5
 3 TO  8 3  6 TO 18
 9 TO 10 3 19 TO 23
11 TO 16 3 24 TO 36
SUBSTRUCTURE DATA
3 1 1 .3 1
BOUNDARY CONDITIONS
 1 TO  2 X 3 Y 2
 3       X 1 Y 2
 4 TO  5 X 3 Y 2
 6 TO 36 X 3 Y 4
19 TO 23 X .333333 END
SUBSIDIARY CONDITIONS
 5  6
18 19
23 24
 1 36
PLOT
GEOMETRY NODES
BOUNDARY DISP 0.1 TY 1
```

Module 1

Module 2

Module 4

Figure 5. Plane stress problem - shear wall

Typical for the problem is the large rigid body motions, *i.e.*, it is in principle a deflection problem. Different types of discretizations have been used, and the results are compared by studying the horizontal deflection at point A (Figure 5). Subsidiary conditions are used to couple the nodes in the corners. Table 3 shows the results; listed is also the CPU-time on an IBM 3033N required for the establishment and solution of the system of equations.

Table 3. FEM- and BEM-results for the shear wall problem

	horizontal deflection at A	CPU-time(sec) establ.	solution
48 finite elements	118.7	1.59	
96 linear elements	117.4	6.55	3.28
64 linear elements	115.4	2.94	1.06
16 linear elements	83.9	0.22	0.03
48 quadratic elements	118.9	3.76	3.32
32 quadratic elements	118.7	1.72	1.06
8 quadratic elements	116.2	0.14	0.03

The finite element calculation was made using 48 (4x12) 8-noded isoparametric elements, and the time given is for both the establishment and solution. Two conclusions can be drawn from this study, namely: 1) Quadratic elements give better results, and 2) They also require much less time for the establishment. Constant elements cannot describe this problem, since they are subparametric. Figure 6 shows plots of node numbers, displacements, and tractions for 16 quadratic elements.

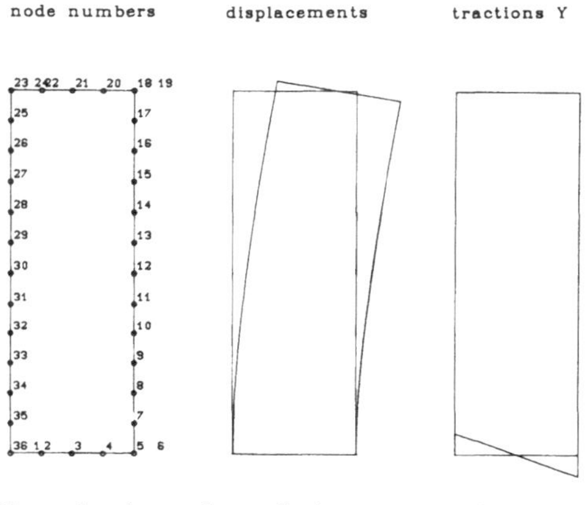

Figure 6. Plots of node numbers, displacements, and tractions

Rectangular plate with a circular hole

The stress concentration near a circular hole in a rectangular plate is studied. Due to symmetry, the problem can be reduced to that shown in Figure 7.

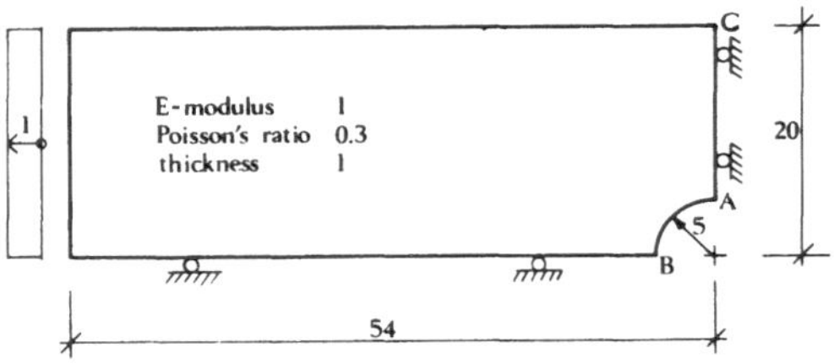

Figure 7. Rectangular plate with a hole

Table 4 shows the calculated results, horizontal stress at A (Figure 7), and vertical stress at B.

Table 4. Stresses in a rectangular plate with a hole

	horizontal stress at A	vertical stress at B	number of equations
69 linear elements	3.25	1.18	148
48 linear elements	3.18	1.17	106
25 linear elements	3.08	1.11	60
40 quadratic elements	3.24	1.18	170
24 quadratic elements	3.21	1.16	106
Wennerström and Petersson (1979), FEM	3.31		214
Roark (1965), empirical	3.235		

Both linear and quadratic elements gave good results in this typical stress concentration problem. Also for constant elements the result converged, but many more elements were needed. However, the quadratic elements are still more attractive, since they require less time for the establishment. During the study of this example, it became obvious that the representation of the geometry of the hole was important (as reported by Wennerström and Petersson). The difference in the results between a polygonial and curved boundary was about 5%. Horizontal tractions are plotted in Figure 8.

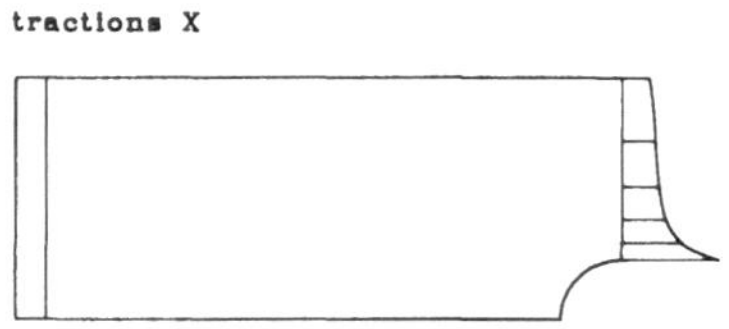

Figure 8. Horizontal tractions

Strip load on a finite layer

Figure 9 shows the problem and the points A and B, where displacement and traction, respectively, are studied. The problem has been calculated with Poisson's ratio equal to 0 and 0.5. Quadratic elements have been used.

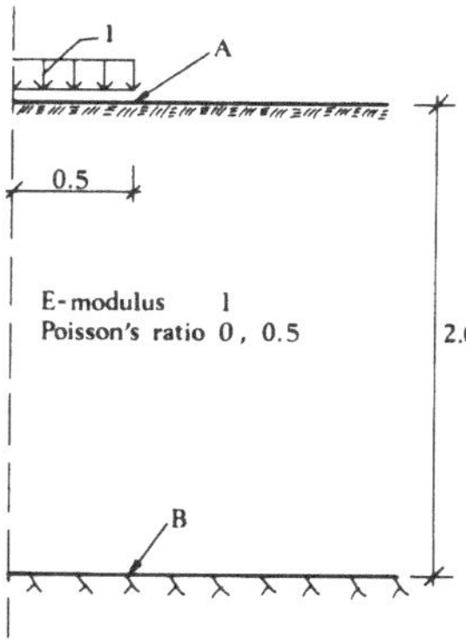

Figure 9. Strip load on a finite layer

Figure 10 shows the discretization with 18 elements and Table 5 calculated and analytical results.

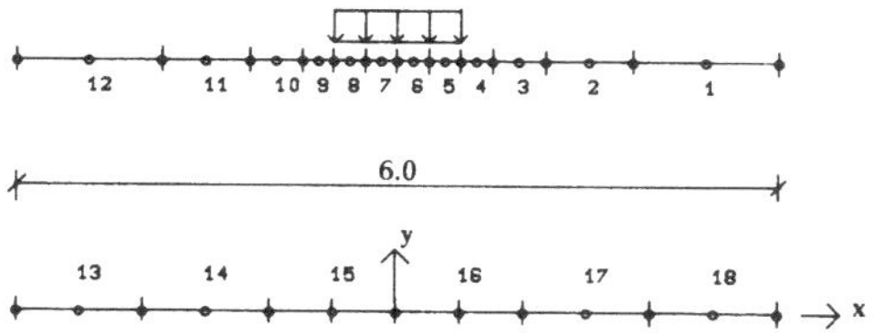

Figure 10. BEM-discretization with 18 quadratic elements

Table 5. Analytical and calculated results for a strip load on a finite layer

	$\nu = 0.5$		$\nu = 0$	
	analytical	calculated	analytical	calculated
vertical stress at B	0.35	0.348	0.35	0.350
vertical displ. at A	-0.84	-0.866	-0.38	-0.378
horizontal displ. at A	-0.38	-0.370	0.06	0.070

The analytical results are given with two decimals only, due to the accuracy of the diagrams in Poulos and Davis (1974). In Figure 11 the vertical stress distribution at the bottom surface and the displacements of the upper surface are shown. Two conclusions can be drawn from this example, namely: 1) Incompressible materials can be treated, and 2) Semi-infinite domains can be approximated with open element models. The CPU-time required for the establishment and solution of the system of equations was 0.62 and 0.23 seconds, respectively.

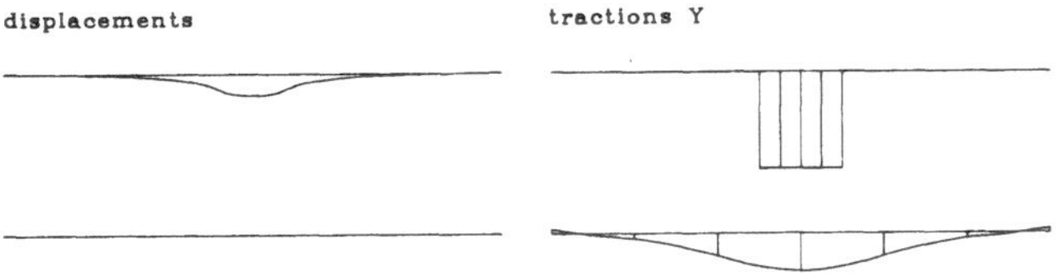

Figure 11. Vertical stress and displacements

Strip foundation on a semi-infinite mass

The contact pressure distribution beneath a smooth strip subjected to a uniform pressure is studied (Figure 12.) This example was calculated using 20 four-node finite elements and 50 linear boundary elements. The discretization was refined at the strip ends.

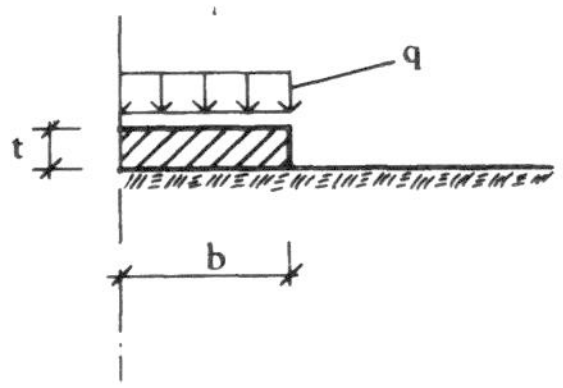

Figure 12. Strip foundation on a semi-infinite mass

Two calculations, with different relative stiffnesses, have been made. The relative stiffness is defined below. Figure 13 shows plots of the displacement profile and the contact pressure distribution for the two calculations. Borowicka (1939) has presented solutions for the distribution of contact pressure p beneath a strip subjected to a uniform pressure q, and some of his results are indicated in the plots. No difference between the plots below and plots of Borowicka's result can be observed. All calculations were performed assuming plane strain conditions.

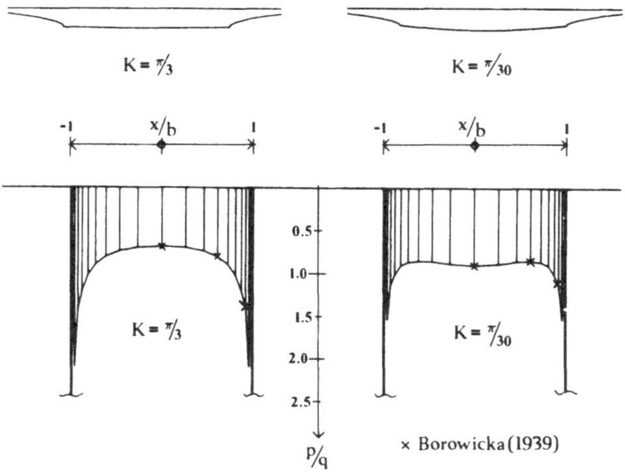

Figure 13. Displacements and contact pressure beneath uniformly loaded smooth strip

The relative stiffness, K, is defined as

$$K = \frac{1}{6} \frac{(1-\nu_s^2)}{(1-\nu_p^2)} \frac{E_p}{E_s} \left(\frac{t}{b}\right)^3 \quad (64)$$

where E_p is the elastic modulus of the strip, ν_p Poisson's ratio of the strip, E_s the elastic modulus of the mass, ν_s Poisson's ratio of of the mass, t the strip thickness, and b the half width of the strip. This example demonstrates one of the features of coupling BEM and FEM, namely, the possibility to treat soil-structure interaction problems.

CONCLUSIONS

BEMSTAT, a boundary element program for two-dimensional elasticity problems, has been presented. The program is easy to use, it requires a small amount of input data, the output facilities are extensive, and it is built up of well-separated modules (allowing restart). Due to the new philosophy applied when integrating and establishing element matrices, the computer time required has been reduced considerably. Infinite and semi-infinite problems can be treated by open element models, since the strongly singular integrals are calculated analytically. Non-homogeneous materials can be handled by dividing the body into a number of homogeneous substructures. Coupling with FEM can be made, which makes the program well suited to solve soil-structure interaction problems. Practical rules for discretization and integration are given in the paper.

The examples studied show that typical deflection problems are best solved with quadratic elements, and few elements are needed. Constant elements cannot describe these types of problems. In stress concentration problems, quadratic elements are still superior to the linear ones, but not as much as for the deflection problems. Constant elements can describe stress concentration problems if a large number of elements are used.

ACKNOWLEDGEMENT

The support received from IBM Sweden, in the form of computer time, is greatly appreciated.

REFERENCES

Alarcon, E., Martin, A., and Paris, F. (1979), Boundary elements in potential and elasticity theory, Computers and Structures, 10, pp. 351-362.

Banerjee, R.K. and Butterfield, R. (1979), Developments in Boundary Element Methods, Appl. Science Publishers Ltd, London.

Borowicka, H. (1939), Druckverteilung unter elastischen Platten, Ingenieur Archiv, vol. X, No. 2, pp. 113-125.

Brebbia, C.A. and Walker, S. (1980), Boundary element techniques in engineering, Newnes-Butterworths, London.

Hartmann, F. (1980), Computing the C-matrix in non-smooth boundary points, Proc. of the Second Int. Sem. on Recent Advances in Boundary Element Methods, held at Univ. of Sothampton, March 1980.

Jaswon, M.A. and Symm, G.T. (1977), Integral Equation Methods in Potential Theory and Elastostatics, Academic Press, New York.

Jeng, G and Wexler A (1977), Isoparametric, finite element variational solution of integral equations for three-dimensional fields, Int. J. for Numerical Methods in Eng., 11, pp. 1455-1471.

Kelly, D.W., Mustoe, G.W, and Zienkiewicz, O.C. (1979), Coupling Boundary Element Methods with Other Numerical Methods, Developments in Boundary Element Methods - 1 (editors: Banerjee and Butterfield), Appl. Science Publishers Ltd, London.

Nedelec, J.C. (1977), Cours de l'Ecole d'Ete' d'Analyse Numerique, C.E.A., I.R.I.A., E.P.F..

Poulos and Davis (1974), Elastic solutions for soil and rock mechanics, John Wiley & Sons, New York.

Roark, R.S. (1965), Formulas for stress and strain, McGraw Hill, New York.

Watson, J.O. (1980), Advanced implementation of the boundary element method for two- and three-dimensional elastostatics, Developments in Boundary Element Methods - 1 (editors: Banerjee and Butterfield), Appl. Science Publishers Ltd, London.

Wendland, W.L., Stephan, E., and Hsiao, G.C. (1979), On the integral equation method for the plane mixed boundary value problem of the Laplacian, Math. Meth. in Appl. Sci., 1, pp. 265-321.

Wennerström and Petersson (1979), GENFEM-3, Verification Manual, Publ. 79:5, Dept of Structural Mechanics, Chalmers University of Technology, Göteborg, Sweden.

APPENDIX A - DEFINITIONS OF MATRICES USED IN BEM FOR TWO-DIMENSIONAL ELASTOSTATICS

A.1 *Fundamental matrices*

The fundamental matrices for two-dimensional plane strain elasticity are defined as

$$U(x,\xi) = \frac{1+\nu}{4\pi E(1-\nu)}\begin{bmatrix} (3-4\nu)\ln\frac{1}{r} + r_x^2 & r_x r_y \\ \text{SYM} & (3-4\nu)\ln\frac{1}{r} + r_y^2 \end{bmatrix} \tag{65}$$

$$T(x,\xi) = \frac{-1}{4\pi(1-\nu)}\frac{1}{r}\begin{bmatrix} \frac{\partial r}{\partial n}((1-2\nu) + 2r_x^2) & 2\frac{\partial r}{\partial n}r_x r_y + (1-2\nu)(r_x n_y - r_y n_x) \\ 2\frac{\partial r}{\partial n}r_x r_y + (1-2\nu)(r_y n_x - r_x n_y) & \frac{\partial r}{\partial n}((1-2\nu) + 2r_y^2) \end{bmatrix} \tag{66}$$

where the pole is placed in ξ, and

$$r = |x-\xi| \qquad \nabla r = [\partial r/\partial x \quad \partial r/\partial y]^T = [r_x \quad r_y]^T \tag{67}$$

The fundamental matrix for the tractions can be derived in the following manner:

$$t(x) = \tilde{n}^T(x)\sigma(x) = \tilde{n}^T S^d \varepsilon = \tilde{n}^T S^d \tilde{\nabla}_x u(x) = \tilde{n}^T S^d \tilde{\nabla}_x U(x,\xi)e(\xi) = T(x,\xi)e(\xi) \tag{68}$$

$$T(x,\xi) = \tilde{n}_x^T S^d \tilde{\nabla}_x U(x,\xi) \tag{69}$$

where the unit normal vector is defined by

$$\tilde{n} = \begin{bmatrix} n_x & 0 & n_y \\ 0 & n_y & n_x \end{bmatrix}^T \tag{70}$$

A.2 *Gradients of fundamental matrices*

When the strains and stresses are to be calculated the gradients of U and T must be derived. It should be noted that the gradient operator worked on the free argument, not the pole coordinate, when the T-matrix was derived. Here, the gradient operators operate on the pole-argument. By using the following identities:

$$\tilde{\nabla}_x = -\tilde{\nabla}_\xi, \quad \nabla_x = -\nabla_\xi \tag{71}$$

the old definitions of gradients of r, Equation 67, can be kept, so (ξ is the pole)

$$\tilde{\nabla}_\xi U = \frac{-A}{r}\begin{bmatrix} r_x(2(1-r_x^2)-B) & r_y(1-2r_x^2) \\ r_x(1-2r_y^2) & r_y(2(1-r_y^2)-B) \\ r_y(1-B-4r_x^2) & r_x(1-B-4r_y^2) \end{bmatrix} \tag{72}$$

where

$$A = \frac{1+\nu}{4\pi E(1-\nu)} \qquad B = 3-4\nu \tag{73}$$

and

$$\nabla_\xi T^T = \frac{-C}{r^2}\begin{bmatrix} (2r_x^2+D)n_x + (2-D-4r_x^2)2\frac{\partial r}{\partial n}r_x & (2r_x^2-1)Dn_y + 2(1-D)r_y r_x n_x + (1-4r_x^2)2\frac{\partial r}{\partial n}r_y \\ (2r_y^2-1)Dn_x + 2(1-D)r_x r_y n_y + (1-4r_y^2)2\frac{\partial r}{\partial n}r_x & (2r_y^2+D)n_y + (2-D-4r_y^2)2\frac{\partial r}{\partial n}r_y \\ 2(1+D)r_x r_y n_x + (2D+(1-D)2r_x^2)n_y + (1-D-8r_x^2)2\frac{\partial r}{\partial n}r_y & (2D+(1-D)2r_y^2)n_x + 2(D+1)r_y r_x n_y + (1-D-8r_y^2)2\frac{\partial r}{\partial n}r_x \end{bmatrix} \quad (74)$$

where

$$C = \frac{-1}{4\pi(1-\nu)} \qquad D = 1-2\nu \quad (75)$$

A.3 *Exact calculation of the free term matrix for non-smooth boundaries*

The exact calculation of the free term matrix for two-dimensional elastostatics was shown by Hartman (1980). With the definitions in Figure 14, the matrix is

$$D(x) = \frac{1}{4\pi(1-\nu)}\begin{bmatrix} (4\pi-2(\alpha_1-\alpha_2))(1-\nu)+\frac{1}{2}(\sin2\alpha_1-\sin2\alpha_2) & \sin^2\alpha_1-\sin^2\alpha_2 \\ \text{SYM} & (4\pi-2(\alpha_1-\alpha_2)(1-\nu)-\frac{1}{2}(\sin\alpha_1-\sin\alpha_2) \end{bmatrix} \quad (76)$$

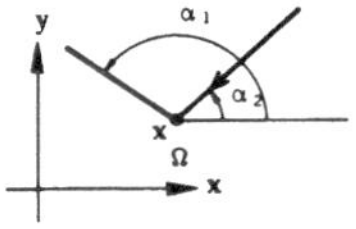

Figure 14. Definition of positive rotation and angles involved

A.4 *Exact integration of the singular integrals for straight line elements*

When the boundary is approximated with straight line elements, it is possible to integrate the singular terms exactly. Here, the result of the interior integration of an element with linear shape functions is shown. The collocation point is in node I, where I = 1 or 2 for linear shape functions:

$$G^e = A\begin{bmatrix} (\frac{DX}{L})^2 + B(2{,}5-I+\ln L) & \frac{DXDY}{L^2} & (\frac{DX}{L})^2 + B(I-0{,}5+\ln L) & \frac{DXDY}{L^2} \\ \frac{DXDY}{L^2} & (\frac{DY}{L})^2 + B(2{,}5-I+\ln L) & \frac{DXDY}{L^2} & (\frac{DY}{L})^2 + B(I-0{,}5+\ln L) \end{bmatrix} \quad (77)$$

$$\bar{H}^e = C\begin{bmatrix} 0 & 1-I+(2-I)\ln L & 0 & 2-I+(1-I)\ln L \\ -(1-I+(2-I)\ln L) & 0 & -(2-I+(1-I)\ln L) & 0 \end{bmatrix} \tag{78}$$

where

$$A = \frac{1+\nu}{4\pi E(1-\nu)}\,\frac{L}{2} \qquad B = 3-4\nu \qquad C = \frac{-1}{4\pi(1-\nu)}(1-2\nu) \tag{79}$$

$$DX = \tilde{x}_2 - \tilde{x}_1 \qquad DY = \tilde{y}_2 - \tilde{y}_1 \tag{80}$$

and L is the element length. Notice that the contribution from the free term matrix is not included in $\bar{H}^e$.

Section VI

Coupling of Boundary and Finite Element Methods

INTERFACING FINITE ELEMENT AND BOUNDARY ELEMENT DISCRETIZATIONS

Carlos A. Felippa
Applied Mechanics Laboratory
Lockheed Palo Alto Research Laboratory
Palo Alto, California 94304, USA

INTRODUCTION

This paper reviews critically various implementation methods for coupling a mechanical finite element model to an infinite external acoustic domain discretized through boundary-element techniques. The associated physical problem is that of a three-dimensional structure submerged in an acoustic fluid, and impinged by a pressure shock wave.

Three coupling methods for advancing the dynamic calculations are described: field elimination, simultaneous integration, and partitioned integration. Variants of these techniques have been tried on the case problem over the past seven years.

The three methods are assessed from gained experience, and their advantages and disadvantages noted. Some generalizations to more general FE/BE coupling scenarios are then offered.

THE PROBLEM

The specific problem used as case study in this paper is illustrated in Figure 1. A linear or nonlinear three-dimensional structure is submerged in an infinite acoustic fluid. A pressure shock wave propagates through the fluid and impinges on the structure. The structure and fluid are discretized through finite-element (FE) and boundary-element (BE) methods, respectively.

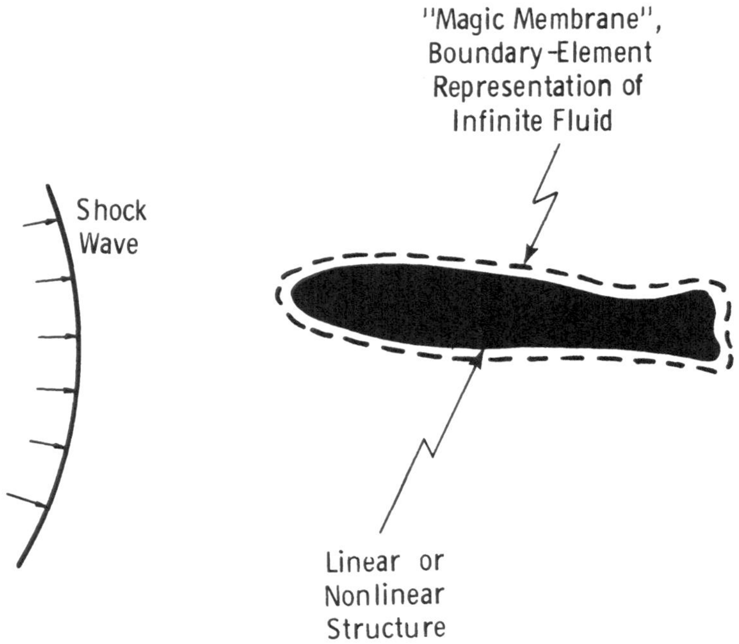

Figure 1 Structure submerged in an acoustic fluid

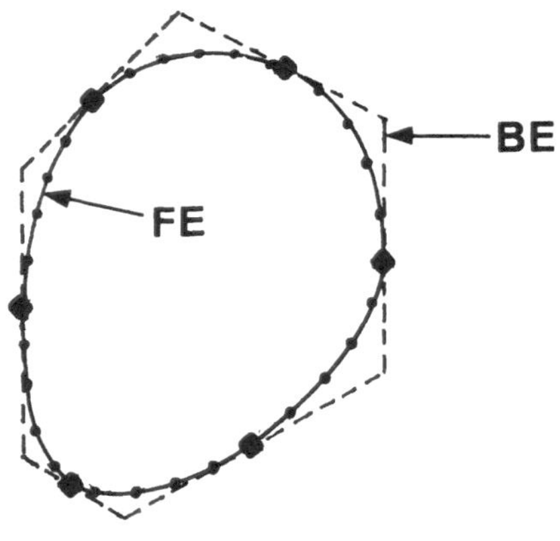

Figure 2 Layout of BE and FE meshes on wet surface
(◆ = BE control points; • = FE node points)

Before proceeding to the governing equations, two practical considerations of relevance to this problem should be mentioned.

First, the structural response (and most especially the structure's survivability) is of primary concern, whereas what happens in the fluid is of little interest.

Second, the FE and BE meshes on the "wet surface" are not necessarily in one-to-one correspondence, as illustrated in the two-dimensional sketch of Figure 2. Rather, a "fluid BE" typically overlaps several structural elements. This ties up with the first consideration in the sense that determination of structural deformations and stresses demands a finer subdivision.

Structural response equations

The governing matrix equation of motion for the dynamic response of a discrete structure is

$$\underset{\sim}{M}_s \, \ddot{\underline{x}} + \underset{\sim}{C}_s \, \dot{\underline{x}} + \underset{\sim}{K}_s \, x = f + \underline{N} \tag{1}$$

where $\underline{x} = \underline{x}(t)$ is the structural displacement vector, $\underset{\sim}{M}_s$, $\underset{\sim}{C}_s$ and $\underset{\sim}{K}_s$ are the structural mass, damping and stiffness matrices, respectively, $\underline{f}$ is the external force vector, $\underline{N} = \underline{N}(\underline{x})$ is a nonlinear psudo-force vector, and a dot denotes temporal differentiation.

For excitation of a submerged structure by an acoustic wave, $\underline{f}$ is given by

$$\underline{f} = -\underset{\sim}{G} \, \underset{\sim}{A}_f \, (\underline{p}_I + \underline{p}_S) \tag{2}$$

where $\underline{p}_I$ and $\underline{p}_S$ are nodal pressure vectors for the wet-surface fluid mesh pertaining to the (known) incident wave and the (unknown) scattered wave, respectively, $\underset{\sim}{A}_f$ is the diagonal area matrix associated with elements in the fluid mesh, and $\underset{\sim}{G}$ is the transformation matrix that relates the structural and fluid nodal forces. Introduction of this matrix takes care of the FE/BE "mesh-mismash" noted previously (cf. Figure 2).

Fluid equations

The response of the fluid is modelled by the Doubly Asymptotic Approximation (DAA) of Geers (1971,1978)

$$\underset{\sim}{M}_f \, \dot{\underline{p}}_S + \rho c \, \underset{\sim}{A}_f \, \underline{p}_S = \rho c \, \underset{\sim}{M}_f \, \underline{u}_S \tag{3}$$

where $\underline{u}_S$ is the vector of scattered-wave fluid-particle velocity normal to the structure's wet surface, ρ and c are the density and sound velocity of the fluid, respectively, and $\underset{\sim}{M}_f$ is the symmetric fluid mass matrix for the wet-surface fluid mesh. This matrix is produced by a boundary-element treatment of Laplace's equation for the irrotational flow generated in an infinite, inviscid, incompressible fluid by motions of the structure's wet surface; it is fully populated with non-zero matrix elements. When transformed into structural coordinates, the fluid mass matrix yields the added mass matrix, which, when combined with the structural mass matrix, yields the virtual mass matrix for motions of a structure submerged in an incompressible fluid. Details of the calculation procedure may be found in a paper by DeRuntz and Geers (1978).

The approximate pressure-velocity relation (3) is called "doubly asymptotic" because it approaches exactness in both the high-frequency (early-time) and low-frequency (late-time) limits. For high-frequency motions, $|\dot{\underline{p}}_S| \gg |\underline{p}_S|$, so that (3) approaches the relation $\underline{p}_S = \rho c \, \underline{u}_S$, which is the correct limit for short acoustic wavelengths. For low-frequency motions, $|\dot{\underline{p}}_S| \ll |\underline{p}_S|$, so that (3) approaches the incompressible-flow relation $\underset{\sim}{A}_f \, \underline{p}_S = \underset{\sim}{M}_f \, \dot{\underline{u}}_S$, which is the correct limit for long acoustic wavelengths.

For excitation by an incident acoustic wave, $\underline{u}$ is related to structural response by the kinematic compatibility relation

$$\underset{\sim}{G}' \, \dot{\underline{x}} = \underline{u}_I + \underline{u}_S \tag{4}$$

where the prime superscript denotes matrix transposition. Equation (4) expresses the constraint that normal fluid-particle velocity match normal structural velocity on the wet surface of the structure. The fact that the transformation matrix relating these velocities is $\underset{\sim}{G}'$ follows from the invariance of virtual work with respect to either of the wet-surface coordinate systems.

Generally, $\underset{\sim}{G}$ is a rectangular matrix whose height greatly exceeds its width, inasmuch as the number of structural DOF usually exceeds considerably the number of fluid DOF, as noted previously. Typical numbers: 5000 structural DOFs and 160 fluid DOFs.

Interaction Equations

The introduction of Equation (2) into (1) and (4) into (3) yields the interaction equations

$$\begin{aligned} \mathbf{M}_s \ddot{\underline{x}} + \mathbf{C}_s \dot{\underline{x}} + \mathbf{K}_s \underline{x} + \underline{N} &= -\mathbf{G}\, \mathbf{A}_f (\underline{p}_I + \underline{p}_S) \\ \mathbf{M}_f \dot{\underline{p}}_S + \rho c\, \mathbf{A}_f\, \underline{p}_S &= \rho c\, \mathbf{M}_f\, (\mathbf{G}'\, \ddot{\underline{x}} - \dot{\underline{u}}_I) \end{aligned} \tag{5}$$

The computational structure of these coupled systems is very different. As can be expected, the FE (structural) system is usually large but sparse. The BE (fluid) system is typically small but dense. It is therefore of interest to design solution methods that exploit these attributes to maximum advantage.

SOLUTION APPROACHES

Three approaches to solving the coupled FE/BE system (4) are reviewed here. They are presented in chronological order, i.e., in roughly the same sequence as they were tried and evaluated over the past seven years.

Field Elimination

The first approach tried is now known as "field elimination", where the term "field" refers to one of the physical components of the coupled system (structural and fluid in our case). As noted previously, the structural response is of primary interest. It is therefore natural to think of eliminating the scattered-pressure vector $\underline{p}_S$ from the coupled equations of motion (5). If $\mathbf{G}$ is the identity matrix, this results in the following third-order ODE system for the structural displacements $\underline{x}$:

$$\begin{aligned} \mathbf{M}_s \dddot{\underline{x}} + [\mathbf{C}_s + \mathbf{A}_f \mathbf{M}_f^{-1} \mathbf{M}_s]\, \ddot{\underline{x}} + \mathbf{K}_s \dot{\underline{x}} + \\ + \mathbf{A}_f \mathbf{M}_f^{-1} \mathbf{K}_s \underline{x} = \underline{r}\, (\mathbf{A}_f, \mathbf{M}_f, \underline{p}_I, \dot{\underline{u}}_I, \underline{N}) \end{aligned} \tag{6}$$

where the RHS vector $\underline{r}$ accomodates incident pressure and incident fluid-particle velocity boundary conditions, and nonlinear effects. If $\mathbf{G}$ is not the identity, as invariable happens in realistic three-dimensional problems, the coefficient matrices in Equation (6) become considerably more complex, because the generalized inverse of $\mathbf{G}$ enters the elimination process.

The structural response $\underline{x}(t)$ can now be determined by numerically integrating Equation (6).

This was in fact the first approach tried to tackle the time-integration of the coupled system; cf. Felippa et al. (1974). Although moderately sucessful for the first problem series (submerged shells of revolution, linear structural behavior), from current perspective it can be properly characterized as a poor strategy that eventually leads to a "computational horror show" for more general problems. Why?

1. The order of the reduced differential system is raised (in this example, from two to three). The appearance of higher derivatives can be the source of many difficulties, the worst of which is noted next.

2. Proper treatment of initial conditions is complicated by the increased ODE order. In our case study, it turned out that (6) had to be integrated once (yielding an integrodifferential system) so as to regularize the treatment of wavefront-induced singularities. Time integrals of forcing terms had then to be carrying along in the calculations -- a grievous burden.

3. Sparseness and symmetry attributes of the original matrices is adversely affected by the elimination process, as can be observed in (6) for the identity $\underset{\sim}{G}$. For a general transformation matrix, all left-hand matrices become unsymmetric and dense.

4. The development of specialized software is required. For example, available software for dealing with the uncoupled problems (structural dynamics and acoustic shocks) separately is not likely to be of much use in solving the reduced system (6).

Simultanous integration

In this approach Equations (5) are viewed as a single second-order system

$$\begin{bmatrix} \underset{\sim}{M}_s & \underset{\sim}{0} \\ \underset{\sim}{0} & \underset{\sim}{0} \end{bmatrix} \begin{Bmatrix} \ddot{\underline{x}} \\ \ddot{\underline{p}} \end{Bmatrix} + \begin{bmatrix} \underset{\sim}{C}_s & \underset{\sim}{G}\,\underset{\sim}{A}_f \\ \rho c\,\underset{\sim}{M}_f \underset{\sim}{G}' & \underset{\sim}{M} \end{bmatrix} \begin{Bmatrix} \dot{\underline{x}} \\ \dot{\underline{p}} \end{Bmatrix} + \begin{bmatrix} \underset{\sim}{K}_s & \underset{\sim}{0} \\ \underset{\sim}{0} & \rho c\,\underset{\sim}{A}_f \end{bmatrix} \begin{Bmatrix} \underline{x} \\ \underline{p} \end{Bmatrix} = \begin{Bmatrix} -\underset{\sim}{G}\,\underset{\sim}{A}_f \underline{p}_I \\ -\rho c\,\underset{\sim}{M}_f \underline{u}_I \end{Bmatrix} \quad \ldots (7)$$

This approach removes many of the objections raised against the field elimination technique. Inasmuch as the ODE order is not raised, initial-condition difficulties do not arise and better use can be made of existing software for dealing with second-order ODE systems.

The computational burden for realistic three-dimensional problems can be prohibitive, however. Note that the coefficient matrix of higher-derivative terms is singular, which means that implicit integration is required to construct a marching scheme in time.

But the assembly and formation of the implicit coefficient matrix was found to pose enormous computational demands because of presence of BE/FE coupling terms that can extend across thousands of equations.

For example, it was estimated that the factorization of that matrix for a 5000-DOF problem would require 3 hours on a Cyber 175 computer. Carrying out a nonlinear transient- response analysis of a realistic model was then adjudged infeasible.

Partitioned integration

In the partitioned integration approach, the solution state is advanced over each of the two subsystems: FE structural model and BE fluid model, in a staged fashion. Interaction terms are treated as "forcing" actions that have to be judiciously extrapolated.

What is now called the staggered solution procedure is a specific partitioned-integration method originally formulated for the system (5) by Park, Felippa and DeRuntz (1977). A version of this procedure was implemented in a production- level computer program described by DeRuntz et al. (1978).

Success of this method led to further applications and eventually the development of a general theory of partitioned time integration; see Park (1980), Park and Felippa (1980). A state-of-the-art review of formulation aspects is provided in a recent survey by Felippa and Park (1980).

The staggered solution procedure was found to offer two important advantages: enhanced software modularity and computational efficiency.

The first advantage accrues from the fact that relatively few modifications to programs available for processing the uncoupled systems are necessary. Given current costs in software development, augmentation and maintenance, this is indeed an important virtue of this approach. For our specific problem, a BE fluid analysis module was written, and data-coupled to existing large-scale structural analysis codes such as NASTRAN, SPAR and STAGS.

An obvious advantage of "plug-in" modularity is the freedom afforded the analyst as regards the selection of a structural analysis code that best fits the problem at hand; for example, the nonlinear analyzer STAGS when plasticity or finite displacements had to be considered. Moreover, if there is a choice among structural analyzers that can do just about the same thing, the user can select the one he or she is most comfortable with.

As regards computational efficiency, the cost per time step is roughly the same as adding up those incurred in processing the FE and BE models as isolated entities. This is because the overhead introduced by the flow of information (which consists primarily of computational vectors) among the two analysis modules becomes comparatively insignificant in large-scale problems. It follows that the staggered integration procedure appears as economically attractive should time stepsize considerations be excluded from consideration.

Unfortunately, the latter assumption was not easy to realize in practice. The high computational efficiency per time step is counteracted by the fact that satisfactory numerical stability properties are hard to achieve; in fact, the practical feasibility of this technique hinges almost entirely on the stability analysis. The reader is referred to the cited sources for additional details. Suffice here to say that a specific integration algorithm of unconditional stability was found for (7) when the BE fluid was suitably modified through an "augmentation" technique.

FINAL REMARKS

It is time now to summarize some lessons gained from this case study of FE/BE interfacing, and to venture various generalizations. This will be done following a question-and-answer style.

Why was a dynamic problem chosen as case study?

We have used boundary-element techniques primarily for dynamic problems; in particular, wave propagation, impact and reentry studies. This has provided a body of experience "complementary" to that gathered by investigators dealing with static problems.

Moreover, the computer implementation of dynamic and nonlinear-static analysis share many common facets. In fact, using the dynamic relaxation concept, the latter can be always viewed as pseudo-dynamic systems whose steady-state solution is sought, and much of the discussion on time-marching algorithms apply.

How about BE for interior problems?

Our experience has been that BE techniques are primarily useful for discretizing unbounded homogeneous domains governed by linear equations. For linear problems in bounded homogeneous domains, they are not competitive with properly written finite element codes, and the latter are far easier to extend to nonhomogeneous regions and nonlinear problems. Advertised reductions in mesh preparation efforts are largely illusory in these days of powerful pre-processors complemented by inexpensive interactive graphics.

But boundary-element methods come on their own for treating unbounded linear media, notably when their "interior" response is of little interest. The model-description effort is greatly reduced, and the construction of "quite boundaries" is simplified.

How "tight" should the FE/BE interface be?

As "loose" as possible. This goes along with the philosophy of maximizing software modularity. One facet of this philosophy says that the analyst ought to have the freedom of selecting FE and BE meshes independently according to the physics of the problem and response-resolution requirements.

In our problem, for instance, it would be plain silly to force the BE fluid mesh to be constrained by the presence of internal structural stiffeners.

The analysis of fluid-solid impact and similar problems requiring moving and sliding interfaces provide further ammunition on the argument for a high degree of independence in FE/BE mesh definitions.

What's a most compelling reason for modularity?

Keeping complexity down and analyst's sanity up. Nontrivial FE and BE programs tend to be fairly complex beasts even when taken separately. And the complexity of a monolithic marriage can easily escape anybody's control.

There are many things that can go wrong and (true to Murphy's law) will: modelling, numerics, machine problems, data management, result interpretation. Nonlinear dynamics problems, for example, are particularly vulnerable to many trouble sources. Experience has shown that keeping modular interfaces not only reduces the chances for troubles, but also makes their resolution prompter.

Do modularity requirements impact solution strategies?

Yes. And that's a major reason for the advent and high degree of success of partitioned analysis procedures.

Is success or failure of partitioned integration procedures contingent upon implementation details?

Very much so. The survey paper by Felippa and Park (1980) highlights this theme.

ACKNOWLEDGEMENT

Preparation of this paper was supported by the Independent Research Program of Lockheed Missiles and Space Co., Inc.

REFERENCES

DeRuntz, J. A. and Geers, T. L. (1978) Added Mass Computation by the Boundary Integral Method, Int. Journal Numerical Methods in Engineering, 12, 531-550.

DeRuntz, J. A., Geers, T. L., and Felippa, C. A. (1978) The Underwater Shock Analysis (USA) Code: A Reference Manual, Report DNA-4524F to Defense Nuclear Agency, Lockheed Palo Alto Research Lab., Palo Alto, California.

Felippa, C. A., Geers, T. L., and DeRuntz, J. A. (1974) Response of a Ring-Stiffened Cylindrical Shell to a Transient Acoustic Wave, Report LMSC-D403671 to Office of Naval Research, Structural Mechanics Laboratory, Lockheed Palo Alto Research Lab., Palo Alto, California

Felippa, C. A. and Park, K. C. (1980), Staggered Solution Transient Analysis Procedures for Coupled Mechanical Systems: Formulation, Computer Methods in Applied Mechanics and Engineering, 24, 61-111.

Geers, T. L. (1971) Residual Potential and Approximate Methods for Three-Dimensional Fluid-Structure Interaction, Journal of the Acoustical Society of America, 45, 1505-1510.

Geers, T. L. (1978) Doubly Asymptotic Approximations for Transient Motions of Submerged Structures, Journal of the Acoustical Society of America, 64, 1500-1508.

Park, K. C. (1980) Partitioned Transient Analysis Procedures for Coupled-Field Problems: Stability Analysis, ASME Journal of Applied Mechanics, 47, 370-376.

Park, K. C. and Felippa, C. A. (1980) Partitioned Transient Analysis Procedures for Coupled-Field Problems: Accuracy Analysis, ASME Journal of Applied Mechanics, 47, 919-926.

Park, K. C., Felippa, C. A. and DeRuntz, J. A. (1977) Stabilization of Staggered Solution Procedures for Fluid-Structure Interaction Analysis, in: T. Belytschko and T. L. Geers (eds.), Computational Methods for Fluid-Structure Interaction Problems, ASME Applied Mechanics Symposia Series, AMD-Vol. 26, 94-124.

THE DERIVATION OF STIFFNESS MATRICES FROM INTEGRAL EQUATIONS

Friedel Hartmann

University of Dortmund, Dept. of Structural Mechanics

Introduction

Besides Finite Elements in the domain we apply today also integral equation methods on the boundary to problems in elasticity. This boundary element method (B.E.M.), as it is called, leads to non-symmetric and fully populated matrices and it is not easy to convince engineers, that anything good can come out of such strange matrices.

Immagine these same engineers when they discover that the stiffness matrices which they derive from these linear systems turn out to be non-symmetric... . God knows there is not much engineers believe in, but they certainly believe that stiffness matrices have to be symmetric and they are therefore deeply irritated, finding themselves in a situation where two fundamental beliefs clash (Mathematics and Mechanics) and either one is equally convincing.

But naturally the situation is not that dramatic and we will see, as you might have guessed, that also B.E.M. lead to symmetric stiffness matrices, though not so easily as F.E.M. To see how this is achieved and to justify the modifications we must necessarily bring about, we make a small detour by studying instead of the original equation $Hu = Gt$ the equivalent integral equation $G^{-1}Hu = t$, where $G^{-1}H$ is a self-adjoint operator. If the equation $G^{-1}Hu = t$ is solved with Galerkin's method or by minimizing a potential $\Pi(u)$, we obtain indeed a stiffness matrix, which has all three properties $P1^*$ (Rigid body movement), $P2^*$ (Equilibrium) and $P3^*$ (Symmetry), commonly associated with such a matrix. So there is nothing wrong with integral equation methods.

But solving the equation $G^{-1}Hu = t$ on a finite set of boundary elements amounts to substituting for the operator $G^{-1}H$ a modified operator $PG^{-1}HP$, where P stands for a collocation or an orthogonal L_2-projection onto the set of nodal shape functions. While the L_2-operator $PG^{-1}HP$ (i.e. if P is the L_2-projection) preserves the properties, which guarantee a correct stiffness matrix, the collocation operator $PG^{-1}HP$ is no longer self-adjoint and looses also the feature that guarantees the property $P2^*$ (Eq.) in its stiffness matrix.

These remarks are first hints at the problems we encounter if we do numerical computations. But prior to computations, we have to cope with the fact that an expression for the operator $G^{-1}H$ is unknown and we therefore have to return to the original equation $Hu = Gt$ and use the matrices $\underline{\underline{PHP}}$, $\underline{\underline{PGP}}$ (the operators PHP, PGP etc. coincide with matrices) to approximate the matrix $\underline{\underline{PG^{-1}HP}}$ with the product $\underline{\underline{PGP}}^{-1}\underline{\underline{PHP}}$. Using in practical computations numerical integration amounts furthermore to approximating the product $\underline{\underline{PGP}}^{-1}\underline{\underline{PHP}}$ with the product $\underline{\underline{P'G'P'}}^{-1}\underline{\underline{P'H'P'}}$, where the primes denote the modified operators. So we end up with the chain:

$$G^{-1}H \quad \rightarrow \quad PG^{-1}HP \quad \rightarrow \quad PGP^{-1}PHP \quad \rightarrow \quad P'G'P'^{-1}P'H'P'$$

The only operator at hand is the one last in line, i.e. the matrix $\underline{\underline{P'G'P'}}^{-1}\underline{\underline{P'H'P'}}$ and neither the collocation nor the L_2-version of it preserves the three properties P1, P2, P3 (see below), which originally in $G^{-1}H$ or in $PG^{-1}HP$ (if P is the L_2-projection) guaranteed a correct stiffness matrix.

Being confronted with the task to obtain a correct stiffness matrix from $\underline{\underline{P'G'P'}}$ and $\underline{\underline{P'H'P'}}$ we can now, on going backwards, analyze where we loose the single properties needed for a correct stiffness matrix and modify the approximation $\underline{\underline{P'G'P'}}^{-1}\underline{\underline{P'H'P'}}$ accordingly.

Our main concern is the symmetry of the stiffness matrix and this is automatically achieved, if we calculate the stiffness matrix by minimizing the potential $\Pi(u)$, but we could also make the L_2-operator $P'G'P'^{-1}P'H'P'$ self-adjoint, (because $PG^{-1}HP$ has this property) and then use Galerkin's method.

With respect to the properties $P1^*$ (Rigid body movement) and $P2^*$ (Eq.) we must either modify the matrices to ensure these properties or hope that the defect is negligible. This should occur rather in the L_2-matrices than in the collocation matrices. In the following we will discuss all this in more detail.

1. Finite elements in the domain

We study in the bounded domain B with smooth boundary S the 2nd b.v.p. ("Neumann problem") for the vector valued displacement u(x)

$$L\,u = 0 \qquad x \in B \qquad\qquad \lim_{i}\ \tau(u) = \tilde{t} \quad \text{on} \quad S \tag{1}$$

where L is the differential operator of classical elastostatics

$$L = \mu\,\Delta + \frac{\mu}{1-2\nu}\ \text{grad div} \tag{2}$$

and

$$\tau(\ ) = 2\,\mu\,\frac{\partial}{\partial n}\cdot + \frac{2\mu}{1-2\nu}\ n\ \text{div} + \mu\ (n \wedge \text{rot}) \qquad (3)^{\dagger}$$

is the traction operator.

We approximate the solution u on a set of nodal shape functions $\{\phi_1, \phi_2, \phi_3, \dots \phi_N\} \subset H^1(B)$ by a function

$$Pu = \sum_r u_r\,\phi_r(x) = \underset{\sim}{\Phi}^T\,\underset{\sim}{\tilde{u}}$$

and the traction $\tilde{t}$ by the function

$$P\tilde{t} = \sum_r t_r\,\tilde{\phi}_r(x) = \underset{\sim}{\tilde{\Phi}}^T\,\underset{\sim}{\tilde{t}}$$

(It is $\underset{\sim}{\Phi} = [\phi_1, \phi_2, \dots \phi_N]^T$, $\underset{\sim}{u} = [u_1, u_2, \dots u_N]^T$, $\underset{\sim}{\tilde{t}} = [\tilde{t}_1, \tilde{t}_2 \dots \tilde{t}_N]^T$ and it is $\tilde{\phi}_i = \phi_i(x)$, $x \in S$. Note that the "vectorcomponents" u_i etc. are themselves vectors, i.e. $\underset{\sim}{u}$ etc. represent "hypervectors" and consequently are all the matrices "hypermatrices", i.e. their elements K_{ij} etc. are themselves (3x3) matrices).

and we choose the coefficients u_i with Galerkin's method (see ref.[1], p. 94 and 95)

$$(LPu, \phi_i) = \int_B E(Pu)\cdot C\cdot E(\phi_i)\,dv - \int_S P\tilde{t}\cdot\tilde{\phi}_i\,ds = 0 \tag{4}$$

$$i = 1, 2 \dots$$

which yields the linear system

$$\underline{\underline{K}}\,\underset{\sim}{u} = \underline{\underline{F}}\,\underset{\sim}{t} = \underset{\sim}{f} \tag{5}$$

---

† n is the outward normal and $\mu > 0$, $0 \leqslant \nu < 1/2$ are the elastic constants.

with

$$K_{ij} = \int_B E(\phi_i)\cdot C\cdot E(\phi_j)\; dv \qquad F_{ij} = \int_S \tilde{\phi}_i \cdot \tilde{\phi}_j \; ds = (\tilde{\phi}_i, \tilde{\phi}_j) \quad (6)$$

or we choose as solution the approximation $Pu = \sum_r u_r \; \phi_r$, which minimizes the functional

$$\Pi(u) = \frac{1}{2} \int_B E(u)\cdot C\cdot E(u)\; dv - \int_S P\tilde{t}\cdot\tilde{u} \;\; ds \qquad (7)$$

which yields again Eq.(5). The vector $\underset{\sim}{f}$ in Eq. (5) represents the so called "equivalent nodal forces".

2. Boundary elements on the surface

If $\tilde{t}$ is smooth enough, then the solution u of Eq. (1) is regular, $u \in C^2(B) \cap C^1(\bar{B})$, and the Somigliana identity holds on S

$$C(x)\tilde{u}(x) + \int_S T(y,x)\tilde{u}(y)ds_y = \int_S U(y,x)\tilde{t}(y)ds_y \qquad (8)$$

or shortly

$$H\tilde{u} = G\tilde{t}$$

with $\tilde{t} = \lim_i \tau_e(u)$. The operator G is self-adjoint, while H is not.

In boundary element methods we use Eq.(8) to determine the unknown trace u(x) approximately by a projection method, which yields a linear equation on R^{3N}

$$\underline{\underline{PHP}}\; \underset{\sim}{\tilde{u}} = \underline{\underline{PGP}}\; \underset{\sim}{\tilde{t}} \qquad (9)^{\dagger}$$

where $\underset{\sim}{u}$ and $\underset{\sim}{t}$ are the vectors of (generalized) nodal values with respect to a projection P

$$\tilde{u} \rightarrow P\tilde{u} = \underset{\sim}{\Phi}^T \underset{\sim}{\tilde{u}} \qquad \tilde{t} \rightarrow P\tilde{t} = \underset{\sim}{\Phi}^T \underset{\sim}{\tilde{t}} \qquad (10)$$

A formal manipulation of Eq.(9) leads then to the following equation between the nodal displacements u_i and the equivalent nodal forces f_i :

$$\underline{\underline{F}}\;\underline{\underline{PGP}}^{-1}\underline{\underline{PHP}}\;\underset{\sim}{\tilde{u}} = \underline{\underline{F}}\;\underset{\sim}{\tilde{t}} = \underset{\sim}{f} \qquad (11)$$

---

† see ref.[2] for a definition of C(x), T(y,x) and U(y,x).

where the (usually) non-symmetric matrix $\underline{\underline{F}}\ \underline{\underline{PGP}}^{-1}\underline{\underline{PHP}}$ appears in the role of a stiffness matrix which irritates engineers.

In the following we want to discuss what the transition from Eq.(9) to Eq.(11) means and show how we can use Eq.(9) to derive a symmetric stiffness matrix.

Suppose we study instead of the problem

$$H\,\tilde{u} = G\,\tilde{t} \tag{12}$$

the problem

$$G^{-1}H\,\tilde{u} = \tilde{t} \tag{13}$$

$G^{-1}H$ is a linear operator on $C^{1,\gamma}(S)$ with the properties

(P1,Ker) $\quad G^{-1}H\,\tilde{u} = \tilde{0} \qquad \forall\ \tilde{u} = \tilde{a} + \tilde{b} \wedge \tilde{x}$ (rigid body movement)

(P2,Eq.) $\quad (G^{-1}H\,\tilde{u},\ \tilde{a} + \tilde{b} \wedge \tilde{x}) = 0 \qquad \forall\ \tilde{u}$ (equilibrium condition)

(P3,Adj.) $\quad (G^{-1}H\,\tilde{u}^1,\ \tilde{u}^2) = (\tilde{u}^1,\ G^{-1}H\,\tilde{u}^2) \qquad (A_{1,2} = A_{2,1})$

and it is with Korn's second inequality, see ref.[1] p. 381,

$$(G^{-1}H\,\tilde{u},\tilde{u}) = (\,\tau(u),\ \tilde{u}) = \int_B E(u)\cdot C\cdot E(u)\,dv \ > \ c\ \ ||u||_{H^1(B)} - ||u||^2_{L^2}$$

if u is the solution, $L\,u = 0$, with the trace $\tilde{u} = \lim u$. Property (P3, Adj.) is a consequence of the Betti equation

$$\int_B [u^1\cdot Lu^2 - u^2\cdot Lu^1]\,dv = \int_S [\tau(u^1)\cdot\tilde{u}^2 - \tau(u^2)\cdot\tilde{u}^1]\,ds \tag{15}$$

and the fact, that for any pair of functions $\tilde{u}^i \in C^{1,\gamma}(S)$, $\tilde{t}^i \in C^{o,\gamma}(S)$, which stand in the relation

$$H\,\tilde{u}^i = G\,\tilde{t}^i \tag{16}$$

there exists a function $u^i \in C^2(B) \cap C^1(\bar{B})$, $Lu^i = 0$, with $\lim\limits_i u^i = \tilde{u}^i$ and $\lim\limits_i \tau(u^i) = \tilde{t}^i$, see ref.[2], theorem 8.1 and see Kupradze's (et alia) result on the regularity of elastic potentials, ref.[3], theorem 5.2, p. 313 and theorem 6.2, p.315.

---

† The Galerkin method is identical with an orthogonal L_2-projection. Eq.(9) represents therefore both, the collocation method and the Galerkin method. For a definition of the two matrices $\underline{\underline{PGP}}$, $\underline{\underline{PHP}}$ see Eq.(27 a,b) (Collocation) and Eq.(28a,b) (Galerkin).

Furthermore, if u satisfies $L\,u = 0$, then the value of $\Pi(u)$ at u is

$$\Pi(u) = \frac{1}{2}\int_S \tau(u)\,\tilde{u}\,ds - \int_S \tilde{t}\,\tilde{u}\,ds =$$

$$\frac{1}{2}\int_S G^{-1}H\,\breve{u}\;\tilde{u}\,ds - \int_S \tilde{t}\,\tilde{u}\,ds \tag{17}$$

and the trace $\tilde{u}$, i.e. the value of u on the surface S, minimizes, as u minimizes $\Pi(u)$ in Eq.(7) on $H^1(B)$, the functional $\Pi(u)$ in Eq.(17) on $H^{1/2}(S)$.

The Galerkin method on a set of functions

$$\{\,\phi_1, \phi_2, \ldots\ \phi_N\,\} \subset H^{1/2}(S) \cap C^{1,\gamma}(S)$$

$$(G^{-1}H\,P\tilde{u} - P\tilde{t},\ \phi_i) = 0 \qquad i = 1,2\ldots N \tag{18}$$

yields the same linear equations for $\tilde{u}$ as the problem to minimize $\Pi(u)$

$$\Pi(P\tilde{u}) = \frac{1}{2}\int G^{-1}H\,P\tilde{u}\cdot P\tilde{u}\;ds - \int P\tilde{t}\cdot P\tilde{u}\,ds \rightarrow \text{Min} \tag{19}$$

namely

$$\underline{\underline{K}}\,\underset{\sim}{\tilde{u}} = \underline{\underline{F}}\,\underset{\sim}{\tilde{t}} \tag{20}$$

with

$$K_{ij} = (G^{-1}H\,\phi_j,\ \phi_i) \tag{21}$$

The matrix $\underline{\underline{K}}$ in Eq.(20) is a true stiffness matrix, i.e. it has all the properties, commonly associated with such a matrix.

3. A modification

To steer into the direction of Eq.(11) we modify the Galerkin method. We do not solve Eq.(18), but the problem

$$(PG^{-1}HPu - Pt,\phi_i) = (\,\underset{\sim}{\Phi}^T\ \underline{\underline{PG^{-1}HP}}\ \underset{\sim}{u} - \underset{\sim}{\Phi}^T\ \underset{\sim}{t},\ \phi_i) = 0 \tag{22}$$ †

$$i = 1,2\ldots N$$

† We delete in the following text the upper tilde, which we used to designate the trace.

and the same holds for the functional Π where we solve instead of Eq.(19) the problem

$$\Pi(Pu) = \frac{1}{2} \int_S PG^{-1}HPu \cdot Pu \, ds - \int_S Pt \cdot Pu \, ds =$$

$$\frac{1}{2} \int_S \underset{\sim}{\Phi}^T \underline{\underline{PG^{-1}HP}} \, \underset{\sim}{u} \cdot \underset{\sim}{\Phi}^T \underset{\sim}{u} \, ds - \int_S \Phi^T \underset{\sim}{t} \cdot \underset{\sim}{\Phi}^T \underset{\sim}{u} \, ds \rightarrow \text{Min} \tag{23}$$

i.e. we restrict the procedures to the projections of $G^{-1}H\ Pu$ onto the space H_N spanned by the functions ϕ_i.

We use either a collocation projection (we assume $\phi_i(x^j) = \delta_{ij}$)

$$P = P_c \qquad u \rightarrow \underset{\sim}{\Phi}^T \underset{\sim}{u} \qquad \text{with} \quad u_i = (u, \delta(x-x^i))$$

or an orthogonal L^2- projection (i.e. $(u - \underset{\sim}{\Phi}^T \underset{\sim}{u}, \phi_i) = 0$)

$$P = P_{L^2} \qquad u \rightarrow \underset{\sim}{\Phi}^T \underset{\sim}{u} \qquad \text{with} \quad \underset{\sim}{u} = \underline{\underline{F}}^{-1} \underset{\sim}{w}(u) \quad \text{and } w_i(u) = (u, \phi_i)$$

On $H^{1/2}(S)$ the operator $PG^{-1}HP : u \rightarrow PG^{-1}HPu$ is defined by a matrix,

$$u \rightarrow \underset{\sim}{\Phi}^T \underline{\underline{PG^{-1}HP}} \, \underset{\sim}{u} \tag{24}$$

with $PG^{-1}HP_{ij} = (\delta(x-x^i), G^{-1}H\phi_j) = \tau(u(\phi_j))(x^i)$ if $P = P_c$ or

$$PG^{-1}HP_{ij} = \sum_k^N F_{ik}^{-1}(G^{-1}H\phi_k, \phi_j)$$

if $P = P_{L^2}$.

The introduction of the L^2-projection is camouflage, because the thus "modified" problems, Eq.(22) and Eq.(23), have the same solutions as the original problems, Eq.(18) and Eq.(19), but we hoped to gain some consistency in the derivation by doing so.

Note, that the projection of a function that is contained in the set H_N coincides with the function itself.

Keeping this remark in mind, it follows easily, that the L^2-operator $PG^{-1}HP$, (i.e. if P is the L^2-projection), has all three properties (P1), (P2) and (P3), if H_N contains the traces of the six basic rigid body movements, while the collocation operator has only the property (P1,Ker).

The loss of (P2,Eq.) and (P3,Adj.) under the collocation projection is naturally due to the fact that the basis functions ϕ_i and the weight functions $\delta(x-x^i)$ differ, but with N and (consequently) $\phi_i \rightarrow \delta(x-x^i)$, we might expect to regain these properties in the limit.

With this modification, i.e. replacing $G^{-1}HP$ by $PG^{-1}HP$ we are free to couple the two methods with either a L^2 - or a collocation projection.

Method	Projection
Galerkin	L^2
Π - Min	Collocation

which yields essentially only three different procedures to solve $G^{-1}H\,u = t$ because (Galerkin x L^2) and (Π- Min x L^2) coincide, see below. The combination (Galerkin x Collocation) does not lead to a symmetric stiffness matrix and so we are left with the combinations (Galerkin x L^2) and (Π - Min x Collocation).

The Galerkin vector $\underset{\sim}{u}$, the solution of Eq. (22), satisfies the equation

$$\underline{\underline{F}}\,\underline{\underline{M}}\,\underset{\sim}{u} = \underline{\underline{F}}\,\underset{\sim}{t} \tag{25}$$

while the Π - Min vector $\underset{\sim}{u}$, the solution of Eq.(23), satisfies the equation

$$\frac{1}{2}\,(\underline{\underline{F}}\,\underline{\underline{M}} + \underline{\underline{M}}^T\underline{\underline{F}})\,\underset{\sim}{u} = \underline{\underline{F}}\,\underset{\sim}{t} \tag{26}$$

(it is $\underline{\underline{F}} = \underline{\underline{F}}^T$)

Both vectors coincide, if $\underline{\underline{PG^{-1}HP}} = \underline{\underline{M}}$ is symmetric, i.e. if P is the L^2-projection and it easily follows, then that the stiffness matrix $\underline{\underline{K}} = \underline{\underline{F}}\,\underline{\underline{M}}$ has the properties

P1*: $\quad \underline{\underline{K}}\,\underset{\sim}{s} = \underset{\sim}{0}$ $\qquad$ P2* : $\quad \underset{\sim}{s}^T\,\underset{\sim}{f} = \underset{\sim}{s}^T\,\underline{\underline{K}}\,\underset{\sim}{u} = 0 \quad \forall \quad \underset{\sim}{u},\ \underset{\sim}{s}$

P3*: $\quad \underset{\sim}{f}_1^T\,\underset{\sim}{u}_2 = \underset{\sim}{f}_2^T\,\underset{\sim}{u}_1$ $\qquad$ if $\underline{\underline{K}}\,\underset{\sim}{u}^i = \underset{\sim}{f}^i \qquad i = 1, 2$

because $PG^{-1}HP$ has the properties (P1), (P2) and (P3) or, what is equivalent, the matrix $\underline{\underline{M}} = \underline{\underline{PG^{-1}HP}}$ satisfies the Eqs.(33) - (35), see below. Analogously the stiffness matrix

$$\underline{\underline{K}} = \frac{1}{2}\,(\underline{\underline{F}}\,\underline{\underline{M}} + \underline{\underline{M}}^T\underline{\underline{F}})$$

with $\underline{\underline{M}} \neq \underline{\underline{M}}^T$ has the properties P1*, P2* and P3* if $PG^{-1}HP$ has the properties (P1) and (P2).

4. A substitution

But now the operator $G^{-1}H$ is unknown and we therefore use the operators G and H from Eq.(8), $H\,u = G\,t$, and their modifications with respect to P

$$PHP: \qquad u \rightarrow \underset{\sim}{\Phi}^T\,\underline{\underline{PHP}}\,\underset{\sim}{u}$$

$$PGP: \qquad t \rightarrow \underset{\sim}{\Phi}^T\,\underline{\underline{PGP}}\,\underset{\sim}{t}$$

where $\underline{\underline{PGP}}$ and $\underline{\underline{PHP}}$ are the matrices

$$P = P_c \qquad PHP_{ij} = (\delta(x-x^i),\ H\phi_j) \qquad PGP_{ij} = (\delta(x-x^i),\ G\phi_j) \tag{27a,b}$$

$$P = P_{L^2} \qquad PHP_{ij} = \sum_k^N F_{ik}^{-1}(H\phi_k,\phi_j) \qquad PGP_{ij} = \sum_k^N F_{ik}^{-1}(G\phi_k,\phi_j) \tag{28a,b}$$

to substitute for the matrix $\underline{\underline{PG^{-1}HP}}$ the matrix $\underline{\underline{PGP}}^{-1}\underline{\underline{PHP}}$

$$\underline{\underline{PG^{-1}HP}} \cong \underline{\underline{PGP}}^{-1}\underline{\underline{PHP}} = \underline{\underline{\tilde{M}}} \tag{29}$$

This substitution rests on the following argument: consider t fixed and let u be the function that satisfies $PHPu = PGPt$ or equivalent $\underline{\underline{PHP}}\ \tilde{u} = \underline{\underline{PGP}}\ \tilde{t}$ and suppose the same u satisfies $PG^{-1}HPu = Pt$, or equivalently $\underline{\underline{PG^{-1}HP}}\ \tilde{u} = \tilde{t}$, then follows Eq.(29). But this assumption with respect to $\tilde{u}$ holds naturally only for the original equations, i.e. $Hu = Gt \Longrightarrow G^{-1}Hu = t$. Note that

$$(Gt,t) = \int_{R^3} E(u)\cdot C\cdot E(u)\ dv \tag{30}$$ †

where

$$u = Gt = \int_S U(y,x)t(y)ds_y \qquad x \in R^3 \tag{31}$$

and therefore $(Gt,t) = 0$ yields $u = a + b \wedge x$ and due to

$$0 - 0 = \lim_i\ \tau_e(u) - \lim_e \tau_e(u) = \frac{1}{2}\,t + {}_S\!\int T^*t\,ds_y - (-\frac{1}{2}\,t + {}_S\!\int T^*t\,ds_y) =$$

also $t = 0$, which means, that the matrix $\underline{\underline{F}}\ \underline{\underline{PGP}} = [(G\phi_j,\phi_i)]$ is non-singular, which should also hold, if N is large enough, for the collocation matrix $\underline{\underline{PGP}} = [(G\phi_j)(x^i)]$ due to $Gt=0 \Longleftrightarrow t=0$.

5. Some corrections

In practical computations we do not approximate $\underline{\underline{PG^{-1}HP}}$ with $\underline{\underline{PGP}}^{-1}\underline{\underline{PHP}}$, but, due to numerical quadrature, rather with the product

$$\underline{\underline{P'G'P'}}^{-1}\underline{\underline{P'H'P'}} = \underline{\underline{\tilde{M}'}} \tag{32}$$

where a prime denotes the modified operator due to numerical quadrature. The properties (P1), (P2) and (P3) read, if we replace $\underline{\underline{PG^{-1}HP}}$ by $\underline{\underline{PGP}}^{-1}\underline{\underline{PHP}}$ (we delete the primes)

$$\text{(P1,Ker)} \qquad \underline{\underline{PGP}}^{-1}\underline{\underline{PHP}}\ \tilde{s} = \tilde{0} \qquad \forall\ \tilde{s} \tag{33}$$

$$\text{(P2,Eq.)} \qquad \int \tilde{\Phi}^T\ \underline{\underline{PGP}}^{-1}\underline{\underline{PHP}}\ \tilde{u}\cdot\tilde{\Phi}^T\tilde{s}\ ds = 0 \qquad \forall\ \tilde{u},\ \tilde{s} \tag{34}$$

---

† See the appendix for a proof of Eq.(30).

¶ The index e on $\tau_e(\ )$ means, that the normal points into the exterior.

or equivalently

$$\underset{\sim}{s}^T \underline{\underline{F}}\ \underline{\underline{PGP}}^{-1}\underline{\underline{PHP}} = \underset{\sim}{0}^T \qquad \forall\ \underset{\sim}{s} \tag{34a}$$

(P3,Adj.)

$$\underline{\underline{PGP}}^{-1}\ \underline{\underline{PHP}} = (\underline{\underline{PGP}}^{-1}\underline{\underline{PHP}})^T \tag{35}$$

where $\underset{\sim}{s}$ is the vector of nodal values of a rigid body movement $a + b \wedge \tilde{x}$ with respect to the functions ϕ_i.

The collocation matrix $\underline{\underline{M}}'$ has none of the properties (P1), (P2) and (P3). While the property (P1,Ker) gets lost under numerical integration, the lack of properties (P2,Eq.) and (P3,Adj.) is due to the collocation projection itself and has nothing to do with substituting $\underline{\underline{PGP}}^{-1}\underline{\underline{PHP}}$ for $\underline{\underline{PG^{-1}HP}}$.

The L^2-matrix $\underline{\underline{\tilde{M}}}'$ also looses (P1,Ker) under numerical quadrature, while the lack of (P2,Eq.) and (P3,Adj.) is due to the substitution, but we might hope to miss these three properties only by a narrow margin.

The properties $P1^*$, $P2^*$ and $P3^*$ are considered essential for a stiffness matrix and depend, as we have seen, on the properties P1, P2 and P3 of $\underline{\underline{M}}$.

Introducing approximations $\underline{\underline{\tilde{M}}}'$ for $\underline{\underline{M}} = \underline{\underline{PG^{-1}HP}}$ we have, therefore, to ensure that the properties P1, P2 and P3 are met by the substitute $\underline{\underline{\tilde{M}}}'$. Consequently we replace the L^2-matrix $\underline{\underline{\tilde{M}}}'$ by the symmetrix matrix

$$\underline{\underline{\tilde{M}}}'' = \frac{1}{2}(\underline{\underline{\tilde{M}}}' + \underline{\underline{\tilde{M}}}'^T) \tag{36}$$

and modify $\underline{\underline{\tilde{M}}}''$ to guarantee (P2,Eq.) and (P1,Ker).

Analogously we correct the collocation matrix $\underline{\underline{\tilde{M}}}'$ to ensure (P1,Ker) and furthermore modify it, to obtain property (P2,Eq.), which is originally no property of $\underline{\underline{M}}$, but will be needed to obtain a correct stiffness matrix. Naturally there is no need to symmetrize the collocation matrix, if we couple it with the Π - Min method.

With these corrections we obtain from (Galerkin x L^2) and (Π - Min x Collocation) correct stiffness matrices.

6. Stiffness matrices for exterior unbounded regions

Let us denote for clarity by $G^{-1}H_I = G^{-1}H$ the operator of interior regions which we studied up to now.

The approximation and derivation of stiffness matrices for exterior regions, i.e. for the operator $G^{-1}H_E$ proceeds as before and we therefore restrict the discussion to an analysis of the operator $G^{-1}H_E$.

We study the b.v.p.

$$L\,u = 0 \qquad x \in B_e \qquad \lim_e \tau_i(u) = \tilde{t} \tag{37}$$

and call u a solution, if u satisfies the equations above and belongs to $H^1(B_e)$.

We know, that if $\tilde{t}$ is smooth enough, then there exists a unique regular solution $u \in C^2(B_e) \cap C^1(\bar{B}_e) \cap H^1(B_e)$, which satisfies the Somigliana identity

$$C_e(x)u(x) + \int_S T_i(y,x)u(y)ds_y = \int_S U(y,x)t(y)ds_y \tag{38}$$

or shortly

$$H_E u = G\ t \tag{39}$$

In interior regions we encountered the operator $H = H_I$, namely

$$H_I u = C_i(x)u(x) + \int_S T_e(y,x)u(y)ds_y \tag{40}$$

and it follows with

$$C_i(x)s(x) + \int_S T_e(y,x)s(y)ds_y = 0 \qquad \forall \quad s = a + b \wedge x \tag{41}$$

and

$$C_i(x) = I - C_e(x) \qquad\qquad T_e(y,x) = T_i(y,x)\cdot(-1) \tag{42}$$

that

$$-(H_E - I)s = H_I s = 0 \qquad \forall \quad s \tag{43}$$

or equivalently

$$H_E s = s \qquad \forall \quad s \tag{44}$$

The operator G^{-1} is defined by $G^{-1}G\ t = t$ and due to

$$G\ t = \int_S U(y,x)t(y)ds_y \tag{45}$$

it follows, with the well known jump of the traction of the potentials of the first kind, that

$$G^{-1}(\ ) = \lim_i \tau_e(\ ) - \lim_e \tau_e(\ ) \tag{46}$$

Note that this definition makes sense only for those $u \in H^{1/2}(S)$, whose continuation into B_i and B_e are known. The rigid body movement belongs certainly to this class, and so we obtain

$$G^{-1}H_E s = G^{-1}s = \lim_i \tau_e(s) - \lim_e \tau_e(s) = 0 - 0 = 0 \tag{47}$$

i.e. the operator $G^{-1}H_E$ has the property (P1,Ker).

Exterior regular solutions satisfy furthermore the Betti equation in unbounded regions and with theorem 8.2 in ref.[2] it follows as above, that $G^{-1}H_E$ has the property (P3,Adj.). Only (P2,Eq.) is no property of $G^{-1}H_E$. Exterior solutions must not satisfy the equilibrium conditions on the surface S. (This is no longer true for regular exterior solutions in the plane. The equilibrium condition is a necessary condition in the plane. Otherwise the solution would not satisfy a radiation condition.)

Consequently have exterior stiffness matrices the properties $P1^*$ and $P3^*$, (in 3-D), which should make the approximation of these matrices perhaps a little bit easier.

7. Practical computations

In this section we have finally a short look at the numerical effort needed to derive a stiffness matrix with either a L^2- or a collocation projection from the equation $Hu = Gt$.

Let us suppose the surface S to consist of M elements S_m, $m_i = 1, 2 \ldots M$ and we assume furthermore, that there are N nodes x^i on S and that integration is done on a reference element S_o with a numerical integration technique with, (say), K weights w_k:

$$\int_S f(x)ds = \sum_m^M \int_{S_o} f^m(\xi)J^m(\xi)d\xi = \sum_m^M \sum_k^K f^m(\xi^k)J^m(\xi^k)w_k = \sum_\ell f(\underline{x}^\ell)J^\ell w^\ell$$

where $\ell = (m-1)K + k$, $J^\ell = J^m(\xi^k)$, $w^\ell = w_k$ if $\ell = (m-1)K + k$.

The points $\underline{x}^\ell$, $\ell = 1,2\ldots L$ are the integration points on S, w^ℓ are the corresponding weights and J^ℓ is the corresponding value of the Jacobian in x^ℓ. The nodal points usually do not coincide with an integration point. Next we introduce the diagonal matrix $\underline{\underline{W}}_{(LxL)}$ with $W_{ij} = \delta_{ij}J^j w^j$ and introduce N nodal shape functions ϕ_i, which generate the matrix $\underline{\underline{V}}_{(NxL)}$ with $V_{ij} = \phi_i(x^j)$ and with the matrices $\underline{\underline{G}}_{(NxL)}$ and $\underline{\underline{H}}_{(NxL)}$

$$H_{ij} = C(x^i)\delta_{ij}W_{ij}^{-1} + T(\underline{y}^j, x^i) \qquad G_{ij} = U(x^i,\underline{y}^j) = U(x^j,\underline{y}^i)$$

and $\underline{\underline{\bar{G}}}_{(LxL)}$ and $\underline{\underline{\bar{H}}}_{(LxL)}$

$$\bar{H}_{ij} = C(\underline{x}^i)\delta_{ij}W_{ij}^{-1} + T(\underline{y}^j,\underline{x}^i) \qquad \bar{G}_{ij} = U(\underline{x}^i,\underline{y}^j) = U(\underline{x}^j,\underline{y}^i) \quad \dagger$$

we obtain the collocation matrices

$$\underline{\underline{PHP}} = \underline{\underline{H}}\,\underline{\underline{W}}\,\underline{\underline{V}}^T \qquad \underline{\underline{PGP}} = \underline{\underline{G}}\,\underline{\underline{W}}\,\underline{\underline{V}}^T$$

and the L^2- matrices

$$\underline{\underline{PHP}} = \underline{\underline{F}}^{-1}\underline{\underline{V}}\,\underline{\underline{W}}\,\underline{\underline{\bar{H}}}\,\underline{\underline{W}}\,\underline{\underline{V}}^T \qquad \underline{\underline{PGP}} = \underline{\underline{F}}^{-1}\underline{\underline{V}}\,\underline{\underline{W}}\,\underline{\underline{\bar{G}}}\,\underline{\underline{W}}\,\underline{\underline{V}}^T$$

(for completeness we state: $\underline{\underline{F}} = \underline{\underline{V}}\,\underline{\underline{W}}\,\underline{\underline{V}}^T = \underline{\underline{D}}\,\underline{\underline{V}}^T$)
which yields with $\underline{\underline{D}}_{(NxL)} = \underline{\underline{V}}\,\underline{\underline{W}}$ for the collocation matrix $\underline{\underline{\tilde{M}}}'$:

$$\underline{\underline{\tilde{M}}}' = (\underline{\underline{G}}\,\underline{\underline{W}}\,\underline{\underline{V}}^T)^{-1}\,\underline{\underline{H}}\,\underline{\underline{W}}\,\underline{\underline{V}}^T = (\underline{\underline{G}}\,\underline{\underline{D}}^T)^{-1}\underline{\underline{H}}\,\underline{\underline{D}}^T$$

and for the L^2-matrix $\underline{\underline{\tilde{M}}}'$:

$$\underline{\underline{\tilde{M}}}' = (\underline{\underline{V}}\,\underline{\underline{W}}\,\underline{\underline{\bar{G}}}\,\underline{\underline{W}}\,\underline{\underline{V}}^T)^{-1}\underline{\underline{V}}\,\underline{\underline{W}}\,\underline{\underline{\bar{H}}}\,\underline{\underline{W}}\,\underline{\underline{V}}^T = (\underline{\underline{D}}\,\underline{\underline{\bar{G}}}\,\underline{\underline{D}}^T)^{-1}\underline{\underline{D}}\,\underline{\underline{\bar{H}}}\,\underline{\underline{D}}^T$$

where $\underline{\underline{D}}\,\underline{\underline{\bar{G}}}\,\underline{\underline{D}}^T$ is a symmetric matrix, while $\underline{\underline{G}}\,\underline{\underline{D}}^T$ is not.

---

† $\underline{\underline{\bar{H}}}$ and $\underline{\underline{\bar{G}}}$ are not defined on the diagonal, but this is irrelevant here and no major obstacle to the notation.

So, while the L^2-method (or Galerkin method) calls for a higher resolution of $G \to \underline{\underline{\overline{G}}}_{(LxL)}$ and $H \to \underline{\underline{\overline{H}}}_{(LxL)}$ and involves two more matrix multiplications

$$(\underline{\underline{D}}\ \underline{\underline{G}}\ ..)^{-1}\underline{\underline{D}}\ \underline{\underline{H}}\ ..$$

the matrix to invert, $\underline{\underline{D}}\ \underline{\underline{G}}\ \underline{\underline{D}}^T$, is symmetric and as approximation of $\underline{\underline{F}}\ \underline{\underline{PGP}} = [(G\phi_i,\phi_j)]$ probably positive definite.

Furthermore we might expect that the defect of $\underline{\underline{\tilde{M}}}'$ with respect to (P1,Ker) and (P2,Eq.) becomes smaller, if we use the L^2-projection, which should give the L^2-method a fair chance in comparison with the collocation projection.

Having obtained a symmetric stiffness matrix with the means proposed above, we now turn to the modifications we must bring about to ensure the properties (P1,Ker) and (P2,Eq.).

The property (P2,Eq.) seems to be of considerable influence on the exactness of a solution. This we conclude from similar experience with the Laplace operator Δ, where the analogue property is $(\partial u/\partial n, 1) = 0$ and remarks in ref.[5], which confirm this view.

The necessary modifications would be (relatively) simple if we neglect possible unbalanced couples and satisfy only the equilibrium condition with respect to translations. The following three simple vectors are a basis of the set of all possible translations:

$$\underset{\sim}{s}^{(1)} = [1,\ 0,\ 0,\ 1,\ \ldots \qquad \ldots\ 0]^T$$

$$\underset{\sim}{s}^{(2)} = [0,\ 1,\ 0,\ 0,\ 1,\ \ldots \qquad \ldots\ 0]^T$$

$$\underset{\sim}{s}^{(3)} = [0,\ 0,\ 1,\ 0,\ 0,\ 1,\ \ldots \qquad \ldots\ 1]^T$$

Let us denote by $\underset{\sim}{\varepsilon}^{(k)}$ the residuum of the translation $\underset{\sim}{s}^{(k)}$

$$\underset{\sim}{s}^{(k)\ T}\ \underline{\underline{F}}\ \underline{\underline{\tilde{M}}}' = \underset{\sim}{\varepsilon}^{(k)\ T} \tag{48}$$

The residua $\varepsilon^{(k)}$ would be zero vectors, if $\underline{\underline{F}}\ \underline{\underline{\tilde{M}}}'$ was a correct matrix, see Eq.(34a). The modified matrix $\underline{\underline{F}}\ \underline{\underline{M}}$ has then the components

$$[FM]_{ij} = [F\tilde{M}]_{ij} - \sum_{k=1}^{3} \varepsilon_j^{(k)}\ s_i^{(k)} \qquad i,\ j = 1,\ 2\ \ldots\ 3N \tag{49}$$

The property (P1,Ker) should give us the least trouble. There is a technique to calculate the Cauchy principal value of the kernel $T(y,x)$ with a rigid body movement. If we do this, then the matrix $\underline{\underline{\tilde{M}}}'$ automatically has the property (P1,Ker). If we do not use this technique, the property (P1,Ker) might not be obtained exactly, but the defect should be very small. Otherwise our numerical quadrature needs mending.

8. Conclusion

Summing it all up, we state:
To derive from the equation

$$\underline{\underline{PHP}}\,\underset{\sim}{u} = \underline{\underline{PGP}}\,\underset{\sim}{t}$$

a stiffness matrix with the properties $P1^*$, $P2^*$ and $P3^*$ proceed as follows:

If P is the L^2-projection, calculate $\tilde{\underline{\underline{M}}}'$, see Eq.(29), symmetrize it, see Eq.(36), and modify the new matrix $\tilde{\underline{\underline{M}}}''$ to ensure property (P1,Ker), calculate the stiffness matrix $\underline{\underline{K}} = \underline{\underline{F}}\,\tilde{\underline{\underline{M}}}''$ according to Eq.(24), and modify it to guarantee property (P2,Eq.).

If P is the collocation projection calculate $\tilde{\underline{\underline{M}}}'$, see Eq.(29), modify it to ensure (P1,Ker) and modify $\underline{\underline{F}}\,\tilde{\underline{\underline{M}}}'$ to satisfy Eq.(34) and then calculate the stiffness matrix $\underline{\underline{K}} = 1/2\,(\underline{\underline{F}}\,\tilde{\underline{\underline{M}}}' + \tilde{\underline{\underline{M}}}'^T\underline{\underline{F}})$.

These matrices are the stiffness matrices, if we solve the equation $G^{-1}Hu = t$ with Galerkin's method or by minimizing a potential in connection with a projection method and if we approximate $\underline{\underline{PG^{-1}HP}}$ with $\underline{\underline{PGP}}^{-1}\underline{\underline{PHP}}$.

APPENDIX

Proof of Eq.(30)

The function $u(x) = \int U(y,x)t(y)\,ds_y$ satisfies $Lu = 0$ in any point $x \in R^3 = B_i \cup B_e$ and has a finite strain energy in the continuum, i.e.

$$\int_{R^3} E(u)\cdot C\cdot E(u)\,dv \quad < \quad \infty \tag{a}$$

Due to $Lu = 0$ it follows, see ref. [1], p.95,

$$(\tilde{t}^i, \tilde{u}) = \int_{B_i} E(u)\cdot C\cdot E(u)\,dv \tag{b}$$

$$(\tilde{t}^e, \tilde{u}) = \int_{B_e} E(u)\cdot C\cdot E(u)\,dv \tag{c}$$

and by adding Eq.(b) to Eq.(c) we obtain

$$(\tilde{t}^i + \tilde{t}^e, \tilde{u}) = \int_{R^3} E(u)\cdot C\cdot E(u)\,dv \tag{d}$$

where

$$\tilde{t}^i = \lim_i \tau_e(u) = \frac{1}{2}t + \int T^* t\,ds_y \tag{e}$$

$$\tilde{t}^e = \lim_e \tau_i(u) = -\left(-\frac{1}{2}t + \int T^* t\,ds_y\right) \tag{f}$$

Substituting for $\tilde{t}^i$ and $\tilde{t}^e$ the expression in Eqs.(e) and (f) we obtain with $\tilde{u} = Gt$ finally the result:

$$(Gt,t) = \int_{R^3} E(u)\cdot C\cdot E(u)\,dv \tag{g}$$

References

[1] Flügge, S.(Ed.), Encyclopedia of Physics, Volume VIa/2, Mechanics of Solids II, (Vol. Ed. Truesdell), Springer Verlag 1972

[2] Hartmann, F. Elastische Potentiale in Gebieten mit Ecken. Dissertation, University of Dortmund, 1980.

[3] Kupradze, Gegelia, Bashelishivili, Burchuladze. Three dimensional problems of the mathematical theory of elasticity and thermoelasticity. North-Holland Publishing Company, Amsterdam, New York, Oxford, 1979.

[4] Brebbia, C.A. and Georgiou, P. Combination of boundary and finite elements in elastostatics, Appl. Math. Modelling 1979, Vol.3

[5] Zienkiewicz, O.C., D.W. Kelly, and P.Bettess. Marriage à la mode - The Best of Both Worlds (Finite Elements and Boundary Integrals), in Energy methods in F.E. analysis, ed. by Glowinski, Rodin and Zienkiewicz, John Wiley & Son, 1979.

[6] Mattioli, F. Numerical instabilities of the integral approach to the interior boundary value problem for the two-dimensional Helmholtz equation. Int. Journal for Numerical Methods in Eng. Vol. 15, 1303 - 1313 (1980).

[7] Weyl, H., Symmetry. Princeton, Princeton University Press 1952.

ACKNOWLEDGEMENT

This work was completed while the author was visiting the School of Engineering at the University of California, Irvine. The financial support of the NATO Science Committee (Grant 490/420/585/0) is gratefully acknowledged.

THREE DIMENSIONAL SUPER-ELEMENT BY THE BOUNDARY INTEGRAL EQUATION METHOD FOR ELASTOSTATICS

F. Volait

CETIM - France

ABSTRACT

In this paper, we describe the formation of a stiffness matrix using a boundary integral method for three dimensional elastostatics problems.

This technique is based upon energy minimization and leads directly to a symmetric form. A special discretization is adopted in order to avoid problems due to eventual discontinuities of the traction vector (geometrical boundary discontinuities). Such a matrix enables a coupling with a finite element discretization or an incorporation within a standard finite element software. Numerical aspects induced by the integration are discussed.

INTRODUCTION

For an homogeneous isotropic medium V of boundary S, a linear elasticity problem can be formulated by using Navier's equation :

$$\Delta^{*} u_i(y) + f_i(y) = 0 \text{ for } y \in V \qquad (1)$$

and classical boundary conditions

$$u_i(y) = \overline{u_i}(y) \quad \text{for} \quad y \in S_u \qquad (2)$$

$$t_i(y) = \overline{t_i}(y) \quad \text{for} \quad y \in S_t \qquad (3)$$

where : $t_i(y) = \sigma_{ij}(y)\, n_j(y)$, $S = S_u \cup S_t$

$n_j(y)$ is the unit outward normal vector at point y

CLASSICAL BOUNDARY ELEMENT METHOD

Numerical solution of this problem can be computed by using classical boundary element method (see ref [1], [2])
This solution derives from an equivalent formulation of the first problem :

$$\int_V U_{ij}\,(x-y)\,\left[\Delta^{*} u_j\,(y) + f_j\,(y)\right] dV = 0 \qquad (4)$$

for $x \in V$

with boundary conditions : $u_i\,(y) = \overline{u_i}\,(y)$ on S_u

$t_i\,(y) = \overline{t_i}\,(y)$ on S_t

In equation (4), $U_{ij}\,(x-y)$ is the elementary solution of the Navier's equation which is nothing but the Kelvin's tensor.

In most of numerical formulations, equation (4) is transformed by using Betti's theorem and becomes :

$$c_{ij}\,(x)\,u_j(x) + \int_S T_{ij}\,(x-y)\,u_j\,(y)\,dS = \int_S U_{ij}\,(x-y)\,t_j\,(y)\,dS + \int_V U_{ij}\,(x-y)\,f_j(y)dV \qquad (5)$$

which, in the case of zero body forces, is reduced to a boundary integral equation. Then, boundary conditions are taken into account by replacing the functions u_i (resp. t_i) by their known values $\overline{u_i}$ (resp. $\overline{t_i}$)

Finally, we obtain an equation where unknowns are of two kinds, displacements and tractions, which can be resolved by analytical or numerical methods (collocation method for example).

A SYMMETRICAL BOUNDARY INTEGRAL EQUATION

In the previous formulation, we obtained a relation between u_i and t_i (Eq 5) in which we imposed boundary conditions.

The initial problem can be now formulated in such a way that displacements are principal unknowns. This approach, described by Zienkiewicz et al. in [6], [7] is based on the minimization of the energy functional.

$$\Pi\,(u) = \frac{1}{2}\int_V \sigma_{ij}\,\varepsilon_{ij}\,dV - \int_V u_i f_i dV - \int_{S_t} u_i \overline{t_i} dS \qquad (6)$$

with $u_i = \overline{u_i}$ (7)

If we now impose as an extra condition that :

$$\Delta^{*} u_i + f_i = 0 \tag{8}$$

the expression of the functional (6) becomes :

$$\Pi(u) = \frac{1}{2} \int_S u_i\, t_i\, dS - \int_{S_t} u_i\, \bar{t}_i dS + \frac{1}{2} \int_V u_i f_i\, dV \tag{9}$$

which is to be minimized with both conditions (7) and (8).

The solution of the initial problem is then achieved by calculating a stationary value of Π , which is given by :

$$\delta\Pi\ (u) = 0 \tag{10}$$

Calculating the expression of $\delta\Pi$ we obtain :

$$\delta\Pi(u) = \frac{1}{2} \int_S \left[u_i\, \delta t_i + t_i\ \delta u_i \right] dV - \int_{S_t} \delta u_i\, \overline{t_i}\, dS + \frac{1}{2} \int_V \delta u_i\ f_i\ dV = 0 \tag{11}$$

with $u_i = \overline{u_i}$ on S_u

and with the equilibrium condition $\Delta^{*} u_i + f_i = 0$ which is equivalent to the integral equation (5). It is to be noticed that this integral equation gives a relation between tractions and displacements which will be used to eliminate tractions from the expression (11)

NUMERICAL TREATMENT

A classical discretization procedure can be applied to this problem and, assuming that $f_i = 0$ in V, this discretization is reduced to the boundary [7]. Coordinates, displacements and tractions are approximated by a set of polynomial functions. A collocation procedure is used to obtain a linear system from integral equation (5)

$$[A]\ \{u\} = [B]\ \{t\} \tag{12}$$

In most situations, this technique gives a well conditio - ned solvable system of equations and tractions may be expressed as a linear function of displacements.

However, an important feature, which has been related by M. Chaudonneret [3], and by the author in previous papers [4], [5], is the possible discontinuity of the tractions which cannot be taken into account by a classical discretization.

In order to avoid introducing numerical errors, the discretization procedure is modified. Nodal values of the tractions are now chosen inside the element as shown in Figure 1. In this way, discontinuities of the traction vector occurring on the outline of the elements may be precisely approximated.

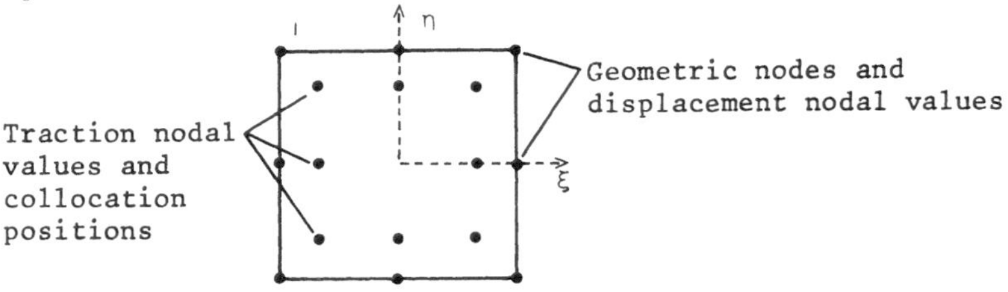

Figure 1 - Discretization scheme

However, such a discretization increases the number of tension nodal values and consequently the number of columns of the matrix B. This leads to a non square system and therefore a non solvable one.
To overcome this problem, the collocation procedure is modified so that the collocation positions coincide with the traction nodal value positions.
It is to be noticed that the displacements interpolations can be kept continuous.

Finally, the integral equation (5) leads to the relation :

$$\{t\} = [B^{-1}] \; [A] \; \{u\} \tag{13}$$

By using the same discretization in equation (11) a matrix form is obtained :

$$\frac{1}{2} < \delta u > [C] \{t\} + \frac{1}{2} < t(\delta u) > [C]^T \{u\} - < \delta u > \{F\} = 0 \tag{14}$$

$$\text{with} \quad [C] = \int_S N^T M \, dS$$

$$\text{and} \quad \{F\} = \int_S N^T \, \bar{t} \, dS$$

in these expressions N (resp. M) denotes the interpolant functions of u (resp. t).
Applying the relation (13) to the displacements {u} and {δ u} , the equation (14) can now be expressed in terms of displacements only so that :

$$\frac{1}{2} < \delta u > \Big[[E] + [E]^T \Big] \{u\} - < \delta u > \{F\} = 0 \tag{15}$$

$$\text{where} \quad [E] = [C] \, [B]^{-1} \, [A]$$

Numerical solutions of the initial problem are then achieved by solving the approximate problem :

$$[K] \{u\} - \{F\} = 0 \qquad (16)$$

with $u_i = \overline{u_i}$ on S_u

As has been pointed out in /6/ , this method leads to a symmetric linear system (16) which can be assembled to a system of finite element equations as the contribution from a new element. The compatibility is ensured if the boundary interpolant for displacement is chosen to be identical to the adjacent finite element interpolation on the interface boundary.

NUMERICAL INTEGRATION

One of the most important problems to solve when computing the coefficients of the matrices [A] and [B] is certainly the numerical integration accuracy. This problem has been solved by Lachat and Watson /7/ in the three dimensional elastostatics program E.I.T.D. * which was based on a classical boundary integral equation formulation.

However, due to the new discretization, we had to modify it.

Coefficients to be computed are of the kind :

$$I(x) = \int_{-1}^{+1} \int_{-1}^{+1} K(x, y^e(\xi, \eta)) N^c(\xi, \eta) J^e(\xi, \eta) d\xi d\eta \qquad (17)$$

where K(x,y) is a singular kernel depending on the distance r (x,y) ; $y^e(\xi, \eta)$ represents the global coordinates of a point y on the element e and $J^e(\xi, \eta)$ is the Jacobian of the geometrical transformation.

In most cases, an approximate value can be achieved by a Gauss numerical integrating with a (3 x 3) points scheme. However, when the distance becomes small compared with the size of the element, this formula has to be modified by dividing the element and by increasing the number of integration points.

* E.I.T.D. CETIM'S Program running on a C.D.C. 7600

When the collocation point belongs to the element over which the integration has to be performed, the integral becomes improper and has to be evaluated with particular care.

. For $K(x,y) = U_{ij}(x,y)$, the kernel contains a singularity of the order $\frac{1}{r}$ and good accuracy is obtained by dividing the element as shown in figure 2. Here, the number of integration points is chosen according to the value of θ.

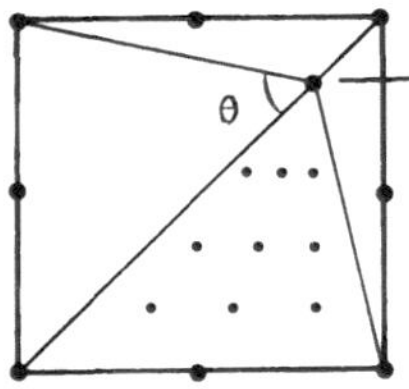

Figure 2 - Integration scheme on singular element

This integration scheme was tested on the function $\frac{1}{r}$ which is simpler than $U_{ij}(x, y)$ but singular. Results were compared with analytical calculations and the error was only of 0.03 % .

. For $K(x,y) = T_{ij}(x,y)$, the kernel varies as $\frac{1}{r^2}$ and the integral has to be interpreted in the sense of the Cauchy principal value. Numerical evaluation of this integral is very difficult.

To overcome this problem, we tried to use the solution described by Lachat and Watson in [7].

This consists of decreasing the order of the singularity by using displacement interpolant vanishing when x = y. The product of both functions $T_{ij}(x,y(\xi, \eta)$ and $N^c(\xi, \eta)$ is reduced to a $\frac{1}{r}$ singularity (Extended calculation is related in [5]), for the bidimensional case.

However, due to the collocation position (inside the element), we had to modify the discretization of the displacements and use the same interpolation functions as the ones used for the traction vectors. In this way, collocation positions coincide with displacements nodal value and the integrating scheme previously described may be used. After computing the matrices A^* and B (A^* corresponding to displacements u^* - fig. 3) and in order to keep the compatibility with a possible finite element zone, we have to return to the previous interpolation. This is simply done by expressing the inside nodal values in terms of outline nodal values (see fig. 3)

$$\{u^*\} = [M]\{u\}$$

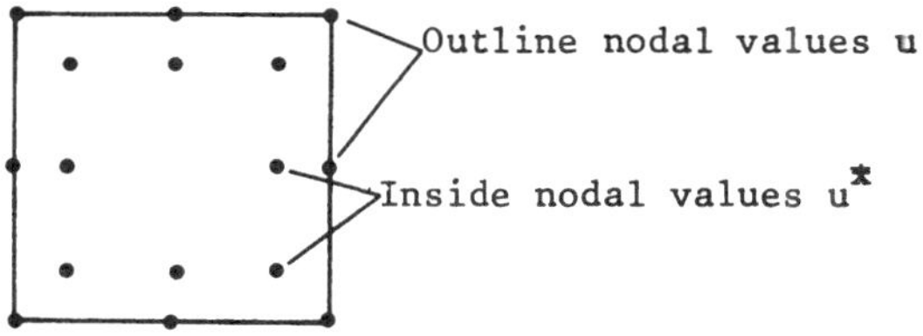

Figure 3 - Position of the displacement nodal values

CONCLUSION

In this paper, we have emphasized the numerical treatment of the symmetrical boundary element method applied to three dimensional elasticity problems.

At present, a computer program is being developed at CETIM and numerical results should be available in a short time. However, the results obtained with the two dimensional program [5] showed a good accuracy. But, due to the computation of the inverse matrix, computing time was considerably increased with respect to a classical formulation.

ACKNOWLEDGEMENTS

This work was carried out with the support of the "Direction des Recherches et Etudes Techniques" . Thanks are due to Prof. O.C. Zienkiewicz, Drs. A. Chaudouet, D.W. Kelly, G.G.W. Mustoe and G. Loubignac for their help and encouragement.

BIBLIOGRAPHY

/1/ Brebbia, C.A., Walker, S., Boundary Element Technique in engineering , Newnes - Butterworths,1980.

/2/ Lachat, J.C., A further development on the Boundary Integral technique for elastostatics, Ph. D. Thesis, University of Southampton, 1975.

/3/ Chaudonneret, M. , On the discontinuity of the stress vector in the boundary integral equation method for elastic analysis. Int. Symp. on recent advances in boundary element methods, Southampton. 1978.

/4/ Mustoe, G.G.W., Volait,F.,A symmetric direct boundary integral equation method for two dimensional elastostatics. Int. Symp. on recent advances in boundary element methods, Southampton. 1980.

/5/ Volait, F., Calcul d'un super élément par une méthode d'équation intégrale. Thèse de 3ème cycle - Université de Technologie de Compiègne. 1980.

/6/ Zienkiewicz, O.C., Kelly, D.W. and Bettess, P. Marriage a la mode - The best of both worlds (finite elements and boundary integrals - Int. Symp. on innovative Numerical analysis in applied engineering Science, Versailles France. (1977)

/7/ Lachat, J.C. and Watson, J.O. Effective numerical treatment of boundary integral equations : A formulation for three - dimensional elastostatics.
Int. J. for num. meth. in eng., Vol 10, 1976.

THE COUPLING OF BOUNDARY AND FINITE ELEMENT METHODS FOR INFINITE DOMAIN PROBLEMS IN ELASTO- PLASTICITY

G. Beer and J.L. Meek

Dept. of Civil Engineering, University of Queensland, Australia

ABSTRACT

The implementation of a coupled analysis capability into an existing Finite Element computer program is discussed. The coupled analysis is then applied to a circular excavation in a infinite domain where the region of plasticity is confined to the Finite Element mesh. Further potential usage of the coupled analysis is then discussed in relation to mine design.

INTRODUCTION

The coupling of Boundary Element and Finite Element methods was first discussed in a general context by Zienkiewicz et al., (1977) although some more specific applications appeared earlier (Chen 1974). The method was applied to a number of field problems and problems in fluid mechanics (Kelly 1979). Application to problems in elasto-statics have appeared more recently (Brebbia 1979 and Mustoe 1980). None of these, however, deal with the 'exterior' problem i.e. one involving an infinite domain. The principle approach of the above methods is to obtain a stiffness matrix of the region which is bounded by Boundary Elements. The stiffness matrix is then made symmetric by an energy approach or by minimising the errors in the non-symmetric terms. This method will be called the Symmetric Direct Boundary Element method.

At the same time as the above developments took place, a completely different approach was presented by Ungless for elasto-static problems and by Bettess for field problems and problems in fluid mechanics. This approach used special Finite Elements which extended to infinity in one or more directions. The functions or displacements were assumed to decay in the infinite direction exponentially or inversely proportional to the distance $(\frac{1}{R})$ from a decay origin. The authors have later shown that a decay of $\frac{1}{R}$ gives good results for underground excavation problems. In contrast with the Boundary Elements, these Elements still require a volume integration which is carried out numerically in a finite mapped region of the infinite domain.

These two approaches will be examined herein in more detail. Both methods have been programmed, incorporated in an existing Finite Element program, and run on a PDP-11 mini computer at the University of Queensland.

SYMMETRIC DIRECT BOUNDARY ELEMENT METHOD

Boundary integral equations

The derivation of Boundary Integral equations for elastostatics is well known (Rizzo 1967) and will not be repeated here. This paper follows the notation by Watson. For the exterior problem, the following integral equation is obtained (see Fig. 1):

$$c_{ij}(x)u_j(x) + \int_S T_{ij}(x,y)u_j(y)\, dSy = \int_S U_{ij}(x,y)t_j(y)\, dSy \quad (1)$$

where

$$c_{ij} = \lim_{\varepsilon\to 0} \int_{S(x,\varepsilon)} T_{ij}(x,y)\, dSy \quad (2)$$

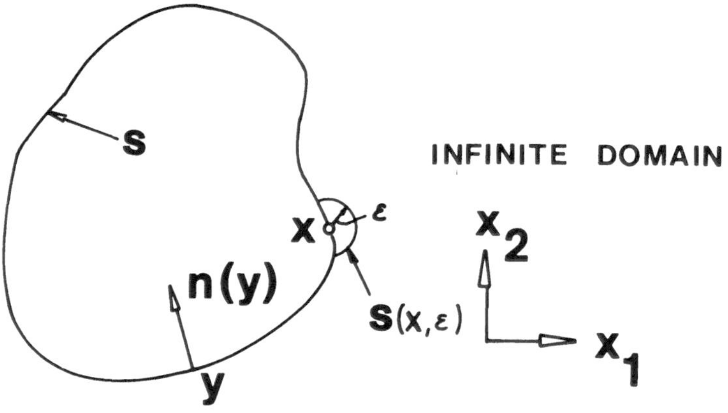

Figure 1. Opening in an infinite domain

In the above equation, x is a point on the surface and y the integration variable; u_i and t_i are the displacements and tractions at the boundary. The kernels $U_{ij}(x,y)$ and $T_{ij}(x,y)$ for the case of plane strain are:

$$U_{ij}(x,y) = C_1 \{C_2 \delta_{ij} \ln(\frac{1}{r}) + \frac{(x_i - y_i)(x_j - y_j)}{r^2}\} \tag{3}$$

and

$$T_{ij}(x,y) = C_3 \cdot \frac{1}{r} \left\{ C_4 \left[n_i(y) \frac{(x_j - y_j)}{r} - n_j(y) \frac{(x_i - y_i)}{r} \right] + \left[C_4 \delta_{ij} + 2 \frac{(x_i - y_i)(x_j - y_j)}{r^2} \right] \cdot n_s(y) \frac{(x_s - y_s)}{r} \right\} \tag{4}$$

where

$$C_1 = \frac{(1+\nu)}{4\pi E(1-\nu)}$$

$$C_2 = 3 - 4\nu$$

$$C_3 = \frac{1}{4\pi(1-\nu)}$$

$$C_4 = 1 - 2\nu$$

In the above, x_i, y_i are the coordinates of points x and y at the surface, r is the distance between the points and $n_i(y)$ is the vector normal to the surface at point y pointing inward; δ_{ij} is the Kroneker Delta.

Discretisation of boundary integral equations

Here a point collocation procedure is used to discretize the Boundary Integral Equation described previously. This means that equation 1 is written for n points along the Boundary. These points are the nodal points of Boundary Elements. For each Boundary Element, the geometry and functions are described by

$$x_i(\xi) = N^k(\xi)\, x_i^k \tag{5}$$

and

$$u_i(\xi) = N^k(\xi)\, u_i^k \tag{6}$$

$$t_i(\xi) = N^k(\xi)\ t_i^k \tag{7}$$

In the above, $N^k(\xi)$ are suitable shape functions of the local (element) coordinate ξ, and the superscript refers to the kth node of an Element.

Two forms of the shape functions will be examined later, namely parabolic and linear. The parabolic function involves 3 element nodes and the serendipity shape function can be written as:

$$N^1(\xi) = \frac{1}{2}\xi(\xi+1) \tag{8}$$

$$N^2(\xi) = 1 - \xi^2 \tag{9}$$

$$N^3(\xi) = \frac{1}{2}\xi(\xi - 1) \tag{10}$$

The linear variation involves only 2 nodes and the corresponding shape functions are:

$$N^1(\xi) = \frac{1}{2}(\xi + 1) \tag{11}$$

$$N^2(\xi) = \frac{1}{2}(\xi - 1) \tag{12}$$

As with Finite Elements, a global numbering system is used for the nodes. Let d(b,c) be the global number of node c of Element b. The discretized form of equation 1 becomes:

$$c_{ij}(x^a)\ u_j\ (x^a) + \sum_{b=1}^{p}\sum_{c=1}^{n(b)} u_j\ (x^{d(b,c)}) \int_{S_b} T_{ij}(x^a,y(\xi))$$

$$N^c(\xi)\ J(\xi)\ d\xi = \sum_{b=1}^{p}\sum_{c=1}^{n(b)} t_j(x^{d(b,c)}) \int_{S_b} U_{ij}(x^a,y(\xi))$$

$$N^c(\xi)\ J(\xi)\ d\xi \tag{13}$$

In Equation 13, the summation is over all elements p and nodal points n(b) of each Element and $J(\xi)$ is the Jacobian of Equation 5.

Numerical Integration of Boundary Elements

Due to the complexity of the integrands, the integration has to be carried out numerically. The following scheme is used:

For d(b,c) ≠ a, i.e. where the point a does not lie within an Element, a 4 point Gauss-Legendre integration for a parabolic shape function and a 2-point scheme for a linear shape function are used.

When d(b,c) = a, the product $U_{ij}(x^a, y(\xi)) \cdot N^c(\xi)$ tends to infinity and requires special attention. For i = j the integrand becomes:

$$\int_{S_b} C_1 \left\{ C_2 \ln\left(\frac{1}{r}\right) + \frac{(x_i - y_i)^2}{r^2} \right\} N^c(\xi) \quad J(\xi) \quad d\xi \qquad (14)$$

The integral

$$\int_{S_b} C \ln\left(\frac{1}{r}\right) N^c(\xi) \quad J(\xi) \quad d\xi \qquad (15)$$

will be examined in more detail. Substituting $\bar{r} = \frac{r}{R}$ where R is the length of Boundary Element b Equation 15 can be rewritten as:

$$\int_{S_b} C\, N^c(\xi) \ln\left(\frac{1}{\bar{r}R}\right) J(\xi)\, d\xi = \int_0^1 C\, N^c(\xi(\bar{r})) \ln\left(\frac{1}{\bar{r}}\right) \frac{\partial \xi}{\partial \bar{r}} J(\xi) \, \overline{dr} + \int_{-1}^{1} C\, N^c(\xi) \ln\left(\frac{1}{R}\right) J(\xi)\, d\xi \qquad (16)$$

The first part of the integral above can now be evaluated numerically using the integration formulae given by Stroud which integrate

$$\int_0^1 \ln\left(\frac{1}{\bar{r}}\right) f(\bar{r}) \, \overline{dr}$$

whereas, for the second part simple Gauss-Legendre is sufficient. Watson has shown that the integrals $T_{ij}(x^a, y(\xi))$ $N^c(\xi)$ and the free term $c_{ij}(x^a)$ need not be evaluated at points d(b,c) = a. The value can be obtained by summing all off-diagonal terms and changing sign. For the exterior problem, a value of δ_{ij} has to be subtracted before changing sign.

Symmetric stiffness matrix of boundary element region

Using matrix notation equation 13 can be rewritten as:

$$[A]\{a\} = \{B\}[b] \tag{17}$$

where matrices [A] and [B] contain the integrals of the kernel functions and the vectors {a} and {b} list the displacements and tractions at all nodes.

It should be mentioned here that at a node where 2 Elements join the node has two different local numbers, for example, $d(1,3) \equiv d(2,1)$. When constructing the matrices [A] and [B] this causes the addition of some terms. Equation 17 can now be solved to give the tractions as a function of the displacements:

$$\{b\} = [B]^{-1}[A]\{a\} = [C]\{a\} \tag{18}$$

To obtain the stiffness matrix of the Boundary Element region, two approaches have been used which the authors believe to be identical. The first approach is by a minimisation of the potential Energy of the system and has been suggested by Kelly and later used by Mustoe. The potential energy of the exterior region is

$$\pi = \frac{1}{2}\int_{\Omega} \{\sigma\}^T \{\varepsilon\} \, d\Omega - \int_{\Gamma} \{u\}^T \{\bar{t}\} \, d\Gamma \tag{19}$$

where $\{\sigma\}$ and $\{\varepsilon\}$ are the stress and strain vectors and $\{\bar{t}\}$ are tractions applied at the boundary. The strain energy term may be transformed into a boundary integral and the total potential energy can be written as

$$\pi = \frac{1}{2}\int_{\Gamma} \{u\}^T \{t\} \, d\Gamma - \int_{\Gamma} \{u\}^T \{\bar{t}\} \, d\Gamma \tag{20}$$

Upon substitution of equation 6 and 7 the discrete form of equation 20 is:

$$\pi = \frac{1}{2}\{a\}^T [M]\{b\} - \{a\}^T \{F\} \tag{21}$$

Where the matrix [M] contains the integrals of shape function products, for example:

$$M_{1,2} = \int_{S_1} N^1(\xi)N^2(\xi)\ J(\xi)\ d\xi \quad \text{etc.} \tag{22}$$

The vector {F} contains integrals of the product of shape functions and traction values, for example:

$$F_1 = \int_{S_1} N^1(\xi)\cdot \bar{t}_1(\xi)\ \ J(\xi)d\xi \quad \text{etc.} \tag{23}$$

Upon substitution of equation 18 relationship 21 can be rewritten as:

$$\pi = \frac{1}{2}\ \{a\}^T\ [M]\ [C]\ \{a\}\ -\ \{a\}^T\{F\} \tag{24}$$

Setting the variation to {a} zero gives

$$\frac{\delta\pi}{\delta\{a\}} = \frac{1}{2}\left\{[M]\ [C]\ +\ ([M]\ [C])^T\right\}\{a\} - \{F\} = 0 \tag{25}$$

where

$$[K] = \frac{1}{2}\left([M]\ [C]\ +\ ([M]\ [C])^T\right) \tag{26}$$

is the stiffness matrix.

In the second approach, Brebbia uses virtual work principles to obtain

$$[K]_{us} = [M]\ [C] \tag{27}$$

which is not in general symmetric.The matrix is made symmetric by the substitution

$$[K]_s = \frac{1}{2}\left([K]_{us} + [K]^T_{us}\right) \tag{28}$$

which is identical to the previous approach.

Coupling of boundary and finite element meshes

Once the stiffness matrix of the Boundary Element region has been found, the analysis of coupled problems proceeds as in the standard Finite Element process. All stiffness matrices are assembled and the system of equations solved for applied forces. Since the stiffness matrix of the Boundary Superelement is symmetric, standard Finite Element software with symmetric equation solvers can be used.

It is needless to say that to ensure compatability between the Finite and Boundary Element meshes, the interpolation functions for the displacements must be identical. Thus, parabolic Boundary Elements should be combined with Finite Elements which have a parabolic variation of the displacements at the connecting boundary although Brebbia apparently obtained good results by coupling constant Boundary Elements with parabolic Finite Elements

Tests on accuracy - choice of shape functions

Tests were made on a circular opening in an infinite medium to determine the accuracy obtainable with the symmetric direct Boundary Element method for infinite domain problems. The virgin stress field was assumed to be $\sigma_H = -0.4$ in horizontal direction and $\sigma_v = -1.0$ in vertical direction. A Poisson's ratio of 0.2 was used. The problem was solved by using equation 17, that is, a conventional Boundary Element method and using equation 25, that is, the symmetric Direct Boundary Element method.

In the first series of runs parabolic Boundary Elements and two different meshes were used. The displacements along the boundary are plotted in Figure 2 together with the theoretical values by Pender (continuous curve). It can be seen that although the conventional Boundary Element method gives excellent results with relatively few parabolic elements, there is an appreciable loss in accuracy where a symmetric stiffness matrix is used. The symmetrising causes minor oscillations of the displacements. No reason could be found except that the stiffness matrix [K] may be not well conditioned. The tests were made on a PDP 11/34 mini computer with a wordlength of 32 bits. Mustoe apparently has not experienced any problems with interior regions using parabolic Elements although he used a CDC 7600 for his analyses. This computer has a wordlength of 60 bits.

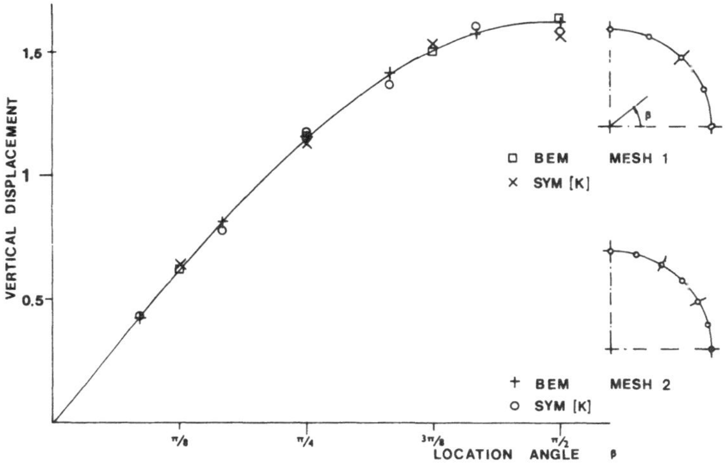

Figure 2. Plot of $u_2 E/\sigma_v$ vs. β along the surface of opening using parabolic Boundary Elements

When using Boundary Elements with a linear variation of displacements and tractions, the problem disappears and results by the conventional and symmetric Boundary Element method are practically identical. This is shown in Figure 3 where 2 different meshes of linear Boundary Elements are used. Obviously, a finer mesh is required to obtain results similar to the ones obtained with parabolic Boundary Elements. It is interesting to note here that Brebbia has also reported no loss in accuracy between conventional Boundary Element method and symmetric Boundary Element method for constant Boundary Elements.

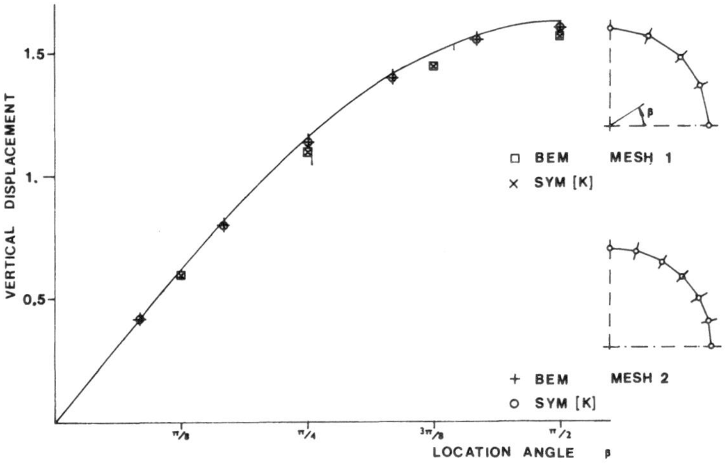

Figure 3. Plot of $u_2\ E/\sigma_v$ vs. β along the surface using linear Boundary Elements

Coupled analysis of circular opening

After the previous tests, it was decided to use linear Boundary Elements and couple them with the variable number of node quadilateral Finite Elements available in the existing program. This Element allows the choice of either a linear or parabolic variation of displacement along an element boundary depending on the presence or omission of the midside node.

The mesh used for the coupled analysis is shown in Figure 4 where also the displacements along the tunnel periphery are shown. It can be seen that there is good agreement with the theory. There is also good agreement with the theoretical stresses.

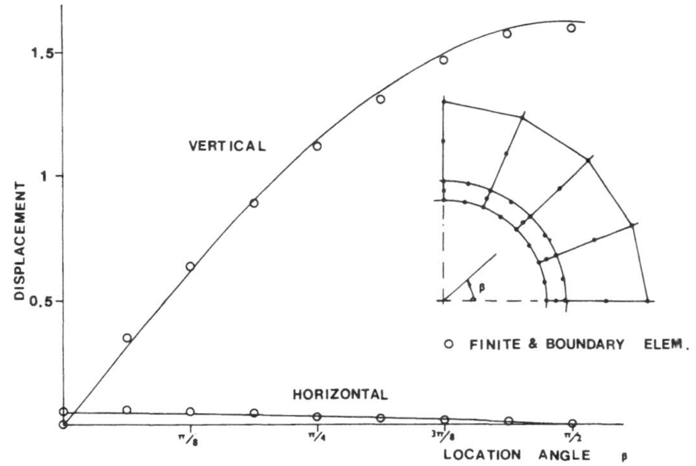

Figure 4. Results of Coupled analysis

INFINITE ELEMENTS

The theory of the Infinite Element has been presented elsewhere (Beer 1980) and here only a brief summary will be given.

The infinite Element is a special Finite Element which is of infinite extension in one direction. The geometry of the element is described by

$$x_i = N^k(\xi_1)\ {}^{\infty}N^k(\xi_2)\ x_i^k \tag{29}$$

where ${}^{\infty}N^k$ are singular shape functions of the local coordinate ξ_2. The variation of the displacement within the Element is

$$u_i = \frac{1}{r} N^k\ (\xi_1)\ u_i^k \tag{30}$$

where r is the distance from a decay origin in this case the centre of the opening.

To obtain the stiffness matrix of the infinite element a volume integration is still needed. The accuracy of the solution now depends not only on the fineness of the mesh in the finite direction but also on the remoteness of the Element from the excavation surface because the variation of displacements is based on a far field solution. Thus, these elements can not in general be used on the free surface. However, recent tests have shown (Beer 1981) that convergence of the element is good and in most cases only 2 or 3 layers of Finite Elements between the surface and the infinite Element are needed to give good results. The main advantage of the infinite Elements is that they are easy to program and that they do not destroy the banded nature of the Finite Element equations thus making the equation solving more efficient. The accuracy and computational efficiency of the two methods is compared in Table 1 for the circular opening.

TABLE 1

	Max. Displacement	Execution time PDP 11/34
Theory	1.632	-
Finite Elements and Boundary Elements	1.6007	6:05
Finite Elements and Infinite Elements	1.6077	3:19

It can be seen that whereas the accuracy is practically the same, the coupled approach requires significantly more computing time.

APPLICATIONS OF COUPLED ANALYSIS

As mentioned in the Introduction, a coupled analysis could be applied to infinite domain problems in elasto-plasticity. Figure 5 shows such as application where a deep circular opening in a visco plastic soil is analysed. The plasticity parameters where such that the region of inelastic deformation was confined to the Finite Element mesh. For this particular example, a similar result can be obtained using infinite Elements.

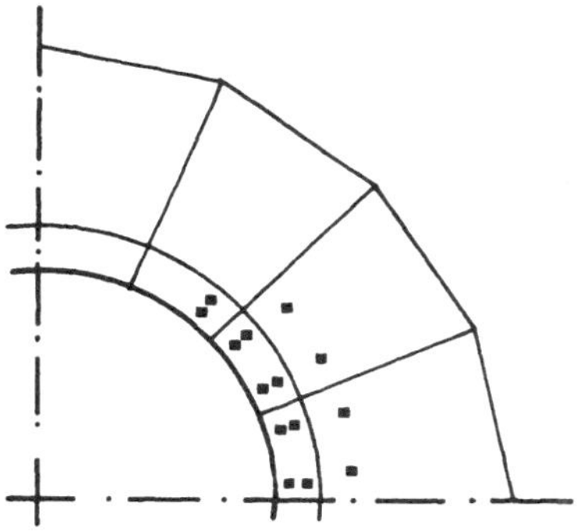

Figure 5. Region of plasticity for angle of friction 40° and cohesion of 0.5

Another application of the coupled analysis which occurs in underground mining is the study of pillar behaviour. Figure 6 shows a typical cross-section of a mine opening. The aim of the study is to observe the non-linear behaviour in the pillar. This is essentially a parameter study and may have to be carried out a large number of times with different material properties, failure criteria and geometry.

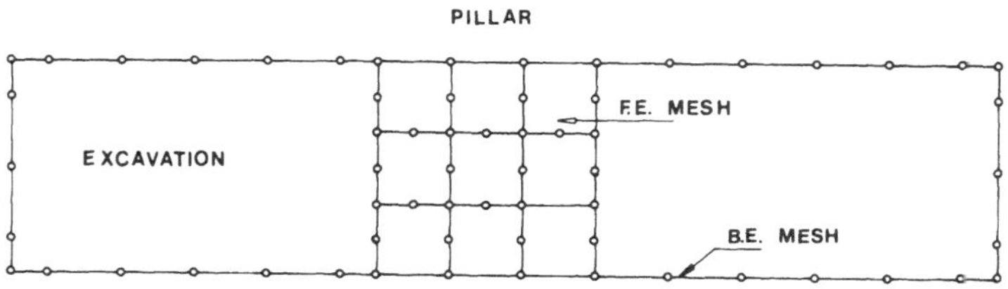

Figure 6. Mine pillar problem

For such a study, it is convenient to discretize the opening into Boundary Elements and the pillar into Finite Elements. There only the stiffness matrix of the nodes connecting to the Finite Element mesh is needed and the symmetrisation process has to be modified slightly. The vector {a} is partitioned into vector $\{a_1\}$ which lists displacements of the nodes not connected to the Finite Element mesh and a vector $\{a_2\}$. Similarly, the vector {b} is partitioned into $\{b_1\}$ and $\{b_2\}$. Equation 17 is then rewritten as:

$$\begin{bmatrix} [A]_1 & [A]_2 \end{bmatrix} \begin{Bmatrix} \{a_1\} \\ \{a_2\} \end{Bmatrix} = \begin{bmatrix} [B]_1 & [B]_2 \end{bmatrix} \begin{Bmatrix} \{b_1\} \\ \{b_2\} \end{Bmatrix} \qquad (31)$$

and rearranged:

$$\begin{bmatrix} [A]_1 & [B]_2 \end{bmatrix} \begin{Bmatrix} \{a_1\} \\ -\{b_2\} \end{Bmatrix} = \begin{bmatrix} [B]_1 & [A]_2 \end{bmatrix} \begin{Bmatrix} \{b_1\} \\ -\{a_2\} \end{Bmatrix} \qquad (32)$$

Multiplying both sides with the inverse of the matrix on the left hand side gives:

$$\begin{Bmatrix} \{a_1\} \\ -\{b_2\} \end{Bmatrix} = [E] \begin{Bmatrix} \{b_1\} \\ -\{a_2\} \end{Bmatrix} \qquad (33)$$

Where the new matrix [E] is the inverse of the matrix on the left hand side, multiplied by the matrix on the right hand side in (32). The tractions on the Boundary Finite Element interface can be computed from (33) by partitioning:

$$\{b_2\} = [E]_{22}\ \{a_2\} - [E]_{21}\ \{b_1\} \qquad (34)$$

In the above $\{b_1\}$ are the (known) tractions at the Boundary Element surface and $\{a_2\}$ the displacement of the Boundary-Finite Element interface. Equation 34 can then be substituted into (21):

$$\pi = \frac{1}{2}\ \{a\}^T\ [\bar{M}] \left([E]_{22}\ \{a_2\} - [E]_{21}\ \{b_1\} \right) \qquad (35)$$

where $[\bar{M}]$ now only involves the Boundary Elements connected to the Finite Element mesh. Minimisation of the potential energy yields:

$$\frac{1}{2}\left[[\bar{M}][E]_{22} + ([\bar{M}][E]_{22})^T\right]\{a_2\} - [\bar{M}][E]_{21}\{b_1\} = 0 \quad (35)$$

The stiffness matrix of the interface nodes is:

$$K = \frac{1}{2}\left[[\bar{M}][E]_{22} + ([\bar{M}][E]_{22})^T\right] \quad (36)$$

and the nodal force vector becomes:

$$\{F\} = [\bar{M}][E]_{21}\{b_1\} \quad (37)$$

CONCLUSIONS

This paper has concentrated on the implementation of a coupled Finite Element Boundary Element program. The attention was focused on the exterior problems of a deep underground excavation although the computer code also allows the treatment of interior domains. Details of implementation of a Symmetric Direct Boundary Element method were discussed and test runs made to assess the accuracy and computing time of a coupled analysis. The implementation of the coupled analysis capability into an existing general purpose Finite Element code only required minor modifications to the program and the addition of 12 Subroutines and a total of 500 Fortran statements.

In the paper, it was also shown that the coupled analysis can be applied successfully to the excavation of a circular tunnel with localised plasticity. It was also shown that a completely different approach using infinite Elements gives similar answers with greater economy. Infinite Elements, however, can not be used on the free surface internal boundary. Thus, potential application of the coupled Finite-Boundary Element analysis is for parameter studies on rock pillars in the mining industry. In this application, the excavation surface is discretized into Boundary Elements and the pillar region into Finite Elements. The Boundary problem is solved giving the stiffness matrix of the elastic rock mass only at the interface nodes. Once this has been done, a large number of non-linear analyses can be carried out (changing the material models, pillar dimensions, properties etc.) on a relatively small system. Work is currently underway in this direction.

ACKNOWLEDGEMENT

The work reported herein was carried out as part of the research project 'Computer Based Excavation Design Techniques for Multiple Orebodies in Jointed Rock' which is sponsored by the Mining Research section of Mount Isa Mines Ltd., Australia.

REFERENCES

Anderson, D.L. and Ungless, R.L. (1979) Infinite Finite Elements. Int. Symposium on Innovative Numerical Analysis in Applied Engineering Science, Versailles, France.

Beer, G. (1981) On the Finite Element Analysis of Underground Excavations. Submitted to Int. Jnl. Anal. and Num. Meth. in Geomechan.

Beer, G. and Meek, L.L. (1980) A Boundary Finite Element for underground mining applications. in New Developments in Boundary Element Methods, CML Publications, Southampton, U.K. pp. 281-294.

Bettess, P. (1977) Infinite Elements. Int. Jnl. Num. Meth. Engng., Vol. 11: 53-64.

Brebbia, C.A. and Georgiou, P. (1979) Combination of Boundary and Finite Elements in elastostatics. Appl. Math. Modelling, Vol. 3.

Chen, H.S. and Mei, C.C. (1974) Oscillations and Wave Forces in a Man-made Harbour. 10th Naval Hydro Symp., Dept. of Civil Eng., M.I.T., Cambridge, U.S.A.

Kelly, D.W., Mustoe, G.G. and Zienkiewicz, O.C. (1979) Coupling Boundary Element methods with other numerical methods. Ch. 10 in Developments in Boundary Element Methods-1, Applied Science Publishers Ltd., London.

Mustoe, G.G. and Volait, F. (1980) A Symmetric Direct Integral Equation method for two-dimensional elastostatics. Paper presented at 2nd Int. Seminar on Boundary Element methods, Southampton.

Pender, M.T. (1980) Elastic Solutions for a deep circular tunnel, Geotechnique XXX, 2: 216-222.

Rizzo, F.J. (1967) An integral equation approach to boundary value problems of classical elasto-statics. Quart. Appl. Math., 25: 83-95.

Stroud, A.J. and Secrest, D. (1966) Gaussian Quadrature Formulas, Prectice-Hall.

Watson, T.O. (1979) Advanced Implementation of the Boundary Element method for two and three dimensional elastostatics. Ch. 3 in Developments in Boundary Element methods-1, Banerjee ed., Applied Science Publ.

Zienkiewicz, O.C., Kelly, D.W. and Bettess, P. (1977) Marriage a la mode - The best of both worlds (Finite Elements and Boundary Integrals). Int. Symp. on Innovative Num. Anal. in Appl. Eng. Science, Versailles, France.

Zienkiewicz, O.C., Kelly, D.W. and Bettess, P. (1977) The Coupling of Finite Element and Boundary Solution Procedures, Int. J. Num. Meth. Engng., Vol. 11: 355-376.

A FINITE ELEMENT - BOUNDARY INTEGRAL SCHEME TO SIMULATE ROCK-EFFECTS ON THE LINER OF AN UNDERGROUND INTERSECTION

B. A. Dendrou[1] and S. A. Dendrou[2]

[1]Agbabian Asociates, El Segundo, California
[2]Camp Dresser & McKee, Annandale, Virginia

INTRODUCTION

The collapsing mechanism of an underground facility in a blocky jointed rock is generally associated with rock wedges in loosened conditions along preexisting surfaces of discontinuities. During the movement of the rock fallout the initial stress field is released, resulting in a new state of equilibrium that affects the liner of the underground opening.

A three-dimensional finite element analysis is generally required to simulate the excavation process and the installation of the liner. The analytical description of the rock fallout occurring after construction of the liner, however, requires either a great computational effort, if one uses "joint" elements, or an over-simplification if only the gravity load of the rock wedge is applied on the liner. A viable alternative is an iterative procedure that progressively loads the finite element discretization of the liner through the use of a rock fallout and that takes into consideration the confinement of the rock medium that results from the excavation process.

In this study the main effort is placed in the coupling of the above finite element and boundary integral procedures, as applied in the study of rock fallouts affecting an underground intersection. The results of this scheme are compared with the results of an oversimplified approach in which the gravity loads of the rock wedge are applied directly on the liner. This reveals that the proposed methodology offers both accuracy and cost efficiency.

FORMULATION OF THE FINITE ELEMENT-BOUNDARY INTEGRAL SCHEME

The possibility of rock fallout following an excavation stage is of primary concern for the designer of the liner of an underground facility. The conventional method consists of assuming that the rock medium around the opening is in a loosening state of stress, which allows large movements of rock blocks along preexisting discontinuous surfaces. These rock blocks are free to transmit all or part of their weight to the supporting liner and affect the redistribution of the stresses in the continuous rock medium. Clearly the rock block represents a collapsing mechanism and the liner should be designed with respect to this ultimate load. This important fact is also recognized by Cording (1976) who provided design recommendations that were based on possible configurations of rock fallout. At the present state of knowledge, location of the initiation of rock failure is not known, and consequently it is not possible to simulate the propagation of failure for a known pattern of discontinuities. Therefore, a possible approach is a postfailure analysis in which the failure surface is assumed known and its effect on the conventional analytical model is estimated. Many investigators, Ghaboussi (1973), Goodmann (1968), and Zienkiewicz (1970) have implemented this concept of slip surfaces in a finite element computer code through the use of element representations. These "joint elements" have met with moderate success because of the required large computational effort resulting from the increase of the degrees of freedom that describe the sliding friction. The proposed technique is conceptually simple and is based on the premises that the slip surface between the rock medium and the rock fallout is a stress free area described by a series of adjacent nodal points. The described solution is obtained by prescribing nodal pressures at the slip nodes in such a manner that the friction law is satisfied. In what follows the governing equations for a finite-element model of a liner and a boundary-integral formulation for the rock-fallout-liner intersection are developed. These equations are then combined to form the proposed iterative scheme.

The finite element equilibrium equation for the liner structure is given, in matrix form as follows

$$\underset{\sim}{K}_S \cdot \underline{u} = \underline{W}_G + \underline{F}_E \qquad (1)$$

where $\underset{\sim}{K}_S$ is the stiffness matrix, $\underline{u}$ is the structural displacement vector, $\underline{W}_G$ is the vector associated with the gravity load of the liner and $\underline{F}_E$ is the force vector associated with the rock fallout action on the liner. The stiffness matrix and vector $\underline{W}_G$ are easily obtained from a Finite Element Solution. The force vector F_E on the other hand is the resultant of two vectors

$$\underline{F}_E = \underline{F}_{RG} - \underline{F}_I \qquad (2)$$

where $\underline{F}_{RG}$ is the load vector for a completely rigid liner, and $\underline{F}_I$ is the correction vector that accounts for the rock/liner interaction. This vector is defined as:

$$\underline{F}_I = \underset{\sim}{T}^T \underset{\sim}{K}_R \underset{\sim}{T} \underline{u} \tag{3}$$

where $\underset{\sim}{K}_R$ is the stiffness of the rock medium and $\underset{\sim}{T}$ is a transformation matrix that selects the structural degrees of freedom used to define the rock/liner interface. Substituting equation (3) in equation (1) we obtain the expression:

$$(\underset{\sim}{K}_S + \underset{\sim}{T}^T \underset{\sim}{K}_R \underset{\sim}{T}) \underline{u} = \underline{W}_S + \underline{F}_{RG} \tag{4}$$

Matrix $\underset{\sim}{K}_R$ and vector $\underline{F}_{RG}$ are now determined through the application of the boundary-integral technique as follows.

The fundamental boundary-integral equation is given by Brebbia (1978). Neglecting body forces and assuming a virtual state (denoted by superscript *) on the domain Ω of the rock medium with a boundary S, and applying the Betti-Maxwell reciprocity principle, the following relation is obtained:

$$\int_S t_i \, u_i^* \, dS = \int_S t_i^* \, u_i \, dS \tag{5}$$

where u_i and t_i are respectively the displacement and traction on boundary S.

Furthermore, if we assume the virtual state to be the fundamental solution state provided by Kelvin's problem then the virtual displacements are given by:

$$u_i^* = \frac{1}{16\,\pi G(1-\nu)r} \left\{ (3-4\nu)\,\delta_{ij} + \frac{\partial r}{\partial x_i} \cdot \frac{\partial r}{\partial x_j} \right\} e_j \tag{6}$$

where r is the distance from the point of application of a unit load to the point of evaluation of the virtual displacement. G is the shear modulus of elasticity and ν is Poisson's ratio of the rock medium, δ_{ij} is Kronecker's delta and e_j the unit vector in the direction of the unit load.

Similarly, the tractions at a point on the surface with a normal direction η are given by:

$$t_i^* = \frac{1}{8\pi(1-\nu)r^2} \left\{ \frac{\partial r}{\partial \eta} \left[(1-2\nu)\,\delta_{ij} + 3\,\frac{\partial r}{\partial x_i} \frac{\partial r}{\partial x_j} \right] \right.$$

$$\left. + (1-2\nu)\left(\eta_j \frac{\partial r}{\partial x_i} - \eta_i \frac{\partial r}{\partial x_j}\right) \right\} e_j \tag{7}$$

Through the division or rock's external boundaries into a series of boundary elements, equation (5) may be expressed in matrix notation as:

$$\underset{\sim}{H}\ \underline{u} = \underset{\sim}{G}\ \underline{t} \tag{8}$$

In the present implementation the displacement and traction shape functions are assumed to be constant over each boundary element. Once the matrices in equation (8) are computed, the rock stiffness matrix $\underset{\sim}{K}_R$ can be obtained from the following expression:

$$\underset{\sim}{K}_R\ \underline{u} = \underset{\sim}{G}^{-1}\ \underset{\sim}{H} \cdot \underline{u} \tag{9}$$

Attention should be given to the fact that the derived stiffness matrix is not symmetric and should be symmetrized in order to use it in expression (4).

Only a brief presentation of the boundary integral formulation is given here since an extensive coverage of the subject is available in the literature. Rather, in what follows the effort is placed in the description of the iterative procedure adopted for the simulation of the rock fallout. If Ω_2 represents the rock medium and Ω_1 represents the rock fallout (Figure 4), then our main concern is the description of the sliding of Ω_1 with respect to Ω_2 during the loosening process. The phenomenon of sliding friction is not well understood at the present time but the adopted friction law is representative of many experimental observations, Jaeger (1971).

Two are the important observations: (a) the frictional resistance of two surfaces in contact is proportional to the normal force across the friction surface, and (b) the "kinetic" friction is smaller than the "static" friction. In this application the former case is considered. Then the frictional resistance prior to slippage obeys the following relationship

$$f_T = k\ f_N \tag{10}$$

where f_T is the tangential force acting at a particular slip node f_N is the normal force acting at a particular slip node and k is the coefficient of friction.

In a typical computational procedure, the slip nodes are first assumed to be attached to the adjacent material surface. The inception of slippage occurs when the tangential force exceeds the frictional resistance given in expression (10). This is expressed as

$$|f_T^M| > f_T \tag{11}$$

where f_T^M is the force existing on the rock medium, Figure 4. After sliding begins the adhesion of the frictional surface is lost and at this point the tangential force acting at the slip node is given by

$$|f_T^M| = \alpha\ k\ f_N^M \tag{12}$$

where α equals $\pm$ 1, so as to ensure that the frictional force acts always in a direction opposite to that of slippage. Application of this frictional force implies a new equilibrium state in which adjacent slip nodes can move only in the tangential direction as shown in Figure 4. The new equilibrium state is identified using the boundary integral technique and the compatibility of the displacements of the slip nodes in the normal direction is assured by a correction displacement vector that has to be incorporated in the subsequent computational step. In a formal way, the above methodology is presented by the following iterative steps:

(i) Define the boundaries of the rock fallout and compute tangential and normal forces acting on the boundaries based on the computed stress field of the three-dimensional finite element analysis.

(ii) Identify boundary elements where frictional resistance is exceeded and apply corresponding tangential forces.

(iii) Compute loading vector and displacements of the liner according to expression (4).

(iv) Compute the displacements of the slip nodes resulting from the tangential forces and the response of the liner.

(v) Check the relative normal displacements of the slip nodes. If compatibility is violated impose a correction loading vector and return to step (iii).

APPLICATION OF THE PROPOSED SCHEME-DISCUSSION AND RESULTS

The case of an underground intersection and in particular the study of the rock-fallout effects on the liner at the intersection are used as an application of the above described methodology. Two are the main issues associated with the application of the above methodology, namely (a) the evaluation of the stress field of the rock medium at the final excavation stage, before initiation of the loosening of the rock fallout, and (b) the comparison of the proposed approach with other simplified methods.

The initial stress field for the application of the finite-element Boundary Integral method is evaluated by a three-dimensional finite element BMINES code, Agbabian (1976). The three-dimensional model of the intersection illustrated in Figure 1 includes 490 brick elements. Thirteen excavation stages disposed in three consecutive horizontal layers are used to simulate the removal of the rock. The liner is modeled with a grid of beams that are adequately calibrated to respond in a similar way as the real liner made of shotcrete and steel ribs.

The corresponding vertical stress fields at stages 8, 9, and 13, and for the diagonal section shown in Figure 1, are illustrated in Figure 2. The stress field at stage 13 is retained as the initial stress field of the rock fallout study. Interestingly, it can be observed that loosening conditions are seen to develop close to the base of the rock wedge, Figure 2, stage 13. The geometric configuration of the assumed failure surface that defines the rock wedge is given in Figure 3, a. Investigation of the effect of the above defined rock fallout on the liner requires the retention of a smaller domain of analysis as shown in Figure 4. If a finite element model is used in conjunction with the simulation of the relative displacements of the rock wedge with respect to the rock medium using "joint elements," an estimated 1200 degrees of freedom would be necessary to describe accurately the physical phenomenon. Retaining the same discretization of the boundaries as in the finite element approach, but using the boundary integral formulation, reduces the degrees of freedom to 180, resulting in a more tractable analysis.

The initial stresses and displacements of the boundaries of the domains Ω_1 and Ω_2 are obtained by the previously described finite element model. This requires compatibility between domains Ω_1 and Ω_2 at the interfacing boundary, achieved by using spatial 1-D elements. Sliding between domains Ω_1 and Ω_2 is modeled to be initiated at the base of the interface surface, and is simulated by releasing the tangential frictional resistance, Figure 4. The resulting displacements at the interface between Ω_1 and Ω_2 obtained by the boundary integral model used assuming a rigid liner are shown in Table 1. The equality between normal and tangential stresses in iteration 1, Table 1, is due to the 45° slope of the assumed discontinuity surface. The displacement fields in Ω_1 and Ω_2 are seen to be incompatible along the interface. Relaxing the assumption of rigidity of the liner, the vertical displacement field of the liner is obtained by use of a simplified finite element discretization of the structural components of the liner, namely the steel ribs and the shotcrete. This technique was originally applied by Brierley (1975). In a second iteration, the incompatibilities of the normal displacements at the interfacing boundary are removed resulting in an adjusted tangential displacement field and normal force field along the interface, Table 1. Reduction of the normal force incompatibilities indicate smooth convergence of the iterative scheme. In iteration 2, incremental tangential stresses are computed as resulting from the adjustment of the relative normal displacement, Table 1. To this displacement field corresponds a new set of displacements of the liner shown in Table 2. The displacement field of the final iteration is added to the displacement field obtained from the finite element model of the excavation process. This total displacement field corresponds to the stresses of the steel ribs and the stresses of the shotcrete shown in Table 4.

The above results are compared to a simplified model in which the action of the rock fallout is simply simulated by the action of the gravity load. This comparison is shown in Tables 2 and 3. Interestingly, the difference between the above two approaches is larger at the springline of the intersection, the simplified model showing smaller displacements at that location. Thus, the proposed Finite Element Boundary Integral method can produce a more realistic analysis. However, attention should be given to the importance of the initiation of the loosening conditions. In the above application, loosening was assumed to be initiated at the springline of the liner.

In fact many causes can initiate loosening conditions, such as, for example, the alteration of physical properties of the rock mass due to the presence of water. Finally, concerning the slippage conditions at the discontinuity surface, kinetic friction and dynamic effects ought to be included in the model if blasting operations are to be conducted at a nearby construction site.

ACKNOWLEDGMENTS

The investigation reported above was partially supported by NSF, grant No. DAR76-80044.

Agbabian Assoc. (1976) User's Guide for a Computer Program for Analytical Modeling of Rock/Structure Interaction, U-7638-3-4183. El Segundo, CA:AA, Jun.

Brebbia, C.A. (1978) "The Boundary Element Method for Engineers," Pentech Press.

Brierley, G.S. (1975) "The Performance During Construction of the Liner for a Large, Shallow Underground Opening in the Rock." Ph.D. Dissertation, Univ. of Illinois, Urbana.

Brown, E.T. and Hocking, G. (1976) "The Use of Three Dimensional Boundary Integral Equation Method for Determining Stresses at Tunnel Intersections," in Proc. 2nd Australian Tunnelling Conf. Melbourne, 1976, pp. 55-64.

Cording, E.J.; Mathews, A.A.; and Peck, R.B. (1976) Design Criteria for Permanent Structural Linings for Station Excavations in Rock, Washington Metropolitan Area Transit Authority. Unpublished draft report prepared for DeLeuw, Cather and Co., General Engineering Consultant, Washington Metropolitan Area Transit Authority, Washington, D.C., Jun.

Cruse, T.A. and Rizzo, F.J., eds. (1975) Boundary Integral Equation Method: Computational Applications in Applied Mechanics, AMD-Vol. 11, ASME, New York.

Ghaboussi, J.; Wilson, E.L.; and Isenberg, J. (1973) "Finite Element for Rock Joints and Interfaces," J. Soil Mech. Found. Div., ASCE Vol. 99, No. SM10.

Goodman, R.E.; Taylor, R.L.; and Brekke, T.L. (1968) "A Model for the Mechanics of Jointed Rock," J. Soil Mechanics and Foundations Div., ASCE, Vol. 94, No. SM3.

Jaeger, J.C. (1971) "Friction of Rocks and Stability of Rock Slopes," Geotechnique, Vol. 21.

Zienkiewicz, O.C. et al. (1970) "Analysis of Nonlinear Problems in Rock Mechanics with Particular Reference to Jointed Rock Systems," Proc: 2nd Cong. of Int. Soc. for Rock Mechn., Belgrade.

Table 1. Computed Displacements and Pressures at the Interface Between Ω_1 and Ω_2

		Normal Direction						Tangential Direction					
		Normal Displacement [in.]			Normal Traction [psi]			Tangential Displacement [in.]			Tangential Traction [psi]		
Iterations	Locations*	Ω_1	Ω_2	Relative Value	Ω_1	Ω_2	Relative Value	Ω_1	Ω_2	Relative Value	Ω_1	Ω_2	Relative Value
					Given						Given		
1	a	-.0124	-.0113	.0011	50	-50		.0800	.0908	.0108	-50	+50	
	b	-.0130	-.0117	.0013	50	-40		.0935	.0974	.0039	-40	+40	
	c	-.0132	-.0107	.0025	35	-35		.0970	.0978	.0009	-35	+35	
		Given											
2	a	.0005	-.0005		3.3	-3.5	.20				0.8	-0.7	
	b	.0006	-.0006		4.	-4.3	.30				1.2	-1.	
	c	.0012	-.0012		5.	-5.2	.20				1.3	-1.5	

*Locations are indicated in Figure 4.

Table 2. Computed Vertical Displacements (in inches)

Steel Ribs	Nodes	Displ. at Final Excavation	F.E. Gravity Model		B.I. - F.E. Model	
			Partial Displ.	Total Displ.	Partial Displ.	Total Displ.
	1	-.076	-.0137	-.0897	-.014	-.090
3*	2	-.07	+.0074	-.064	+.0082	-.078
	3	-.082	+.0119	-.071	-.00203	-.084
	4	-.096	-.0024	-.098	-.003	-.099
	1	-.08	-.0182	-.0982	-.026	-.106
4	2	-.068	-.0208	-.0888	-.029	-.097
	3	-.073	-.0018	-.074	-.0025	-.075
	4	-.099	+.0023	-.101	-.003	-.132
	1	-.085	-.0530	-.138	-.054	-.139
5	2	-.017	-.0991	-.170	-.098	-.169
	3	-.073	-.0465	-.119	-.049	-.122
	4	-.09	-.0104	-.100	-.015	-.105
	1	-.085	-.0543	-.139	-.055	-.140
6	2	-.07	-.0771	-.149	-.078	-.148
	3	-.075	-.0325	-.107	-.035	-.110
	4	-.094	-.0098	-.103	-.012	-.106

*Numbers of steel ribs are given in Figure 5.

Table 3. Compressive Stresses in Steel Ribs (in psi)

Steel Ribs	Members	Final Excavation		F.E. Gravity Model		B.I. - F.E. Model	
		Strains	Stresses	Strains	Stresses	Strains	Stresses
	1	1.53 E-3	46.10	2.43 E-3	70.0	2.50 E-3	72.0
3*	2	1.57 E-3	45.70	2.42 E-3	70.0	2.48 E-3	72.0
	3	1.47 E-3	42.70	1.51 E-3	44.0	2.0 E-3	58.0
	1	1.61 E-3	46.6	2.40 E-3	69.0	2.45 E-3	70.0
4	2	1.62 E-3	47.0	2.39 E-3	69.0	2.50 E-3	71.0
	3	1.52 E-3	44.10	2.23 E-3	65.0	2.48 E-3	72.0
	1	1.23 E-3	35.50	1.64 E-3	47.0	1.7 E-3	49.0
5	2	1.36 E-3	39.40	2.39 E-3	69.0	2.4 E-3	69.0
	3	0.88 E-3	25.64	2.3 E-3	67.0	2.4 E-3	70.0
	1	2.3 E-3	67.28	0.89 E-3	26.0	1.0 E-3	29.0
6	2	2.31 E-3	67.10	1.25 E-3	36.0	1.28 E-3	38.0
	3	0.55 E-3	16.10	1.36 E-3	39.0	1.4 E-3	40.0
	1	0.83 E-3	24.10	1.29 E-3	37.0	1.3 E-3	38.0
7	2	0.67 E-3	19.37	0.79 E-3	23.0	0.8 E-3	23.0
	3	1.01 E-3	29.50	1.08 E-3	31.0	1.42 E-3	32.0

*Numbers of steel ribs are given in Figure 5.

Table 4. Stresses in Shotcrete at Four Gaussian Points (in psi) for Panel A Shown in Figure 5

	Gauss Point	σ_x	σ_y	τ_{xy}	Principal Stresses σ_1	σ_2
Final Excavation	1	-666.0	-327.	895.	-1407.0	419.0
	2	-1130.0	-316.	666.	-1504.0	58.0
	3	-444.0	-340.	190.	-1099.0	345.0
	4	-471.0	-353.	283.	-918.0	94.0
F.E. Gravity Model	1	-691.0	70.0	371	-842	222
	2	-1150.0	-10.0	33	-1151	9
	3	-467.0	51.0	745	-997	581
	4	-486.0	-96.0	482	-881	229
B.I. - F.E. Model	1	-670.0	-108.0	700		
	2	-1140.0	-60.0	350		
	3	-470.0	-150.0	260		
	4	-490.0	-408.0	200		

(-) sign indicates compression.

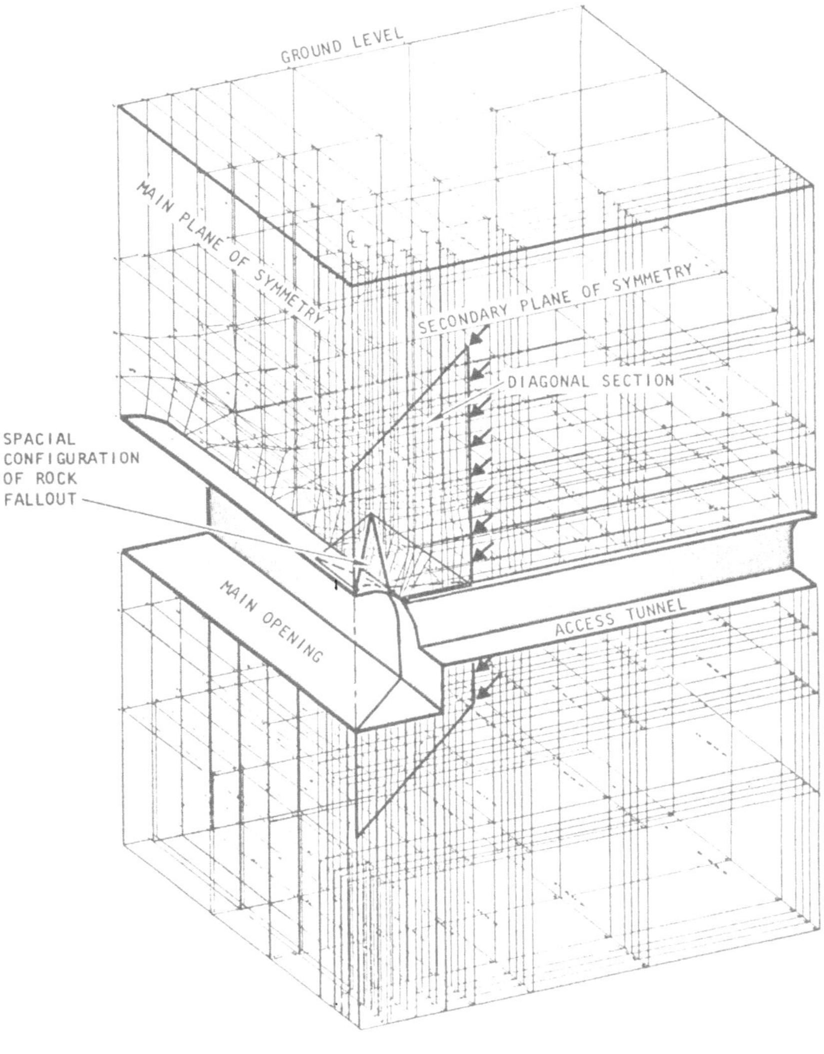

FIGURE 1. THREE-DIMENSIONAL CONFIGURATION OF POSSIBLE ROCK FALLOUT AT INTERSECTION WITH RESPECT TO FINITE ELEMENT DISCRETIZATION

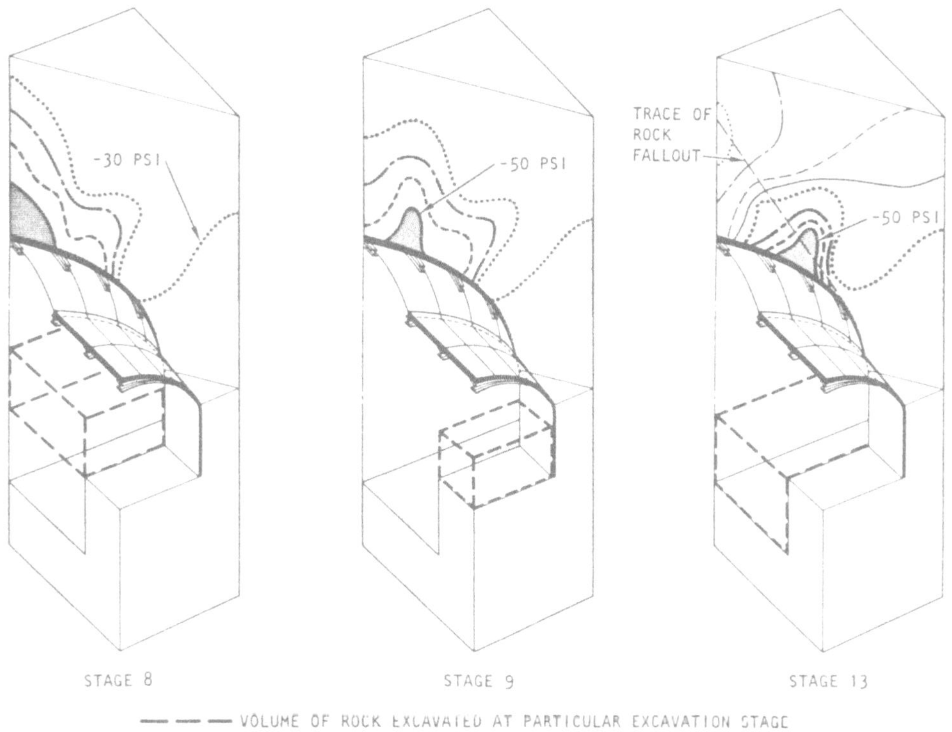

FIGURE 2. FINITE ELEMENT ESTIMATION OF VERTICAL STRESS DISTRIBUTION IN ROCK MEDIUM AT DIAGONAL SECTION OF INTERSECTION

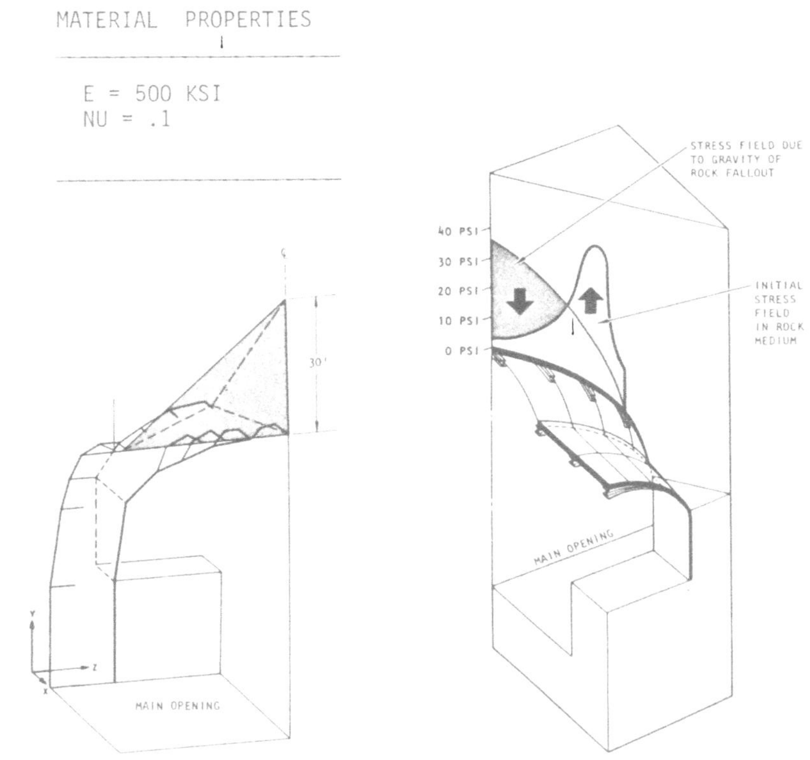

(a) Geometric configuration of rock fallout

(b) Stress field applied to liner

FIGURE 3. GEOMETRIC CONFIGURATION AND INITIAL VERTICAL STRESS OF ASSUMED ROCK FALLOUT

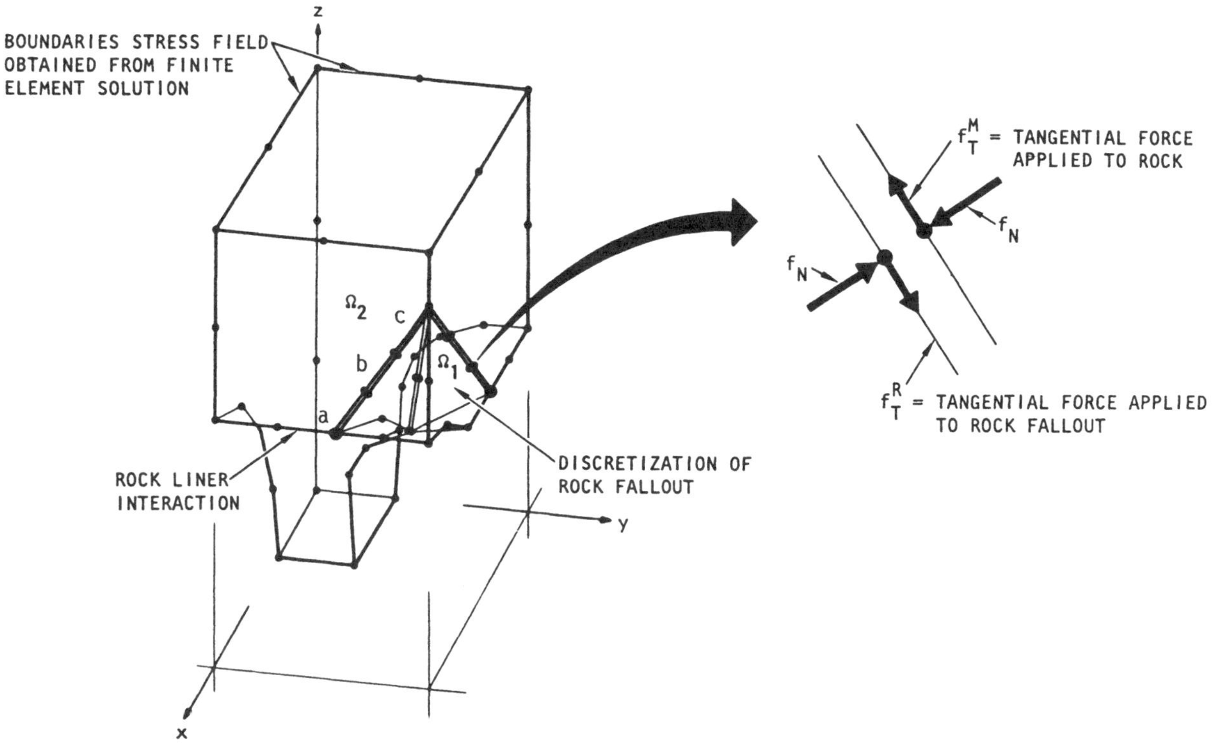

FIGURE 4. DISCRETIZATION OF BOUNDRIES RETAINED FOR STUDY OF ROCK FALLOUT EFFECTS

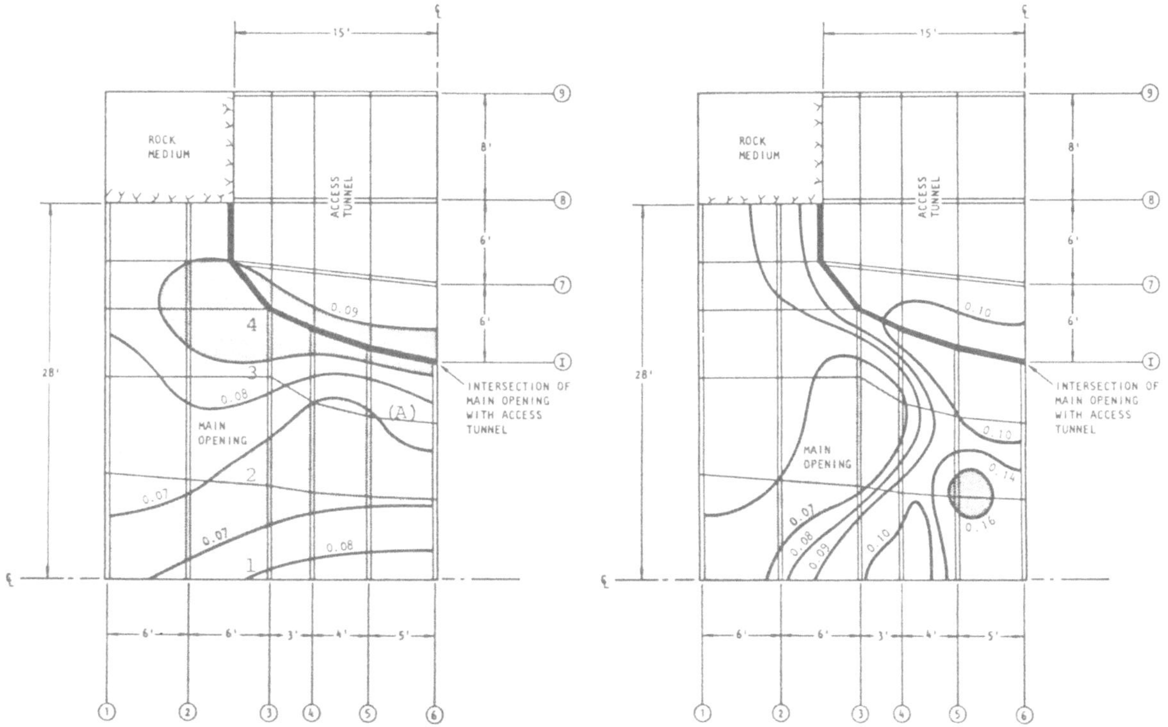

(a) Displacements at excavation stage 13

(b) Displacements due to rock fallout

FIGURE 5. CONTOURS OF COMPUTED DISPLACEMENTS AT UNDERGROUND INTERSECTION

THE USE OF GREEN'S FUNCTIONS IN THE NUMERICAL ANALYSIS OF POTENTIAL, ELASTIC AND PLATE BENDING PROBLEMS

C. Katz
Technical University Munich, Germany

INTRODUCTION

Neither the Boundary Element Method (BEM) nor the Finite Element Method (FEM) is the unique technique to solve engineering problems. While the first method satisfies the differential equation exactly and approximates the boundary conditions, the latter does the reverse. This implies in general that a solution obtained by the BEM will not fulfil global conditions as equilibrium or balance of fluxes in potential theory.

But this is not the only reason to link the BEM and the FEM. An effective computer code exists for the FEM, but on the other hand, the modelling of infinite regions is difficult in FEM and the advantage of the BEM to discretize the boundary only follows the thinking of practical engineers.

Linking procedures will in general try to calculate a stiffness matrix (symmetric) using the BEM. Zienkiewicz et al. (1977) showed a way to do this by using a variational principle and the direct integral equation approach. Their method has two disadvantages, first it implies the inversion of a fully populated nonsymmetric matrix, second the independent interpolation of displacements and stresses or potential and fluxes introduces some errors.

The next step in a new direction was taken by Margulies (1980). He showed that using the Green's function for a special region a symmetric stiffness matrix can be calculated. This approach however has the handicap that this particular Green's function has to be known and the severe restriction that the kernels of his equations are strongly singular and any numerical integration is impossible for the diagonal coefficients.

This paper pushes forward Margulies' approach and shows some new ways to get a stiffness matrix by the BEM.

THE METHOD OF MARGULIES

Margulies started with the fundamental equation of the BEM which gives the values of the potential in an observation point z_o by the boundary values $\phi(z)$ and $\frac{\partial \phi(z)}{\partial n}$.

$$\phi(z_o) = \oint_{z\in\Gamma} G(z,z_o)\cdot \frac{\partial \phi(z)}{\partial n}\, ds - \oint_{z\in\Gamma} \frac{\partial G(z,z_o)}{\partial n}\cdot \phi(z)\, ds \qquad (1)$$

In general G is the free space Green's function which is the potential due to a unit source in z and is symmetric with respect to z and z_o. Changing G to the real Green's function, that is the solution of the unit source in z imposed to the boundary condition

$$G(z,z_o) = 0 \qquad \text{for } z \text{ or } z_o \in \Gamma \qquad (2)$$

equation (1) simplifies to

$$\phi(z_o) = -\oint_{z\in\Gamma} \frac{\partial G(z,z_o)}{\partial n}\cdot \phi(z)\, ds \qquad (3)$$

which is an explicit representation for the solution given by the boundary values $\phi(z)$.

Using this expression for the variational formulation of the problem which is for an interior problem

$$\pi = \frac{1}{2}\int_{z_o\in\Gamma_o} \phi(z_o)\cdot \frac{\partial \phi(z_o)}{\partial n_o}\, ds_o \qquad (4)$$

we obtain

$$\pi = -\frac{1}{2}\oint_{z_o\in\Gamma_o} \phi(z_o)\cdot \oint_{z\in\Gamma} \frac{\partial^2 G(z_o,z)}{\partial n \partial n_o}\cdot \phi(z)\, ds\, ds_o \qquad (5)$$

and using the interpolation trial functions as usual for ϕ

$$\pi = -\frac{1}{2}\oint_{\Gamma_o} N_i(z_o)\cdot\phi_i \oint_{\Gamma} \frac{\partial^2 G(z_o,z)}{\partial n \partial n}\cdot N_j(z)\cdot\phi_j\, ds\, ds_o \qquad (6)$$

From this this the symmetric stiffness matrix is deduced

$$a_{ij} = - \oint_{\Gamma_o} \oint_{\Gamma} N_i(z_o) \cdot \frac{\partial^2 G(z_o,z)}{\partial n \partial n_o} N_j(z) \, ds \, ds_o \tag{7}$$

Margulies has calculated the stiffness matrix for the cases when Γ is a circle or a sphere and for constant or linear trial functions. However, the integration of the diagonal coefficients has to be performed analytically or by some additional constraints. In the following I restrict myself to the potential problem. The extension to elastic problems will be touched on shortly afterwards and will be reported in detail elsewhere.

FIRST EXTENSION: GREEN'S FUNCTIONS FOR OTHER REGIONS

Though the Green's Function for the circle of radius a is

$$G = - \frac{1}{4\pi} \ln \frac{(r/a)^2 + (a/r_o)^2 - 2(\frac{r}{r_o}) \cos(\alpha - \alpha_o)}{1 + (r/r_o)^2 - 2(\frac{r}{r_o}) \cos(\alpha - \alpha_o)} \tag{8}$$

and the derivatives can be calculated from there, it is much easier to use the complex representation. Each harmonic function is the real part of an analytic function. For the free space function the Green's function is the real part of

$$GG(z,z_o) = \frac{-1}{2\pi} \ln (z - z_o) \tag{9}$$

The imaginary part of (9) is a harmonic function too and is called the stream function.

To obtain the analytic Green's function for the circle three components are needed

- a source in z_o in the interior of the circle
- a sink in the inverse point $a^2 / \overline{z_o}$
- a constant factor to shift the potential $\ln (\overline{z_o}/a)$

The complex Green's function is then given by

$$GG = \frac{-1}{2\pi} \ln \frac{a(z - z_o)}{(a^2 - z\overline{z}_o)} \tag{10}$$

The real part of (10) is

$$G = \frac{-1}{4\pi} \ln \frac{a^2(z - z_o)\cdot(\overline{z} - \overline{z}_o)}{(a^2 - z\overline{z}_o)\cdot(a^2 - \overline{z}z_o)} \tag{11}$$

and is equivalent to (8).

It is worth noting that z is a complex variable x + i y while $\bar{z}$ is x - i y but is independent of z when calculating complex derivatives.

Similar the complex Green's function for the half plane is given by

$$GG = -\frac{1}{2\pi} \ln \frac{z - z_o}{z - \bar{z}_o} \qquad (12)$$

For other simply connected regions the complex Green's function is given by the function for the unit circle or half plane and the conformal mapping from the region to the circle or half plane. Let ζ represent a point in the mapped plane, z in the original one, then the conformal mapping is

$$\zeta = \omega(z) \qquad (13)$$

and for the complex Green's function the expression holds:

$$GG_z(z, z_o) = GG_\zeta(\omega(z), \omega(z_o)) \qquad (14)$$

The proof can be found in Kantorowitsch/Krylow (1964).

If the region to be mapped is the unit square, the conformal mapping to the unit circle is given by Gaier (1964)

$$\zeta = a\cdot z + \frac{a^5}{10} \cdot z^5 + \frac{a^9}{120} \cdot z^9 + \frac{11a^{13}}{15600} \cdot z^{13} + \cdots$$

$$a = 0.92703733865 \qquad (15)$$

The mapping is of high quality as is shown in the following figure

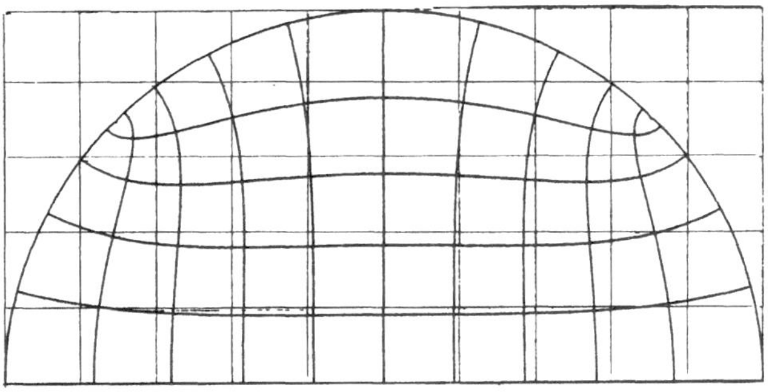

Figure I Conformal mapping of the unit square (15) using 4 coefficients

This function is used to calculate the Green's function for the unit square. The following figure gives the lines of equal potential for this function.

Figure II Equipotentials, Green's function for a square

To calculate the derivatives the following expressions can be used

$$2 \cdot \frac{\partial G}{\partial z} = \frac{\partial G}{\partial x} - i \cdot \frac{\partial G}{\partial y} \qquad (16)$$

$$2 \cdot \frac{\partial G}{\partial \bar{z}} = \frac{\partial G}{\partial x} + i \cdot \frac{\partial G}{\partial y} \qquad (17)$$

when the normal and tangent of the boundary are given by

$$n = e^{i\theta} = \cos\theta + i \cdot \sin\theta \qquad (18)$$
$$s = i \cdot n$$

the derivatives in the direction of n and s will be

$$\frac{\partial G}{\partial n} = e^{-i\theta} \frac{\partial G}{\partial \bar{z}} + e^{i\theta} \frac{\partial G}{\partial z} \qquad (19)$$

$$i \frac{\partial G}{\partial s} = e^{-i\theta} \frac{\partial G}{\partial \bar{z}} - e^{i\theta} \frac{\partial G}{\partial z} \qquad (20)$$

and from equation (19) follows

$$\frac{\partial^2 G}{\partial n \partial n_o} = e^{i\theta} e^{i\theta o} \frac{\partial^2 G}{\partial z_o \partial z} + e^{-i\theta} e^{-i\theta o} \frac{\partial^2 G}{\partial \bar{z}_o \partial \bar{z}}$$

$$= e^{-i\theta} e^{i\theta o} \frac{\partial^2 G}{\partial z_o \partial \bar{z}} + e^{i\theta} e^{-i\theta o} \frac{\partial^2 G}{\partial \bar{z}_o \partial z} \qquad (21)$$

SECOND EXTENSION: NUMERICAL EVALUATION OF THE INTEGRALS

Though the kernel of the stiffness matrix is now known for all regions which can be mapped on the unit circle or the half plane the numerical evaluation of integral (7) is impossible. The original Green's function is of logarithmic order, the second derivative is of order $(1/r^2)$ and this can not be integrated in a double integral without special tricks.

This is a problem which seems spurious because there is no physical reason for it. Indeed the difficulties come from the singularity of the unit potential on the boundary, while the potential in nature has no jumps at all. In the following figures the flow-field for a unit potential and the real potential on the boundary is shown. It is evident that in the former case the fluxes will be infinite, while the latter case has finite fluxes.

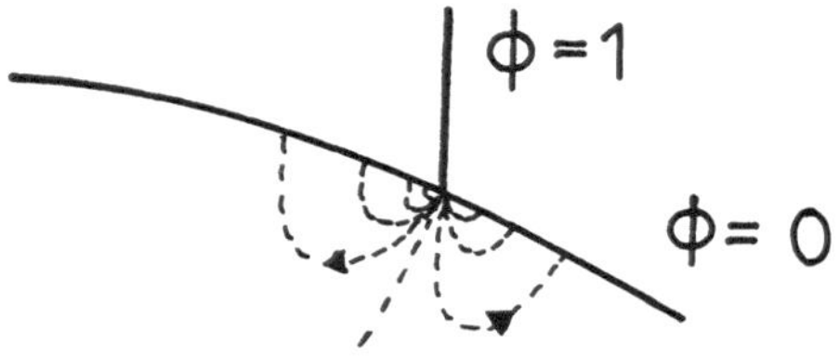

Figure III Unit Potential

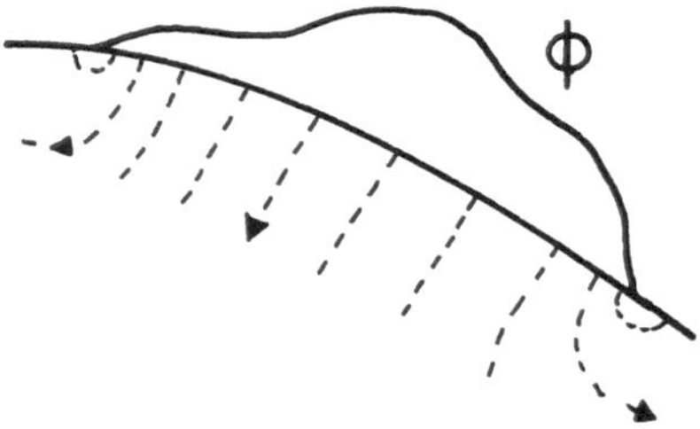

Figure IV Distributed Potential (real case)

Moreover, the nature of the trial functions is that they are bounded in some parts of the boundary and vanish on the rest of the boundary.

All this leads to the idea of integrating by parts. This is first performed on equation (3) which gives the value of ϕ at z_o.

$$\phi(z_O) = - \oint_\Gamma \frac{\partial G(z,z_O)}{\partial n} \cdot \phi(z)\, ds$$

$$= - \int \frac{\partial G}{\partial n} ds \cdot \phi(z) + \oint_\Gamma \cdot \frac{\partial \phi(z)}{\partial s} \cdot \int \frac{\partial G}{\partial n} ds\, ds \qquad (22)$$

The first part will vanish if we integrate in a closed contour or if we consider only one trial form. However, there might be some additional values due to the multi-valued functions but this is discussed later. So we have the new function

$$G^*(z,z_o) = \int \frac{\partial G(z,z_O)}{\partial n} ds \qquad (23)$$

which I call the <u>Green's function of the second kind</u>. With this function the expression for an interior value will be

$$\phi(z_O) = \oint_\Gamma \frac{\partial \phi(z)}{\partial s} \cdot G^*\, ds - \phi(z1) \cdot \oint_\Gamma G^*\, ds$$

$$= \oint_\Gamma \frac{\partial Ni}{\partial s} \cdot G^*\, ds \cdot \phi_i + \phi_o \qquad (24)$$

Similar a second partial integration is performed on the other part of the boundary Γ_o. This leads to the expression for the stiffness matrix

$$a_{ij} = - \oint_\Gamma \oint_\Gamma \frac{\partial Ni}{\partial s_i} \cdot G^{**}(z,z_O) \cdot \frac{\partial Ni}{\partial s_j} ds_i\, ds_j \qquad (25)$$

in which

$$G^{**}(z,z_o) = \iint \frac{\partial^2 G(z,z_O)}{\partial n \partial n_O} ds\, ds_o \qquad (26)$$

is called the <u>Green's function of the third kind</u>.

Both new functions remain of logarithmic order and can, therefore, be integrated easily using any quadrature formula. As one can see, a constant potential gives no value to (24) than ϕ_o and not any to the stiffness matrix. Indeed this means that the row-sums of the stiffness matrix are exactly zero. This even holds for each stage of the integration because the feature of the trial functions (their sum is one in every point) yields that their sum of the derivatives will vanish in every point. So far this approach seems very easy. To get the Green's functions of the second or third kind, a normal complex integration would be very cumbersome. But there is some more physical meaning in these functions.

To show this, let us first assume that our Green's function of the first kind G can be represented as the sum of four analytic functions as follows

$$G(z,z_o) = \Phi(z,z_o) + \overline{\Phi(z,z_o)} + \psi(z,\bar{z}_o) + \bar{\psi}(\bar{z},z_o) \qquad (27)$$

This is true in all cases because G is the real part of an analytic function and is harmonic in z and z_o. Using (27) the two derivatives will become

$$\frac{\partial G}{\partial n} = e^{-i\theta} \cdot \left(\frac{\partial\bar{\Phi}}{\partial\bar{z}} + \frac{\partial\bar{\psi}}{\partial\bar{z}}\right) + e^{i\theta}\left(\frac{\partial\Phi}{\partial z} + \frac{\partial\psi}{\partial z}\right) \qquad (28)$$

and

$$\frac{\partial^2 G}{\partial n\partial n_o} = e^{-i\theta}e^{-i\theta_o}\frac{\partial^2\bar{\Phi}}{\partial\bar{z}\partial\bar{z}_o} + e^{i\theta}e^{i\theta_o}\frac{\partial^2\Phi}{\partial z\partial z_o} + e^{-i\theta}e^{i\theta_o}\frac{\partial^2\bar{\psi}}{\partial\bar{z}\partial z_o} + e^{i\theta}e^{-i\theta_o}\frac{\partial^2\psi}{\partial z\partial\bar{z}_o} \qquad (29)$$

I claim now that the functions G^* and G^{**} are given by

$$G^* = \frac{1}{i}\left|\Phi(z,z_o) - \bar{\Phi}(\bar{z},\bar{z}_o) + \psi(z,\bar{z}_o) - \bar{\psi}(\bar{z},z_o)\right| \qquad (30)$$

$$G^{**} = \left|-\Phi(z,z_o) - \Phi(\bar{z},\bar{z}_o) + \psi(z,\bar{z}_o) + \bar{\psi}(\bar{z},z_o)\right| \qquad (31)$$

The proof is given by the fact that

$$G^* = \int \frac{\partial G}{\partial n} ds + \text{Const} \iff \frac{\partial G^*}{\partial s} = \frac{\partial G}{\partial n} \tag{32}$$

and

$$G^{**} = \iint \frac{\partial^2 G}{\partial n \partial n_o} ds ds_o + \text{Const} \iff \frac{\partial^2 G^{**}}{\partial s \partial s_o} = \frac{\partial^2 G}{\partial n \partial n_o} \tag{33}$$

and applying the following rules to (30) and (31)

$$i \frac{\partial G^*}{\partial s} = e^{-i\theta} \frac{\partial G^*}{\partial \bar{z}} - e^{i\theta} \frac{\partial G^*}{\partial z} \tag{34}$$

and

$$i^2 \frac{\partial^2 G^{**}}{\partial s \partial s_o} = e^{i\theta} e^{i\theta_o} \frac{\partial^2 G^{**}}{\partial z \partial z_o} + e^{-i\theta} e^{-i\theta_o} \frac{\partial^2 G^{**}}{\partial \bar{z}_o \partial \bar{z}} - e^{i\theta} e^{-i\theta_o} \frac{\partial^2 G^{**}}{\partial z \partial \bar{z}_o} - e^{-i\theta} e^{i\theta_o} \frac{\partial^2 G^{**}}{\partial \bar{z} \partial z_o} \tag{35}$$

Looking at (30) and (31) closer one can see that the Green's function of the second kind is nothing else but the imaginary part of the complex Green's function or, if one prefers, the so called stream function. Green's function of the third kind can be interpreted as complement function to the one of the first kind because the two singularities now have the same sign.

SPECIAL NUMERICAL CONSIDERATIONS

Let us start with the Green's function of the second kind. For the unit circle it will be

$$G^* = \frac{1}{2\pi} \cdot \text{Im} \left[\ln \frac{z - z_o}{1 - z\bar{z}_o} \right] = \frac{1}{2\pi} \cdot \arctan \left\{ \frac{z - z_o}{1 - z\bar{z}_o} \right\} \tag{36}$$

when going round about the circle, this function will jump by the value 1. So ϕ_o in equation (24) has the value of the potential where the jump occurs. It is convenient to shift this jump to a node to avoid difficult interpolations. Normally the function will jump when the argument is - a. (This is dependent on the computer). This will be when

$$\frac{z-z_o}{1-z\bar{z}_o} = \frac{\bar{z}-\bar{z}_o}{1-\bar{z}z_o} < 0 \qquad (37)$$

Noting that z is on the unit circle and

$$z = \sigma = 1/\bar{\sigma} \qquad (38)$$

one gets a quadratic equation with the following solutions

$$\sigma_1 = -\frac{1-z_o}{1-\bar{z}_o} \qquad \sigma_2 = \frac{1+z_o}{1+\bar{z}_o} \qquad (39)$$

where σ_1 is the right one.

The evaluation of the integral (24) starts at the chosen node counterclockwise. When position σ_1 is passed the value 1 has to be added to G*.

To calculate the fluxes in some point z_o equation (24) is differentiated with respect to $\bar{z}_o$ which gives

$$\frac{\partial\phi}{\partial x_o} + i\frac{\partial\phi}{\partial y_o} = 2\oint_{z\in\Gamma} \phi_i \frac{\partial N_i}{\partial s} \frac{\partial G^*}{\partial\bar{\zeta}_o} \overline{\omega'(z_o)}ds \qquad (40)$$

This is the only formula where the derivative of the conformal mapping occurs. In all other relations only the coordinates of z and z_o or of their images are needed. This is caused by the fact that the solution of the potential problem and finding the conformal mapping are equivalent.

For the calculation of the stiffness matrix only the upper half has to be calculated. It is useful to interprete the double integral as two dimensional integral and using an isoparametric concept. So we have to integrate in a unit square where each dimension represents one boundary element and can use n by n Gauss points:

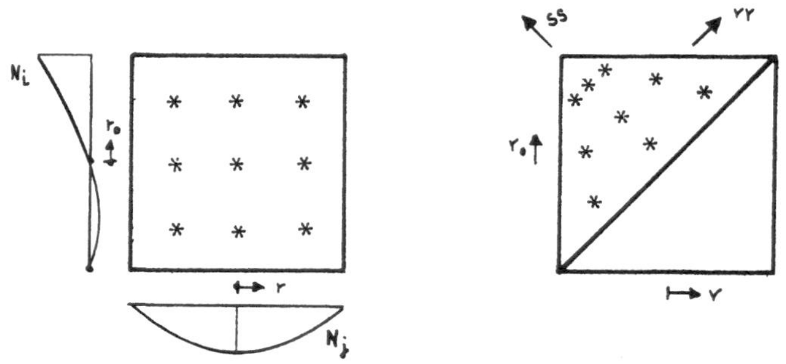

Figure V Integration schemes

If the two elements coincide, the logarithmic singularity is on the diagonal of the square. As we calculate only the half of the stiffness matrix we have to restrict ourselves to one half of the square. Therefore, we can integrate in one triangle only and use special quadrature formulas or transform our n by n Gauss points by

$$rr = \frac{1}{2}\left[r\ (1-r_o) - (1 + r_o)\right]$$

$$ss = \frac{1}{2}\left[r\ (1-r_o) + (1 + r_o)\right] \qquad (41)$$

The determinant of this transformation will be

$$\det |J| = \frac{1}{2}\ (1 - r_o) \qquad (42)$$

with which the integrand has to be multiplied.

If desired, the conformal mapping can be improved by using our knowledge that the images of the boundary points are located on the unit circle resp. the real axis and applying the following corrections to the images

$$\text{circle} \qquad \zeta := \frac{\zeta}{|\zeta|} \qquad (43)$$

$$\text{half plane}\ \zeta := \mathrm{Re}[\zeta] \qquad (44)$$

NUMERICAL EXAMPLE

To demonstrate the new approach we consider a square with prescribed potential at two opposite sides. The fluxes will be constant, the potential will vary linearly. Each side is divided in three linear elements with four Gauss points for each element.

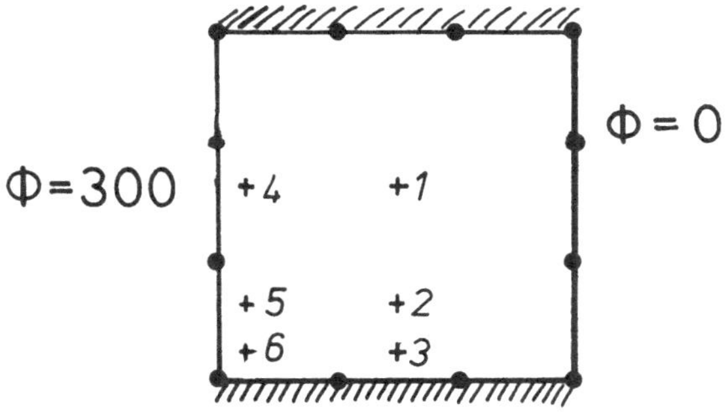

Figure VI Constant flow example

As we use only a trial form for the potential no corner problem exists besides that in the conformal mapping.

First we calculate some values in interior points applying the Green's function of the first and the second kind (equations 3,24,40) prescribing the exact solution of our problem at the boundary. The conformal mapping was done by equation (15) using four coefficients. The following table gives the calculated values in the points marked in Figure VI. The exact values of the fluxes are $\phi_{,x} = 5.000$ and $\phi_{,y} = 0.000$.
As one can see, the values obtained by the Green's function of the second kind are of much higher quality than the original ones. The fluxes in the points near the boundary are very poor when one uses the Green's function of the first kind.

Table I Values at internal points

Point no.	exact value	Values calculated with Green's function 1st kind			Values calculated with Green's function 2nd kind		
	Φ	Φ	$\Phi_{,x}$	$\Phi_{,y}$	Φ	$\Phi_{,x}$	$\Phi_{,y}$
1	150	150.0	4.999	0.000	150.0	4.999	0.000
2	150	149.9	5.017	0.065	150.0	4.995	0.000
3	150	146.2	5.378	3.036	150.0	4.872	0.000
4	275	267.4	-1.070	0.000	275.0	5.001	0.000
5	275	267.4	-1.203	0.204	274.9	5.011	0.022
6	275	280.0	2.502	3.210	274.8	4.981	0.014

Then a stiffness matrix was calculated following equation (25). This matrix is fully populated but exactly symmetric and has the eigen-value zero i.e. the row sums are zero within the computer precision. This matrix was used to solve the given problem. After a solution is obtained the original matrix is multiplied with the potential to give the nodal sources.

$$Q_i = a_{ij} \cdot \phi_j \qquad (45)$$

The next table gives the potential in the two nodes on the upper resp. lower side and the nodal sources in the nodes with prescribed potential. The values do not converge to the exact solution but to that one which is prescribed by the conformal mapping.

Table II Values obtained by the stiffness matrix

Number of Gauss points/element	Φ_1	Φ_2	Q_1	Q_2	sum of Q
exact value	200.00	100.00	5.000	10.000	30.00
4	199.90	100.10	4.633	10.143	29.55
5	199.98	100.02	4.714	10.142	29.71
6	200.03	99.97	4.832	10.064	29.80
8	200.17	99.83	4.882	10.069	29.90

EXTENSION TO ELASTIC PROBLEMS

The technique for elastic systems is similar but much more difficult. First and most severe restriction is the Green's function.

Elastic problems are prescribed by the biharmonic equation. The use of conformal mapping is therefore not easy and a simple substitution does not give the Green's function for the mapped region. Moreover, Green's functions for even very simple regions are rather complicate. The only way I can see to obtain those functions is the method of Muskelishvilli (1951). He used the complex representation of a biharmonic equation called the formula by Goursat:

$$\Delta\Delta G = 0 \iff G = \mathrm{Re}\left[\bar{z}\cdot\varphi(z) + X(z)\right] \qquad (46)$$

where φ and X are analytic functions.

The partial integration is more complicated than the previous one but facilitated by the use of equation (46).

The stretching of plates is described by a logarithmic kernel. So the technique remains quite similar to that of the potential problem. The kernel for bending of plates however is of order $r^2 \ln r$ and here higher derivatives occur. Despite of this, it is possible to keep in the logarithmic order integrating the rotational degrees by part once and the deflections twice. The research on these problems is still going on and will be reported elsewhere.

CONCLUSION

A technique was developed in this paper to calculate exact stiffness matrices for a given region using Boundary Elements and the complex representation of harmonic or biharmonic functions. A numerically stable approach is given which makes use of a continuous potential or displacement at the boundary. New types of Green's functions are introduced, the so called Green's functions of the second and third kind.

REFERENCES

Banerjee, P.K., Butterfield, R. (1979) Developments in Boundary Element Methods I, Applied Science Publishers Ltd., London.

Bettess, P. (1977) Infinite Elements, Int. Journ. for num. Meth. in Engineering Vol 11, 53-64.

Brebbia, C.A. (1978) The Boundary Element Method for Engineers. Pentech Press, Plymouth.

Gaier, D. (1964) Konstruktive Methoden der konformen Abbildung. Springer Verlag, Berlin.

Jaswon, M.A., Symm, G.T. (1977) Integral Equation Methods in Potential Theory and Elastostatics, Academic Press, London.

Kantorowitsch /Krylow (1964) Approximative Methods of Higher Analysis, Noordhof Ltd., Groningen.

Margulies, M. (1980) Exact treatment of the exterior problem in the combined FEM-BEM. New Developments in Boundary Element Methods (C. Brebbia Edit.) CML Publications.

Morse, P., Feshbach, H. (1953) Methods of Theoretical Physics. McGraw-Hill, New York.

Muskhelishvilli, N.I. (1951) Some basic problems of mathematical theory of elasticity. Noodrhof Ltd., Groningen.

Mustoe, G.G.W., Volait, F. (1980) A Symmetric Direct Boundary Integral Equation Method for two Dimensional Elastostatics.

Zienkiewicz, O.C., Kelly, D.W., Bettess, P. (1977) The Coupling of the Finite Element Method and Boundary Solution Procedures. Int. Journal Num. Meth. in Engineering Vol 11 pp 355-375.